Introduction to Forensic Chemistry

Introduction to Forensic Chemistry

Kelly M. Elkins

CRC Press
Taylor & Francis Group
Boca Raton London New York

CRC Press is an imprint of the
Taylor & Francis Group, an **informa** business

CRC Press
Taylor & Francis Group
6000 Broken Sound Parkway NW, Suite 300
Boca Raton, FL 33487-2742

Library of Congress Cataloging-in-Publication Data

Names: Elkins, Kelly M., author.
Title: Introduction to forensic chemistry / Kelly M. Elkins.
Description: Boca Raton, FL : CRC Press/Taylor & Francis Group, [2019] |
Includes bibliographical references and index.
Identifiers: LCCN 2018015615| ISBN 9781498763103 (hardback : alk. paper) |
ISBN 9780429454530 (ebook)
Subjects: LCSH: Chemistry, Forensic.
Classification: LCC RA1057 .E46 2019 | DDC 614/.12--dc23
LC record available at https://lccn.loc.gov/2018015615

**Visit the Taylor & Francis Web site at
http://www.taylorandfrancis.com**

**and the CRC Press Web site at
http://www.crcpress.com**

To my husband, Tim, and our children, Madeleine, Katie, and Sara

Contents

List of Figures   xiii
List of Tables   xxi
Acknowledgments   xxiii
Author   xxv
List of abbreviations   xxvii

1   An introduction to forensic chemistry and physical evidence   1
     Learning objectives   1
     Questions   11
     Bibliography   11

2   Chemical tests   15
     Learning objectives   15
     Colorimetric tests for drugs   16
     Chemical tests for poisons   24
     Colorimetric tests for explosives   24
     Microcrystalline tests for drugs   26
     Microcrystalline tests for explosives   26
     The future of chemical tests   27
     Questions   27
     Reference   27
     Bibliography   27

3   The microscope   33
     Learning objectives   33
     Early microscopes   34
     Parts of the microscope   35
     Light   36
     Magnification and resolving power   37
     Stereomicroscopy   38
     Compound light microscopy   38
     Koehler illumination   40
     Polarized light microscopy   40
     Phase contrast microscopy   42
     Fluorescence microscopy   43
     Microspectrophotometer   43
     Comparison microscopy   43
     Scanning electron microscopy   45
     Transmission electron microscopy   46
     Questions   47
     Bibliography   48

4   Light spectroscopy   51
     Learning objectives   51
     Ultraviolet-visible spectroscopy   53
     Fluorescence spectroscopy   58
     Infrared spectroscopy   62
     Raman spectroscopy   68

Microspectrophotometry 70
Questions 71
Reference 72
Bibliography 72

5   Advanced spectroscopy 77
Learning objectives 77
Mass spectrometry 79
Nuclear magnetic resonance spectroscopy 87
Questions 94
Bibliography 95

6   Chromatography 99
Learning objectives 99
Thin-layer chromatography 101
Paper chromatography 103
Column chromatography 103
High-performance liquid chromatography 105
Ultra-performance liquid chromatography 106
Gas chromatography 106
Other separation methods 112
Questions 112
Bibliography 113

7   Inorganic poisons and contaminants 115
Learning objectives 115
Flame test 117
Emission spectrograph 117
Thin-layer chromatography 118
UV-Vis spectroscopy 118
IR spectroscopy 120
Raman spectroscopy 120
X-ray fluorescence 121
Atomic absorption 122
Inductively coupled plasma-mass spectrometry 123
X-ray crystallography and x-ray diffraction 124
Neutron activation analysis 125
Scanning electron microscopy 125
Questions 125
References 126
Bibliography 126

8   Controlled substances 129
Learning objectives 129
Control and scheduling 130
Classes of drugs 130
Stimulants 131
Depressants and antianxiety drugs 134
Hallucinogens 135
Opiates/opioids 138
Anabolic steroids 140
Other drugs abused in sports 141
New psychoactive substances 141

Chemical analysis: Identification and quantitation | 142
Questions | 148
Bibliography | 148

9 Toxicology | **153**
Learning objectives | 153
Questions | 163
References | 164
Bibliography | 164

10 Trace evidence | **167**
Learning objectives | 167
Glass | 169
Soil | 172
Paint | 173
Polymers | 175
Hair | 176
Fibers | 178
Other trace materials | 184
Questions | 185
References | 186
Bibliography | 186

11 Questioned documents and impression evidence | **191**
Learning objectives | 191
Questioned documents | 192
Physical analysis | 192
Chemical analysis of inks and paper | 193
Impression evidence | 199
Questions | 206
Bibliography | 206

12 Latent print development | **209**
Learning objectives | 209
Questions | 218
References | 219
Bibliography | 219

13 Firearms | **223**
Learning objectives | 223
Firearms | 224
Handguns | 225
Rifles | 226
Shotguns | 226
Assault rifles | 226
Firearms manufacturing methods | 226
Explosives and propellants | 227
Ballistics | 227
Firearms evidence handling and labeling | 228
Firearms comparisons | 229
Gunshot residue analysis | 231
Questions | 231
Bibliography | 232

14 Fire, arson, and explosives   **235**

Learning objectives   235
Fire   236
Arson and accelerants   240
Explosives   244
Incendiary weapons   245
Identification of explosives   246
Questions   247
Bibliography   248

15 Chemical, biological, radiological, nuclear, and explosives (CBRNE)   **251**

Learning objectives   251
Weapons of mass destruction   251
Chemical weapons   253
Blood agents   253
Pulmonary/choking agents   254
Blister agents   255
Nerve gases   256
Nettle or urticant agents   257
Incapacitating agents   257
Vomiting agents   258
Riot/tear agents   258
Trends in chemical characteristics of chemical warfare agents   259
Toxic industrial chemicals   259
Detection and identification methods   259
Biological weapons   262
History of biological weapons use   262
Modern threat classification   263
Recent biological weapons cases and ongoing threats   264
Bacteria   265
Fungi   265
Viruses   266
Protein toxins   267
Small molecule toxins   271
Methods of detection and identification   272
Nuclear weapons   273
Nuclear chemistry reactions and types of radiation   274
Radioactive decay and half-life   275
History of nuclear chemistry and radioactivity   276
Modes of radioactive decay   277
Isotopes and nuclear reactions   278
Uses of radionuclides in bombs   278
Radioactivity units   279
Detection and identification of radioactive material   280
Cases of accidental poisoning with radiological material   281
"Dirty bombs" and accessibility of radioactive material   282
Treatment of radiation poisoning   282
Use of radioactive isotopes in nuclear power plants   283
Identification of a nuclear bomb detonation versus a nuclear power plant meltdown   284
Dual use research: Impacts and publication   284
Emerging threats and designer weapons   286
Questions   287
Bibliography   288

16   **Environmental forensics**   **295**
      Learning objectives   295
      Pesticides   297
      Herbicides   300
      Fungicides   301
      Antimicrobials   302
      In the environment   304
      Examples of detection, identification, and quantification of herbicides and pesticides   306
      Questions   308
      Bibliography   309

Index   311

List of Figures

Figure 1.1   Methamphetamine lab crime scene.   5

Figure 1.2   Evidence collection tools.   5

Figure 1.3   Forensic analysis from the crime scene to the lab.   7

Figure 1.4   NFPA diamond with meanings of scale (a); NFPA diamond for methanol (b).   8

Figure 1.5   Accuracy and precision.   8

Figure 2.1   Morphine and formaldehyde combine to form an oxocarbenium salt in the Marquis reaction.   17

Figure 2.2   Ehrlich reagent complex with LSD.   18

Figure 2.3   Fast Blue B salt (KN test) reaction with two molecules of THC.   19

Figure 2.4   Proposed Dille–Koppanyi reaction product coordination complex.   19

Figure 2.5   Complex formed with cobalt thiocyanate test reagent.   20

Figure 2.6   Mecke test reaction: oxidation of morphine by sulfuric acid.   20

Figure 2.7   Chen test complex with ephedrine molecules.   21

Figure 2.8   Ferric chloride test product with 5-sulfosalicylic acid.   21

Figure 2.9   Apomorphine formed using the potassium permanganate reagent.   22

Figure 2.10   Ammonium phosphomolybdate.   22

Figure 2.11   Zimmerman's test for benzodiazepines.   23

Figure 2.12   Selected colorimetric spot tests for drugs on a spot plate.   24

Figure 2.13   Griess test chemical reaction.   25

Figure 2.14   Crystals formed with microcrystalline tests.   26

Figure 3.1   Magnifying glass.   34

Figure 3.2   Real versus virtual image.   35

Figure 3.3   Compound light microscope.   36

Figure 3.4   Wavelength and amplitude.   36

Figure 3.5   Michel-Levy chart.   37

Figure 3.6   Resolution of two objects.   38

Figure 3.7   Stereoscope.   38

Figure 3.8   Light reflected off bullets.   39

Figure 3.9   Objective lenses with specifications.   39

Figure 3.10   Polarized light microscope.   40

Figure 3.11   A rayon fiber viewed using a polarizing light microscope (a), with a retarding analyzer (b), and with a compensator retardation plate inserted (c).   41

Figure 3.12   Fluorescence microscope.   42

Figure 3.13   Infrared microspectrophotometer.   43

Figure 3.14   Comparison microscope.   44

Figure 3.15   Comparison ballistics microscope.   45

Figure 3.16   Two fibers in a split field of view recorded using a comparison microscope.   45

Figure 3.17   Two cartridge cases in a split field of view recorded using a comparison microscope.   46

Figure 3.18   Scanning electron microscope.   46

Figure 3.19   TEM of gold nanoparticles.   47

Figure 4.1   Electromagnetic spectrum.   53

Figure 4.2   Schematic of the light path in a UV-Vis spectrometer.   54

Figure 4.3   UV-Vis cuvette.   54

Figure 4.4   Chemical structures of chromophores FD&C Yellow 5 and Red Dye 40.   55

Figure 4.5   Excitation of an electron from the ground state to a higher energy level.   55

Figure 4.6   UV-Vis absorption from the HOMO to the LUMO energy level.   55

Figure 4.7   UV-Vis absorption spectrum of DNA.   56

Figure 4.8   UV-Vis absorption spectra of red food coloring at dilutions of 1:10 and 1:1000.   56

Figure 4.9   Dilution series concentration standard red food dye.   57

Figure 4.10   Beer's Law calibration curve of red food dye dilution series.   57

Figure 4.11   NanoDrop UV-Vis spectrophotometer.   58

Figure 4.12   UV-Vis spectrometer.   58

Figure 4.13   UV-Vis absorption spectrum of bovine hemoglobin.   59

Figure 4.14   UV-Vis absorption spectrum of salicylic acid.   59

Figure 4.15   Fluorescence spectrometer.   60

Figure 4.16   Fluorescence emission of light energy after excitation of an electron to an excited state.   60

Figure 4.17   Schematic of a fluorescence spectrometer.   61

Figure 4.18   Excitation-emission plot of bovine serum albumin bound to gold nanoparticles in preparation for conjugation to an antibody for lifestyle analysis of fingerprints.   62

Figure 4.19   Structures of fluorophores ethidium bromide, fluorescein, and SYBR Green I.   62

Figure 4.20   Quantitation of *E. coli* standard DNA using real-time PCR with SYBR Green I fluorescence.   63

Figure 4.21   Infrared vibrations.   64

Figure 4.22   Schematic of ATR FT-IR diamond crystal and evanescent wave.   66

Figure 4.23   ATR FT-IR spectrometer.   66

Figure 4.24   A polystyrene thin film standard in the sample holder in an infrared spectrophotometer.   67

Figure 4.25   ATR FT-IR spectrum of red and green food dyes.   67

Figure 4.26   ATR FT-IR spectrum of salicylic acid.   68

Figure 4.27	Jablonski diagram of Raman spectroscopy as compared to other spectroscopy types.	68
Figure 4.28	Raman spectrometer.	69
Figure 4.29	Raman spectrum of salicylic acid.	69
Figure 5.1	Mass spectra of (a) acetone and (b) propionaldehyde.	80
Figure 5.2	Mass spectra of (a) 1-propanol and (b) 2-propanol.	82
Figure 5.3	Chemical ionization of methane gas.	83
Figure 5.4	Molecular ion formation using ionized methane gas.	83
Figure 5.5	Mass spectrum of phentermine.	85
Figure 5.6	Mass spectrum of methamphetamine.	85
Figure 5.7	Mass spectrum of salicylic acid.	86
Figure 5.8	Collected mass spectrum of caffeine and library search result overlay (center).	86
Figure 5.9	400 MHz JEOL NMR spectrophotometer.	88
Figure 5.10	An NMR sample tube.	88
Figure 5.11	^{1}H NMR spectrum of lidocaine in D_2O.	89
Figure 5.12	^{13}C NMR ephedrine in D_2O.	91
Figure 5.13	^{1}H NMR ephedrine in D_2O.	92
Figure 5.14	^{1}H NMR α-PVT alone and cut with powdered sugar in D_2O.	93
Figure 6.1	A silica TLC plate in a developing chamber.	101
Figure 6.2	A silica TLC plate used to separate inks.	102
Figure 6.3	Paper chromatography used to separate inks.	103
Figure 6.4	FPLC instrument with fraction collector.	104
Figure 6.5	Elution profile of proteins separated using FPLC.	105
Figure 6.6	Steel-jacketed HPLC C18 column.	105
Figure 6.7	HPLC instrument.	106
Figure 6.8	HPLC elution profile of protein separation.	107
Figure 6.9	UPLC instrument.	107
Figure 6.10	Schematic of GC-MS.	108
Figure 6.11	GC-MS instrument.	108
Figure 6.12	GC-MS vials with insert (left) and without (right).	109
Figure 6.13	Salicylic acid and caffeine are separated by GC. Their boiling points are 211°C and 178°C, respectively.	110
Figure 6.14	GC-FID used in determination of blood alcohol content.	111
Figure 6.15	A capillary electrophoresis instrument.	112
Figure 7.1	Principle of emission spectroscopy.	117
Figure 7.2	Student spectroscope.	118
Figure 7.3	Line spectra of elements (a) neon, (b) argon, and (c) nitrogen.	119

Figure 7.4   Quantitation of iron using UV-Vis spectroscopy and Beer's Law on complexation
with salicylic acid.   119

Figure 7.5   ATR FT-IR spectrum of the inorganic pigment sodium nitroprusside.   120

Figure 7.6   Raman spectrum of the inorganic pigment sodium nitroprusside.   120

Figure 7.7   XRF instrument.   121

Figure 7.8   XRF spectrum.   122

Figure 7.9   Atomic absorption spectrometer.   123

Figure 7.10   ICP-MS instrument.   124

Figure 8.1   Drug seizure of cannabis, cocaine, and MDMA.   131

Figure 8.2   Chemical structures of controlled substances amphetamine, heroin, fentanyl, THC,
phenobarbital, JWH-018, and testosterone.   132

Figure 8.3   Seized marijuana plant material submitted to the Maryland State Police crime lab for analysis.   136

Figure 8.4   Kratom capsule and contents.   142

Figure 8.5   Acidic (top row: aspirin, ibuprofen, and cocaine hydrochloride) and basic (bottom row:
caffeine, nicotine, and crack cocaine) drugs.   145

Figure 8.6   UV-Vis spectrum for methamphetamine collected using a NanoDrop UV-Vis
spectrophotometer.   145

Figure 8.7   UV-Vis spectrum for phentermine collected using a NanoDrop UV-Vis spectrophotometer.   146

Figure 8.8   Mass spectrum for methamphetamine collected using an Agilent 5975C MSD.   146

Figure 8.9   Mass spectrum for phentermine collected using an Agilent 5975C MSD.   147

Figure 8.10   ATR FT-IR spectra of methamphetamine and phentermine.   147

Figure 9.1   Structure of ethanol.   155

Figure 9.2   BAC over time in hours with and without food intake.   155

Figure 9.3   ALDH enzymatic reaction to convert ethanol to formaldehyde.   156

Figure 9.4   ClustalW 2.1 ALDH1A1 and ALDH2 protein sequence alignment with Glu504Lys
mutation in yellow.   156

Figure 9.5   Crystal structure of human ALDH2 (3sz9.pdb) bound to 1-(4-ethylbenzene)
prop-2-en-1-one inhibitor.   157

Figure 9.6   Relative risk increases as blood alcohol concentration increases.   157

Figure 9.7   Screenshot of BAC Free app blood alcohol determination for a male (left) and female (right) of
the same weight and alcohol consumed.   158

Figure 9.8   ATR FT-IR spectrum of ethanol.   159

Figure 9.9   Heroin metabolism reaction.   161

Figure 9.10   Human carboxylesterase 1 (hCE1) bound to naloxone methiodide, a heroin
analogue (1MX9.pdb).   161

Figure 9.11   Chemical structures of poisons including CO_2, CO, CN, HCl, Cl_2.   162

Figure 9.12   Structure of human hemoglobin (1gzx.pdb), a hetero-tetramer consisting of two alpha and two beta
chains. The ribbons, which represent the alpha helices, are colored by the order of the secondary structure elements
and trace the backbone structure of the protein chains. The oxygens are shown as four white diatomic balls, each in

one quadrant of the structure, and are tethered by the heme moieties shown in stick form. The proximal and distal histidine residues that are essential for oxygen binding are shown in purple. Poisons including diatomic gases CO, CN, HCl, and Cl_2 can bind at the oxygen binding site in the protein. Carbon dioxide (CO_2) binds at the allosteric site to induce a conformational change in the protein to promote oxygen release into the cell.

Figure 10.1   Locard's exchange principle.   169

Figure 10.2   Determination of glass density.   170

Figure 10.3   ATR FT-IR spectra of nail polish paint of different colors.   174

Figure 10.4   ATR FT-IR spectra of clearcoat from three automobiles.   175

Figure 10.5   Chemical structures of plastic polymers.   176

Figure 10.6   ATR FT-IR spectra of selected materials from the six recycling codes.   176

Figure 10.7   Schematic and cross section of hair.   178

Figure 10.8   Images of human Caucasian (left) and Negroid (right) hairs under a comparison microscope.   179

Figure 10.9   Micrograph of red fox (left) and deer (right) hairs.   181

Figure 10.10   Micrograph of horse (left) and rabbit (right) hairs.   181

Figure 10.11   Micrograph of acrylic (left) and acetate (right) fibers.   182

Figure 10.12   ATR FT-IR spectra of synthetic fibers acrylic, Antron nylon, acetate, and Dacron polyester.   183

Figure 10.13   ATR FT-IR spectra of natural fibers cotton, coconut husk, silk, and viscose.   183

Figure 10.14   ATR FT-IR spectra of hairs from human, dog, cat, rabbit, camel, musk ox, yak, llama, sheep, and goat.   184

Figure 10.15   Micrograph of tape weave and pattern matching.   185

Figure 11.1   US $50 viewed under white light.   194

Figure 11.2   US $50 viewed using an ALS shows the fluorescent fibers in the paper.   194

Figure 11.3   US $50 viewed under UV light shows the security strip.   195

Figure 11.4   Microwriting is visible on a US $50 bill on magnification.   195

Figure 11.5   Original numbers in red ink (top left); original numbers viewed under the ALS (top right); altered numbers in red ink (bottom left); and altered numbers viewed under the ALS (bottom right).   196

Figure 11.6   IR spectra of color paper.   196

Figure 11.7   TLC plate containing separations of inks from four red pens.   197

Figure 11.8   TLC of ink from four red pens excited using a UV light.   198

Figure 11.9   UV-Vis spectra of pen inks from various manufacturers.   198

Figure 11.10   ATR FT-IR spectra of red pen inks from various manufacturers.   199

Figure 11.11   ATR FT-IR spectra of blue pen inks from various manufacturers.   199

Figure 11.12   ATR FT-IR spectra of black pen inks from various manufacturers.   200

Figure 11.13   ATR FT-IR spectrum of crystal (Gentian) violet dye.   200

Figure 11.14   Raman spectrum of crystal (Gentian) violet dye.   201

Figure 11.15   West Midlands Police preserve footprint evidence.   201

Figure 11.16   Suspect footprint impression evidence in the snow (simulated evidence).   202

Figure 11.17   Cast of a suspect's shoe (simulated evidence) in dental stone.   202

Figure 11.18   Biofoam cast of a reference shoe (simulated evidence).   203

Figure 11.19   Inked suspect shoe prints (simulated evidence).   203

Figure 11.20   Inked tire tread prints.   204

Figure 11.21   Tool impressions in wood.   205

Figure 11.22   Impressed serial number in metal.   205

Figure 12.1   Three men share remarkably similar facial characteristics but their fingerprints are different.   211

Figure 12.2   Fingerprints with a (left to right) double loop, whorl, and single loop.   212

Figure 12.3   Fingerprint minutiae.   212

Figure 12.4   Fingerprint points for analysis.   213

Figure 12.5   Fingerprint dusted with black magnetic fingerprint powder.   215

Figure 12.6   Fingerprint developed on currency by dusting with fluorescent fingerprint powder and viewing under UV light.   215

Figure 12.7   Superglue fuming humidity chamber.   216

Figure 12.8   Fingerprint developed using superglue fuming.   216

Figure 12.9   Fingerprint developed using ninhydrin.   217

Figure 12.10   Comparison of fingerprints using a comparison microscope.   218

Figure 13.1   Components of a bullet (1-bullet; 2-casing; 3-propellant; 4-rim; 5-primer).   225

Figure 13.2   Revolver handgun (0.357 Magnum).   225

Figure 13.3   Pistol.   225

Figure 13.4   Rifle.   226

Figure 13.5   Shotgun.   227

Figure 13.6   AR-15 rifle.   227

Figure 13.7   Lands and grooves in a barrel.   228

Figure 13.8   Structures of primer explosives.   228

Figure 13.9   Structures of gunpowder propellants.   229

Figure 13.10   Selected cartridges with bullets (left to right): (1) .17 HM2; (2) .17 HMR; (3) .22LR; (4) .22 WMR; (5) .17 SMc; (6) 5 mm/35 SM4; (7) .22 Hornet; (8) .223 Remington; (9) .223 WSSM; (10) .243 Winchester; (11) .243 Winchester Improved (Ackley); (12) .25–06; (13) .270 Winchester; (14) .308; (15) .30–06; (16) .45–70 Govt; (17) .50–90 Sharps.   229

Figure 13.11   Micrograph of lands, grooves, twists, and striations on 7.62 × 51 mm NATO bullets and an unfired cartridge, with counterclockwise rifle marks.   230

Figure 13.12   Dermal nitrate test.   231

Figure 14.1   Fire triangle.   237

Figure 14.2   Fire "V" charring pattern.   239

Figure 14.3   Developmental stages of a fire.   239

Figure 14.4   Ignitable liquids: methyl ethyl ketone, charcoal lighter fluid, and turpentine.   240

Figure 14.5   Gas chromatogram of 1-butanol. The sample was run on an Agilent 7890A and 0.2 µL of a 1 µL/mL solution prepared in methanol was injected at 280°C and cooled to 50°C for 2 minutes prior to separation with a 50°C–280°C temperature ramp at a rate of 60°C/min. The final temperature was held for 4 minutes; the run time was 9.83 min.   242

Figure 14.6   Gas chromatogram of unburned lighter fluid.   242

Figure 14.7   Gas chromatogram of unburned gasoline.   243

Figure 14.8   Gas chromatogram of unburned turpentine.   243

Figure 14.9   Gas chromatogram of unburned kerosene.   243

Figure 14.10   Gas chromatogram of unburned lamp oil.   243

Figure 14.11   Mass spectrum of dodecane found in kerosene.   244

Figure 14.12   Chemical structures of explosives.   245

Figure 14.13   Molotov cocktail.   246

Figure 14.14   Improvised explosive device.   246

Figure 14.15   ATR FT-IR spectra of explosives NH_4NO_3, KNO_3, $NaNO_3$, $KClO_4$, and $NaClO_4$.   247

Figure 15.1   Chemical structures of blood agents.   254

Figure 15.2   Chemical structures of pulmonary/choking agents.   254

Figure 15.3   Chemical structures of blister agents.   255

Figure 15.4   Chemical structures of nerve agents.   256

Figure 15.5   Chemical structure of a nettle/urticant agent.   258

Figure 15.6   Chemical structure of an incapacitating agent.   258

Figure 15.7   Chemical structures of vomiting agents.   259

Figure 15.8   Chemical structures of riot/tear agents.   259

Figure 15.9   *Bacillus anthracis* stained with the Gram stain.   266

Figure 15.10   *E. coli* culture.   267

Figure 15.11   Castor beans containing ricin.   268

Figure 15.12   Ricin A chain structure (1ift.pdb).   269

Figure 15.13   Abrin A chain structure (1abr.pdb).   269

Figure 15.14   Methods and chemicals used in ricin extraction and crude and analytical purification schemes.   270

Figure 15.15   Botulinum toxin (1bta.pdb).   271

Figure 15.16   Diphtheria toxin (1f0l.pdb).   271

Figure 15.17   Structures of biological toxins.   272

Figure 15.18   Uranium cake, a form of uranium oxide.   277

Figure 15.19   23 kiloton XX-34 BADGER 1953 US nuclear bomb test at the Nevada Test Site.   279

Figure 15.20   Geiger counter.   281

Figure 15.21   Chernobyl Unit 4 reactor damaged by nuclear meltdown.   284

Figure 15.22   Fukushima Dai-ichi Nuclear Power Plant near reactor Unit 3 damaged by a 2011 earthquake and 15-meter tsunami.   285

Figure 16.1   Structures of pesticides by class: organophosphates, pyrethroids, carbamates, and organochlorines.   297

Figure 16.2   Structure of DDT.   298

Figure 16.3   Structures of organochlorine pesticides.   298

Figure 16.4   Structures of organophosphate pesticides.   299

Figure 16.5   Structures of carbamates.   299

Figure 16.6   Structures of pyrethrins.   300

Figure 16.7   Herbicide products.   301

Figure 16.8   Structures of chlorinated herbicides.   301

Figure 16.9   Structures of sulfonylurea herbicides.   302

Figure 16.10   Structures of glyphosate and fluazifop-p-butyl herbicides.   302

Figure 16.11   Structures of selected agricultural fungicides.   304

Figure 16.12   Structures of selected antifungals.   304

Figure 16.13   Structures of selected antibiotics.   305

Figure 16.14   Fluorescence EEM plot for the SRFA-Al3+-DCPPA ternary solution prepared at pH 4.0 in 0.1 M NaClO$_4$ buffer containing 15 mg/L SRFA, 0.300 mM Al3+, and 0.715 mM 2,4-dichlorophenoxypropionic acid. The solution was incubated in the dark for 24 hours, then excited from 300 to 360 nm (y-axis), and detected fluorescence emission was plotted from 380 to 500 nm (x-axis). The relative fluorescence units (color) are plotted from 0 to 100 (z-axis).   307

Figure 16.15   Mass spectra f 2,4-D (a) and 2,4,5-T (b).   308

List of Tables

Table 1.1	Brief history of some notable advances in forensic chemistry	3
Table 1.2	Units of forensic laboratories that use forensic chemistry	4
Table 1.3	Examples of class and individual characteristics for physical evidence types	6
Table 1.4	SWGDRUG categories of analytical techniques by discriminating power (Category A > Category B > Category C)	9
Table 2.1	Positive results with color tests for explosives	25
Table 4.1	Nobel Prize Laureates in the field of spectroscopy	54
Table 4.2	Fluorophores and their excitation and emission maxima	61
Table 4.3	Infrared spectrum ranges	63
Table 4.4	IR correlation chart	65
Table 4.5	Raman versus infrared spectroscopy: A contrasting view	70
Table 4.6	Correlation chart for Raman spectroscopy	71
Table 5.1	Nobel Prizes awarded for advances in mass spectrometry	79
Table 5.2	Commonly observed fractionation ions in mass spectrometry	81
Table 5.3	Nobel Laureates in the field of NMR	87
Table 5.4	Chemical shift for several proton types	90
Table 5.5	Chemical shift by carbon type	91
Table 6.1	Gas chromatography stationary phases and their applications	109
Table 6.2	Boiling point and order of elution of hydrocarbons in fuels on a GC equipped with a polydimethylsiloxane column	110
Table 6.3	Gas chromatography detectors and their applications	111
Table 7.1	Flame colors of metals	118
Table 8.1	Controlled Substance Act drug schedule	131
Table 8.2	Banned exogenous anabolic androgenic steroids as of January 1, 2012. Related drugs with the same parent structure and biological activity are also banned	140
Table 8.3	Drugs and their chemical and spectroscopic features	143
Table 9.1	Drugs and secondary metabolites	160
Table 9.2	Retention time in urine for several drugs	162
Table 10.1	Chemical composition of six common glass types	169
Table 10.2	Densities of various types of glass	170
Table 10.3	Refractive indices for liquids and oils	171
Table 10.4	Refractive indices for several types of glass	172

Table 10.5   Soil particles by size   173

Table 10.6   Plastic polymers by uses and recycling code   175

Table 10.7   Characteristics of human head hairs   178

Table 10.8   Fibers and their uses   180

Table 10.9   Fiber characteristics   182

Table 11.1   Fluorescence observed using an ALS on several red inks   195

Table 12.1   Latent print development methods   214

Table 14.1   Flash point and autoignition temperature for selected ignitable liquids   237

Table 14.2   Heats of combustion for various fuels   238

Table 14.3   Boiling point and molecular mass of common ignitable liquid accelerants or major components   241

Table 14.4   Ignitable liquids in the light, medium, and heavy classes   242

Table 14.5   Major mass spectrometry ion peaks for different hydrocarbon compound cation types. More masses are included in Table 5.2   244

Table 15.1   Chemical properties of chemical weapons   260

Table 15.2   US CDC Category A, B, and C agents   264

Table 15.3   Categories of bacteria by Gram stain   266

Table 15.4   Viruses by type and lethal dose (where reported)   267

Table 15.5   Sizes and toxicity of protein toxins (in kilo Daltons, kDa)   268

Table 15.6   Methods and chemicals used in ricin extraction and crude and analytical purification schemes   270

Table 15.7   Radioactive particles   274

Table 15.8   Radiation emitted by commonly encountered radionuclides; all nuclides with more than 83 protons are radioactive   275

Table 15.9   Nobel prizes awarded in nuclear chemistry   276

Table 16.1   Classification of herbicides by inhibition effect   303

Table 16.2   Classification of pesticides based on toxicity by the EPA   305

Acknowledgments

I was first asked to teach an undergraduate forensic chemistry course a dozen years ago. It was a team-taught course in which I focused most of my efforts on the forensic biology portion. Even so, forensic science intrigued me and I was hooked. I stayed only pages ahead of my students that semester and was grateful to have Richard Saferstein's book *Criminalistics: An Introduction to Forensic Science* (Pearson, 2003, 8th ed.) to teach me what my academic training in chemistry and biochemistry had not.

I am grateful to Todd Hizer for giving me the chance to teach forensic chemistry the first time. I am fortunate to have had excellent mentors including Chris Tindall, Jr., now retired, who tasked me with creating a new forensic biology course and also gave me the opportunity and tools to teach courses in forensic chemistry and drug analysis. Thanks also to Tom Brettell and Larry Quarino who have been very helpful in guiding me to become a forensic scientist. Thanks to Sue Schelble for offering me so many opportunities and inviting me to collaborate to bring nuclear magnetic resonance spectroscopy to undergraduates. Thank you to Tim Brunker for teaching me how to use Towson's nuclear magnetic resonance spectrometer, Ellen Hondrogiannis for guiding me on gas chromatography-mass spectrometry, Cindy Zeller for training me on our new capillary electrophoresis instrument, and George Kram for instructing me on the use of the Raman spectrometer. Thanks to Keith Reber for providing use of and tutorials on MNova and Dan Macks for providing use of ChemDraw. Thank you to Beth Kautzman for useful discussions of end of chapter questions. Thanks to Tim Phillips for taking photographs for this book. Thanks to Ashley Cowan for preparing all of the chemical structure figures, performing literature research, reviewing the manuscript, and formatting most of the references for this book. Thanks to Ryan Casey for supporting me writing this book and my forensic chemistry students for their ideas and feedback. Thank you to the Chemistry Department for stimulating lunch conversations. I am glad to work with you every day.

Thank you to my family, friends, and three children who encouraged me in this project and supported my long summer days of writing. Thank you to Karen Pinco for the discussions about writing and sharing in the journey. I am grateful to my early readers and manuscript reviewers who made many excellent suggestions for improvement including Alicia Quinn, Ashley Cowan, Theresa DeAngelo, Mark Profili, and Sue Schelble.

Thanks to Mark Listewnik, acquisitions editor, who approached me about taking on this project. It has been a pleasure to work with you. I truly appreciate the work of Mark Listewnik and Misha Kydd at CRC Press/Taylor & Francis who have guided me in making this a better book. Thank you to Lara Silva McDonnell of Deanta Global for editing services.

I am grateful to everyone who reads and uses my books. I truly appreciate the emails from faculty who have adopted my books and correspondence from my readers.

Author

Kelly M. Elkins earned a BS in biology and BA in chemistry from Keene State College in Keene, New Hampshire. She went on to earn MA and PhD degrees in chemistry from Clark University in Worcester, Massachusetts. She was a Fulbright Scholar in Heidelberg, Germany, from 2001 to 2002 and a Cancer Research Institute postdoctoral fellow in the Biology Department at MIT in Cambridge, Massachusetts, from 2003 to 2004. She was a temporary assistant professor of chemistry at Armstrong Atlantic State University in Savannah, Georgia, for two years and an assistant professor of chemistry at Metropolitan State College of Denver in Denver, Colorado, for five years, where she served as director of the Criminalistics Program (2010–2012), a Forensic Science Education Programs Accreditation Commission (FEPAC)-accredited program, supervised undergraduate research and internships, and developed the curriculum for their Criminalistics II (Forensic Biology) course. She joined the Towson University Chemistry Department and FEPAC-accredited Forensic Sciences program in 2012, where she is currently associate professor of the Chemistry and Graduate Faculty in the Masters of Science in Forensic Science Program. She teaches undergraduate and graduate forensic chemistry courses including Forensic Chemistry, Forensic Serology, and Weapons of Mass Destruction. Her research interests include the development of new methodologies and course materials to support forensic science. Her research focuses on the development of new real-time high-resolution melt polymerase chain reaction assays for species identification of "legal high" plants, controlled species, and food-borne pathogens.

Dr. Elkins is a member of the American Chemical Society (ACS) and the American Academy of Forensic Sciences (AAFS). She was promoted to Fellow of the American Academy of Forensic Sciences in 2018. She is a member of the ACS Ethics Committee (ETHX) and the ACS Exams Institute Diagnostic of Undergraduate Chemical Knowledge (DUCK) 2017 exam committee. She is an elected alternate Councilor of the Maryland sections of the ACS and co-chair of the Women Chemists committee. She is also a member of the Council of Forensic Science Educators (President, 2012). Dr. Elkins has published five book chapters and twenty-five peer-reviewed papers in journals including the *Journal of Forensic Sciences*, *Drug Testing and Analysis*, and *Medicine, Science and the Law*. She has presented her work at national and international conferences in the United States, Canada, Mexico, England, Germany, Russia, and Korea. Her work has been funded by the National Science Foundation, Maryland TEDCO, ACS Project Seed, ACS Petroleum Research Fund, and the Forensic Sciences Foundation. She has been interviewed by NBC 9News Denver, FOX 31 News Denver, ABC 7 News Denver, the *Denver Post*, and *Forensic Magazine* as a forensic expert. Her research was recently highlighted by GenomeWeb.com and she recently contributed an invited article to The *Expert Witness* journal. Her first book is *Forensic Biology: A Laboratory Manual* (2013).

List of abbreviations

AA	atomic absorption
AAFS	American Academy of Forensic Sciences
ABC	American Board of Criminalistics
ABFT	American Board of Forensic Toxicology
ACS	American Chemical Society
ADH	alcohol dehydrogenase
ALDH	acetaldehyde dehydrogenase
ALS	alternate light source
amu	atomic mass units
ANFO	ammonium nitrate fuel oil
ANZFSS	Australian and New Zealand Forensic Science Society
ASCLD	American Society of Crime Lab Directors
ASCLD-Lab	American Society of Crime Laboratory Directors-Laboratory Accreditation Board
ASTM	American Society of the International Association for Testing and Materials
ATCC	American Type Culture Collection
ATF	Bureau of Alcohol, Tobacco, Firearms and Explosives
ATR FT-IR	attenuated total reflectance Fourier transform-infrared spectroscopy
BAC	blood alcohol concentration
Bq	becquerel
BWC	biological weapons convention
BZP	N-benzylpiperazine
CBRNE	chemical, biological, radiological, nuclear, and explosives
CDC	Centers for Disease Control
CE	capillary electrophoresis
CI	chemical ionization
CMYB	cyan-magenta-yellow-black
COSY	correlation spectroscopy
CSA	Controlled Substances Act
CSFS	Chartered Society of Forensic Sciences
DART	direct analysis in real-time
DDT	dichlorodiphenyltrichloroethane
DEAE	diethyl amino ethyl cellulose
DEA	Drug Enforcement Agency
DHS	Department of Homeland Security
DMT	dimethyltryptamine
DNA	deoxyribonucleic acid
DRIFTS	diffuse reflectance infrared Fourier transform spectroscopy
DURC	dual use research of concern
E	energy
ECD	electron capture detector
EDS	energy dispersive spectroscopy
EDTA	ethylene diamine tetra acetic acid
EI	electron impact ionization
ELISA	enzyme-linked immunosorbent assay
EM	electron multiplier
EPA	Environmental Protection Agency

ESI	electrospray ionization
FBI	Federal Bureau of Investigation
FI	field desorption ionization
FID	flame ionization detector
FPLC	fast performance liquid chromatography
FSS	Forensic Sciences Society
GC	gas chromatography
GC-MS	gas chromatography–mass spectrometry
GHB	gamma-hydroxybutyrate
GPCR	G protein-coupled receptor
GSR	gunshot primer residue
HDPE	high-density polyethylene
HGH	human growth hormone
HPLC	high-pressure liquid chromatography (also known as high-performance liquid chromatography)
HMBC	heteronuclear multiple bond correlation
HOMO	highest occupied molecular orbital
HSQC	heteronuclear single quantum coherence
IAFIS	Integrated Automated Fingerprint Identification System
IAI	International Association for Identification
IBIS	Integrated Ballistic Identification System
ICP-MS	inductively coupled plasma emission spectroscopy–mass spectrometry
IDDA	instrumental data for drug analysis
IED	improvised explosive device
IR	infrared spectroscopy
ISO	International Organization for Standardization accreditation
LA	laser ablation
LC	liquid chromatography
LDPE	low-density polyethylene
LUMO	lowest unoccupied molecular orbital
LSD	lysergic acid diethylamide
MALDI	matrix-assisted laser desorption/ionization
MDA	3,4-methylenedioxyamphetamine
MDMA	3,4-methylenedioxymethamphetamine
MMDA	5-methoxy-3,4-methylenedioxyamphetamine
MP	melting point
MRSA	methicillin-resistant *Staphylococcus aureus*
MSDS	material safety data sheet
NAA	neutron activation analysis
NCI	negative chemical ionization
NIBIN	National Integrated Ballistics Information Network
NIH	National Institutes of Health
NIST	National Institutes of Standards and Technology
NFPA	National Fire Protection Association
NHTSA	National Highway Traffic Safety Administration
NMR	nuclear magnetic resonance spectroscopy
NOESY	nuclear overhauser effect spectroscopy and experiments
NPS	new psychoactive substances
OSAC	Organization of Scientific Area Committee
PBI	polybenzimidazole
PCC	1-piperidinocyclohexanecarbonitrile
PCI	positive chemical ionization
PCP	phencyclidine
PCR	polymerase chain reaction

PD	plasma desorption
PDQ	paint data query
PEEK	polyether ether ketone
PEN	polyethylene naphthalate
PETE	polyethylene terephthalate
PETN	pentaerythritol tetranitrate
PFTBA	perfluorotributylamine
PI	photoionization
PID	photoionization detector
PLA	polylactic acid
PP	polypropylene
PPE	personal protective equipment
ppm	parts per million
PS	polystyrene
PSA	polysulfone
PTT	polytrimethylene terephthalate
PVA	polyvinylalcohol
PVC	polyvinyl chloride
QA	quality assurance
QC	quality control
QNB	3-quinuclidinyl benzilate
RDX	Research Department eXplosive
RNA	ribonucleic acid
RFLP	restriction fragment length polymorphism
RFU	relative fluorescence units
RGB	red-green-blue
ROESY	rotating frame nuclear overhauser effect spectroscopy
RUVIS	reflected ultraviolet imaging system
SCAN	scan mode
SERS	signal-enhanced Raman spectroscopy
SEM	scanning electron microscope
SIM	single ion monitoring mode
SOFT	Society of Forensic Toxicologists
SOP	standard operating procedure
SPME	solid phase microextraction
SRM	standard reference material
Sv	sievert
SWGDRUG	Scientific Working Group for the Analysis of Seized Drugs
TATP	triacetone triperoxide
TEM	transmission electron microscope
TEPP	tetraethyl pyrophosphate
THC	tetrahydrocannabinol
TIC	toxic industrial chemicals
TLC	thin-layer chromatography
TMS	tetramethylsilane
TNB	trinitrobenzene
TNT	trinitrotoluene
TOF	time-of-flight
TOCSY	total correlation spectroscopy
TS	thermospray ionization
TTI	transmitting terminal identifier
TWGFEX	Technical Working Group for Fire and Explosions
UMHW	ultra-high-molecular-weight polyethylene

UPLC	ultra-performance liquid chromatography
USDA	United States Department of Agriculture
UV	ultraviolet spectroscopy
UN Manual	United Nations Rapid Testing Method of Drugs of Abuse Manual
UNDOC	United Nations Office of Drugs and Crime
Vis	visible spectroscopy
WADA	World Anti-Doping Agency
WHO	World Health Organization
WMD	weapons of mass destruction
XRF	x-ray fluorescence spectroscopy

CHAPTER 1

An introduction to forensic chemistry and physical evidence

KEY WORDS: forensic science, forensic chemistry, criminalistics, physical evidence, crime scene investigator, chain of custody, class characteristics, individual characteristics, presumptive test, reference samples, comparison standards, safety data sheets, control samples, background controls, positive control, negative control, accuracy, precision, replicates, standard operating procedures, quality control, quality assurance, expert witness

LEARNING OBJECTIVES

- To explain the difference between forensic science, criminalistics, and forensic chemistry
- To understand the historical development of forensic science
- To know the locations and identities of several forensic laboratories
- To list the units of forensic laboratories that use forensic chemistry
- To identify physical evidence in a forensic case
- To differentiate between class and individual characteristics for physical evidence types
- To identify the Scientific Working Group for the Analysis of Seized Drugs (SWGDRUG) categories of analytical techniques by category
- To understand the role of the forensic chemist in the laboratory, in the forensic community, and in court

ALCOHOL POISONING: METHANOL AND OTHER DENATURANTS

A man arrived at the hospital hallucinating. Although not readily apparent, the hallucinations turned out to be a symptom of methanol present in the alcohol he had consumed.

Alcohol, also known as ethanol or ethyl alcohol, is the most widely used legal drug. It is a depressant and affects the central nervous system. At low doses, it can lead to the loss of inhibitions and increased talkativeness. At higher doses, it affects reasoning, behavior, memory, speech, emotion, and abstract thinking. At very high doses, it can lead to a loss of consciousness and death.

Passed in 1919, the 18th Amendment to the US Constitution banned the manufacture, sale, and transportation of alcoholic beverages into the country. Enforcement began with the passage of the Volstead Act on January 1, 1920. Thus began prohibition. As a result, drinkers resorted to drinking wood alcohol and industrial alcohol with severe effects. Although alcohol was illegal to consume as a beverage, it was still used in industry and manufacturing in paint thinners, fuels, and medical supplies, and was also used as a solvent.

On September 7, 1919, the *New York Times* reported an increase in the numbers of deaths from people drinking wood alcohol as a substitute for grain alcohol. Methanol (methyl alcohol) is found in alcohol produced by distilling wood. The National Committee for the Prevention of Blindness recorded over 1000 reported cases of blindness (across the country) resulting from the consumption of wood alcohol. Dr. Alexander Gettler, a toxicologist with

ALCOHOL POISONING: METHANOL AND OTHER DENATURANTS (continued)

the New York Office of the Chief Medical Examiner and Chemical Laboratory of the Pathological Department, Bellevue and Allied Hospitals, also reported an increase in deaths due to wood alcohol. He reported examining over 700 human organs for alcohol in 1918–1919. As a result, states began to pass laws to regulate and control the sale of wood alcohol.

Beginning in 1906, industrial users could purchase ethanol without paying the tax levied on drinking alcohol. The US government devised a method of making the ethanol deadly to drink—by adding methanol—while leaving the bulk chemical properties unchanged. (Methanol is used today in windshield washer fluid and is poisonous and extremely toxic.) The resultant alcohol was labeled as "denatured" alcohol. Several other denaturing methods followed. Some involved the addition of poisonous metals such as mercury, cadmium, and zinc to the ethanol. Others involved the addition of less lethal but extremely bitter compounds to the ethanol, rendering it undrinkable. Bootleggers hired chemists to distill the alcohol to remove the contaminants and return the ethanol to a composition that was safe to consume. In response, by mid-1927, new denaturants were added to the alcohol including common chemicals such as gasoline, kerosene, chloroform, camphor, ether, formaldehyde, acetone, iodine, and quinine.

Eventually, prohibition was overturned with the ratification of the 21st Amendment and consumption of alcohol was again legalized on December 5, 1933.

BIBLIOGRAPHY

Many Deaths Due to Wood Alcohol, *New York Times*, September 7, 1919.

Gettler, A.O. 1920. Critical study of methods for the detection of methyl alcohol. *J. Biol. Chem.* 42:311–328.

Blum, D. The chemist's war: The little-told story of how the U.S. government poisoned alcohol during Prohibition with deadly consequences. February 19, 2010, www.slate.com/articles/health_and_science/medical_examiner/2010/02/the_chemists_war.html (accessed January 23, 2018).

Forensic science is the application of the scientific method to legal questions. The laws themselves are enforced and upheld by the criminal justice system including federal, state, and local law enforcement agencies and the courts. The goal of the criminal justice system is the establishment of the guilt or innocence of a suspect or suspects accused of a crime.

Forensic chemistry is a subdiscipline of forensic science. Its principles guide the analyses performed in modern forensic laboratories. Forensic chemistry's roots lie in medicolegal investigation, toxicology, and microscopy. Deaths due to tainted food products, new applications of materials in the home, drug use and abuse, and industrial pollution sped up the development of modern forensic science investigations and practices.

Forensic chemistry emerged in Europe in the 1830s with advances by scientists including James Marsh. Marsh was a British chemist who developed a method for testing the presence of arsenic in human tissue that was the first use of toxicology in a jury trial. The Marsh Test (1836), as it is now widely known, employs testing using zinc and sulfuric acid. Arsine gas is formed in the presence of even small amounts of arsenic; the method was used to detect the ingestion of rat poison containing arsenic in cases of suspected poisonings.

Approximately 50 years later, University of Pennsylvania professor Theodore Wormley authored the first American book, *Micro-chemistry of Poisons* (1885), dedicated in the preface to "the study of the chemical properties of poisons as revealed by the aid of the microscope." The United States Pure Food and Drugs Act (1906), which was signed into law by then President Theodore Roosevelt, regulated food and medicines and ultimately paved the way for the modern Food and Drug Administration. The Pure Food and Drugs Act prevented the production and trafficking of poisonous, mislabeled, or adulterated foods as well as pharmaceutical drugs and alcoholic beverages. The American toxicologist Dr. Alexander Gettler was instrumental in advancing forensic chemistry in his work as chief chemist at the New York Medical Examiner's office; he significantly advanced the science through his several publications including his paper "The Toxicology of Cyanide" published with his student J. Ogden Baine in 1938 in the *American Journal of the Medical Sciences*. It documents the case study of Fremont and Annie Jackson who died in 1922 by the inhalation of fumigation products in their Manhattan apartment.

As forensic chemistry is focused on materials analysis, innovations that have advanced the field have been many and varied as shown in Table 1.1. The innovations include the development of new discernible chemical reactions,

Table 1.1   Brief history of some notable advances in forensic chemistry

Year	Advance
1590s	Zacharias Janssen develops first compound light microscope
1784	First use of fracture edge matching/pattern matching in John Toms' case
1810	Konigin Hanschritt document dye analyzed by chemical test
1828	William Nichol invents polarized light microscope
1835	Charles Wheatstone invents emission spectroscopy
1836	James Marsh develops test for arsenic and it is used in a jury trial
1858	Johann Peter Griess develops test for nitrites
1867	Alfred Nobel receives US patent for his invention of dynamite
1880	Henry Faulds suggests using fingerprints on clay and glass to solve crimes
1883	K. Mandelin develops test for strychnine later applied to alkaloids
1885	Theodore Wormley publishes book *Micro-chemistry of Poisons*
1889	Alexandre Lacassagne matches bullets using lands and grooves to a gun barrel
1891	Hans Gross describes the use of physical evidence in solving crimes in his book *Handbuch für Untersuchungsrichter* and coins the term Kriminalistik (Criminalistics)
1892	Francis Galton publishes first book on fingerprints
1894	Alphonse Bertillon's handwriting analysis is used to convict Alfred Dreyfus (falsely)
1898	J. J. Thomson measures mass-to-charge ratio of the electron
1898	Paul Jeserich uses minutiae to individualize bullets
1903	Will West prison case solved using latent fingerprints
1903	M. S. Tswett separates plant pigments using paper chromatography
1906	President T. Roosevelt signs US Pure Food and Drugs Act signed into law
1910	Albert Sherman Osborn publishes *Questioned Documents*
1915	First use of chemical weapons
1919	Francis Aston builds the first fully functional mass spectrometer and later uses it to discover 212 naturally occurring isotopes
1928	Geneva Protocol signed that prohibits use of chemical and biological weapons in war
1928	C. V. Raman develops Raman spectroscopy
1930	Edmond Locard's Principe de l'echange "Exchange Principle" coined
1930s	Pierre Duquenois develops color test for THC
1940	Glenn Seaborg, Jospeh Kennedy, Edwin McMillan, Emilio Segre, and Arthur Wahl discover plutonium-239
1945	First nuclear magnetic resonance spectroscopy (NMR) spectra of liquids and solids by Felix Bloch and Edward Mills Purcell, independently
1948	Founding of the American Academy of Forensic Sciences
1951	Archer John Porter Martin and Richard Laurence Millington Synge invent modern gas chromatography
1955	Modern flame atomic absorption spectrometer developed by Sir Alan Walsh
1962	Rachel Carson publishes book *Silent Spring*
1970	First meeting of the Society of Toxicology on Long Island
1973	GC-MS applications to analysis of drugs and metabolites
1974	Richard Ernst pioneers two-dimensional NMR COSY experiment
1974	SEM-EDX is applied to gunshot residue analysis
1977	Application of FT-IR in forensic science
1988	Franz Hillenkamp and Michael Karas pioneer the matrix-assisted laser desorption ionization-MS technique
1988	Introduction of enzyme-multiplied immunoassay technique (EMIT) in forensic toxicology
1991	Richard Ernst develops high-resolution nuclear magnetic resonance spectroscopy
1992	GC-IR is applied to forensic drug analysis
1996	Raman spectroscopy is introduced to forensic use
1997	Scientific Working Group for the Analysis of Seized Drugs is created by the US National Institute of Standards and Technology
2001	US Federal Bureau of Investigation investigates Amerithrax case of deaths due to mailed letters containing anthrax spores

instrumental tools, books, laws, methods, index cases, and even the development of dual-use materials so often misused by criminals.

While federal, state, and local law enforcement agencies are the primary providers of forensic chemistry services to the criminal justice system, private and university laboratories are also available for this purpose. In the United States, major federal agencies including the Federal Bureau of Investigation (FBI), Drug Enforcement Agency (DEA), Bureau of Alcohol, Tobacco, Firearms and Explosives (ATF), Environmental Protection Agency (EPA), Department of Homeland Security (DHS), and the Postal Service (USPS) have their own labs or contract with outside laboratories to perform forensic testing and research. In addition to the federal labs, states, cities, and counties may have their own forensic labs focused on criminal investigations. University forensic labs are common in Europe and other parts of the world. Forensic chemistry analyses are performed in sections including controlled substances analysis, toxicology, explosives and fire debris, trace evidence, latent prints, firearms, tool marks and impression evidence, and questioned documents. The units of the crime laboratory that utilize forensic chemistry, which will be covered in this book, are listed in Table 1.2. Forensic laboratories also examine environmental samples that may contain pesticides, herbicides, and chemicals used as weapons, and the improper use or disposal of these and other chemicals by individuals and industry. Crime scene investigation and forensic biology are also important sections of forensic laboratories, and while both utilize chemical principles and tests in their evaluation of evidence, they will not be covered in this book.

Criminalistics describes the branch of forensic science focused on evaluating physical evidence collected at crime scenes. Scientists working in the field of criminalistics are termed criminalists and may conduct crime scene investigations, perform analyses in the laboratory, write reports, and testify as expert witnesses in court. Criminalists focus on recognizing, documenting, collecting, preserving, analyzing, and reporting on physical evidence. A methamphetamine drug synthesis crime scene is shown in Figure 1.1. Several evidence items are visible including glassware, drug material or chemical intermediates, containers, and tubing. Notably, discarded matchboxes are visible in the photo; the red phosphorous from the strike pads is used in the synthesis of methamphetamine.

Physical evidence may include any type of physical material found at a crime scene. This type of evidence can include everyday items such as household chemicals, fabrics and fibers, hairs, glass, fingerprints, soil, plant material, handwritten or typed documents, checks, polymers and plastics, inks and dyes, serial numbers, and tools and tool marks.

Table 1.2   Units of forensic laboratories that use forensic chemistry

Unit	Evidence	Methods
Controlled substance analysis	White powders, colored chemicals, botanical material, and other suspected controlled substances or their starting materials or intermediates	Color spot tests, macroscopic tests, stereomicroscopy, microcrystalline tests, FTIR, GC-MS
Toxicology	Blood-alcohol samples Body fluid-drug samples including blood, urine, saliva, stomach contents, and vitreous humor	GC-MS, LC-MS, ELISA
Latent print examination	Latent and visible prints, impression evidence	ALS, photography, fingerprint powder, superglue fuming, chemical latent print development methods, lifts
Questioned documents	Handwritten and typewritten documents including checks, suicide notes, and ransom notes, among others	ALS, stereomicroscopy, TLC, Raman spectroscopy, IR imaging, SEM
Trace evidence	Polymers, paint, glass, hair, fiber, plastic, paper, soil	Stereomicroscopy Compound light microscopy Polarizing light microscopy Microspectrometer (UV-Vis, FTIR) Scanning electron microscope (SEM) Phase contrast microscopy Fluorescence microscopy Solubility testing Hot stage microscopy
Firearm and tool mark examination	Firearms, tools, serial numbers	Stereomicroscopy comparisons
Explosive and fire debris examination	Burned materials, explosives remnants, accelerants	GC, GC-MS, SEM

Figure 1.1 Methamphetamine lab crime scene. (From Nathan Russell, https://www.flickr.com/photos/nathanrussell/2690501345/in/photolist-56Kwje-e2uUsZ-4Shks7-3MuHv7-kAeuvD-6nX5zE-oot25S-eegoz7-7Py3rm-GnhsDD-Ge8kWy-dEWvJE-pyYH16-6Hj3kt-3kQNeH-6Hj3fV-6Ho63U-6Ho5Hh-6Hj3t2-6Hj3zp-6Hj3DR-6Ho6C5-aKcU4F-eegmMG-etR3Qi-6Fmg8i-5Zh4gf-56KwaF-2RRujG-oosCQ1-oot1M6-7fB3wD-4Wtdfk-h4nFNn-2hfrtA-oos7Sg-6Hj48x-5P6iGJ-6Ho6fW-6Ho6u9-6Ho5QC-6Ho6ry-KWAUS-6Hj3LT-6Ho6zh-6Hj3RZ-6Hj3oH-6Ho69m-QXDsxn-QXFtr2.)

Physical evidence may also include narcotics, marijuana, and drug paraphernalia, weapons, ammunition and shell casings, flammable substances and accelerants, explosives, body fluids, impressions such as tire markings, shoe prints, tracks, bite marks, and fabric impressions. Cigarette butts, chewing gum, contact lenses, clothing, rags, plastic bags, sawdust, duct tape, and rope may all be submitted as physical evidence for analysis by the forensic laboratory. The collection of physical evidence is subject to search and seizure laws.

Physical evidence is collected and labeled by a *crime scene investigator* or technician who is trained in forensic science. These specialists are responsible for identifying, photographing, logging, collecting, tagging, and transporting evidence from the crime scene that can be used to gain knowledge of the events, persons, and circumstances surrounding the crime. Each evidence item is logged on an evidence submission form. Care must be taken not to introduce outside contaminants such as DNA, fingerprints, hairs, and clothing fibers to the physical evidence as well as

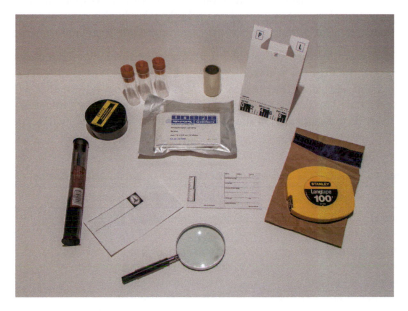

Figure 1.2 Evidence collection tools. (Courtesy of Tim Phillips.)

not to lose trace evidence items. Evidence is packaged based on the type of evidence and its physical characteristics. In order to guard against altering or damaging evidence, each item must be packaged separately. Some evidence collection devices and containers are shown in Figure 1.2. For example, powders and fine particles may be packaged in a folded paper pharmacy fold. Burned materials recovered in an investigation of suspected arson would be collected in a clean paint can and closed tightly with a metal lid. Pills and small pieces of broken glass are collected in a plastic pill container or vial. Clothing with blood or other body fluids are packaged in paper bags to allow the garments to dry out and prevent the growth of mold. Trace fibers may be collected in plastic bags, containers, or small envelopes. Swabs should be placed in separate containers with desiccant. Fingerprints should be placed on fingerprint cards or the item should be collected. Evidence items should be labeled with the technician name or initials/signature, case number, item number, date, and time. Evidence items should be recorded on *chain of custody* forms for tracking the lineage of the sample from collection to testing to admission in court. Every person who examines or handles the evidence is logged on the chain of custody form. Although evidence may be shipped by FEDEX or the regular postal mail, it is often hand-delivered to avoid lengthening the chain of custody.

Physical evidence has several advantages over other types of evidence. It is tangible and can be presented in court as exhibits to the jury. It can be taken by the jury to the jury room for further examination. It cannot be distorted by the defendant and is not subject to the memory loss of a witness. It can be tested by independent contract laboratories. Cases without physical evidence are often weaker. Other types of evidence include direct evidence and circumstantial evidence.

Physical evidence has class and individual characteristics (Table 1.3). *Class characteristics* are characteristics shared by all members of a particular class. While they cannot be attributed to a single source, they can be used to conclusively eliminate association with a suspect or source location. Mass-produced products often only have class characteristics unless a change to the item has been imparted such as a tear, breakage, or wear pattern that would give the item individual characteristics. *Individual characteristics* are unique characteristics that can be used to associate an object to a particular individual, suspect, or place associated with a crime. They can be used to attribute evidence to a common source with an extremely high degree of certainty. These individual characteristics set the object apart from others in the same class. Most physical evidence has both class and individual characteristics that can be associated with it. For example, the drug methamphetamine may be associated with a class of drugs that are all white powders and test positive via an orange-brown color with a chemical test called the Marquis test. However, the particular lot of methamphetamine will also have individual characteristics including trace chemicals used in the production that may not have been fully purified out of the final product and chemical markers associated with various manufacturers. These materials can be teased out using chemical techniques and associated by attribution methods using chemical instrumentation and statistical methods. For firearms, class characteristics of an item may involve a revolver or .38 caliber with six lands and grooves and a left-hand twist impressed on bullets fired with the weapon. Individual characteristics may include unique striated markings that can be observed on the bullets shot by this particular revolver that are imparted by the microgrooving produced in the production of the firearm or random wear characteristics in the barrel.

After evidence is identified, a *presumptive test*, or color test, may be performed to tentatively determine the identity of the substance or to determine the class of materials to which it belongs. If the substance remains of value to the case, the item is packaged and labeled as described earlier. At the laboratory, criminalists will perform further presumptive or screening tests. Based on the results of these tests, the material will be evaluated with spectroscopic and chromatographic tests to confirm the identity of the substance. Additional analysis will be performed to determine

Table 1.3 Examples of class and individual characteristics for physical evidence types

Class characteristics	Individualizing characteristics
Brand of shoe label on sole	Wear pattern on sole
Diameter of firearm barrel	Striations on bullet
Density of broken glass	Fit of broken piece to make whole
Drug identity	Trace metals that attribute its source
Species origin of hair	DNA identification from follicular tag
Make, model, and color of vehicle	Serial number of vehicle
Letters in a typed word	Angle and wear characteristics from typewriter

Figure 1.3 Forensic analysis from the crime scene to the lab.

the quantity of each of the components of the sample, as needed. A sample evidence collection and processing scheme is shown in Figure 1.3.

Criminalists are often asked to determine if two objects originated from the same source. In order to make this determination, they may compare the item in question (Q) to the item of known (K) origin. For example, a comparison may be made between a piece of plastic seemingly from an automobile headlight cover and a broken headlight cover that is still intact in the automobile involved in a crime or incident. Careful comparison under magnification will allow the criminalist to determine if the evidence is associated with or included in the reference material or is not associated with or a match to the source. *Reference samples* are those collected from a different area than the evidence but have a known or authenticated source or composition. These may include hairs from victims/suspects, buccal swabs from victims/suspects, unburned or unstained areas of fibers or fabrics, or unaltered paint from a car involved in a "hit-and-run."

In other cases, reference samples are not always available for a determination of origin. In order to determine the source of the evidence, criminalists rely on databases assembled for this purpose. For example, the Integrated Automated Fingerprint Identification System (IAFIS) can be used to search for a match to a fingerprint of unknown source based on the individual characteristics of the fingerprint in question. *Comparison standards* may also be used to determine the identity of materials. These standards have a known identity and have a trusted source or origin. For example, the National Institutes of Standards and Technology (NIST) produces various standard reference material (SRM) for purchase and come with certification papers. The American Society of the International Association for Testing and Materials (ASTM) is another trusted source of chemical and material standards.

Care must be taken when working with all chemicals, even standard chemicals, as well as physical evidence. Even though standards are of known origin, they, like evidence, have risks associated with them. Unlike evidence, they are delivered with chemical and occupational safety and health information in the form of a *safety data sheet* (SDS). This includes information on the identity of the substance or mixture; supplier details; hazards information including classification of the substance or mixture; composition/ingredients; first aid measures in case of exposure; fire-fighting measures; precautions in case of accidental release; handling and storage; personal protection for handling; physical and chemical properties; stability and reactivity; toxicological information; ecological information; disposal considerations; transport information and precautions; and regulator information. The National Fire Protection Association (NFPA) diamond (Figure 1.4) is often included in the SDS and includes information in colored and numerical code for health, fire, reactivity, and specific hazards. Figure 1.4 also shows the NFPA ratings for methanol, the toxic alcohol discussed in the case study at the beginning of the chapter.

To protect themselves from these items, criminalists wear one or more of the following personal protective equipment (PPE) items: goggles, latex or nitrile gloves, lab coat, Tyvek shoe covers, Tyvek or Kleengard suits, and/or a mask or respirator. Additionally, for the safety of all the scientists working in the laboratory, smoking, eating, drinking, and makeup application should never be performed in the lab. In the case of chemical analysis of samples that may pose a biohazard threat, it is important to decontaminate equipment; use biohazard labels for waste; dispose of contaminated gloves, masks, suits and consumables in red biohazard bags; and work in a lab space that is designated as appropriate for the biohazard level. Clothing and equipment contaminated with nuclear weapon material or fallout should also be removed immediately, labeled appropriately, and decontaminated.

Also used in the evaluation of physical evidence are controls, standards, and blanks. *Control samples* are substances or materials that are tested simultaneously with the item in question. They are used to reveal potential problems with the test or procedure. Background controls, positive controls, and negative controls are all examples of control samples. *Background controls* are run to test that the background is negative or as expected for the test run without a sample. *Positive controls* are often conducted using standards to test and ensure that the method produces the appropriate positive results or responses with a sample. A *negative control* is also included in a well-designed experiment and should not produce any response or a known negative response for the test. Instrumental plus other qualitative and quantitative data should be recorded as reported without rounding so as not to lose *accuracy* (measure of the true

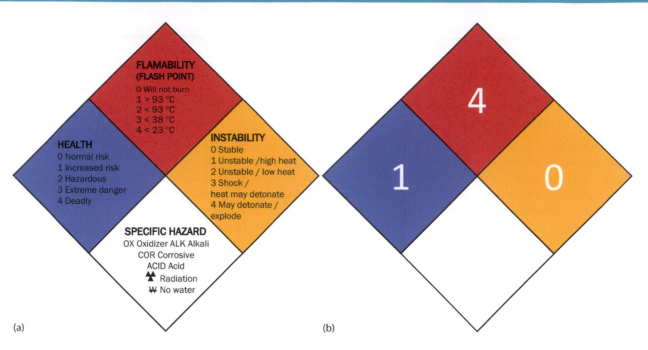

Figure 1.4   (a) NFPA diamond with meanings of scale; (b) NFPA diamond for methanol.

value). *Replicates*, or retests of a part or small amount of the evidence sample, should be performed. Ideally, samples should be tested in triplicate or more to evaluate data *precision* (measure of reproducibility). Accuracy and precision are demonstrated by the targets in Figure 1.5. High accuracy and high precision are characterized by hitting the bull's-eye target in the center repeatedly. High precision but low accuracy is characterized by hitting another position on the target (but not the bull's-eye) repeatedly. Low precision but high accuracy is characterized by hitting the bull's-eye on average, but not directly.

A methodology used in the forensic laboratory must follow *standard operating procedures* (SOPs) that have been validation tested in that lab, by the lab's staff, on the lab's instruments. The SOP often references published methods and procedures for evaluating chemical physical evidence. SOPs include methods for the use of control and standard samples and performing calibration checks of instrumentation and methodology using calibration curves.

In addition to using SOPs, forensic laboratories follow strict quality control and quality assurance methods. *Quality control* (QC) methods include evaluating control samples, blanks, internal standards (spikes), reference samples, and/or standard samples with evidence samples. *Quality assurance* (QA) measures ensure that results are scientifically valid and that expert opinions are concluded only from data and results that are deemed reliable. The QA approach that the lab implements is performed for the purpose of satisfying quality requirements for the data and complying with regulations that surround the testing. This may include staff education requirements, evidence handling procedures, laboratory security procedures, peer review of all results, specific case file documentation, process for distributing reports, audits of testimony, and proficiency tests. Because of their work in the criminal justice systems, criminalists working with evidence can expect thorough background checks, polygraph examinations, screening for illegal drug use, driving record checks, credit checks, residential history checks, and a review of employment history.

Low Accuracy   High Accuracy   High Accuracy
High Precision   High Precision   Low Precision

Figure 1.5   Accuracy and precision.

Several QA groups exist in the forensic community including the American Society of Crime Lab Directors (ASCLD), the Association of Forensic Quality Assurance Managers (AFQAM), the Technical Working Groups (TWGs), the Scientific Working Groups (SWGs), and now the Organization of Scientific Area Committees (OSACs).

Criminalists are often asked to serve as *expert witnesses* for cases in the legal system. Based on their knowledge, skill, education, experience, and training, criminalists must carefully decide on and triage evidence in order of relative value. They must also educate police officers, lawyers, judges, and juries about how the chemical and instrumental tests they perform work and how they arrived at their conclusions from the evidence tests in lay terms. However, criminalists may also be challenged on his or her basic qualifications, methodology used, and ability to give an opinion on the methods, results, and aspects of the case. Criminalists must combine appropriate credentials including academic degrees, successful professional test results (e.g., American Board of Criminalistics), successful proficiency test results, documented professional development and training, and professional experience to be successfully admitted as an expert witness by the judge. For a judge to admit the evidence and the results of the forensic analysis as part of the official record, the methodology must be scientifically sound and reliable and appropriately applied to support the data and conclusions presented. The theory or methodology must have gained general acceptance by other practitioners in the field or the relevant scientific community and/or have been successfully published in a peer-reviewed scientific journal. The judge will also consider the rate of error of the technique and the existence and maintenance of standards for the technique.

The workload of the criminalist—and the capacity of the forensic laboratory—has dramatically increased over the last 60 years, in part due to the increase in drug-related crime. Additionally, in the United States, Supreme Court decisions in the 1960s were responsible for the greater police emphasis on professional investigation and evaluating physical evidence scientifically. Advances in chemical instrumentation and analysis methods have led to improved capabilities for analyzing and discriminating evidence. Research in chemical and material analysis has led to a rich peer-reviewed literature.

The SWGDRUG has categorized modern chemical instrumentation and wet chemical methods into three categories (A, B, and C) by maximum potential discriminating power (Category A is highest) as shown in Table 1.4. For seized drug and other submitted evidence identification analysis, at least two validated techniques must be used if a Category A method is used. If no Category A method is used, at least three different (validated) techniques must be employed including at least two uncorrelated Category B methods. In the case of plant material suspected to be marijuana, Category B macroscopic and microscopic investigation techniques may be used if the submitted evidence has sufficient observable detail under the microscope; the additional technique is frequently the Duquenois–Levine color test. Examinations must lead to reviewable data including printed spectra or chromatograms, photographs or images, thin layer chromatography plates, peer review (for microcrystalline tests), references to published data (for pharmaceutical identifiers), or detailed descriptions of microscopic observations (for marijuana only).

Table 1.4 SWGDRUG categories of analytical techniques by discriminating power (Category A > Category B > Category C)

Category A	Category B	Category C
Infrared spectroscopy	Capillary electrophoresis	Color tests
Mass spectrometry	Gas chromatography	Fluorescence spectroscopy
Nuclear magnetic resonance spectroscopy	Ion mobility spectrometry	Immunoassay
Raman spectroscopy	Liquid chromatography	Melting point
X-ray diffractometry	Macroscopic examination[a]	UV spectroscopy
	Microcrystalline tests	
	Microscopic examination[a]	
	Pharmaceutical identifiers	
	Thin layer chromatography	

Source: SWGDRUG, www.swgdrug.org/Documents/SWGDRUG%20Recommendations%20 Version%207-1.pdf.

[a] Cannabis only; if submitted evidence lacks sufficient observable detail under the microscope, an extract of the plant material is analyzed by another technique. Most commonly, one of the following is employed: thin layer chromatography or gas chromatography coupled with mass spectrometry.

Forensic scientists are scientists and forensic chemists are chemists that apply their knowledge of chemical methods and procedures to solve forensic problems. The work of forensic chemists is varied and challenging but is always based in chemical principles. The forensic chemist must be flexible, resourceful, creative, and persistent. Unlike traditional lab work in chemistry, forensic chemists do not know the origins of their samples, the starting materials used, the manufacturers that provided those materials, what the materials have been mixed with, the environmental conditions to which the evidence was exposed, and so on. Oftentimes, physical evidence samples are junk—and would be considered such in any other setting.

The tools of the forensic chemist include qualitative and quantitative methods of evaluating physical and chemical properties. These tools include measurement, microscopy, chromatography, and spectroscopy techniques, all of which will be described further in the upcoming chapters. Forensic chemists rely heavily on databases and libraries and sophisticated software with pattern-matching algorithms to interpret the data collected from the evidence. However, new, synthetic drug substances pose real challenges to the everyday workflow. These materials are often too new to be included in chemical libraries. Professional forensic chemists meet these challenges using chemistry, through discussions with peers and other experts, using the chemical literature, and using chemical structure determination techniques. Many remain members of—and maintain involvement in—chemical societies related to their expertise and read and contribute to chemistry journals.

Criminalists do not just test evidence with the goal of enforcing current laws. They must also identify and seek to understand materials that are not controlled to help the legal system enact new laws to keep society safe. For example, synthetic cannabinoids and cathinones pose risks to users of these substances. As they are often marketed as "spice," "plant food," "fertilizer," or "bath salts," and labeled "not for human consumption," they pose real risks for users seeking a "legal high" and real challenges to forensic laboratories in analyzing these (often new) unknowns. Coupling methods such as nuclear magnetic resonance spectroscopy, mass spectrometry, melting point, and infrared spectroscopy can yield sufficient chemical information to allow criminalists to solve chemical structures *ab initio*.

The chemical literature is invaluable in solving—and reporting solutions to—these problems. Peer-reviewed journals that are heavily used by forensic chemists include traditional analytical chemistry journals such as *Analytical Chemistry, Journal of Separation Science, Food Chemistry, Journal of Agricultural and Food Chemistry,* and *Journal of Analytical Toxicology,* and criminalistics-specific journals such as the *Journal of Forensic Sciences, Forensic Sciences International, Forensic Science Communications, Australian Journal of Forensic Sciences, Canadian Journal of Forensic Science, Drug Testing and Analysis, Microgram Journal, Journal of Forensic Research, Journal of Forensic and Legal Medicine, Legal Medicine, Science and Justice, Journal of Forensic Identification, Medicine, Science, and the Law,* and *Forensic Science, Medicine and Pathology.* In 2015, the first journal focused specifically on this field, *Forensic Chemistry,* was launched by Elsevier.

Forensic chemists are professionals and many are members of professional organizations including the American Academy of Forensic Sciences (AAFS), Society of Forensic Toxicologists (SOFT), American Board of Forensic Toxicology (ABFT), American Board of Criminalistics (ABC), American Chemical Society (ACS), Australian and New Zealand Forensic Science Society (ANZFSS), British Academy of Forensic Sciences, Canadian Society of Forensic Science, the Chartered Society of Forensic Sciences (CSFS), Forensic Sciences Society, Japanese Society of Legal Medicine, and the International Association for Identification (IAI), among others. Some forensic chemists are also members of ASCLD. These professional organizations have a recognized and accepted authority in the profession. These organizations promote the forensic profession in society, have a code of ethics to which their members uphold, form the basis of a professional network for their members, and serve as the voice of the profession to the government.

The integrity of forensic laboratories is of the utmost importance. The integrity is documented by a combination of several intentional measures including the use of proper tools, validation, following protocols, using standards, criminalist training, and accreditation. There are two main agencies that accredit forensic chemistry laboratories in the United States. These include the American National Accreditation Board (ANAB): American National Standards Institute (ANSI)–American Society for Quality (ASQ), which recently merged their forensic operations and is one of several entities entitled to offer International Organization for Standardization accreditation (e.g., ISO 17025). Another organization, COLA, is a clinical lab accreditation agency.

QUESTIONS

1. Which of the following defines criminalistics?

   a. A branch of forensic science that applies science to law through the recognition, documentation, collection, preservation, and analysis of physical evidence

   b. A branch of science devoted to the application of science to criminal and civil laws that are enforced by police agencies in the criminal justice system

   c. A branch of science devoted to determining the time, manner, and cause of death

   d. A branch of science devoted to determining genetic lineages

   e. All of the above

2. Which of the following items would not be a piece of physical evidence collected by CSIs?

   a. Victim's clothing

   b. Investigator's clothing

   c. Suspect's clothing

   d. Blood splatter

3. For quality control, evidence samples may be split into three parts. These parts are referred to as _____.

   a. Replicates   b. Controls

   c. Calibration standards   d. Reference samples   e. All of the above

4. Establishing the exact whereabouts of an item of evidence and under whose control it was from the time of its collection to its admissibility as evidence in court is known as maintaining the:

   a. Scientific method   b. Quality control

   c. Chain of command   d. Link report   e. Chain of custody

5. A function of a forensic scientist includes:

   a. Furnishing training on the proper collection of physical evidence

   b. Analysis of physical evidence

   c. Providing expert testimony

   d. All of the above

6. Historically, how did forensic chemistry develop? What did it emerge from?

7. List some personal protection equipment to wear when handling physical evidence.

8. Which laboratories perform testing on forensic evidence?

9. Explain the difference between class and individual characteristics. Give an example of each.

10. Define the three categories of SWGDRUG testing.

Bibliography

American Academy of Forensic Sciences. www.aafs.org/about-aafs/ (accessed January 23, 2018).

Amerithrax Investigative Summary. 2010. www.justice.gov/archive/amerithrax/docs/amx-investigative-summary2.pdf (accessed January 23, 2018).

ANSI-ASQ National Accreditation Board. www.anab.org/forensic-accreditation (accessed November 3, 2017).

ASTM International. www.astm.org (accessed November 3, 2017).

Bloch, F., W.W. Hansen, and M. Packard. 1946. Nuclear induction. *Phys Rev.* 69:127.

Bloch, F., W.W. Hansen, and M. Packard. 1946. The nuclear induction experiment. *Phys Rev.* 70:474–485.

Bloch, F. 1952. The principle of nuclear induction. *Nobel Lecture.* pp. 203–215.

Blum, D. 2011. *The Poisoner's Handbook: Murder and the Birth of Forensic Medicine in Jazz Age New York*. New York: Penguin Press.

Brand, W.A. 2004. Chapter 38: Mass spectrometer hardware for analyzing stable isotope ratios. In *Handbook of Stable Isotope Analytical Techniques, Volume I*, ed. P.A. deGroot, 835–856. Amsterdam: Elsevier B. V.

Buchwald, J.Z. and A. Warwick, eds. 2001. *Histories of the Electron: The Birth of Microphysics*. Cambridge: The MIT Press.

Carson, R. 1962. *Silent Spring*. Boston: Houghton Mifflin Company.

Collins, G.P. 1991. Nobel chemistry prize recognizes the importance of Ernst's NMR work. *Physics Today* 44(12):19.

Draffan, G.H., R.A. Clare, and F.M. Williams. 1973. Determination of barbiturates and their metabolites in small plasma samples by gas chromatography-mass spectrometry. Amylorbarbitone and 3′-hydroxyamylobarbitone. *J Chromatogr*. 75(1):45–53.

Elkins, K.M., S.E. Gray, and Z.M. Krohn. 2015. Evaluation of technology in crime scene investigation, *CS Eye*. www.cseye.com/content/2015/april/research/evaluation-of-technology (accessed January 25, 2018).

Ettre, L.S. 1991. 1941–1951: The golden decade of chromatography. *Analyst* 116(12): 1231–1235.

Ettre, L.S. 2003. M.S. Tswett and the invention of chromatography. *LC+GC Europe*. pp 1–7.

Everts, S. 2015. When chemicals became weapons of war. *Chemical and Engineering News*. http://chemicalweapons.cenmag.org/when-chemicals-became-weapons-of-war/ (accessed January 24, 2018).

Faulds, H. 1880. On the Skin-furrows of the Hand. *Nature* 22:605.

Frankel, E. 2008. William Nicol. Complete Dictionary of Scientific Biography. www.encyclopedia.com/people/science-and-technology/metallurgy-and-mining-biographies/william-nicol (accessed January 23, 2018).

Galton, F. 1892. *Finger Prints*. London: MacMillan and Co.

Gettler, A. and J.O. Baine. 1938. The toxicology of cyanide. *Am. J. Med. Sci*. 195(2):182–198.

Gorter, C.J. and L.J.F. Broer. 1942. Negative result of an attempt to observe nuclear magnetic resonance in solids. *Physica* 9:591.

Gross, H. 1893. *Handbuch für Untersuchungsrichter*. Graz: Leuschner & Lubensky.

Harding, L.A. 1892. Forensic microscopy. *Science* 20(508):242–243.

Horning, M.G., J. Nowlin, K. Lertratanangkoon, R.N. Stillwell, W.G. Stillwell, and R.M. Hill. 1973. Use of stable isotopes in measuring low concentrations of drugs and drug metabolites by GC-MS-COM procedures. *Clin. Chem*. 19(8):845–852.

Jacobs, A.D. and R.R. Steiner. 2014. Detection of the Duquenois-Levine chromophore in a marijuana sample. *Forensic Sci. Int*. 239:1–5.

James, F.A.J.L. 1983. The study of spark spectra 1835–1859. *Ambix*. 30(3):137–162.

Kalasinsky, K.S., B. Levine, and M.L. Smith. 1992. Feasibility of using GC/FT-IR for drug analysis in the forensic toxicology laboratory. *J. Anal. Toxicol*. 16(5):332–336.

Katz, H. 2010. *Cold Cases: Famous Unsolved Mysteries, Crimes, and Disappearances in America*. Santa Barbara: Greenwood Publishing Group.

Kean, S. 2011. *The Disappearing Spoon*. New York: Back Bay Books.

Klinger, L.S. 2015. *The Sherlock Holmes Book (Big Ideas Simply Explained)*. London: DK Publishing.

Lewellen, L.J and H.H. McCurdy. 1988. A novel procedure for the analysis of drugs in whole blood by homogeneous enzyme immunoassay (EMIT). *J. Anal. Toxicol*. 12(5):260–264.

Lewis, J.K, J. Wei, and G. Siuzdak. 2000. Matrix-assisted laser desorption/ionization mass spectrometry in peptide and protein analysis. In *Encyclopedia of Analytical Chemistry*, ed. R.A. Meyers, 5880–5894. Chichester: John Wiley & Sons Ltd.

Locard, E. 1930. The analysis of dust traces, Part I. *Am. J. Police Sci*. 1(3):276.

Locard, E. 1930. The analysis of dust traces, Part II. *Am. J. Police Sci.* 1(4):401–418.

Locard, E. 1930. The analysis of dust traces, Part III. *Am. J. Police Sci.* 1(5):496–514.

L'vov, B.V. 2005. Fifty years of atomic absorption spectrometry. *J. Analyt. Chem.* 60(4):382–392.

Lyle, D.P. 2008. *Howdunit Forensics*. Cincinnati, OH: Writer's Digest Books.

May, L. 1930. The identification of knives, tools, and instruments: A positive science. *Am. J. Police Sci.* 1(3): 246–259.

NicDaeid, N. ed. 2010. *Fifty Years of Forensic Science*. West Sussex, UK: Wiley-Blackwell.

Nickell, J. and J.F. Fischer. 1998. *Crime Science: Methods of Forensic Detection*. Lexington: University Press of Kentucky.

Nobel Media AB. 2014. All Nobel prizes. www.nobelprize.org/nobel_prizes/lists/all/ (accessed January 18, 2018).

Nobel Media AB. 2014. List of Alfred Nobel's patents. www.nobelprize.org/alfred_nobel/biographical/patents.html (accessed January 23, 2018).

Osborn, A.S. 1910. *Questioned Documents*. Rochester: The Lawyers' Co-Operative Publishing Co.

Pestaner, J.P., F.G. Mullick, and J.A. Centeno. 1996. Characterization of acetaminophen: Molecular microanalysis with Raman microprobe spectroscopy. *J. Forensic Sci.* 41(6):1060–1063.

Petruzzi, J. 1973. GC/MS in drug analysis. *Anal. Chem.* 45(14): 1213A–1213A.

Poe, C.F. and D.W. O'Day. 1930. A study of Mandelin's test for strychnine. *J. Pharm. Sci.* 19(12):1292–1299.

Purcell, E.M., H.C. Torrey, and R.V. Pound. 1946. Resonance absorption by nuclear moments in a solid. *Phys. Rev.* 69:37–38.

Purcell, E.M. 1952. Research in nuclear magnetism. *Nobel Lecture*. pp. 219–231.

Rabi, I.I. 1937. Space quantization in a gyrating magnetic field. *Phys. Rev.* 51:652–654.

Robinson, M. S., F. Stoller, M. Costanza-Robinson, and J.K. Jones. 2008. *Write Like a Chemist: A Textbook and Resource*. Oxford: Oxford University Press.

Saferstein, R. 2015. *Criminalistics: An Introduction to Forensic Science*. 11th ed. Boston: Pearson.

Seaborg, G.T. 1985. Man's first glimpse of plutonium. *New York Times*. www.nytimes.com/1985/07/16/science/man-s-first-glimpse-of-plutonium.html (accessed January 23, 2018).

Seaborg, G.T. 2001. *Adventures in the Atomic Age*. New York: Farrar, Straus and Giroux.

Shaler, R.C. 2012. *Crime Scene Forensics: A Scientific Method Approach*. Boca Raton: CRC Press.

Shen, Y., Q. Zhang, X. Qian, and Y. Yang. 2015. Practical assay for nitrite and nitrosothiol as an alternative to the Griess assay or the 2,3-diaminonaphthalene assay. *Anal. Chem.* 87(2): 1274–1280.

Singh, R. 2002. C.V. Raman and the discovery of the Raman effect. *Phys. Perspect.* 4(4):399–420.

Society of Forensic Toxicologists, Inc. www.soft-tox.org/society (accessed January 23, 2018).

Starmans, B.J. 2017. CSI – The prologue: The history of forensic science. *The Social Historian*. www.thesocialhistorian.com/csi-the-prologue/ (accessed January 23, 2018).

Stuart, J.H., J. J. Nordby, and S. Bell. 2014. *Forensic science: An Introduction to Scientific and Investigative Techniques*. 4th ed. London: Taylor & Francis.

SWGDRUG. 2016. Scientific Working Group for the Analysis of Seized Drugs Recommendations Version 7.1. www.swgdrug.org/Documents/SWGDRUG%20Recommendations%20Version%207-1.pdf (accessed October 10, 2017).

SWGDRUG History. 2017. www.swgdrug.org/history.htm (accessed January 23, 2018).

Taudte, R.V., A. Beavis, L. Blanes, N. Cole, P. Doble, and C. Roux. 2014. Detection of gunshot residues using mass spectrometry. *BioMed Research International* http://dx.doi.org/10.1155/2014/965403.

The laying on of hands for fingerprints: Woman expert thinks system will not be confined to criminals, but will become universal—Chinese used it for identification sixteen centuries ago. *New York Times*. Jun 29, 1919. p. 80.

United States House of Representatives. The Pure Food and Drug Act (1906). http://history.house.gov/HistoricalHighlight/Detail/15032393280 (accessed January 23, 2018).

UNODA. 1925. Geneva Protocol. www.un.org/disarmament/wmd/bio/1925-geneva-protocol/ (accessed January 23, 2018).

Vision Engineering Ltd. Hans and Zacharias Jansen: A complete microscope history. www.history-of-the-microscope.org/hans-and-zacharias-jansen-microscope-history.php (accessed January 23, 2018).

What, when, how: in depth tutorials and information. *Forensic Science Analysis*. http://what-when-how.com/forensic-sciences/analysis-2/ (accessed November 3, 2017).

Wormley, T.G. 1885. *Micro-Chemistry of Poisons, Including Their Physiological, Pathological, and Legal Relations; With an Appendix on the Detection and Microscopic Discrimination of Blood; Adapted to the Use of the Medical Jurist, Physician, and General Chemist*. Philadelphia: J. B. Lippincott.

CHAPTER 2

Chemical tests

KEY WORDS: chemical test, presumptive test, color test, Marquis test, nitric acid test, Duquenois–Levine test, Mayer's test, Van Urk's test, Ehrlich test, Fast Blue B salt test, Dille–Koppanyi test, Scott test, cobalt thiocyanate test, HCl test, Liebermann test, Mandelin test, Mecke test, La Fon's test, Chen's test, ferric chloride test, Vitali–Morrin test, Wagner test, Zwikker test, ferric hydroxamate test, ferric sulfate test, Gallic acid test, permanganate test, phosphate test, Froehde test, silver nitrate test, Simon test, Simon with acetone test, sodium nitroprusside test, sulfate test, Zimmerman test, Reinsch test, Griess test, diphenylamine test, alcoholic KOH test, Nessler's test, anthrone spot test, barium chloride test, microcrystalline tests, Cropen test

LEARNING OBJECTIVES

- To explain the difference between presumptive and confirmatory drug tests
- To describe the uses of color tests for drugs at the crime scene and in the forensic lab
- To recognize the major chemical color tests for drugs
- To describe chemical color tests and microcrystalline tests
- To explain the chemical mechanisms of several color tests

FALSE POSITIVES AND THE DRAWBACKS OF COLOR TESTS

Janet Lee was a first-year college honor student at Bryn Mawr traveling home for winter break in 2003. She was arrested at Philadelphia International Airport; three condoms containing a white powder were found in her carry-on bag by airport screeners. Upon testing on-site with the cobalt thiocyanate test, the white powder was determined to be presumptively positive for cocaine. The powder also tested presumptively positive for opium. Lee responded that the condoms were stress reliever hand toys that contained flour that were used at her women's college dorm for reliving stress during exam periods. Nonetheless, she was arrested based on the results of the color test and she was jailed for three weeks on drug trafficking charges in lieu of a $500,000 bond while she awaited further processing. Conviction of the charges could have led to up to 20 years in prison.

Color tests are presented for use as presumptive tests, the results of which can be used to show probable cause to get a warrant for a broader search. Color tests are useful screening tests as they are quick and portable for crime scene testing. However, color tests probe chemical functional groups and are not specific or conclusive. Laboratory tests must be used to confirm the identity and quantity of a substance. However, color tests have been used to arrest and convict people of drug crimes without additional laboratory testing including spectroscopic and chromatographic tests and microscopic analyses.

Lee had (previously) performed volunteer work and a jail guard recognized her. The guard contacted a volunteer group that found her an attorney. The attorney sought additional testing that confirmed that the white powder was flour and showed she was telling the truth. Lee sued the city and 2 years later settled for $180,000 before the case went to trial. Lee later told the *Philadelphia Inquirer* that she didn't know that drug dealers hid drugs in condoms and that she just brought the stress toys home to show friends because she thought they were funny (Shiffman 2005).

FALSE POSITIVES AND THE DRAWBACKS OF COLOR TESTS (continued)

REFERENCE

Shiffman, J. 2005. Flour in condoms sent her to jail. 29 December, *The Philadelphia Inquirer*. http://www.mapinc.org/drugnews/v05/n2019/a01.html.

BIBLIOGRAPHY

Feature: Citing startling research on false positive drug tests, researchers call for moratorium on field drug test kit testing. http://stopthedrugwar.org/chronicle/2009/mar/06/feature_citing_startling_researc (accessed January 23, 2018).

Kelly J. 2008. *False Positives Equal False Justice*. Washington, DC: The Mintwood Media Collective.

Woman jailed over flour-filled condoms settles suit. www.freerepublic.com/focus/f-chat/1762297/posts (accessed January 23, 2018).

Chemical tests are wet chemical tests used to determine the chemical identity of a compound or functional group with the aid of a specific reagent. Chemical tests are used for screening substances including illicit drugs, explosives residue, and toxicology specimens. Chemical tests include color tests and microcrystalline tests. *Color tests* are tests that are interpreted by noting the color produced when a reagent or reagents are added in specific proportions to an evidence sample. These tests are most often used to detect the presence of a substance in a sample but can also be used to quantify the substance. Advantages of color tests are their low cost, ease of use, rapid results, and portability to the crime scene. They can be used for elimination purposes and to obtain a warrant for further search. *Microcrystalline tests* analyze for the formation of microscopic crystals. Microcrystalline tests were deemed to reliably produce crystals of a characteristic form and "habit."

The drug substances may include amphetamines, heroin, fentanyl, cocaine, marijuana, lysergic acid diethylamide (LSD), phencyclidine (PCP), ketamine, and steroids, among others. The categories, structures, sources, and physical and chemical properties of these compounds will be discussed in Chapter 8. Explosives include chlorate and nitrate-containing compounds that will be covered further in Chapter 14. Nitrocellulose, nitroglycerin, pentaerythritol tetranitrate (PETN), trinitrotoluene (TNT), Research Development Explosive (RDX), and tetryl are some commonly employed explosives. Chemical tests are used to presumptively identify drug and explosive substances and are the focus of this chapter.

COLORIMETRIC TESTS FOR DRUGS

More than 30 color tests for drugs are included in the United Nations Rapid Testing Method of Drugs of Abuse Manual (UN Manual). However, crime labs typically routinely perform only a small (6–10) subset of these tests including the Marquis, Duquenois–Levine, Nitric acid, Mayer's, Scott, cobalt thiocyanate, and Mecke tests. Several tests may be needed to make a tentative identification. These chemical tests are often able to indicate only the class of compounds to which a drug belongs. The reagents act by targeting functional groups contained within the compounds including aromatic rings; phenols; alcohols; carboxylic acids; and primary, secondary, and tertiary amines. The detection limits are typically 1–50 mcg. Criminalists need to perform positive and negative controls along with the evidence samples to determine if the reagents are working properly and to ensure that the results will be reliable. The colors produced will be influenced by the concentration of the sample and reagent, adulterants and diluents present in a sample, age of the reagent, recipe used to prepare the reagent, and the length of time the reaction is observed. Color interpretation and recording may vary based on the visual physiology, skill, and color terminology of the examiner.

Color tests are performed at the lab in ceramic or disposable single-use spot plates. A small amount of the suspected drug is placed in the spot plate and the indicated reagent(s) is(are) added as instructed by the manual or standard operating procedure. The color change, if any, is noted. Many of the color tests are commercially packaged and marketed for field use. Glass ampoules containing the prepared reagents are packaged in thick plastic pouches with clips for sealing. Crime scene investigators or law enforcement personnel add a small amount of the seized substance (using a disposable paper applicator estimating the necessary quantity) to the pouch and systematically break the ampoules in order, as directed in the instructions. The pouches are labeled with color swatches of a few commonly encountered substances (or classes of substances) that the test can differentiate. A common commercial product is the NIK® Polytesting System; the reagents are labeled by letters including A, B, D, E, G, H, K, L, M, N, P, Q, R, and U. Another system is Narcotest®. As some drugs react similarly with the color test reagents, confirmation of the suspected substance will need to be performed using other methods including microscopy and instrumental techniques. These will be covered in later chapters in this book beginning with microscopy in Chapter 3.

The first screening test that is performed is often the *Marquis test* (the "A" test in the NIK® system) because it reacts with a wide variety of substances and can place them into chemical classes by the color produced. To a small amount of the sample, one drop of Reagent 1 (0.1 mL of 37% formaldehyde in 4 mL glacial acetic acid) is added, followed by two drops of Reagent 2 (concentrated sulfuric acid). The Marquis test is used to screen for amphetamines, opiates, 3,4-methylenedioxymethamphetamine (MDMA), 3,4-methylenedioxyamphetamine (MDA), and mescaline, among other drugs, and allows the examiner to determine the best pathway for further testing. The deep-purple color change for morphine is proposed to be the result of two molecules of morphine and two molecules of formaldehyde condensing to the dimeric product that is protonated to the oxoniumcarbenium salt as shown in Figure 2.1. Other opiates including heroin and codeine also result in a purple color while amphetamines yield an orange color. Fentanyl results in a slow-developing brown-yellow color and mescaline results in a deep-orange color. Fentanyl, however, is difficult to detect using the available color tests, as it is routinely cut with other drugs (e.g., quinine, caffeine, and sugars) that contribute to the resulting color of the test. Meperidine results in a yellow hay color. LSD yields an olive-black color while MDA hydrochloride results in a black color. A deep-red color is formed with aspirin and a blackish-red color is formed with doxepin hydrochloride.

The second screening test that is performed is often the *nitric acid test* (the "B" test in the NIK® system) because it also reacts with a wide variety of substances and can guide further testing. This oxidizing acid hydrolyzes amines and esters. The nitric acid test (one drop of concentrated nitric acid) is used to discriminate between morphine and heroin; these drugs result in the same color with the Marquis test. Morphine dissolves and forms a blood-red color that slowly changes to an orange-yellow color while heroin results in a pale-yellow color. LSD forms a strong brown color and mescaline hydrochloride results in a dark-red color with the test. Codeine and MDA hydrochloride both result in a light greenish-yellow color. Pentylone results in a light-yellow color. A variation of the nitric acid test is the nitric-sulfuric acid test (0.12 mL of concentrated nitric acid mixed with 4 mL of concentrated sulfuric acid). Two drops of the reagent are added to a small amount of material in a spot plate. Morphine forms a reddish-violet color that changes over time to a blood red or yellowish red with this test. Another screening test employs the use of two drops of concentrated sulfuric acid added to a small amount of material in the spot plate. Amphetamine and methamphetamine form a light-brown or grayish-yellow color with that test.

The *Duquenois–Levine test* (the "E" test in the NIK® system) was developed in the 1930s by Pierre Duquenois and was subsequently modified. It is used to detect Δ-9-tetrahydrocannabinol (THC), a cannabinoid produced by the marijuana (*Cannabis sativa*) plant. Fifty drops (2 mL) of Reagent 1 (0.8 g of vanillin in 40 mL of 95% ethanol and 1 mL of acetaldehyde) are added to a small amount of sample in a test tube. After shaking for a minute, 2 mL of Reagent 2 (concentrated hydrochloric acid) is added and the sample is again shaken. After letting the solution stand for a few minutes, and as a color is formed, 2 mL of Reagent 3 (chloroform) is added. With gentle shaking, the color should transfer to the organic layer if the cannabinoid is present. A positive result for THC is the presence of a purple color

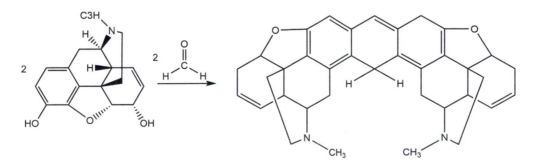

Figure 2.1 Morphine and formaldehyde combine to form an oxocarbenium salt in the Marquis reaction. (Structure prepared by Ashley Cowan.)

transferred to the organic layer on the addition of chloroform. In 1969, a UK scientist reported that 25 plants other than marijuana give a result similar to *Cannabis*. Another study showed that false positives were observed with a variety of vegetable extracts. A 2000 NIST study by O'Neil et al. showed that mace, nutmeg, and tea resulted in very light purple–strong reddish purple, pale reddish purple–light gray purplish red, and light yellow-green, respectively.

Mayer's test (1.36 g of mercuric chloride and 5.00 g potassium iodide dissolved in 100.0 mL of deionized water) is used to detect alkaloid drugs by the formation of a white or cream-colored precipitate. For this test, results are best observed in a black spot plate. It can also be used to test for the presence of drugs comingled with marijuana.

Psilocin and psilocybin compounds produced by so-called magic mushrooms and drugs dimethyltryptamine (DMT) and LSD are detected by *Van Urk's test* (also known as *Ehrlich's* and *p-DMBA [DMAB] test*) (two drops of the reagent consisting of 0.4 g of 4-dimethylaminobenzaldehyde [DMAB] in 4 mL of methanol with 4 mL of concentrated ortho-phosphoric acid). This test identifies indoles, ergoloids, and trypamines by a pink-purple color that results when the p-DMBA complexes with the indole. A deep-purple color indicates the presence of psilocin and/or psilocybin due to the addition of the drug to the p-DMBA reagent (Figure 2.2). Opiates may give false positives due to the presence of tryptophan groups in proteins co-purified with the drug compounds that react with the reagent.

The *Fast Blue B Salt test* (also known as the KN test at the Kanto-Shin'etsu Narcotics Control Office, Japan) is used to detect psilocin and psilocybin compounds and THC through the use of this diazonium salt that adds to aldehydes. This test is conducted in a test tube rather than a spot plate. To a small amount of material to be tested, a small amount of the Fast Blue B salt Reagent 1 (1 g of Fast Blue B salt mixed with 40 g of anhydrous sodium sulfate), 25 drops of Reagent 2 (chloroform), and 25 drops of Reagent 3 (0.16 g sodium hydroxide in 40 mL deionized water aqueous solution) are added with mixing steps of 1 and 2 minutes, respectively, after the addition of Reagents 2 and 3. Figure 2.3 shows two molecules of THC bound to the salt. Psilocin and psilocybin are red when reacted with Fast Blue B and turn blue when the acid is added. The complex is formed from two molecules of the drug for each molecule of Fast Blue B at the nitrile. Nutmeg and mace result in false positives with this test.

The *Dille–Koppanyi test* (the "C" test in the NIK® system) (three drops each of Reagent 1 [0.04 g cobalt (II) acetate tetrahydrate in 40 mL absolute methanol and 0.08 mL glacial acetic acid] and Reagent 2 [2 mL isopropylamine in 38 mL absolute methanol]) is used to test for barbiturates including amobarbitol, pentobarbital, phenobarbital, and

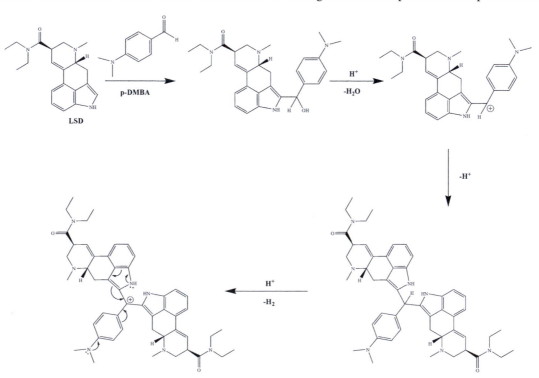

Figure 2.2   Ehrlich reagent complex with LSD. (Structure prepared by Ashley Cowan.)

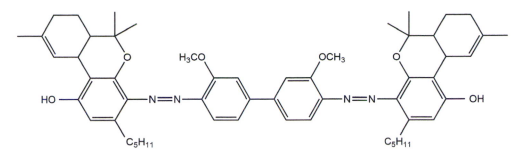

Figure 2.3 Fast Blue B salt (KN test) reaction with two molecules of THC. (Structure prepared by Ashley Cowan.)

secobarbitol. It was developed in the 1930s by Theodore Koppanyi and James M. Dille. The barbiturate nitrogen atoms complex with the cobalt in the reagent to produce a light-purple color. A proposed coordination complex is formed with the cobalt and two barbiturates and two isopropylamine molecules (Figure 2.4).

The *Scott test* (the "G" test in the NIK® system) is used to test for the presence of cocaine in a sample. It was published in 1973. It is a three-step test that is performed in a test tube, like the Duquenois–Levine test. Five drops of Reagent 1 (0.4 g cobalt [II] thiocyanate dissolved in 20 mL of 10% acetic acid mixed with 20 mL of glycerin) are added to the sample. The sample and reagent are mixed by shaking for 10 seconds until it turns blue (or an equal amount of the reagent can be added to drive this color change). On the addition of one drop of Reagent 2 (concentrated hydrochloric acid), a pink color should emerge (or another drop should be added) on shaking prior to the addition of five drops of Reagent 3 (chloroform). The final mixture should be mixed by shaking. A positive for cocaine hydrochloride is denoted by a bright blue color on the surface of the particles (faint blue for cocaine base) after the addition of Reagent 1, a pink color on the addition of Reagent 2, and the formation of a blue color in the organic layer on the addition of Reagent 3. Other drugs that give similar results are lidocaine and diphenhydramine.

An earlier test for cocaine is the *cobalt thiocyanate* test. Also in a test tube, a small amount of evidence material or standard is mixed with one drop of Reagent 1 (16% aqueous hydrochloric acid solution) and one drop of Reagent 2 (1 g of cobalt [II] thiocyanate solution prepared with 40 mL of deionized water), mixing for 10 seconds by shaking after the addition of each reagent. Protonated cocaine hydrochloride samples turn greenish blue on participating in a hexacoordinate complex consisting of cobalt at the center with two molecules of cocaine and thiocyanate molecules as ligands (Figure 2.5). Cocaine base forms a dark gray-blue color. Heroin, meperidine, ephedrine and pseudoephedrine hydrochloride, procaine hydrochloride, and PCP also test positive (greenish-blue color) with this test.

The *HCl test* (two drops of 2 N HCl) is an acid solubility test used to detect the presence of drugs, including cocaine which forms a precipitate. It is a test for amines that are insoluble with 5% NaOH. Diazepam (Valium) forms a light neon-yellow precipitate with this test.

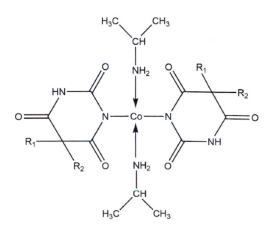

Figure 2.4 Proposed Dille–Koppanyi reaction product coordination complex. (Structure prepared by Ashley Cowan.)

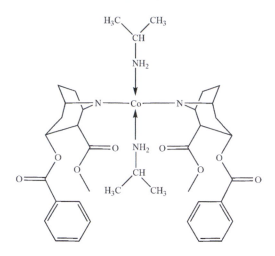

Figure 2.5 Complex formed with cobalt thiocyanate test reagent.

The *Liebermann test* is a test for alkaloids and can be used to detect cocaine, morphine, para-methoxyamphetamine (PMA), and para-methoxy-N-methylamphetamine (PMMA). The test is conducted in a spot plate with a small amount of material to which one drop of the reagent (0.4 g of sodium nitrite in 4 mL concentrated sulfuric acid) is added. Salicylic acid results in an orange-brown color. Meperidine forms an orange color and mescaline turns black with dark-yellow edges.

The *Mandelin test* (the "W" test in the NIK® system) is used to test for alkaloids and results in various colors for various drugs. It was developed in 1883 by Mandelin as a test for strychnine, which turns purple with the test. It is used to test for morphine (grayish-reddish brown), heroin (reddish brown), PCP, methadone (grayish blue), MDA (bluish black), mescaline (dark-yellowish brown), amphetamines (green), Yohimbine (purple), procaine (deep orange), and methalone (orange yellow), as well as LSD, psilocybin, ketamine and PMA. The test is conducted in a spot plate with a small amount of material to which one drop of the reagent (0.4 g of ammonium vanadate in 40 mL concentrated sulfuric acid) is added. To induce the color change, the drugs form a complex with the vanadium metal.

The *Mecke (or La Fon's) test* is also a test for alkaloids and is used to test for opiates, LSD, psilocybin, and mescaline. One drop of the reagent (0.4 g of selenious acid in 40 mL of concentrated sulfuric acid) is added to a small amount of material in a spot plate. Morphine is dehydrated by the sulfuric acid and then undergoes an oxidized re-arrangement as shown in Figure 2.6 in the presence of selenious acid to form a blue-green color. Heroin forms a deep bluish-green color and opium forms an olive-black color. MDA hydrochloride and codeine form a dark bluish-green color. Mescaline hydrochloride and LSD form an olive and greenish-black color, respectively.

Chen's test (Chen-Kao test) is a test for ephedrine, pseudoephedrine, norephedrine, and norpseudoephedrine. It uses two drops each of Reagent 1 (0.4 mL of glacial acetic acid mixed with 40 mL of deionized water), Reagent 2 (0.4 g of copper [II] sulfate dissolved in 40 mL of deionized water), and Reagent 3 (3.2 g of sodium hydroxide in 40 mL of deionized water). Ephedrine turns purple in complex with the copper in the presence of sodium hydroxide. Two molecules of ephedrine are complexed with the copper as shown in Figure 2.7. It is considered "selective" for phenyl-alkylamines with vicinal hydroxyl- and amino-groups. Cathine turns a blue-gray color.

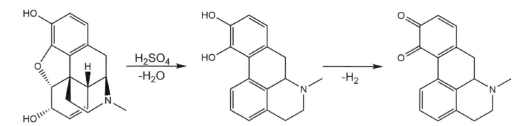

Figure 2.6 Mecke test reaction: oxidation of morphine by sulfuric acid. (Structure prepared by Ashley Cowan.)

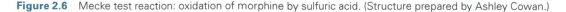

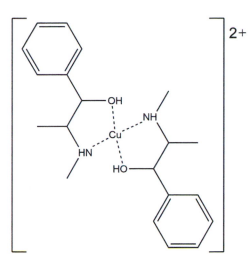

Figure 2.7 Chen test complex with ephedrine molecules. (Structure prepared by Ashley Cowan.)

The *ferric chloride test* (two drops of a solution of 0.8 g of anhydrous ferric chloride dissolved in deionized water) is used to test for morphine and salicylates in urine, but various colors may be observed with different drugs. The ferric chloride test is a test for phenols but enols, hydroxamic acids, oximes, and sulfinic acids also yield positive results. Aldehydes and ketones with significant enolic character and acid chlorides and primary and secondary amines will also form colored complexes. Some phenols may not react. Some of the complexes formed are short-lived and the color fades quickly. The result is a purple color with salicylates, a dark greenish yellow with acetaminophen, and a blue color with morphine. Excedrin forms a purplish-blue color and baking soda forms a deep-orange color. The iron (III) chloride complex formed with 5-sulfosalicylic acid is shown in Figure 2.8.

The *Vitali–Morrin* test is used to test for morphine, heroin, *Salvia divinorum*, and mescaline. A small amount of material is placed in a porcelain spot plate and 0.5 mL of Reagent 1 (concentrated nitric acid) is added and carefully heated over a water bath to dryness. Then 5 mL of Reagent 2 (acetone) and 1 mL of Reagent 3 (0.224 g of potassium hydroxide dissolved in 40 mL of ethanol) are added.

The *Wagner test* is used to detect secondary amine-containing alkaloids and thiols and results in a blue color with MDMA and methamphetamine. It will also oxidize methyl alcohols and methyl ketones to the corresponding carboxylates. It can form iodo compounds (e.g., CHI_3) with terminal methyl groups. In a test tube, five drops of deionized water are added to a small amount of material and mixed by shaking and then two drops of the iodine reagent (0.508 g of iodine and 0.8 g of potassium iodide dissolved in 40 mL of deionized water) are added. Cocaine hydrochloride forms a cloudy brown color with the test.

The *Zwikker test* is used to test for barbiturates: a light-purple color is formed with phenobarbital, pentobarbital, and secobarbitol through complexation with the copper. Reagent 1 consists of 0.2 g of copper (II) sulfate pentahydrate in 40 mL of deionized water and Reagent 2 consists of 2 mL of pyridine in 38 mL of chloroform. In contrast, baking soda forms a light blue and Excedrin forms a light green.

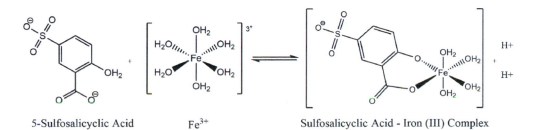

5-Sulfosalicyclic Acid Fe^{3+} Sulfosalicyclic Acid - Iron (III) Complex

Figure 2.8 Ferric chloride test product with 5-sulfosalicylic acid. (Structure prepared by Ashley Cowan.)

The *ferric hydroxamate test* is a test for esters. To perform the test, one drop of Reagent 1 (4 g of hydroxylamine hydrochloride in 40 mL of methanol), three drops of Reagent 2 (0.2 g of ferric chloride in 40 mL of methanol), and one drop of deionized water are added to a spot plate. The reaction with salicylic acid was dark purple.

The *ferric sulfate test* is conducted by probing one drop of the sample drawn from a small amount of material dissolved in three drops of deionized water with one drop of the reagent (2 g of ferric sulfate in 40 mL of deionized water). A positive result for morphine and heroin is a red color. Opium results in a brown color with a red tint.

The *Gallic acid test* (one drop of the 0.2 g Gallic acid reagent dissolved in 40 mL of concentrated sulfuric acid) must be performed using a freshly prepared reagent as it turns the color from clear to brown to purple over time. Addition to diphenhydramine hydrochloride resulted in a bright yellow color.

The *permanganate test* can be used to detect poisons such as methanol. The test uses one drop each of Reagent 1 (3.2 g sodium hydroxide in 40 mL of deionized water) and Reagent 2 (absolute ethanol) that are added to a small amount of material in a spot plate. An alternate recipe employs 3% potassium permanganate followed by the addition of sodium bisulfite and chromotropic acid and then carefully layering on concentrated sulfuric acid. When methanol is oxidized to formaldehyde in the presence of the reagent, a purple color at the acid layer indicates a positive result. Formaldehyde alone will not give a positive result. Morphine is oxidized to apomorphine (ketone form) as shown in Figure 2.9 by potassium permanganate.

The *phosphate test* is used to test for the presence of phosphate ions via the formation of ammonium phosphomolybdate, a bright-yellow precipitate, with the ammonium molybdate reagent as shown in Figure 2.10. The test consists of

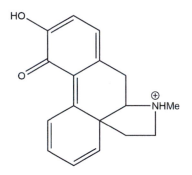

Figure 2.9 Apomorphine formed using the potassium permanganate reagent. (Structure prepared by Ashley Cowan.)

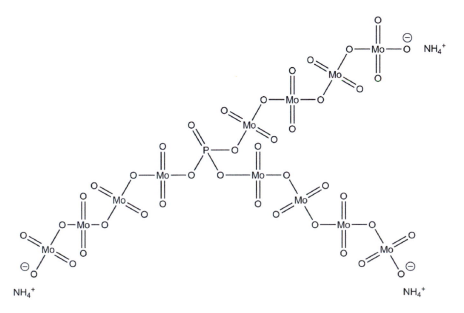

Figure 2.10 Ammonium phosphomolybdate. (Structure prepared by Ashley Cowan.)

two reagents: Reagent 1 (4 g of ammonium molybdate in 40 mL of water) and Reagent 2 (4 mL of concentrated nitric acid to 36 mL of water). Psilocybin will yield a positive result.

A different test also using molybdenum is the *Froehde test*. The test reagent consists of 100 mL of hot, concentrated sulfuric acid to 0.5 g of sodium molybdate and is used to test for alkaloids, specifically opioids. Froehde's test is used to test for opiates, LSD, psilocybin, and mescaline. A deep purplish red is formed with the opiates heroin and morphine and a brownish black with opium. Morphine reacts to form a purplish color that changes to blue-green and then to a faint red color. Aspirin forms a grayish purple. Oxycodone hydrochloride forms a yellow color. Codeine, LSD, and MDA hydrochloride form a very dark green, yellow green, and greenish-black color, respectively.

The *silver nitrate test* is a test for alkenes that bind the silver (I) cation (although other metals may displace the silver), amine-containing alkaloids, and thiols. A positive result is the formation of a precipitate according to solubility rules. Morphine is detected by the presence of a gray precipitate. The test is performed in a test tube using five drops of deionized water, one drop of material, and one drop of the reagent (0.68 g of silver nitrate in 40 mL of deionized water). Nitrates are soluble and will give a negative result.

The *Simon test* is a test for amine-containing alkaloids and the detection of thiols. A dark-blue color is formed in the presence of secondary amines and is observed with MDMA and methamphetamine. A light-purple color is formed with methylphenidate. Secondary amines and acetaldehyde combine to form an enamine that reacts with sodium nitroprusside to form an imine. The immonium salt is hydrolyzed to a bright cobalt-blue Simon-Awe complex. To a small amount of test material on a spot plate, one drop of Reagent 1 (4 mL of acetaldehyde added to 0.36 g of sodium nitroprusside already dissolved in 36 mL of deionized water) and two drops of Reagent 2 (0.8 g of sodium carbonate dissolved in 40 mL of deionized water) are used. A variation on the Simon test is the *Simon test with acetone test* that gives similar results but with primary amines such as amphetamine. For this test, one drop of Reagent 1 (0.4 g of sodium nitroprusside dissolved in 40 mL of 5% aqueous acetone) and one drop of the aforementioned Reagent 2 are added to a small amount of material in a spot plate. Another variation is the *sodium nitroprusside test* that uses two drops each of Reagents 1 (3.2 g of sodium hydroxide in 40 mL of deionized water) and 2 (0.4 g of sodium nitroprusside dissolved in 40 mL of deionized water) added to a small amount of material in a spot plate. The sodium nitroprusside test yields a blue color in the presence of secondary amines such as MDMA and methamphetamine.

The *sulfate test* is used to test for sulfate-containing drugs such as morphine sulfate, quinine sulfate, and so on. The test is prepared by dissolving 2 g of barium sulfate dehydrate in 40 mL of deionized water. Barium sulfates are insoluble and white precipitates are observed.

The *Zimmerman test* is used to test for diazepam and benzodiazepines. It is performed using one drop each of Reagent 1 (0.4 g of 1,3-dinitrobenzene dissolved in 40 mL of methanol) and Reagent 2 (6 g of potassium hydroxide in 40 mL of deionized water) added to a small amount of material in a spot plate. In basic conditions, benzodiazepines are dehydrated and react to add 1,3-dinitrobenzene as shown in Figure 2.11. In a similar reaction, an aqueous solution of sodium carbonate and sodium nitroferricyanide and m-dinitrobenzene in isopropanol turns blue in the presence of ephedrine.

Figure 2.12 shows the results of three color tests with controlled substances. Column one shows the results of the Wagner test with cocaine. Column two shows the results of the cobalt thiocyanate test with cocaine. Column three shows the results of the HCl test with diazepam. For all three tests, the bottom row is a negative control in which only the test reagent and no drug was added.

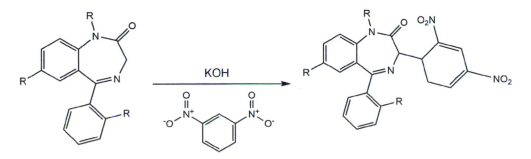

Figure 2.11 Zimmerman's test for benzodiazepines. (Structure prepared by Ashley Cowan.)

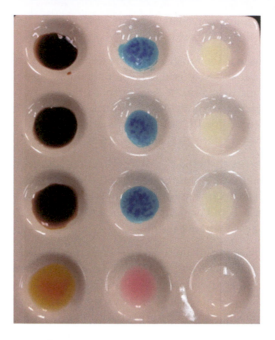

Figure 2.12   Selected colorimetric spot tests for drugs on a spot plate. (Courtesy of Alex Weghorst.)

Other color tests indicate the presence or use of legal drugs such as aspirin and acetylsalicylic acid in pills, white powders, or toxicology samples. The metabolite (and starting material) is salicylic acid. A colorimetric test for determining the presence of salicylates uses an acidic ferric chloride solution (similar to the ferric chloride test previously described) that reacts with salicylates to produce a purple color. Recording the absorbance using a visible spectrometer enables the analyst to determine the concentration to micromolar sensitivity. Other investigations have also focused on quantitating the drug analyte (i.e., methamphetamine) detected by color tests using the smartphone app ColorAssist.

CHEMICAL TESTS FOR POISONS

Chemical tests can also be used in toxicology to detect inorganic, heavy metal poisons including arsenic, mercury, and lead. For example, in the *Reinsch test*, a tissue or body fluid sample is dissolved in hydrochloric acid and a copper strip is inserted into the sample. A chemical reaction with the copper, indicated by a silvery or dark coating on the strip, indicates the presence of a heavy metal.

Methanol can be detected by 3% potassium permanganate followed by the addition of sodium bisulfite and chromotropic acid and then the layering of concentrated sulfuric acid. The methanol is oxidized to formaldehyde and a purple color at the acid-filtrate layer interface indicates a positive result. Formaldehyde in the specimen will result in a false negative test.

COLORIMETRIC TESTS FOR EXPLOSIVES

The Griess, diphenylamine, alcoholic KOH, and Nessler's tests can be used as colorimetric tests for explosives. By running all of these tests on a sample, the laboratory analyst can determine which explosives are present in the sample, differentiate samples containing explosives by the color test results, and determine if the sample contains a mixture of explosives. For example, as shown in Table 2.1, a sample that yields a pink color with the Griess test, a blue color with the diphenylamine test, and no color change with the alcoholic KOH test yields a positive presumptive indication of nitrocellulose.

The *Griess test* is a test for nitrites and the method was first published in 1858 by Peter Griess. It is a two-step test in which the sample is first treated with sodium hydroxide solution prior to adding the Griess reagent (2% sulphanilamide, 0.2% N-alpha-naphthyl-ethylenediamine, 5% phosphoric acid or methanolic N-alpha-naphthyl-ethylenediamine in the presence of sulphanilic acid and zinc dust). The test has been used in forensics to test for nitrite-containing compounds

Table 2.1   Positive results with color tests for explosives

Explosive	Griess	Diphenylamine	Alcoholic KOH	Nessler's
Chlorate				
Nitrate				
Nitrocellulose				
Nitroglycerin				
PETN				
RDX				
TNT				
Tetryl				

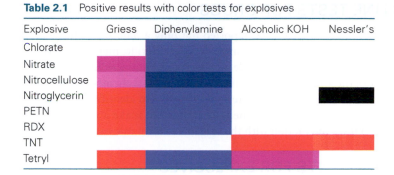

including nitroglycerine. The base breaks down the sample to liberate nitrite ions. As shown in Figure 2.13, the sulphanilic acid of the Griess reagent forms a diazonium salt with the nitrite ion; the diazonium salt reacts with the N-alpha-naphthyl-ethylenediamine or azo dye reagent to produce a pink color. The test is working properly when the color of a sample with the Griess reagent only (no prior addition of base) remains unchanged.

The *diphenylamine test* is a test for oxidizing ions, especially nitrates, but also nitrites, chlorates, and ferric ions. The reagent is a colorless solution of diphenylamine and ammonium chloride in concentrated sulfuric acid. Diphenylamine is oxidized in the presence of nitrates and a blue color is produced. It is used in gunshot residue kits to detect nitroglycerine and nitrocellulose. The alcoholic KOH test is composed of 3% KOH in ethanol. In addition to the colors produced and shown in Table 2.1, 2,4-dinitrotoluene results in a yellow color with the alcoholic KOH test. The aniline hydrochloride test is prepared using aniline acidified with hydrochloric acid and activated with potassium chlorate. It tests for chlorates and chlorites present in explosives. Chlorates are indicated by a blue color. Nessler's reagent (mercuric iodide solution) can be used to detect the presence of ammonium ions. The solution is a faint yellow in color. The formation of an orange-yellow/brown precipitate indicates the presence of ammonium ions.

Another color test that can be used to detect explosives is the *anthrone spot test*. It indicates the presence of carbohydrates including sugars and starches like those in nitrocellulose. Several crystals of anthrone are added to a white spot plate containing the sample and two to three drops of concentrated sulfuric acid are added. The formation of a blue-green color indicates the presence of carbohydrates.

The *barium chloride* and *silver nitrate* spot tests were introduced previously as tests for illicit substances. However, as they are tests for sulfate ions and chloride ions, respectively, they can be used to indicate the presence of explosives containing these ions. A white precipitate indicates the presence of sulfate ions with the barium chloride spot test and the presence of chloride ions with the silver nitrate test. Carbonates will also form precipitates with this test but these can be re-dissolved with acetic acid. Sulfates and other halide ions may also form off-white to yellow precipitates with this test; concentrated ammonium hydroxide will re-dissolve the halide precipitates. The Reinsch test that can be used to detect poisons can also be used to detect inorganics in explosives residue including bismuth, thallium, antimony, and tin.

There are drawbacks to using color tests for explosives. Nonexplosive materials containing nitrites and nitrates, including fertilizers, will also yield a presumptive positive result. All color tests must be confirmed with instrumental chemical tests, which will be covered in upcoming chapters.

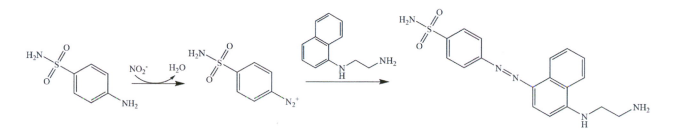

Figure 2.13   Griess test chemical reaction. (Structure prepared by Ashley Cowan.)

MICROCRYSTALLINE TESTS FOR DRUGS

More than 200 microcrystalline tests have been reported but only a handful of these are routinely used. Common microcrystalline tests include bismuth iodide, gold bromide, gold chloride, mercuric chloride, mercuric iodide, platinic bromide, platinic chloride, potassium bismuth iodide, potassium cadmium iodide, potassium tri-iodide, and potassium permanganate. While some of these reagents are readily available in most chemistry labs, others may need to be specially ordered as they contain precious metals. Like color tests, these tests can identify microgram quantities of materials. No single test is specific so at least two microcrystalline tests are used to make a positive identification. Microcrystalline tests are based on the compound's interaction with the reagents to form macroscopic crystals as shown in Figure 2.14.

MICROCRYSTALLINE TESTS FOR EXPLOSIVES

There are also microcrystalline tests for explosives. Microcrystalline tests for explosives include the Cropen microcrystalline test. The *Cropen test* is used to indicate the presence of chlorates and perchlorates. On a microscope slide, one drop of Cropen reagent and one drop of methylene blue solution are added next to each other, but not touching, and the evidence sample is added next to the Cropen reagent. The methylene blue is moved to the Cropen sample area using a glass rod or capillary tube. The results are observed using a microscope under 4–10× magnification. A positive result is indicated by the formation of characteristic insoluble crystals. Perchlorates form blue needle crystals, some with a purplish tinge. They may grow singly or in bundles and exhibit blunt or split ends. Chlorate crystals form blue rosettes or single crystals, some of which have a slight purplish tinge. The chlorate needles are smaller and thinner than those of perchlorates and are slower to develop. High concentration also leads to small crystal formation. Impurities may lead to unusual and distorted crystal formation. A reference standard is used to verify the positive result. Old reagents may not produce the expected crystals.

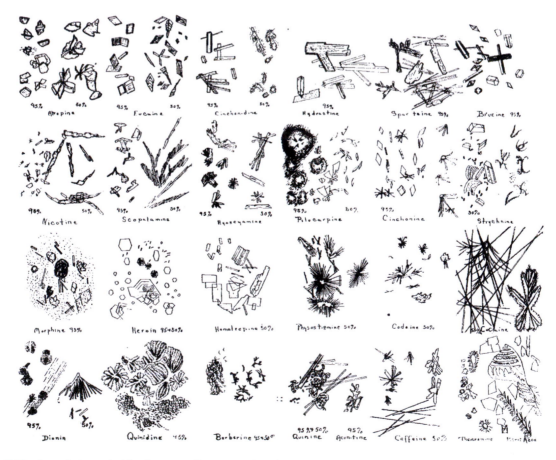

Figure 2.14   Crystals formed with microcrystalline tests. (From Nelson, B.E. and Leonard, H.A. Identification of alkaloids under the microscope from the form of their picrate crystals, *J. Am. Chem. Soc.*, 1922.)

THE FUTURE OF CHEMICAL TESTS

Color tests continue to be used as presumptive tests for drugs, poisons, and explosives at the crime scene and at the lab. Recent research has focused on introducing new reagents for the analysis of designer drugs and synthetic cannabinoid compounds and the automated interpretation of these tests using a smartphone app. The chemical tests rely heavily on investigator experience to interpret and assign colors and crystal shapes observed. The colors are affected by the presence of adulterants or diluents in a sample. Despite their many advantages and uses, color tests cannot be used in court as an identification technique. Microcrystalline tests have not been included in *Clarke's Isolation and Identification of Drugs* since 1986 and many crime labs and police departments have stopped using these tests due to the advent of modern immunoassay, gas chromatography-mass spectrometry (GC-MS), and liquid chromatography-mass spectrometry (LC-MS) techniques. GC-MS and LC-MS methods provide structural and compositional information not provided by these earlier chemical methods.

QUESTIONS

1. The Marquis test turns purple in the presence of _____.
   a. Marijuana   b. Opiates
   c. Cocaine   d. Psilocin   e. Caffeine
2. The Duquenois–Levine test is a screening test for _____.
   a. Heroin   b. Marijuana
   c. Cocaine   d. Psilocin   e. Caffeine
3. Which of the following is a field test for the presence of cocaine?
   a. Dillie–Koppanyi   b. Scott
   c. Duquenois–Levine   d. Marquis   e. Van Urk
4. _____ tests identify drugs by the size and shape of their crystals formed when the drug is mixed with specific reagents.
   a. HPLC   b. Color
   c. TLC   d. Microcrystalline
5. A white precipitate is a positive result with this test used to detect the presence of alkaloids.
   a. Dillie–Koppanyi   b. Scott
   c. Duquenois–Levine   d. Marquis   e. Mayer's
6. Why would an investigator use color tests at a crime scene?
7. Why would an analyst use color tests at a lab?
8. Chemically, how does the Marquis test work? Be specific using morphine as your example.
9. Chemically, how does the Dille–Koppanyi test work? Be specific using barbituric acid as your example.
10. Chemically, how does the Ehrlich/van Urk test work? Be specific using LSD as your example.

Reference

O'Neil, C.L., D.J. Crouch, and A.A. Fatah. 2000. Validation of twelve chemical spot tests for the detection of drugs of abuse. *Forensic Sci. Int.* 109:189–201.

Bibliography

Acharya, S., S. Khatiwada, and K.M. Elkins. 2015. CSI-Pi: A novel automated secure solution to interpret on-site colorimetric tests. *Colonial Academic Alliance Undergraduate Research Journal* 5(2):1–14.

Agg, K., A. Craddock, R. Bos, P. Francis, S. Lewis, and N. Barnett. 2006. A rapid test for heroin (3,6-diacetylmorphine) based on two chemiluminescence reactions. *J. Forensic. Sci.* 51:1080–1084.

Agg, K., N. Barnett, S. Lewis, and J. Pearson. 2007. Preliminary investigations into Tris(2,2′-bipyridyl) ruthenium (III) as a chemiluminescent reagent for the detection of 3,6-diacetylmorphine (heroin) on surfaces. *J. Forensic. Sci.* 52:1111–1114.

ASTM. 2014. ASTM E2329-14: Standard practice for identification of seized drugs. West Conshohocken, PA. www.astm.org/DATABASE.CART/HISTORICAL/E2329-14.htm (accessed May 24, 2018).

Bell, S. 2013. *Forensic Chemistry*, 2nd ed. Upper Saddle River, NJ: Pearson Prentice Hall.

Bell, S. and R. Hanes. 2007. A microfluidic device of presumptive testing of controlled substances. *J. Forensic. Sci.* 52:884–888.

Binette, M.-J. and P. Pilon. 2013. Detecting black cocaine using various presumptive drug tests. *Microgram J.* 10:8–11.

Brandt, S.D., L.A. King, and M. Evans-Brown. 2004. The New drug phenomenon. *Drug Test. Anal.* 6:587.

Choodum, A., K. Parabun, N. Klawach, N. NicDaeid, P. Kanatharana, and W. Wongniramaiku. 2014. Real time quantitative colorimetric test for methamphetamine detection using digital and mobile phone technology. *Forensic Sci. Int.* 235:8–13.

Choodum, A., P. Kanatharana, W. Wongniramaiku, and N. NicDaeid. 2015. A sol-gel colorimetric sensor for methamphetamine detection. *Sensor and Actuators B* 215:553–560.

Choodum, A. and N. NicDaeid. 2011. Digital image-based colorimetric tests for opiate drugs. *Talanta* 86:284–292.

Choodum, A and N. NicDaeid. 2011. Digital image-based colorimetric tests for amphetamine and methylamphetamine. *Drug Test. Anal.* 3:277–282.

Clark, E.G.C. and M. Williams. 1955. Microchemical tests for the identification of alkaloids. *J. Pharm. Pharmacol.* 7(1): 255–262.

Crime Scene Sciences. 2009. *Preliminary Illicit Drug Identification*. Canberra: Australian Federal Police.

Cuypers, E., A.-J. Bonneure, and J. Tytgat. 2016. The use of presumptive color tests for new psychoactive substances. *Drug Testing and Anal.* 8(1):136–140.

Deakin, A.L. 2003. A study of acids used for the acidified cobalt thiocyanate test for cocaine base. *Microgram J.* 1(1–2):40–43.

Dille, J.M. and T. Koppanyi. 1934. Studies on barbiturates. III. Chemical assay of barbiturates. *J. Am. Pharm. Assoc.* 23(11):1079–1084.

Elie, M.P., M.G. Baron, and J.W. Birkett. 2008. Enhancement of microcrystalline identification of gamma-hydroxybutyrate. *J. Forensic Sci.* 53(1):147–150.

Elie, M.P. and L.E. Elie. 2009. Microcrystalline tests in forensic drug analysis. In *Encyclopedia of Analytical Chemistry*, ed. R.A. Meyers, 1–12. New York: John Wiley & Sons, Ltd.

Elkins, K.M., A. Weghorst, A.A. Quinn, and S. Acharya. 2017. Color quantitation for chemical spot tests for a controlled substances presumptive test database. *Drug Test. Anal.* 9:306–310.

Fasanello, J. and P. Higgins. 1986. Modified Scott test for cocaine base or cocaine hydrochloride. *Microgram* 19:137–138.

Federal Bureau of Investigation. 2013. Today's FBI facts and figures 2013–2014. www.fbi.gov/stats-services/publications/todays-fbi-facts-figures/facts-and-figures-031413.pdf/view (accessed May 24, 2018).

Feigl, F. 1966. *Spot Tests in Organic Analysis*. New York: Elsevier Publishing Company.

Feigl, F. and V. Anger. 1972. *Spot Tests in Inorganic Analysis*. New York: Elsevier Publishing Company.

Forrester, D.E. 1997. The Duquenois color test for marijuana: Spectroscopic and chemical studies. Doctoral diss., Georgetown University.

Hanson, A.J. 2005. Specificity of the Duquenois–Levine and cobalt thiocyanate tests substituting methylene chloride or butyl chloride for chloroform. *Microgram* 3:183–185.

Hider, C.L. 1971. The rapid identification of frequently abused drugs. *J. Forensic Sci. Soc.* 11: 257–262.

Hunt, R.W.G. 2004. *The Reproduction of Color*. 6th ed. Chichester, UK: Wiley-IS&T Series in Imaging Science and Technology.

Isaacs, R.C.A. 2014. A structure-reactivity relationship driven approach to the identification of a color test protocol for the presumptive indication of synthetic cannabimimetric drugs of abuse. *Forensic Sci. Int.* 242:135–141.

Jacobs, A.D. and R.R. Steiner. 2014. Detection of the Duquenois–Levine chromophore in a marijuana sample. *Forensic Sci. Int.* 239:1–5.

Johns, S.A., A.A. Wist, and A.R. Najam. 1979. Spot tests: A color chart reference for forensic chemists. *J. Forensic Sci.* 24:631–649.

Jungreis, E. 1996. *Spot Test Analysis: Clinical, Environmental, Forensic and Geochemical Applications*. New York: John Wiley & Sons.

Khan, J., T.J. Kennedy, and C. Christian, Jr. 2012. *Basic Principles of Forensic Chemistry*. New York: Humana Press.

Koppanyi, T., J.M. Dille, W.S. Murphy, and S. Krop. 1934. Studies on barbiturates. II. Contributions to methods of barbital research. *J. Am. Pharm. Assoc.* 23(11):1074–1079.

Kovar, K.A. and M. Laudszun. 1989. Chemistry and reaction mechanisms of rapid tests for drugs of abuse and precursors chemicals. Pharmazeutisches Institut der Universitat Tubingen, Germany. www.unodc.org/pdf/scientific/SCITEC6.pdf (accessed January 18, 2018).

Levine, B. and S.W. Lewis. 2013. Presumptive chemical tests. In *Encyclopedia of Forensic Sciences*, Vol. 3, 2nd ed., ed. J. Siegel, P. Saukko, 616–620. Boca Raton: Elsevier Academic Press.

Levine, B. and S.W. Lewis. 2015. Presumptive chemical tests. In *Forensic Chemistry: Advanced Forensic Science Series*, ed. M.H. Houck, 117–121. San Diego: Elsevier.

Lodha, A, A. Pandya, P.G. Sutariya, and S.K. Menon. 2014. A smart and rapid coloimetric method for the detection of codeine sulphate, using unmodified gold nanoprobe. *RSC Advances* 92:50443–50448.

Logan, B.K., S. Nichols, and D.T. Stafford. 1989. A simple laboratory test for the determination of the chemical form of cocaine. *J. Forensic Sci.* 34(3):678–681.

Lorch, S.K. 1974. Specificity problem with the cocaine-specific field test. ii. non-phenothiazine false positives and the separation of phencyclidine-promazine combinations. *Microgram* 7:129–130.

Manura, J.J., J.-M. Chao, and R. Saferstein. 1978. The forensic identification of heroin. *J. Forensic Sci.* 23(1): 44–56.

Masoud, A.N. 1975. Systematic identification of drugs of abuse I: Spot tests. *J. Pharm. Sci.* 64(5): 841–844.

McMurry, J.E. 2015. *Organic Chemistry*. 9th ed. Pacific Grove: Brooks-Cole Publishing.

Moffat, A.C., D. Osselton, B. Widdop, and J. Watts, eds. 2011. *Clarke's Analysis of Drugs and Poisons*, Vol. 1, 4th ed., 471–495. London: Pharmaceutical Press.

Moffat, A.C., M.D. Osselton, and B. Widdop, eds. 2004. Clarke's analysis of drugs and poisons. In *Pharmaceuticals, Body Fluids and Postmortem Material*. 3rd ed. London: Pharmaceutical Press.

Morris, J.A. 2007. Modified cobalt thiocyante presumptive color test for ketamine hydrochloride. *J. Forensic Sci.* 52(1):84–87.

Munsell, A.H. 1905. *A Color Notation*. Boston: G.H. Ellis Company.

Nagy, G., I. Szöllősi, and K. Szendrei. 2005. Color tests for precursor chemicals of amphetamine-type substances. *United Nations Office on Drugs and Crime SCITEC/20*. www.unodc.org/pdf/scientific/SCITEC20-fin.pdf (accessed January 24, 2018).

National Institute of Justice. 2000. Color test reagents/kits for preliminary identification of drugs of abuse: NIJ Standard-0604.01. www.ncjrs.gov/pdffiles1/nij/183258.pdf (accessed November 3, 2017).

National Research Council. 2009. *Strengthening Forensic Science in the United States: A Path Forward*. Washington, D.C.: The National Academies Press. www.ncjrs.gov/pdffiles1/nij/grants/228091.pdf (accessed January 24, 2018).

Nelson, B.E. and H.E. Leonard. 1922. Identification of alkaloids under the microscope from the form of their picrate crystals. *J. Am. Chem. Soc.* 44(2):369–373.

Newhall, S.M., D. Nickerson, and D.B. Judd. 1943. Final report of the O.S.A. subcommittee on the spacing of the Munsell colors. *J. Opt. Soc. Am.* 33(7):385–418.

NicDaeid, N. and H. Buchannan. 2013. Analysis of controlled substances. In *Encyclopedia of Forensic Sciences*, Vol. 1, 2nd ed., ed. J. Siegel and P. Saukko, 24–28. Boca Raton: Elsevier Academic Press.

Oguri, K., S. Wada, S. Eto, and H. Yamada. 1995. Specificity and mechanism of the color reaction of cocaine with cobaltous thiocyanate. *Jpn. J. Toxicol. Environ. Health* 41: 274–279.

Philp, M., R. Shimmon, M.Tahtouh, and S. Fu. 2016. Development and validation of a presumptive color spot test method for the detection of synthetic cathinones in seized illicit materials. *Forensic Chem.* 1:39–50.

Philp, M., R. Shimmon, N. Stojanovska, M. Tahtouh, and S. Fu. 2013. Development and validation of a presumptive colour spot test method for the detection of piperazine analogues in seized illicit materials. *Anal. Methods.* 5:5402–5410.

Philp, M., Shimmon, R., Tahtouh, M., and Fu, S. 2018. Color spot test as a presumptive tool for the rapid detection of synthetic cathinones. *J Vis Exp.* 2018 Feb 5(132). doi:10.3791/57045.

Pitt, C.G., R.S. Hsia, and R.W. Hendron. 1972. The specificity of the Duquenois color test for marijuana and hashish. *J. Forensic Sci.* 17(4): 693–700.

Poe, C.F. and D.W. O'Day. 1930. A study of Mandelin's test for strychnine. *J. Pharm. Sci.* 19(12):1292–1299.

Rouse, D.B., R.L. Achneider, and E.T. Smith. 2014. Presumptive and confirmatory tests using analogs of illicit drugs: An undergraduate instrumental methods exercise. *Chem. Educator* 19:70–72.

Scott, L.J., Jr. 1973. Specific field test for cocaine. *Microgram* 6:179–181.

Shen, Y., Q. Zhang, X. Qian, and Y. Yang. 2015. Practical assay for nitrite and nitrosothiol as an alternative to the Griess assay or the 2,3-diaminonaphthalene assay. *Anal. Chem.* 87(2): 1274–1280.

Siegel, J.A. 2002. Chapter 4: Forensic identification of illicit drugs. In *Forensic Science Handbook*, Vol. 2, 2nd ed., ed. R. Saferstein. Upper Saddle River, NJ: Pearson Prentice Hall.

Stevens, H.M. 1986. Colour tests. In *Clarke's Isolation and Identification of Drugs*, ed. A.C Moffat, 128–147. London: The Pharmaceutical Press.

Stuart, J.H., J.J. Nordby, and S. Bell. 2014. *Forensic Science: An Introduction to Scientific and Investigative Techniques*, 4th ed. London: Taylor & Francis.

Swiatko, J., P.R. DeForest, and M.S. Zedeck. 2003. Further studies on spot tests and microcrystal tests for identification of cocaine. *J. Forensic Sci.* 48(3):581–585.

Thornton, J. and G. Nakamura. 1972. The identification of marijuana. *J. Forensic Sci. Soc.* 12 (3):461–519.

Toole, K.E., S. Fu, R.G. Shimmon, N. Kraymen, and S Taflaga. 2012. Color tests for the preliminary identification of methcathinone and analogues of methcathinone. *Microgram J.* 9(1):27–31.

Tsujikawa, K., Y.T. Iwata, H. Segawa, T. Yamamro, K. Kuwayama, T. Kanamori, and H. Inooue. 2017. Development of a new field-test procedure for cocaine. *Forensic Sci. Int.* 270:267–274.

Tsumura, Y., T. Mitome, and S. Kimoto. 2005. False positives and false negatives with a cocaine-specific field test and modification of test protocol to reduce false decision. *Forensic Sci. Int.* 155:158–164.

Turk, R.F., H. Dharir, and R.B. Forney. 1969. A Simple chemical method to identify marijuana. *J. Forensic Sci.* 14: 389.

Turvey, B.E. and S. Crowder. 2017. *Forensic Investigations: An Introduction*. Cambridge, MA: Elsevier Academic Press.

U.S. Department of Justice, National Institute of Justice Law Enforcement and Corrections Standards and Testing Program. 2000. Color test reagents/kits for preliminary identification of drugs of abuse: NIJ Standard-0604.01. www.ncjrs.gov/pdffiles1/nij/183258.pdf (accessed January 24, 2018).

U.S. Drug Enforcement Administration. 2016. National forensic laboratory information system special report: 2016 mid-year report. Springfield, VA: U.S. Department of Justice. www.nflis.deadiversion.usdoj.gov/DesktopModules/ReportDownloads/Reports/NFLIS_MidYear2016.pdf (accessed August 7, 2017).

United Nations International Drug Control Programme. 1995. Rapid testing methods of drugs of abuse: Manual for use by national law enforcement and narcotics laboratory personnel. United Nations: Vienna. www.unodc.org/pdf/publications/st-nar-13-rev1.pdf (accessed November 10, 2017).

United Nations Office of Drugs and Crime. 2006. Recommended methods for the identification and analysis of amphetamine, methamphetamine and their ring-substituted analogues in seized materials. United Nations: Vienna. www.unodc.org/pdf/scientific/stnar34.pdf (accessed November 10, 2017).

UNODC. 2013. The challenge of new psychoactive substances. A report from the Global SMART Programme. Vienna: UNODC, Vienna. www.unodc.org/documents/scientific/NPS_2013_SMART.pdf (accessed January 24, 2018).

US Department of Justice. 1978. *NILECJ Standard for Chemical Spot Test Kits for Preliminary Identification of Drugs of Abuse*. p. 12. Washington, D.C.: US Department of Justice.

US Department of Justice. 1981. *NIJ Standard for Color Test Reagents/Kits for Preliminary Identification of Drugs of Abuse*. p. 9. Washington, D.C.: US Department of Justice.

Velapoldi, R.A. and S.A. Wicks. 1974. The use of chemical spot test kits for the presumptive identification of narcotics and drugs of abuse. *J. Forensic Sci.* 19:636–654.

Watanabe, K., G. Honda, T. Miyagi, M. Kanai, N. Usami, S. Yamaori, Y. Iwamuro, et al. 2017. The Duquenois reaction revisited: Mass spectrometric estimation of chromophore structures derived from major phytocannabinoids. *Forensic Toxicol.* 35:185–189.

Wright, F.E. 1916. The petrographic microscope in analysis. *JACS*. 38(9): 1647–1658.

CHAPTER 3

The microscope

KEY WORDS: Microscope, optical parts, illuminator, condenser, iris diaphragm, lens, mechanical parts, base, arm, body tube, turret, stage, fine adjustment, coarse adjustment, working distance, depth of focus, amplitude, wavelength, energy, phase, frequency, refraction, dispersion, polarized, anisotropy, isotropic, birefringence, light retardation, Michel-Levy chart, magnification, resolution, stereomicroscope, compound light microscope, numerical aperture, spherical aberrations, chromatic aberrations, brightfield illumination, dichroism, Koehler illumination, polarized light microscope, extinction, phase contrast microscope, fluorescence microscope, comparison microscope, microspectrophotometer, scanning electron microscope, transmission electron microscope

LEARNING OBJECTIVES

- To identify the parts of the compound light microscope
- To understand wave properties of light
- To define field of view, working distance, magnification, resolving power, numerical aperture, and resolution
- To compare and contrast the different types of microscopes including the stereomicroscope, compound light microscope, polarizing light microscope, scanning electron microscope, and transmission electron microscope in terms of how each microscope works, image formation, magnification, and forensic applications
- To explain the use of a Michel-Levy chart

PETS AT CRIME SCENES: SILENT WITNESSES

Patti Hubbert was murdered by suffocation from a blue towel in September 2006. A single animal hair found on the towel was collected and submitted as forensic evidence in the case. Hubbert's companion, Norman Leighton, and another man, Gerald Morris, were also killed, making the case a triple homicide.

Animal forensics can aid in solving criminal cases. Animals are frequently household pets but may also be zoo animals, wildlife, or animals sought by poachers. In forensic cases, a pet can be a witness, a victim, or a perpetrator. Cases can involve pet-on-pet attacks, pet attacks on humans, human attacks on pets, and crimes in which pets can act as silent witnesses to the case.

Pet hair and fur are common types of physical evidence encountered at crime scenes. This type of trace evidence has been used in forensic investigations since the 1800s. The trace unit of the forensic lab examines the morphology of hairs using compound light microscopes and polarizing microscopes to differentiate and identify animal species and can determine if a hair came from a human or animal. Microscopy is a preferred method in many cases as it is a nondestructive technique and requires no additional costs for consumables or test reagents. Ferret guard hairs can be differentiated from hairs from rabbits, gerbils, degus, and Djungarian hamsters. In this case, pattern differences in the cuticle scales and medulla allowed investigators to discriminate common pet species and similar small mammals using visual comparisons of hairs under the microscope.

PETS AT CRIME SCENES: SILENT WITNESSES (continued)

The hair recovered from the towel was found by Montana State Crime Lab expert trace analyst Alice Amman and was determined by microscopy to be from a dog. She found similar dog hairs on other personal items of the suspect including a jacket and gloves that shared similar characteristics with the hair recovered from the crime scene. The suspect was Richard Covington. Reference samples were collected from the suspect's family dog, Sweet Pea. DNA analysis of the dog hair and Sweat Pea determined that the DNA profile is rare and found in only three of every 1000 dogs.

BIBLIOGRAPHY

Lee, E., Choi, T.-Y., Woo, D., Min, M.-S., Sugita, S., and Lee, H. 2014. Species identification key of Korean mammal hair. *J. Vet. Med. Sci.* 76(5):667–675.

Melton, T. Easy species DNA identification for the forensic laboratory using 12S mitochondrial DNA. *Forensic Magazine*, July 29, 2011. www.forensicmag.com/article/2011/07/easy-species-dna-identification-forensic-laboratory-using-12s-mitochondrial-dna (accessed January 23, 2018).

Sato, I., Nakaki, S., Murata, K., Takeshita, H., and Mukai, T. 2010. Forensic hair analysis to identify animal species on a case of pet animal abuse. *Int. J. Legal. Med.* 124(3):249–256.

Tuttle, G. Dog hairs, seat fibers at triple murder trial. *The Billings Gazette*, March 9, 2010. http://billingsgazette.com/news/local/crime-and-courts/dog-hairs-seat-fibers-at-triple-murder-trial/article_26950e44-2bae-11df-9b02-001cc4c03286.html (accessed January 23, 2018).

After examining items visually, the microscope is often the first tool employed by forensic scientists to evaluate evidence. Microscopes are optical instruments used to magnify evidence, reference, and standard items. The microscope is used to observe and analyze features of evidence items that are not visible or cannot be clearly observed with the naked eye. Microscopy is quick, accurate, and nondestructive and can be used to analyze evidence items and materials *in situ*. Modern microscopes used in forensic science include stereomicroscopes, compound light microscopes, polarizing light microscopes, comparison macroscopes, comparison microscopes, and several types of microspectrophotometers and electron microscopes.

EARLY MICROSCOPES

The earliest and simplest magnification device is made from glass ground into lenses that are used for magnifying items. Objects appear larger than they really are under the magnifying glass due to the refraction, or bending, of light passing through the air into the glass and back through the air to the observer's eye. The observer sees a virtual image as opposed to a real image, depending on whether the object is outside of the focal length of the lens. Magnifying glasses are limited to 5–10× magnification. A magnifying glass is shown in Figure 3.1.

The earliest compound light microscopes were manufactured by Zacharias Janssen of Middleburg, Holland, in the 1590s with help from his father, Hans, although Antoni van Leeuwenhoek of Delft, Holland, is often said to have

Figure 3.1 Magnifying glass. (Courtesy of Tim Phillips.)

invented the microscope after he began grinding lenses and writing letters to the Royal Society of London that were published in the *Philosophical Transactions of the Royal Society*. In 1680, van Leeuwenhoek was elected a full member of the Royal Society. He published drawings of his microscopic work, including the first image of a bacterium, in 1683. Around the same time, Robert Hooke published his famous book *Micrographia* in 1665 concerning his study of thin slices of cork from which he coined the term "cell" using an early compound light microscope. The earliest microscopes achieved 200× magnifications with simple ground glass lenses, although most were of poor quality and allowed for only 20–30× magnification. Although he found microscopes difficult to use because of his poor eyesight, van Leeuwenhoek discovered bacteria, free-living, and parasitic microscopic protists; microscopic nematodes; sperm cells; and blood cells.

Relatedly, an early magnification tool used by artists and scientists was the camera obscura. Using this tool, a right-side-up object placed in front of a pinhole of the camera is projected into an inverted image of the object.

PARTS OF THE MICROSCOPE

Microscopes are made up of mechanical and optical parts. The *optical parts* include a light source (or illuminator), a condenser, and the lenses. The *illuminator* light source may transmit light from the bottom of the microscope up through the item or direct light from above the item that is reflected by the item's surface. The *condenser* collects and concentrates the light onto the evidence item. The most common condenser is the *Abbe condenser*, which has an *iris diaphragm* that opens and closes to control the amount of light that is allowed to pass through. The *lenses* are used for magnification and may include an eyepiece and objective lenses. The object to be magnified is placed under the objective lens and viewed through the upper lens, the eyepiece. The resulting magnified image is a virtual image—upside down and backwards—of the item observed by the eye; a schematic is shown in Figure 3.2.

The *mechanical* parts of the microscope include the base, arm, body tube, stage, and coarse and fine adjustment knobs. The purpose of the mechanical parts is to hold and orient the optical parts in their appropriate positions. The *base* is the square or rectangular foot on which the microscope stands. The *arm* connects the base to the ocular system and is used to carry or move the instrument. The *body tube* is a hollow tube onto which the lenses are mounted on opposite ends. The ocular lens is on one end and the objective lenses are on the other. Multiple objective lenses are mounted on a rotating nosepiece or *turret*. The lenses are positioned so that light may be transmitted through the object or reflected off the object to pass from the objective to the ocular lens to the eye. The *stage* is a horizontal plate that holds the item; it has a hole for the light to pass through, and often contains stage clips to hold items in place during the examination. It may be square or circular, and rotating or stationary. The illuminator is located directly beneath the stage. The *coarse adjustment knob* raises and lowers the stage to produce a "rough" focus. The *fine adjustment knob* also raises and lowers the stage but permits a fine-tuned adjustment to focus the object. These parts can be seen on the compound light microscope shown in Figure 3.3.

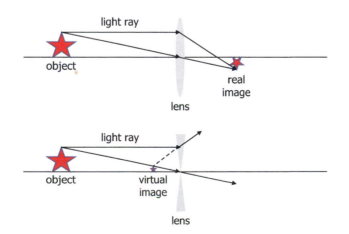

Figure 3.2 Real versus virtual image.

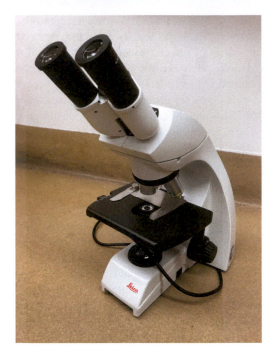

Figure 3.3   Compound light microscope.

The *working distance* of the microscope is the distance between the evidence item on the stage and the closest working portion of the focused objective lens. The *depth of focus* is the distance between two points in an object that is in focus under magnification using defined settings.

LIGHT

Light is a wave phenomenon (Figure 3.4). Each wave has an *amplitude*, a measure of the height of the wave, and a *wavelength*, the distance between the end point of one wave and the start point of the next one. In Figure 3.4, the orange wave has a long wavelength and the blue wave has a short wavelength. Both have the same amplitude.

Light's characteristics include intensity, velocity, wavelength, frequency, amplitude, vibration direction, and phase. The *phase* is the offset or difference between the amplitude or other point on the wave and the next wave. The *frequency* is indirectly proportional to the wavelength. The *energy* (E) of a wave is computed by multiplying Planck's constant (h) times the speed of light (c) and dividing by the wavelength (λ).

$$E = \mathrm{h}c / \lambda \tag{3.1}$$

Light can be refracted, dispersed, and polarized. *Refraction* is a measure of the bending of light: in the visible spectrum, red (shorter wavelength) is least refracted and violet (longer wavelength) is most refracted. *Dispersion* describes the property of separating the wavelengths of light. Normal light is randomly *polarized*—the vibration of the light waves occurs in all (360°) directions—as it is a mixture of many types of waves. However, light can be partially or

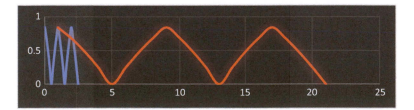

Figure 3.4   Wavelength and amplitude.

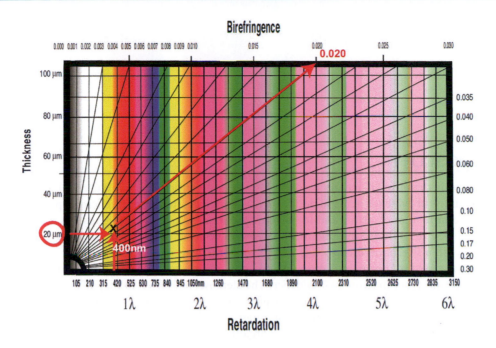

Figure 3.5 Michel-Levy chart. (Used with permission from Petraco, N. and Kubic, T., *Color Atlas and Manual of Microscopy for Criminalists, Chemists, and Conservators*, CRC Press, Boca Raton, FL, 2003.)

totally polarized by reflection, adsorption, or propagation through an anisotropic material. Plane-polarized light has a vibration in only one direction. For example, polarizing sunglasses contain lenses that allow light to pass through only if it is oriented perpendicularly. The lenses contain many tiny vertical scratches so light vibrating in other orientations does not usually pass through and horizontal waves are stopped completely. This allows the wearer to feel a reduction in the glare from the horizontally oriented waves as they are not permitted to pass through the lenses.

Anisotropy is a property of a material in which the light passes is directionally dependent: In anisotropic materials, the velocity of the light varies depending on its direction through the material. Thus, the properties of the material will depend on the direction of the light, its crystalline or other structure, and its thickness. Anisotropic properties are very useful in analyzing materials. In contrast, *isotropic* materials pass light homogenously in all directions. Isotropic crystals refract light at a constant angle and single velocity. Anisotropic crystals can refract incident light in two slightly different directions in a manner dependent on the orientation of the crystal lattice. The refracted light rays are said to be polarized as they have vibration directions at right angles to one another and different velocities. For anisotropic crystals, the difference in their two independent refractive indices, or double or "bi" refraction, is termed *birefringence*. The observed colors' light refracted by anisotropic materials are associated with light retardation.

Light retardation is the difference in the velocity of ordinary and extraordinary rays refracted by the anisotropic crystal. The light retardation is equal to the thickness of the material times its birefringence. A *Michel-Levy Chart* (thickness of thin section in millimeters versus order versus birefringence) is used to quantitate the colors observed in birefringent samples (Figure 3.5). The intersection of the highest order interference colors and known thickness observed in the sample is located on the chart. The birefringence value is determined by the intersecting diagonal lines from the color and thickness.

MAGNIFICATION AND RESOLVING POWER

Magnification is defined as the amount that the evidence item is enlarged by an optical lens system. In computing the total magnification, eyepiece, or ocular lens, magnification is multiplied by the objective lens magnification. So, if the ocular lens is 10× magnification and the objective lens is 10× magnification, the total magnification is 100×.

The *resolving power*, or *resolution*, is the ability of a microscope to distinguish the fine structure and differences in an item or items and to distinguish two small objects from each other (Figure 3.6). Typically, higher magnification yields higher resolving power.

Figure 3.6 Resolution of two objects.

STEREOMICROSCOPY

The *stereomicroscope* is the most frequently used microscope in forensic science (Figure 3.7). It is used to perform preliminary evaluations of evidence items to decide on further testing, to perform macroscopic evaluations, to perform matching or elimination of items, and to perform partitioning of evidence items. As a result, stereomicroscopes are also referred to as dissecting microscopes. As the working stage is relatively large, most items can be initially evaluated with a stereomicroscope. It is also the only method that can be used for opaque evidence items. Stereomicroscopes are used in the analysis of soil, paint chips, glass fractures, matching torn tape edges, plant and mushroom material, and other macroscopic evidence. The stereoscope is used to magnify a paint chip to allow the number of layers to be counted and the color-layer sequence to be determined. Stereomicroscopes can also be used to examine physical features of fibers including crimp, length, color, relative diameter, luster, damage, and adhering debris. Fibers can be tentatively classified as natural, inorganic, or synthetic using this method.

In stereomicroscopes, reflected light is directed on an object from above or from the side, such as the light reflected from the bullets in Figure 3.8. The light source may be internal or external. There is a single-lens system. It is capable of magnification from 2× to 125×. The resulting three-dimensional image observed by the viewer is right side up and correct right to left.

COMPOUND LIGHT MICROSCOPY

In the compound light microscope (CPM), the light source is usually internal and the light is transmitted through the object. This method of illumination is termed *brightfield illumination*. The optical system employs two lenses

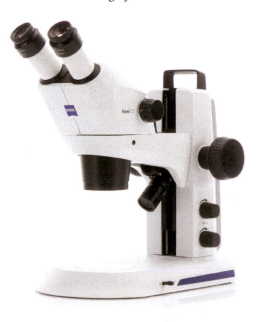

Figure 3.7 Stereoscope. (From ZEISS Microscopy, https://www.flickr.com/photos/zeissmicro/15737691777/in/album-7215762 9398849813/.)

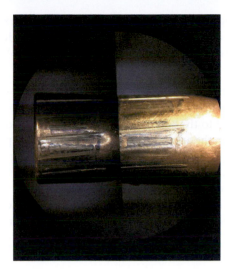

Figure 3.8 Light reflected off bullets.

including an ocular lens at the eyepiece and one or more objective lenses of various magnifications. There are monocular and binocular versions. The resulting image is upside down and backwards to the viewer. The CPM is capable of a magnification of up to 1000×. At high magnification, the depth of focus decreases and the amount of available light is reduced. Brightfield microscopy has several advantages over other microscope methods. The optics do not change the color of the sample. Chemical stains or dyes can be used to visualize the fine structure and specific features of the sample. For example, the Christmas Tree Stain consisting of the nuclear fast red and picroindigocarmine stains is used to stain sperm and epithelial cells. CPM is cheaper than other microscopy methods and does not require additional attachments or features. Thus, it is faster as fewer adjustments to the sample are required prior to observation.

Compound light microscopes are typically equipped with ocular lenses at 10× magnification and objective lenses ranging from 4× to 100× magnification. The lenses are parfocal and remain nearly focused as the lenses are changed. The *numerical aperture* (N.A.) of the lens is notated on the lens tube. This is the ability of a microscope to resolve

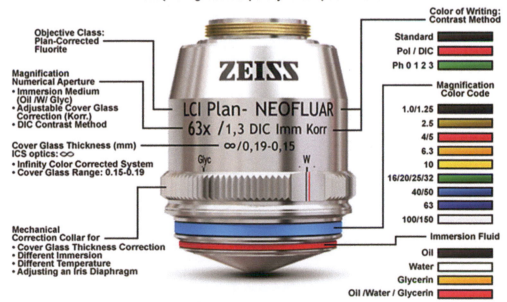

Figure 3.9 Objective lenses with specifications. (From ZEISS Microscopy, https://www.flickr.com/photos/zeissmicro/9662745741/in/album-72157629398849813/.)

two objects as separate images. A higher N.A. has better resolving power. For example, a lens with an N.A. of 1.3 can separate details at half the distance of a lens with an N.A. of 0.65. The theoretical maximum N.A. is 1.4 when using oil immersion and an oil that matches the refractive index of glass. At an N.A. of 1.4, two points only 0.2 microns apart can be resolved.

Modern lenses are also flat field and aberration corrected as noted with other specifications (Figure 3.9). Lens aberrations can include both spherical and chromatic aberrations. *Spherical aberrations* result from light waves that flow through the center of a lens as opposed to the edges, due to being focused at different points. Spherical aberrations can be compensated for by closing down the aperture or by attaching special aspherical lenses. *Chromatic aberrations* are caused by refractive index variation due to wavelength (dispersion) causing the object to be blue in the center and red at the edges. Chromatic aberrations are fixed with achromatic lens systems (Apo). The image in the field of view, the amount of evidence that can be viewed through the eyepiece, may not be in good focus due to the curvature of the lens; this is referred to as field curvature and is fixed using Plan (flat field) objective lenses. Plan Apo lenses are corrected for four wavelengths and give perfect color rendition, image flatness for up to 25 mm diameter fields of view, and the highest numerical aperture.

KOEHLER ILLUMINATION

Koehler illumination is a method of adjusting the light onto an object that eliminates the filament image of the light source on the object while maintaining the quality and uniformity of the projected light. The filament is viewed directly by removing the eyepiece lens. It is used in transmitted and reflected light optical microscopy to generate extremely even illumination and phase contrast, and to ensure that the lighting filament is not in the final image. The aperture diaphragm, condenser, and lens focus are optimized for each specimen.

POLARIZED LIGHT MICROSCOPY

Polarizing light microscopes are special compound light microscopes that are equipped with a polarizer, which is under the stage, and an analyzer, which is located above the nosepiece (Figure 3.10). They also contain an accessory slot for compensators. A compensator can be fixed or variable. A commonly used compensator is a quartz wedge of a full wave (~550 nm) and 1/4 wave (~137 nm) retardation. The *compensator* is an anisotropic material of known

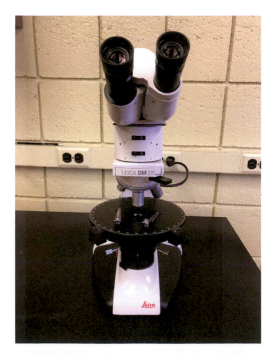

Figure 3.10 Polarized light microscope.

birefringence. It is constructed so that the thickness is controlled and the orientations of vibrations are known. The retardation of the light passing through the compensator can be used to measure the retardation in nanometers.

The polarizer polarizes incoming light to reveal polarizable properties of the material. The light waves then pass through an analyzer before being passed to the eye. The rays that are viewed by the eye are the recombined polarized radiations after passing through the material. The polarizing microscope is used in the analysis of soil, chemical

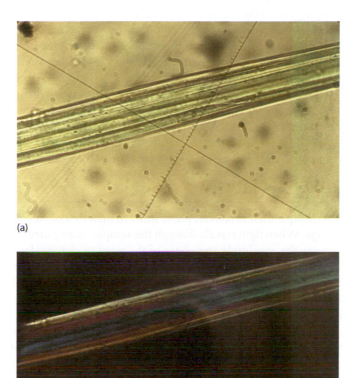

(a)

(b)

(c)

Figure 3.11 A rayon fiber viewed using a polarizing light microscope (a), with a retarding analyzer (b), and with a compensator retardation plate inserted (c).

crystals, ground minerals, and fibers, among other materials, to determine their birefringence. This property can add to the individualization of materials. It can also differentiate natural fibers such as hair as they are not birefringent. The birefringence of an unknown fiber can be estimated when placed in a medium of known refractive index; the birefringence is equal to the parallel refractive index minus the perpendicular refractive index. Isotropic glass and fibers remain black (termed *extinction*) between the crossed polarizer and analyzer. The polarizing microscope can be used to determine the characteristic optical sign for birefringence, order of retardation, and refractive indices of anisotropic materials. For example, if the long axis of a synthetic acetate fiber is oriented northwest-southeast with a first order retardation plate inserted, the color is first order yellow on a magenta background, while a synthetic acetate fiber with the long axis rotated 90 degrees to northeast-southwest is second order blue in color. Thus, the fibers are said to have a positive optical sign for birefringence because the retardation resulted in higher order interference colors. These synthetic fibers exhibit a property termed dichroism. *Dichroism* is the property of exhibiting different colors, especially two different colors, when viewed in polarized light along different axes. A dichroic rayon fiber viewed using a polarizing light microscope, with a retarding analyzer, and with a compensator retardation plate inserted is shown in Figure 3.11. Another use of the polarizing light microscope is to detect a blemish in an LCD screen. It is visible under polarized light but not normal visible light. Some materials, such as natural hairs and animal fibers, are not birefringent, as they are optically isotropic and their refractive index is equal in all directions.

PHASE CONTRAST MICROSCOPY

Phase contrast microscopy is a technique used to convert phase shifts in light that passes through a transparent sample to brightness changes in the image. When light travels through the sample, or any medium other than a vacuum, the interaction with the sample causes the amplitude and phase of the wave to change depending on the properties of the sample or medium. The incident light $[I_o]$ is slowed down while passing through different parts of the sample and becomes out of phase with the transmitted light $[I]$. When the phases of the light are synchronized by an interference lens, a new image is produced with greater contrast. As the phase shifts are invisible, their presence can be detected through the amplitude, or brightness, variations due to scattering and absorption of light. The human eye and microscope cameras are only sensitive to brightness variations. The phase shifts can provide useful information about materials. Some phase contrast materials are basically colorless but can exhibit phase contrast if they cause a phase shift in light within a mounting medium or oil. This can make it possible to visualize structures that are otherwise invisible to the human eye. Other advantages are that stains are not required and the image often appears improved due to the visible fine structure that can be viewed under phase contrast. Phase contrast can be used in the analysis of cotton, flax, asbestos, and other natural fibers including hair as well as synthetic fibers such as Orlon and nylon.

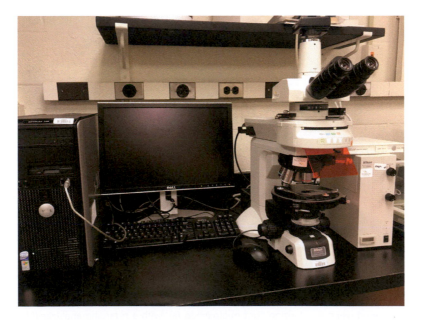

Figure 3.12   Fluorescence microscope.

FLUORESCENCE MICROSCOPY

The compound light microscope can also be equipped with a fluorimeter and camera (Figure 3.12) so that the fluorescence properties of microscopic materials can be examined and recorded. A detailed description of how a fluorimeter works is included in Chapter 4. A fluorimeter consists of an ultraviolet light excitation source, beam splitters, gratings, and excitation and emission barrier filters. The sample can be excited using an ultraviolet light excitation source revealing fluorescence portions of materials so that they can be observed, counted, filtered, sized, and mapped within the whole evidence item. For example, Kevlar fibers exhibit strong fluorescence in complex composite material. Other fluorescent materials include humic and fulvic acid components of soils, some dyed fibers and inks, and stained cellular material.

MICROSPECTROPHOTOMETER

A microscope can be coupled with an ultraviolet, visible, infrared, Raman, or fluorescence spectrophotometer. Spectrophotometry is the primary focus of Chapter 4. The microspectrophotometer can be used to simultaneously evaluate the size, shape, and spectroscopic properties of inks, fibers, soils, paints, and other materials. Spectra are recorded from the top layer of the sample, but samples can be partitioned to expose underlying layers. The microspectrophotometer can be used to differentiate two dyes that may appear identical to the human eye or microscopic fibers contained in fabrics or paints used in art objects. The microspectrophotometer can also be used for quality control of film thickness in materials analysis or to map the color (or fluorescence) along the length of a hair of fiber. The ultraviolet, visible, infrared, Raman, or fluorescence spectra can be differentiated by the peak wavelengths and intensities of each dye. The advantage to this method over traditional spectrophotometry is that the forensic scientist, often confined to limited and small samples, can analyze the heterogeneous microscopic features of materials in a nondestructive manner. An infrared spectrum can determine or help determine the chemical composition of the material. An infrared microspectrophotometer is shown in Figure 3.13.

COMPARISON MICROSCOPY

Comparison microscopes are two compound or stereoscopic microscopes joined by a "bridge" or single eyepiece monocular or binocular system (Figure 3.14). A *comparison macroscope* (ballistics microscope) (Figure 3.15) contains two bridged stereomicroscopes while a *comparison microscope* contains two bridged compound light microscopes. Comparison scopes are invaluable to the forensic scientist, especially for evaluating fibers, hairs, soils, torn items, bullets, and cartridge cases. They allow the scientist to simultaneously view two items, a reference sample and an evidence item, for example, on each of the two stages. The scientist can view the item on the stage of either microscope or show both magnified items simultaneously, splitting the field of view into left and right sides. The items should be viewed under the exact same illumination conditions and light intensity and using the same magnification. A balanced, neutral background is optimal. Comparison microscopy permits a point-by-point side-by-side comparison,

Figure 3.13 Infrared microspectrophotometer.

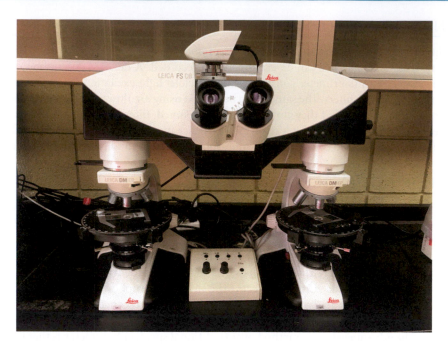

Figure 3.14   Comparison microscope.

which is the most discriminating method of determining if two or more bullets or cartridge cases are consistent with originating from the same source. Cameras are mounted on the bridge to record what the analyst observes in the ocular lenses for both fields of view. Figure 3.16 shows two fibers in a split field of view recorded using a comparison microscope. Figure 3.17 shows two cartridge cases in a split field of view recorded using a comparison microscope.

IDENTIFYING GUNSHOT RESIDUE USING A SCANNING ELECTRON MICROSCOPE

Gunshot residue is the trace material that is transferred to the firing hand, clothing, and immediate area surrounding the shooter, and the skin, clothing, or surface that a bullet enters. A primer is used to ignite gunpowder contained in the body of a cartridge. When the trigger of a firearm is pulled, the firing pin strikes the primer cap at the end of the cartridge causing a spark that ignites the primer. The primer then ignites the gunpowder contained in the body of the cartridge. Products of the chemical reaction include heat and combustion products. The increase in temperature and pressure propels the bullet out of the firearm.

The primer is normally composed of heavy metals including lead (Pb), barium (Ba), and antimony (Sb), but may also consist of tin (Sn) and other elements. Elements from the cartridge or barrel including copper (Cu), zinc (Zn), aluminum (Al), and nickel (Ni) may also be detected. Minor elements may include calcium (Ca), chlorine (Cl), potassium (K), sulfur (S), and silicon (Si). After firing, gunshot residue (the term preferred over primer residue or cartridge discharge residue) is formed from burned molten inorganic primer residues that are released in a vaporous plume from any opening on the weapon. As the plume cools, the metals condense together to form one-, two-, and three-component particles, most of which are 2–10 microns in diameter. One-component particles will consist of only one heavy metal—lead, barium, or antimony—while two-component particles will consist of Pb-Ba, Pb-Sb, or Ba-Sb, and three-component particles will consist of all three elements.

The microscope used to identify gunshot residue is the scanning electron microscope. Gunshot residue examinations are conducted primarily using the scanning electron microscope (SEM) equipped with energy dispersive x-ray spectrometry detection (EDS). With a magnification power of 100,000x, the SEM can be used to detect and identify the elemental composition and morphology of the residue. The three-component Pb-Ba-Sb non-crystalline condensed spheroid particle is characteristic of gunshot residue. Two-component Ba-Sb particles are characteristic of gunshot residue but are also released by some used brake pads. However, in addition to gunshot residue composition, examiners report on morphology, the population of particles, and elemental distribution.

Figure 3.15 Comparison ballistics microscope.

SCANNING ELECTRON MICROSCOPY

The electron microscope utilizes a high-energy electron beam rather than visible light. These instruments are relatively large and are often housed in a dedicated room due to their size and requirements. An SEM is shown in Figure 3.18. SEM focuses a high-energy electron beam (~20 kV) on the surface of the item. Secondary emissions are used to obtain a high-resolution image of the surface and backscattered electrons from the electron beam-surface interaction are used to probe the surface of a specimen. SEM requires samples to be frozen with a cryogen. As a result, the method cannot be used to examine live organisms but can be used to examine insect parts in forensic entomology, as shown in Figure 3.18. SEM is used to examine the external surface morphology of a sample such as gunshot residue or gunshot powder, or the fine structure of extremely small items such as anthrax bacterial spores, pollen, broken cuticles due to damage in hair, cross-sections, or nanoparticles for use in fingerprinting techniques. SEM can also be used to detect if broken glass filaments in light bulbs were hot (on) or cold (off) at the time of an accident. SEM is used for materials analysis including examining polished and etched microstructures, looking for chemical segregation, and detecting contamination. A relatively large sample can be scanned using this method. The resulting image is a three-dimensional image. This microscope yields an extremely high magnification of up to 100,000×. When

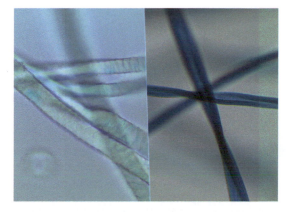

Figure 3.16 Two fibers in a split field of view recorded using a comparison microscope.

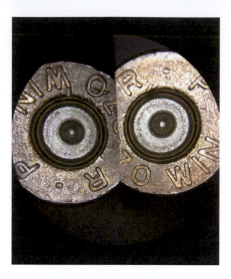

Figure 3.17   Two cartridge cases in a split field of view recorded using a comparison microscope.

coupled with EDS and an x-ray analyzer, it can be used to analyze for chemical composition including trace elements and in the characterization of fibers. The surface morphology can be examined with great depth of field using lower magnifications.

TRANSMISSION ELECTRON MICROSCOPY

Transmission electron microscopy (TEM) is used to transmit a high voltage (~200 kV) beam of electrons through a thin film or transparent surface and then magnify the image. The signals are collected from the transmitted electrons below the sample and analyzed using a quantum wave function. As a result, the sample must be thinly cut. The resulting image is a two-dimensional image. Fine particles can be analyzed under a vacuum. Its uses in forensic science are in materials analysis. It has traditionally been used to study living organisms (e.g., insects) in biology (as opposed to SEM in which samples must be cryo-mounted and therefore dead) to examine their internal structure. Brightfield and darkfield imaging can be used to evaluate stress and other structural defects, magnetic domains, and gain boundaries, and to locate tiny precipitates. It can be used to detect low percent (~10%) components of materials

Figure 3.18   Scanning electron microscope. (From ZEISS Microscopy, https://www.flickr.com/photos/zeissmicro/10710114116/in/photolist-hhEdRg-hhErzb-fHS6HZ-hjq8MW-BBKtfK-hjpFwZ-qPHVfN-hhFoJ8-hjr34g-nYeEad-hjr2SV-pWn8x7-bwu964-mu6Dbv-bwudp4-bwuduk-bwubPT-hhFp2T-bwuiFr-bwucoe-bwuhA8-SUsWws-SUsWXC-hhEfuM-JnLEts-MjAqGv-rLpjW6-rLge4u-vJt-MKU-Nqg2Hc-s3HRDu.)

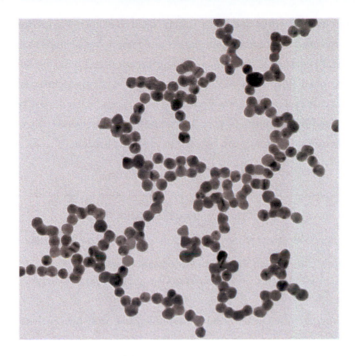

Figure 3.19 TEM of gold nanoparticles. (Courtesy of Pat Bevins.)

that cannot be detected by other techniques. Selected area diffraction can be used to evaluate atomic structure and the crystal structure of the samples. TEM has a much higher resolution than SEM but has a smaller field of view and only slices of a sample can be analyzed at a time. It can be used to analyze nanoparticles as small as 5 nm as shown in Figure 3.19. Particle size can be determined by evaluating the scatter of the electron beam in an atomic plane. The results are shown on fluorescent screens.

QUESTIONS

1. Which of the following terms refers to the separation of light photons into different wavelengths or colors?
   a. Dispersion   b. Birefringence   c. Refraction   d. Absorption
2. The objective lens used was 10×. What is the total magnification from the compound light microscope if the ocular lens is 10×?
   a. 0×   b. 10×   c. 20×   d. 100×
3. The depth of focus under a compound light microscope will be greatest under _____ magnification.
   a. 10×   b. 40×   c. 100×   d. 400×
4. Which of the following microscopes has the lowest resolution?
   a. SEM
   b. Compound light microscope
   c. Stereomicroscope
   d. Polarized light microscope
5. The polarizing compound light microscope would be useful for differentiating:
   a. Synthetic fibers
   b. Soil matter
   c. Crystalline materials
   d. All of the above

6. Briefly explain how a stereoscope works and how it can be used to examine evidence.

7. Briefly explain how a compound light microscope works and how it can be used to examine evidence.

8. Define the terms numerical aperture, resolution, and magnification and describe how they impact the image.

9. Describe the process of Koehler illumination. What is the purpose of this technique?

10. Briefly explain how phase contrast microscopy works and how it can be used to help examine evidence.

11. Briefly explain how polarized light microscopy works and how it can be used to examine evidence.

12. Briefly explain how SEM works and how it can be used to examine evidence.

Bibliography

Bell, S. and K. Morris. 2009. *An Introduction to Microscopy*. Boca Raton: CRC Press.

Bernhard, W.R. 2000. Paint and tape: Collection and storage of microtraces of paint in adhesive tape. *J. Forensic Sci.* 45(6):1312–1315.

Bottrell, M.C. 2009. Forensic glass comparison: Background information used in data interpretation. *Forensic Sci. Commun.* 11(2). https://archives.fbi.gov/archives/about-us/lab/forensic-science-communications/fsc/april2009/review/2009_04_review01.htm (accessed January 24, 2018).

Cengiz, S., A. Cengiz Karaca, I. Cakir, H. Bülent Uner, and A. Sevindik. 2004. SEM-EDS analysis and discrimination of forensic soil. *Forensic Sci. Int.* 141(1):33–37.

Deedrick, D.W. and S.F. Koch. 2004. Microscopy of hair part I: A practical guide and manual for human hairs. *Forensic Sci. Commun.* 6(1) https://archives.fbi.gov/archives/about-us/lab/forensic-science-communications/fsc/jan2004/research/2004_01_research01b.htm (accessed January 24, 2018).

Deedrick, D.W. and S.F. Koch. 2004. Microscopy of hair part II: A practical guide and manual for human hairs. *Forensic Sci. Commun.* 6(3) https://archives.fbi.gov/archives/about-us/lab/forensic-science-communications/fsc/july2004/research/2004_03_research02.htm (accessed January 24, 2018).

Encyclopedia Britannica. 2017. *Antonie van Leeuwenhoek*. www.britannica.com/biography/Antonie-van-Leeuwenhoek (accessed January 23, 2018).

Gaudette, B.D. 1999. Evidential value of hair examination. In *Forensic Examination of Hair*, ed. J.R. Robertson, 243–257. London: Taylor and Francis.

Harding, L.A. 1892. Forensic microscopy. *Science* 20(508):242–243.

Hooke, R. 1665. *Micrographia*. London, UK: Jo. Martyn and Ja. Allestry, printers to the Royal Society.

Hopen, T.L., C. Taylor, L. Peterson, and W. Rantanen. 2007. The Forensic Examination and Analysis of Paper Matches. *NSFTC*. http://projects.nfstc.org/trace/docs/Revised%20Papers/Forensic%20Examination%20and%20Analysis%20of%20Paper%20Matches083007.doc (accessed January 23, 2018).

Köhler, A. 1893. Ein neues Beleuchtungsverfahren für mikrophotographische Zwecke. *Zeitschrift für wissenschaftliche Mikroskopie und für Mikroskopische Technik* 10 (4):433–440.

Köhler, A. 1894. New method of illumination for photomicrographical purposes. *J. Royal Microsc. Soc.* 14:261–262.

Kolowski, J.C., N. Petraco, M.M. Wallace, P.R. DeForest, and M. Prinz. 2004. A comparison study of hair examination methodologies. *J. Forensic Sci.* 49:1253–1255.

Lane, N. 2015. The unseen world: Reflections on Leeuwenhoek (1677) "Concerning little animals." *Philos. Trans. Royal Soc. B* 370(1666). doi:10.1098/rstb.2014.0344.

McCabe, K.R., F.A. Tulleners, J.V. Braun, G. Currie, and E.N. Gorecho. 2013. A quantitative analysis of torn and cut duct tape physical end matching. *J. Forensic Sci.* 58(S1):S34–S42.

McCrone Research Institute. www.mcri.org/ (accessed November 11, 2017).

Meloan, C.E., R.E. James, and R. Saferstein. 2004. Experiment 17. In *Lab Manual-Criminalistics: An Introduction to Forensic Science*. 8th ed. Upper Saddle River, NJ: Pearson Prentice Hall.

Mitosinka, G.T., J.I. Thornton, T.L. Hayes. 1972. The examination of cystolithic hairs of Cannabis and other plants by means of the scanning electron microscope. *J. Forensic Sci. Soc.* 12(3):521–529.

Oien, C.T. 2009. Forensic hair comparison: Background information for interpretation. *Forensic Sci. Commun.* 11(2) https:// archives.fbi.gov/archives/about-us/lab/forensic-science-communications/fsc/april2009/review/2009_04_review02.htm (accessed January 24, 2018).

Parry-Hill, M., R.T. Sutter, and M.W. Davidson. 2017. Microscope alignment for Köhler illumination. www.microscopyu.com/ tutorials/kohler (accessed November 10, 2017).

Petraco, N. and T. Kubic. 2003. *Color Atlas and Manual of Microscopy for Criminalists, Chemists, and Conservators*. Boca Raton: CRC Press.

Reffner, J.A. and P.A. Martoglio. 1995. Uniting microscopy and spectroscopy. In *Practical Guide to Infrared Microspectroscopy*, ed. H.J. Humecki, 41–84. New York: Marcel Dekker, Inc.

Saferstein, R. 2007. *Criminalistics: An Introduction to Forensic Science*. 9th ed. Upper Saddle River, NJ: Pearson Prentice Hall.

Summary of Current Researches Related to Microscopy. 1894. *J. Royal Microsc. Soc.*, p. 261. http://books.google.com/ books?id=0O4BAAAAYAAJ&dq=journal%20of%20the%20royal%20microscopical%20society%201894&pg=PA261#v=onepag e&q&f=false (accessed January 23, 2018).

SWGMAT. 1999. Forensic fiber examination guidelines. *Forensic Sci. Commun.* 1(1). https://archives.fbi.gov/archives/about-us/ lab/forensic-science-communications/fsc/april1999/houcktoc.htm (accessed January 24, 2018).

SWGMAT. 2005. Forensic human hair examination guidelines. *Forensic Sci. Commun.* 7. https://archives.fbi.gov/archives/about- us/lab/forensic-science-communications/fsc/april2005/standards/2005_04_standards02.htm (accessed January 24, 2018).

SWGMAT. 2013. Guideline for assessing physical characteristics in forensic tape examinations. *JASTEE* 5(1):34–41. www.united- statesbd.com/images/unitedstatesbdcom/bizcategories/2961/files/JASTEE_2014_5_1_3.pdf (accessed January 24, 2018).

Taupin, J.M. 2004. Forensic hair morphology comparison—A dying art or junk science? *Sci. Justice* 44(2):95–100.

Tulleners, F.A., J. Thornton, and A.C. Baca. 2013. Determination of unique fracture patterns in glass and glassy polymers. NIJ Report. www.ncjrs.gov/pdffiles1/nij/grants/241445.pdf

Tulleners, F.A. and J.V. Braun. 2011. The statistical evaluation of torn and cut duct tape physical end matching. NIJ Report. www. ncjrs.gov/pdffiles1/nij/grants/235287.pdf

van Oijen, T.A. and J. van der Weerd. 2015. Spectrometric imaging of polarization colors and its application in forensic fiber analysis. *Appl. Spectrosc.* 69(6):773–782.

Vision Engineering Ltd. Hans and Zacharias Jansen: A complete microscope history. www.history-of-the-microscope.org/hans- and-zacharias-jansen-microscope-history.php (accessed January 23, 2018).

CHAPTER 4

Light spectroscopy

KEY WORDS: electromagnetic spectrum, ultraviolet-visible spectroscopy, chromophore, highest occupied molecular orbital, lowest unoccupied molecular orbital, excited state, Beer's Law, fluorescence spectroscopy, excitation, emission, infrared spectroscopy, Raman spectroscopy, Jablonski diagram, correlation chart, microspectrophotometry

LEARNING OBJECTIVES

- To appreciate the historical contributions of scientists that led to the development of spectroscopy
- To diagram the parts of a simple absorption spectrometer
- To understand absorption and emission
- To explain the uses and limitations of ultraviolet-visible, fluorescence, infrared, and Raman spectroscopy
- To compare and contrast infrared spectroscopy and Raman spectroscopy
- To be able to use correlation charts to assign infrared and Raman spectra

DETECTING CARBON MONOXIDE POISONING USING ULTRAVIOLET-VISIBLE SPECTROSCOPY AND CHEMICAL TESTING

A patrolman watched as a man carried a bundle to the river wharf in Brooklyn, New York, dropped it down, and kicked it into the water. It was December 1926. The patrolman chased the man down and brought him to the police precinct. He was a middle-aged man and his name was Francesco Travia. His shoes were soiled and his socks were bloody. Arriving at the apartment, the police found a dead woman on the kitchen floor. She had been cut in half. A butcher's knife and a chisel were found on the table. Only the upper part of her body—her torso, arms, and head—remained in the bloody mess. Travia was arrested on charges of murder and dismemberment. At the city morgue, the woman was identified as Anna Fredericksen. Fredericksen also lived in Travia's building—she ran a rooming house around the corner from his apartment. Her family was horrified as they considered Travia a friend and they didn't think he or anyone else would want to murder her.

Carbon monoxide is a colorless and odorless gas by-product of incomplete combustion reactions such as fuel burning in faulty home furnaces, gas and wood-burning stoves, cigarettes, or gas-powered engines, among others. Carbon monoxide poisoning can also be caused by fires. Carbon monoxide poisoning leads to symptoms including tiredness, headache, dizziness, confusion, shortness of breath, nausea, vomiting, chest pain, impaired judgment, memory impairment, and loss of consciousness. Carbon monoxide can bind competitively to hemoglobin in the blood used to deliver oxygen to the body's organs—hemoglobin binds carbon monoxide over 200 times stronger than oxygen. Hemoglobin has four binding sites for oxygen and exhibits cooperative binding of the gas. Typically, hemoglobin is fully loaded at approximately 96% oxygen and only unloads to 64% bound oxygen, meaning that most of the oxygen remains bound. The binding of carbon monoxide causes the hemoglobin-oxygen binding to curve to the left; the hemoglobin holds the remaining oxygen even more tightly and delivers less of it throughout the body. Carbon monoxide remains bound to hemoglobin for 320 minutes

DETECTING CARBON MONOXIDE POISONING USING ULTRAVIOLET-VISIBLE SPECTROSCOPY AND CHEMICAL TESTING (continued)

(over 5 hours) on average if a person is transferred to clean air. The recovery time is greatly reduced if hyperbaric oxygen is administered. Carboxyhemoglobin (COHb) levels can be used in medicolegal investigations to determine the cause of death. The Centers for Disease Control reports that more than 500 deaths occur annually due to unintentional carbon monoxide poisoning. COHb levels can be detected and quantitated using post-mortem blood samples and whole blood samples drawn in occupational testing to determine exposure to carbon monoxide (CO) in forensic cases. COHb poisoning causes blood to be characteristically red and causes victims to have a pink face and lips.

Dr. Charles Norris, New York City Medical Examiner, investigated the case. He accompanied the police to Travia's apartment. There he observed that the blood pooled around the body was a bright cherry-red color. The dead woman's face was flushed pink even though she'd lost a lot of blood. Immediately, he knew the cause and manner of death, and he told the policeman to release Travia. Carbon monoxide poisoning prior to death causes the blood to be pink in color; a murder victim with adequate blood oxygen levels that bled to death would be porcelain pale. Through chemical testing, Dr. Alexander Gettler, a forensic chemist and toxicologist from the New York Office of the Chief Medical Examiner, proved that Anna Fredericksen was poisoned by carbon monoxide from a stove and was not murdered. Gettler extracted some of her blood from her heart, deposited it in a porcelain dish, added lye, and stirred. When normal blood is tested using this method, it turns dark and gelatinous with dark green-brown layers in the light. Because Frederickson's blood was saturated with carbon monoxide, it was crimson red.

Travia explained to the police that Frederickson had come to him for whiskey as her usual bootlegger was out. They drank his together at his table. He thought they may have argued when he asked her to leave but he couldn't remember and had fallen asleep at the table. When he awoke, he found that she was lying on his floor—stiff and cold to the touch. His head felt dizzy and slow. He had presumed he had killed her in an argument—by shaking, strangulation, or something. He decided he had to get rid of her body or risk being charged with her murder. Travia's case went to trial. With the testimony from the building owner that an empty coffee pot was discovered on Travia's stove and was found to have boiled over thus extinguishing the flame and letting the gas fill the apartment, and with Dr. Gettler's chemical experiments pointing to the cause of death as carbon monoxide poisoning, Travia was acquitted of murder in March 1927. He was still convicted on the dismemberment charge for which he went to prison.

In April 2015, Rodney Todd, a 36-year-old father, and his seven children, aged 5–15, were found dead in their beds in their home in Princess Anne, Maryland. After Todd didn't come to work for several days, his supervisor called the police. She reported that she had not seen Todd since March 28, 2015.

Gettler's protégé created the Office of the Chief Medical Examiner (OCME) in Baltimore, Maryland, which opened in 1890. In 1939, Maryland became the first state to establish a statewide medical examiner system. Medical examiners are medical doctors who have completed a pathology residency and fellowship or additional training and are hired to work for the state. In contrast, coroners in many states are elected by their constituents; the only requirements are to be 18 years of age and have a high-school diploma.

In a published method, oxyhemoglobin (O_2Hb) and methemoglobin are reduced using sodium hydrosulfite solution and the COHb can be detected using ultraviolet-visible (UV-Vis) spectroscopy using an absorbance ratio of 432 nm to 420 nm. A calibration curve is prepared using calibrants from non-CO exposed individuals' blood. A 100% O_2Hb calibrant was created by bubbling ultra-pure O_2 through a blood sample to release any bound CO. Likewise, a 100% COHb calibrant was created by bubbling CO through a blood sample. The limit of quantitation was determined to be 10% COHb. Normal COHb levels are less than 5% and up to 9% in smokers. A COHb level above 25% is associated with serious toxicity and above 70% is fatal. The 100% oxygen-saturated O_2Hb samples are very stable when tightly sealed and refrigerated although the 100% COHb samples increased at the lower end and decreased at the higher end. An alternative to using a conventional UV-Vis spectrophotometer is the use of a specialized automated CO-oximeter, where available. The Baltimore OCME employs the Avoximeter® to obtain rapid blood oxygenation levels.

The electricity to Todd's home had been shut off on March 25. When police investigated the report, they found a generator was being used to power heat and lights in the home and it and an empty gas tank were found inside the house. Generators should only be used outside and at least 15 feet from the building. The deaths of Todd and his children were determined to be accidental and the cause was carbon monoxide poisoning.

DETECTING CARBON MONOXIDE POISONING USING ULTRAVIOLET-VISIBLE SPECTROSCOPY AND CHEMICAL TESTING (continued)

BIBLIOGRAPHY

Anderson, J., C. Campbell, and C. Rentz. Carbon monoxide blamed in deaths of father, 7 children in Princess Anne. *The Baltimore Sun*, April 7, 2015, www.baltimoresun.com/news/maryland/bs-md-carbon-monoxide-20150407-story.html (accessed January 24, 2018).

Blum, D. 2011. *The Poisoner's Handbook: Murder and the Birth of Forensic Medicine in Jazz Age New York*. New York: Penguin Press.

Clinical Guidance for Carbon Monoxide (CO) Poisoning After a Disaster. www.cdc.gov/disasters/co_guidance. html (accessed January 24, 2018).

Luchini, P.D., J.F. Leyton, M. de L.C. Strombech, J.C. Ponce, M. das G.S. Jesus, and V. Leyton. 2009. Validation of a spectrophotometric method for quantification of carboxyhemoglobin. *J. Anal. Toxicol.* 33(8):540–544.

Spectroscopic methods investigate the interaction of chemical materials with electromagnetic radiation as a function of wavelength. In 1666, Sir Isaac Newton coined the term "spectrum" to describe the dispersion of white light. Using spectroscopy, forensic scientists can screen evidence items, determine structural features including chemical functional groups and local environments (e.g., amino acids in proteins), and quantitate drugs and metabolites.

Spectroscopic methods are nondestructive and highly sensitive. The instruments are referred to as spectrophotometers, spectrometers, and spectrographs, depending on the type of radiation (e.g., gamma, x-ray, ultraviolet, visible, near infrared, infrared, terahertz, and microwave) employed and the instrument used. A schematic of the electromagnetic spectrum is shown in Figure 4.1. A prism was used to disperse sunlight in early spectrophotometers but modern instruments employing high-power lasers and gratings and light-emitting diodes (LEDs) are used to produce the desired wavelength(s) of radiation.

Absorption is a term used to describe radiation absorbed by a material while *transmission* is the amount of radiation that is passed through the material. Spectroscopic instruments employed in forensic laboratories include UV-Vis, fluorescence, Fourier transform-infrared spectrometers (FT-IR), and Raman spectrophotometers and microspectrophotometers. Owing to their high sensitivity and low cost, many instruments, including microscopes (discussed in Chapter 3) and chromatography instruments (to be discussed in Chapter 6), are equipped with UV-Vis detectors, although other detectors such as IR detectors are also used.

The development of the optical interferometer by Albert Abraham Michelson laid the groundwork for modern spectrophotometers. It employs a light source that is split into two beams using a beamsplitter and the light beams are reflected back toward the beamsplitter. The amplitudes are then recombined. Michelson was awarded the 1907 Nobel Prize in Physics for his work. Another significant advance in spectroscopy was Sir Chandrasekhara Venkata Raman's discovery of the spectroscopic technique now named after him. Raman was awarded the Nobel Prize in Physics in 1930 "for his work on the scattering of light and for the discovery of the effect named after him." These Nobel Prize winning contributions are tabulated in Table 4.1.

ULTRAVIOLET-VISIBLE SPECTROSCOPY

UV-Vis is a very inexpensive and widely accessible tool used to qualitatively fingerprint and quantitate micromolar concentration ranges of organic and biological molecules such as paint, drugs, dyes, metabolites, DNA, toxins, and

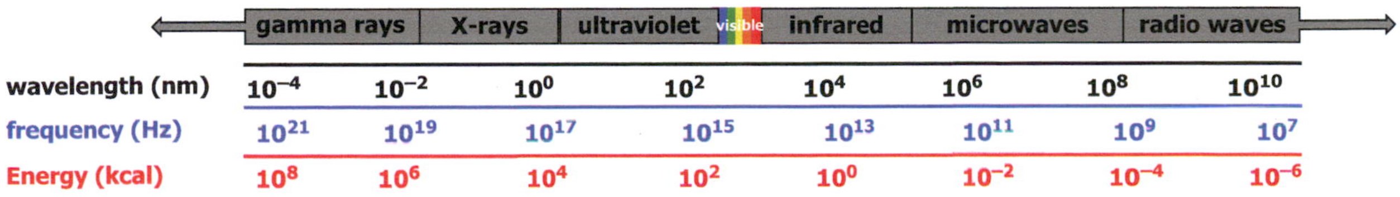

	gamma rays	X-rays	ultraviolet	visible	infrared	microwaves	radio waves
wavelength (nm)	10^{-4}	10^{-2}	10^{0}	10^{2}	10^{4}	10^{6}	10^{8}	10^{10}
frequency (Hz)	10^{21}	10^{19}	10^{17}	10^{15}	10^{13}	10^{11}	10^{9}	10^{7}
Energy (kcal)	10^{8}	10^{6}	10^{4}	10^{2}	10^{0}	10^{-2}	10^{-4}	10^{-6}

Figure 4.1 Electromagnetic spectrum.

Table 4.1   Nobel Prize Laureates in the field of spectroscopy

Award recipient	Nobel prize, year	Noted contribution
Albert A. Michelson	Nobel Prize in Physics, 1907	"for his optical precision instruments and the spectroscopic and metrological investigations carried out with their aid"
C.V. Raman	Nobel Prize in Physics, 1930	"for his work on the scattering of light and for the discovery of the effect named after him"

Source: www.nobelprize.org/.

proteins. UV-Vis spectroscopy measures the amount of electromagnetic radiation absorbed by a solution in the 190–800 nanometer (nm) range. The ultraviolet range spans 190–300 nm and the longwave UV range spans 300–400 nm. The visible range spans 400–800 nm. The excitation light source may be a laser or a light-emitting diode. A grating is used to filter out only the laser wavelengths selected for the sample irradiation or excitation. Samples can be placed in quartz cuvettes that are placed in a cell holder in the instrument for analysis. A schematic of a UV-Vis spectrometer is shown in Figure 4.2. Typical UV-Vis cuvettes hold 1 mL while sub-micro cuvettes can hold as little as 10 μL. UV-Vis cuvettes may be translucent on two parallel sides as the excitation source is parallel to the detector used to record absorbance as shown in Figure 4.3.

Molecules that absorb in the visible region of the electromagnetic spectrum are referred to as *chromophores*. Chromophores include dyes used in paint and food coloring including FD&C Yellow 5 and Red Dye 40 as shown in Figure 4.4. The excitation of an electron from the ground state to a higher energy level from the absorption of light is shown in Figure 4.5. Longwave and lower energy light will promote electrons to lower energy levels than shortwave and higher energy light.

The groups most likely to absorb light in the UV-Vis region in the chromophores are π-electrons in double bonds and heteroatoms with non-bonding valence shell electrons, as shown with the black arrows in Figure 4.6. Transitions between n to σ^* energy levels may also be observed with energy in the 200–800 nm range. When an evidence sample is subjected to an input of energy (such as in the form of light), electrons (from the *highest occupied molecular orbital*, or HOMO) are promoted to higher energy levels (*lowest unoccupied molecular orbital*, or LUMO) altering the electronic structure of the molecules in the material and resulting in an *excited state* species. Each compound will have a select few wavelengths that it will absorb in preference to other wavelengths depending on its molecular structure and the energy differences between the HOMO and LUMO energy levels. The observed absorption peaks reflect the

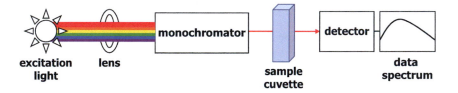

Figure 4.2   Schematic of the light path in a UV-Vis spectrometer.

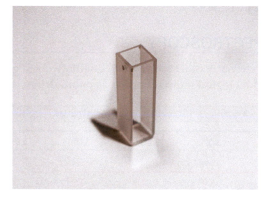

Figure 4.3   UV-Vis cuvette. (Courtesy of Tim Phillips.)

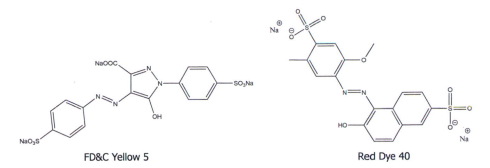

FD&C Yellow 5   Red Dye 40

Figure 4.4 Chemical structures of chromophores FD&C Yellow 5 and Red Dye 40. (Structure prepared by Ashley Cowan.)

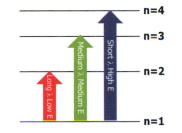

Figure 4.5 Excitation of an electron from the ground state to a higher energy level.

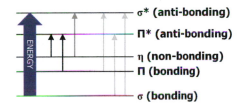

Figure 4.6 UV-Vis absorption from the HOMO to the LUMO energy level.

additive contributions of the UV-Vis active bonds and atoms. For example, in the simple alkene ethene, the energy difference between the π and π* molecular orbital energy levels is 173 kcal/mol, which requires 165 nm radiation to promote an electron from the HOMO to the LUMO energy level. The energy gap between the HOMO and LUMO is smaller in 1,3-butadiene, a conjugated diene. In 1,3-butadiene, UV radiation of 217 nm (a longer wavelength and lower energy) is needed to promote an electron from the HOMO to the LUMO energy level. The addition of successive conjugated double bonds progressively narrows the gap between HOMO and LUMO and lowers the energy of the radiation required for the transition.

DNA has an absorbance maxima of 260 nm, on average. The absorption spectrum of a DNA oligonucleotide is shown in Figure 4.7. Proteins have an absorption maxima of 280 nm, on average, especially if they contain tryptophan residues. UV-Vis is frequently used to quantitate DNA and RNA using their absorbance at 260 nm. Their quality can be determined using the ratio of their absorbance at 260 nm to 280 nm (high-quality DNA has a ratio of 1.8–2.0). When DNA is systematically titrated with protein, absorption of UV light at 260 nm decreases, absorption at 280 nm increases, and an isosbestic point where their spectra cross is observed at approximately 240 nm.

The amount of light absorbed in UV-Vis spectroscopy is directly proportional to the concentration of the compounds in the solution. Thus, UV-Vis spectroscopy is routinely used to quantify drugs or other molecules present in a sample. The relationship between concentration (c) and absorbance (A) is known as *Beer's Law* and is given as:

$$A = a\,b\,c \tag{4.1}$$

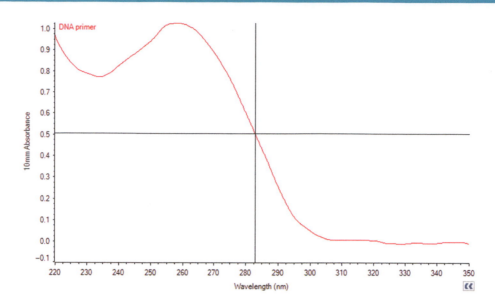

Figure 4.7 UV-Vis absorption spectrum of DNA.

The symbol (a) represents the solution's absorptivity coefficient or constant and (b) the path length that the light travels through the solution, usually 1 cm. Absorbance (data recorded from the spectrophotometer) does not have units. When (b) is in centimeters and (c) is in moles per liter, (a) has the unit liters per moles centimeter ($L*mol^{-1}*cm^{-1}$). The value of the molar absorptivity constant is unique for each absorbing species. Its value can be derived using Beer's Law or from a Beer's Law calibration curve (plot of absorbance vs. concentration).

By carefully selecting a wavelength where one compound absorbs intensely (Figure 4.8), such as 495 nm for red dye, it is often possible to measure the concentration of one compound in the presence of several others. Beer's Law calibration curves are made experimentally by preparing a series of solutions, each with a known concentration of the absorbing species in an appropriate solvent. These solutions are referred to as calibration standards and one microliter of several ten-fold dilution standards is shown in Figure 4.9. The absorbance of each standard is measured at the same wavelength and a calibration curve can be prepared. In addition to determining the value of the molar absorptivity coefficient, this plot can be used to determine the unknown concentration of a solution by measuring the

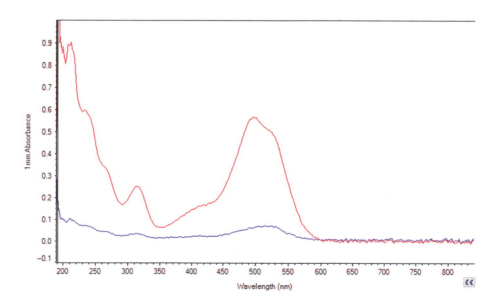

Figure 4.8 UV-Vis absorption spectra of red food coloring at dilutions of 1:10 and 1:1000.

Figure 4.9   Dilution series concentration standard red food dye.

absorbance of an unknown at the same wavelength used to prepare the standards (Figure 4.10). UV-Vis spectroscopy is sensitive to parts per million (ppm) or micromolar (10^{-6} M) concentrations.

Inexpensive and portable UV-Vis instruments include the Ocean Optics spectrophotometers and the Thermo Scientific NanoDrop shown in Figure 4.11. For research purposes, benchtop spectrophotometers such as the Agilent/ Hewlett Packard 8453 or Cary 100 are often used as they are more sensitive and provide results with a higher resolution than portable instruments. Figure 4.12 shows a benchtop UV-Vis spectrophotometer. Drawbacks to the research-grade instruments include the high cost and the larger sample volume required. The NanoDrop requires only one microliter of sample and is ideal for small sample sizes.

Visible spectroscopy can be used to determine the oxygenation level of hemoglobin (Figure 4.13) and the presence and concentration of salicylates in urine or drugs in a body fluid sample (Figure 4.14) using a wavelength of 240 nm or 300 nm (two of the local maxima), especially following separation with high-performance liquid chromatography

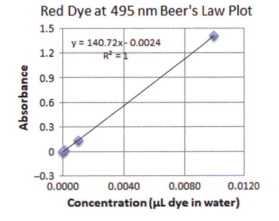

Figure 4.10   Beer's Law calibration curve of red food dye dilution series.

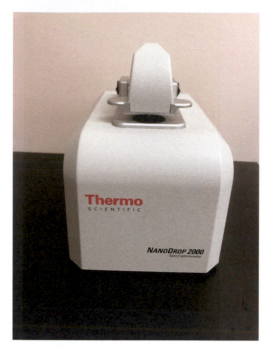

Figure 4.11 NanoDrop UV-Vis spectrophotometer.

(HPLC). The HPLC separated components of the body fluid or other sample flow past the UV-Vis detector as they proceed to collection tubes or a waste container.

FLUORESCENCE SPECTROSCOPY

Fluorescence spectroscopy is used frequently in forensic science to detect and quantitate DNA in a polymerase chain reaction (PCR), DNA typing, and DNA-sequencing applications. Figure 4.15 shows a fluorescence spectrometer. Like UV-Vis spectroscopy, fluorescence can be used to qualitatively fingerprint evidence items for identification and quantitate the amount of a fluorescent substance present. Although fluorescence spectroscopy was once used more regularly through the 1970s in drug identification, infrared (IR) spectroscopy (discussed in the next section) is now used for this purpose. *Fluorescence spectroscopy* detects molecules that not only absorb radiation causing electrons to be boosted to an excited state, but subsequently emit a photon of energy on return to the ground state (Figure 4.16). Thus, it is more specific than UV-Vis spectroscopy as not all molecules that absorb UV-Vis radiation fluoresce. However, some molecules that absorb UV-Vis radiation will fluoresce when the excited electron(s) relax to the ground state. In many cases, the "ground state" is a vibrationally excited state on the potential energy surface of the ground

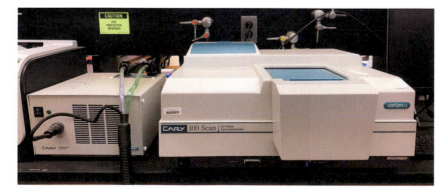

Figure 4.12 UV-Vis spectrometer.

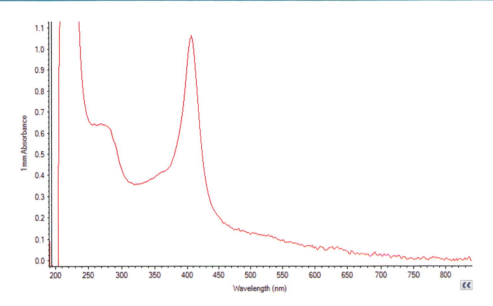

Figure 4.13   UV-Vis absorption spectrum of bovine hemoglobin.

electronic state. The fluorescence emission maxima is usually a longer wavelength than the excitation maxima as energy is lost due to non-radiative transitions.

Spectra can be recorded from samples in quartz cuvettes placed in the cuvette holder. Standard cuvettes can hold volumes up to 4 mL while sub-micro cuvettes can hold as little as 10 µL. Cuvettes used for fluorescence measurements are transparent on all four sides as the emission is recorded perpendicular to the excitation light source as described next and shown in Figure 4.17.

In a fluorescence spectrometer, the excitation light source is typically an ultraviolet or visible light source, but it can also be an x-ray source. Excitation sources include lasers, LEDs, xenon arc lamps, and mercury-vapor lamps. While lasers emit light of only a very narrow wavelength and do not require filters, xenon arc lamps emit light from 300 to 800 nm at constant intensity and sufficient radiation down to 200 nm to allow measurements to that wavelength. The excitation light is passed through a monochromator or filter, allowing only selected light to hit the sample. On absorbing a portion of the light, some of the molecules emit light known as fluorescence. In a

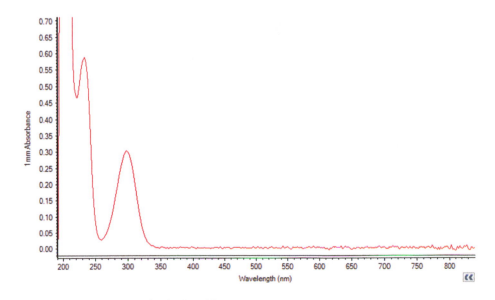

Figure 4.14   UV-Vis absorption spectrum of salicylic acid.

Figure 4.15 Fluorescence spectrometer.

spectrofluorometer, a diffraction grating is used to isolate the excitation and emission light. The emitted energy is of lower energy than the excitation wavelength as some energy is lost as heat due to collisions with other molecules. The fluorescence passes through a second monochromator and filter, which are typically placed at a 90° angle from the excitation source to reduce the reflected or transmitted incident light reaching the detector and improve the signal-to-noise ratio by a factor of 10,000 over 180° detector placement. The cuvette is housed in a dark box to also reduce detection of ambient light. Typically, the emission monochromator is used to scan the emission spectrum but the emission wavelength can also be kept constant so that the excitation spectrum can be scanned. However, in synchronous mode, the analyte is exposed to a range of excitation wavelengths and an emission map is collected for a range of emission wavelengths. The difference between the excitation and emission maxima is referred to as the *Stokes shift*. Table 4.2 lists the excitation and emission maxima for several fluorophores used to label or detect DNA in DNA-typing applications. The Stokes shift for SYBR Green I is a narrow 23 nm.

Three-dimensional (3D) topography contour maps that plot the excitation wavelengths on the y-axis, the emission wavelengths on the x-axis, and the intensity on the z-axis can be used in data analysis. These 3D maps are especially useful for determining the excitation and emission maxima for the substance. Two-dimensional spectra highlight either the excitation range for a given emission wavelength or the emission range for a given excitation wavelength. The fluorescence emission is plotted on the y-axis in relative fluorescence units (RFU). 3D excitation-emission spectra maps have been used to examine metal ion binding to natural organic matter in water and soil and bovine serum albumin with gold nanoparticles prior to conjugation to an antibody for the lifestyle analysis of fingerprints (Figure 4.18), among other uses. In the 3D plots, the excitation is plotted on the y-axis, the emission is plotted on the x-axis, and the z-axis reflects the intensity in RFUs.

Proteins such as hemoglobin and myoglobin can be probed using fluorescence spectroscopy to determine poisoning with the gases carbon dioxide, carbon monoxide, cyanide, and nitrogen; heavy metals including arsenic, lead, and mercury; and volatiles such as methane, ethane, and propane. All are known to bind to the proteins; the protein structure is altered by the binding that is detectable by shifts in emission maxima. Tryptophan, an amino acid in proteins, has an excitation maxima of 279.8 nm and an emission maxima of 348 nm at a neutral pH. Due to its strong quantum yield, it is the predominant fluorescent moiety in proteins and changes in its local environment lead

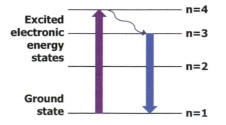

Figure 4.16 Fluorescence emission of light energy after excitation of an electron to an excited state.

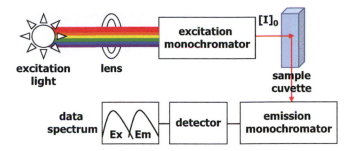

Figure 4.17 Schematic of a fluorescence spectrometer.

to shifts in the emission maxima when the ligand changes and the protein structure is perturbed. In a recent study, carbon monoxide was detected using fluorescence spectroscopy and a naphthofluorescein probe (Yan et al. 2016).

Phosphorescent molecules differ from fluorescent molecules in that they do not immediately emit the absorbed radiation. Rather, the radiation from "forbidden" energy state transitions is slowly emitted at a lower intensity for up to several hours after excitation.

Concentration can be determined using fluorescence spectroscopy with a calibration curve and Beer's Law as described previously. For example, concentration can be determined using fluorescence dyes that are covalently bound to the analyte or that interact via hydrophobic interactions such as π-π stacking and van der Waals interactions. The limit of detection using fluorescence spectroscopy is parts per billion or nanomolar (10^{-9} M) concentrations.

Fluorescence detectors are routinely employed in capillary electrophoresis and real-time PCR instruments used in forensic labs. Fluorescence is used to detect and quantify the analytes. Fluorophores such as SYBR Green I or ethidium bromide (Figure 4.19) can be added to chemical reactions and used to detect and quantitate DNA as it is being amplified in the PCR reaction directly through the PCR tube or plate. SYBR Green I's excitation maxima is 497 nm; its emission maxima is 522 nm. Figure 4.20 shows the quantitation curve of the foodborne pathogen *E. coli* standard DNA using real-time PCR and SYBR Green I and fluorescence spectroscopy.

Other fluorescent dyes, such as fluorescein and its derivatives, can be covalently attached to their targets; this method is used to label PCR primers applied in short tandem repeat DNA-typing methods and nucleotide bases are tagged in DNA sequencing reactions. The PCR-amplified products can be detected and identified as they pass by the detector by the time it takes for them to separate in a capillary or on a gel. These methods are covered in detail in forensic biology textbooks including those by John Butler and Richard Li and *Forensic Biology: A Laboratory Manual*, which was written by the author of this book.

Table 4.2 Fluorophores and their excitation and emission maxima

Fluorophore	Excitation maxima (nm)	Emission maxima (nm)
Ethidium bromide	493	605
SYBR Green I	497	520
ROX	575	602
LIZ	570	591
JOE	520	548
Fluorescein	495	518
FAM	495	516
VIC	538	554
NED	546	575
TAMRA	555	580

Source: ThermoFisher.

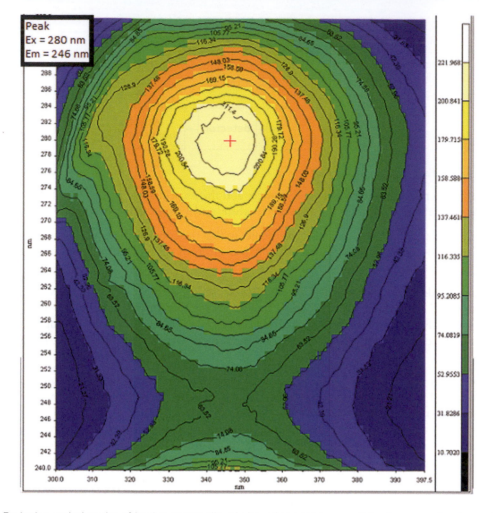

Figure 4.18 Excitation-emission plot of bovine serum albumin bound to gold nanoparticles in preparation for conjugation to an antibody for lifestyle analysis of fingerprints.

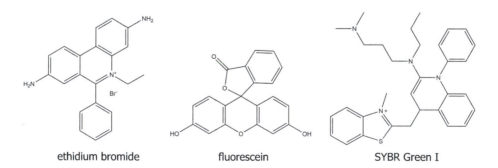

ethidium bromide   fluorescein   SYBR Green I

Figure 4.19 Structures of fluorophores ethidium bromide, fluorescein, and SYBR Green I. (Structure prepared by Ashley Cowan.)

INFRARED SPECTROSCOPY

Infrared spectroscopy is a prevalent tool used to fingerprint and identify organic materials using a library. Forensically relevant materials include paint, drugs, dyes, explosives, plastics, fibers, soil organic matter, and body fluids. IR light spans the region from 0.8 to 1000 micrometers in the electromagnetic spectrum; however, the region that is most commonly used is 2.5 to 50 micrometers. The unit that is commonly used for IR spectroscopy is the wavenumber

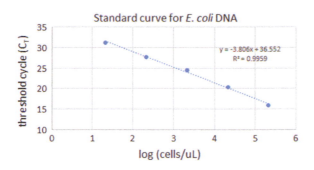

Figure 4.20 Quantitation of *E. coli* standard DNA using real-time PCR with SYBR Green I fluorescence.

(with units of reciprocal centimeters, cm^{-1}). The corresponding range is 4000–400 cm^{-1}. The wavenumber is proportional to energy and frequency and reciprocal with wavelength. Its wavelength is longer than visible light waves but shorter than radio waves. Like these forms of electromagnetic radiation, it is a low-energy form of radiation.

The IR spectrum is divided into three regions: near infrared, middle infrared, and far infrared, as shown in Table 4.3. The near infrared is closest to the visible light spectrum range. The IR absorbance properties are used to probe the structural features of a molecule; a peak in the IR can be linked to the structure and composition of the molecule.

When IR-active molecules are exposed to IR light, the light energy is absorbed by those bonds with the capacity to absorb the same frequency of incoming radiation causing the atoms and bonds of the molecules to vibrate (Figure 4.21). If the frequency of vibration of IR light matches the frequency of vibration of atoms within a molecular bond, an absorbance will occur. Only molecules that have a dipole moment absorb IR radiation. The vibration alters the molecule's dipole moment (the magnitude of the positive and negative charge between two bonded atoms) and causes the absorption. Thus, gas molecules including the homonuclear oxygen (O_2), hydrogen (H_2), chlorine, and (Cl_2) do not absorb IR radiation due to symmetry. Monatomic ions, individual atoms, and the noble gases are also IR inactive. Gases such as sulfur dioxide (SO_2) and carbon dioxide (CO_2) have a dipole moment and will absorb IR light. As carbon dioxide is a prominent component of air, its spectra is recorded in the air background and subtracted from spectra taken in the air.

The frequency of the absorbed vibration will depend on the quantity of each type of atom present and the length and strength of their bonds. Just as a change in the mass of an object on a spring changes the frequency of vibration in the spring, a change in the mass of the atoms found between a chemical bond changes the vibrational frequency of the bond. Typically, as the mass of an atom increases, the frequency absorbed decreases. In addition, the strength of the bond influences the vibrational frequency: as the bond strength increases (from single to double to triple bonds), the frequency also increases. Polar bonds (e.g., O-H, C=O, C=N, etc.) with their strong bond diploes absorb the radiation strongly while nonpolar bonds (e.g., C-H) absorb the radiation weakly. Overlapping bands (e.g., C-H) also tend to result in strong absorptions so molecules that contain many hydrocarbons tend to have strong absorptions. Trends of the vibration frequency of atoms are also observed. For example, light atoms or functional groups (e.g., -CH, -NH, -OH) and multiple bonds (e.g., alkynes and nitriles) vibrate more rapidly than heavy atoms or functional groups (-Cl, -F) and single bonds (e.g., alkanes). The vibration intensity trends follow the change in dipole caused by the vibrating atoms. There are numerous types of vibration that can occur in molecules. These include symmetric and asymmetric stretching vibrations and in-plane rocking, in-plane scissoring, out-of-plane (OOP) wagging, and out-of-plane twisting vibrations, as shown in Figure 4.21.

Each functional group has its own characteristic IR absorption. IR is often used with other instrumental methods to elucidate chemical structures of materials and chemicals not in the library. As molecules increase in complexity, so too do their IR absorption. By carefully analyzing a molecule's IR absorption, forensic scientists can identify the

Table 4.3 Infrared spectrum ranges

Range	Near infrared	Middle infrared	Far infrared
Wavelength (μm)	0.8–2.5	2.5–50	50–1000
Wavenumber (cm^{-1})	12,500–4000	4000–200	200–10
Energy (eV)	1.55–0.5	0.5–0.025	0.025–0.0012

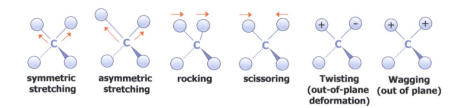

Figure 4.21 Infrared vibrations.

functional groups in a molecule's structure based on the frequency and strength of the absorption. There are four predominant analysis regions: 4000–2500 cm^{-1}, 2500–2000 cm^{-1}, 2000–1500 cm^{-1}, and 1500–400 cm^{-1}, corresponding to the absorption of bonds involving a hydrogen atom (e.g., C-H, N-H, O-H), triple bonds (e.g., C≡C, C≡N), double bonds (e.g., C=O, C=C, C=N), and other bonds and heavy atoms (e.g., C-Cl), respectively. The lowest energy region, 1500–400 cm^{-1}, is referred to as the fingerprint region. Many bands are observed in this region, all resulting from angle bending and heavy atom stretching. Although this region is difficult to interpret in order to identify functional groups present in the molecule, it is widely used in pattern analysis and band-to-band comparisons or "molecular fingerprinting." A high-quality correlation chart can be used to assign the most common functional groups (Table 4.4).

IR samples can be analyzed as gases, liquids, solids, and thin film, although solids are most commonly analyzed in forensics. The samples must be highly concentrated in order to record a high-quality spectrum. To record an IR spectrum, IR radiation is directed at the sample and the amount of energy absorbed is recorded based on how much is not absorbed (transmitted) to the detector.

Depending on the instrument available at the laboratory, data for IR absorbance (plotted as % T) can be recorded using the conventional method, diffuse reflectance, or attenuated total reflectance (ATR). The experiments are not the same and the spectra will be different. It is important, especially when comparing recorded spectra of unknowns, to ensure that the comparison spectra in the library were recorded using the same experimental method. In traditional instruments, a monochromator is used to select the desired frequency at each time. In an FT-IR, the monochromator is replaced with an interferometer. The spectrum is referred to as an interferogram. In some instruments, a series of mirrors are used to split the beam and direct the radiation onto two samples, an unknown and reference, simultaneously. The advantage of this technique is that sample and reference samples can be analyzed simultaneously. In diffuse reflectance infrared Fourier transform spectroscopy (DRIFTS), a powder sample is analyzed by placing it into a sample cup and collecting the IR spectrum on the surface of the bulk sample. The IR light is focused on the powder (sometimes diluted in a non-absorbing matrix) and collected and relayed to the IR detector. This method is ideal for high-surface-area powders and can be used for heterogeneous catalysis, where the temperature and environment of the catalyst can be controlled *in situ* in the DRIFTS cell. It can also be used for drugs, polymers (including foams, powders, fibers, and composites), mineral powders, and matte or dull surfaces such as fabrics and filled plastics. ATR spectra are recorded using a special accessory that allows liquids and solids to be measured directly; the resulting data provides reflectance spectra, as the name suggests. The sample is placed in direct contact with a crystal where a very small amount of light makes its way to the sample. Some of the energy of the evanescent wave is absorbed by the sample and the reflected radiation is returned to the detector (Figure 4.22). FT-IR equipped with ATR accessories are referred to as ATR FT-IR spectrometers; these are very common in modern forensic laboratories (Figure 4.23).

Solids are ground with a mortar and pestle to homogenize the sample and yield a very fine powder. On an ATR instrument, the sample is analyzed directly without any other preprocessing. On a traditional instrument, the solid is mixed with liquid paraffin to make a paste. Several drops of the paste are applied to two sodium chloride (NaCl) plates (or potassium bromide [KBr] plates as these do not absorb in the recording region) and the salt plates are placed in the instrument vertically in a holder so that the light can be efficiently passed through the sample. Samples dissolved in nonpolar solvents can also be applied to circular salt plates in the liquid phase to produce a thin capillary film. Alternatively, a small amount of the homogenized solid (approximately 1–2 mg) can be mixed with pure KBr (approximately 200 mg), which are ground together well and then formed into a pellet using a mechanical die press. The resulting pellet is referred to as a KBr pellet. This too can be introduced to the instrument by placing the pellet vertically in a v-shaped holder in the path of the IR light. Other sampling methods include thin film. The IR spectrum

Table 4.4   IR correlation chart

Wavenumber (cm^{-1})	Group	Type of vibration
3920	O-H in H_2O	Stretching
3650–3590	O-H	Stretching
3550–3500	Ph-O-H	Stretching
3500–3300	N-H	Stretching
3300–3250	-C≡C-H (acetylenes)	Stretching
3490	O-H in H_2O	Stretching
3280	O-H in H_2O	Stretching
3100–3000	-C=C-H	Stretching
3040–3010	C-H (olefins)	Stretching
2960	C-H	Stretching (sym)
2960–2500	C-H	Stretching
2870	C-H	Stretching (asym)
2780	N-CH$_2$	Stretching
2600–2500	S-H	
2260–2215	C≡N	
2260–2100	C≡C	Stretching
1825–1650	C=O	Stretching
1750–1730	-C(O)-O-C (ester)	Stretching
1745–1725	-CBr-C(O)-	Stretching
1740–1720	-C(O)H (aldehyde)	Stretching
1730–1705	-Ph-C(O)-O- (ester)	Stretching
1725–1705	-CH$_2$-C(O)-CH$_2$- (ketone)	Stretching
1720–1680	-Ph-C(O)H (aldehyde)	Stretching
1700–1680	-Ph-C(O)- (ketone)	Stretching
1685–1665	-CH=CH-C(O)- (ketone)	Stretching
1680–1620	C=C	Stretching
1670–1660	-Ph-C(O)-Ph- (ketone)	Stretching
1645	H-O-H in H_2O	Bending
1640-1550	N-H	Bending
1615	N-H	Bending
1470	C-H	Bending (asym)
1430	C-O-H	Bending (in-plane)
1400	-C=C-H	Bending (in-plane)
1400–1000	C-F	
1380	C-H	Bending (sym)
1360–1250	-Ph-C-N	Stretching
1310–1250	-Ph-C-O (ester)	Stretching
1270–1060	C-O	Stretching
1275–1000	C-H	Bending (in-plane)
1240	C-O	Stretching
1220–1020	C-N	Stretching
1100	C-O-C	Stretching
1000–600	-C=C-H	Bending (oop)
930	C-O-H	Bending (oop)
900–690	C-H	Bending (oop)
850–750	NH2	Wagging and twisting
800–600	C-Cl	
150–650	N-H, 2° amide	Wagging
715	N-H	Wagging
700–600	C≡C-H	Stretching
600–500	C-Br, C-I	

Sources: Pimentel, G.C. *J. Chem. Educ.* 37(12):651–657, 1960. Stuart, B. *Infrared Spectroscopy: Fundamentals and Applications.* John Wiley & Sons, 2004.

Note: asym: asymmetric; oop: out of plane.

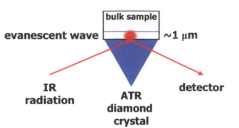

Figure 4.22   Schematic of ATR FT-IR diamond crystal and evanescent wave.

of a thin film sample can be obtained directly when the sample is placed in a holder (typically cardboard) with a slot cut out as a sample window and placed in the sample holder (Figure 4.24). This method is also used for routine calibration of the instrument using a polystyrene film as the band produced by this material is accurately known. The IR spectrum of gases is recorded by adding the gas via an inlet port to a cylindrical gas cell with windows on each end. The cell is composed of an IR-inactive material such as KBr, NaCl, or CaF_2.

Spectra are typically recorded from 4000 to 400 cm^{-1} for 8 or 16 scans, although 128 scans—or more—can be recorded. Typically, a background using air, or the substrate on which the sample is deposited, is recorded and subtracted from the sample spectrum. The detector may be a charge-coupled device (CCD), indium gallium arsenic (InGaAs), or lead sulfide (PbS). The interferometer spectra undergo Fourier transform (FT). Although the molecules absorb the IR light, spectra are typically plotted for display not as absorbance versus wavenumber but percent transmittance (% T) versus wavenumber. The scans are averaged to yield a final spectrum with a low signal-to-noise ratio.

While forensic scientists may assign the absorbance bands to determine the functional groups or bonds present in the molecule, more commonly the spectrum will be identified through a point-by-point comparison with a library of expected substances using a computer algorithm. For example, the IR spectrum of red and green food dyes contains peaks at 3300, 1640, and 1040 cm^{-1} correlated to the O-H and aromatic stretches, aromatic, and sulfoxide stretches, respectively (Figure 4.25). The IR spectrum of salicylic acid is shown in Figure 4.26. Note how many more peaks are present in this spectrum as compared to the UV-Vis spectrum of salicylic acid shown earlier in the chapter. In forensics, the IR spectrometer is an invaluable technique for distinguishing most paint binder formulations, adding further significance to a forensic paint comparison.

The FT-IR is one of the most commonly used instruments in the crime laboratory. Because the structures of most molecules differ, virtually all compounds have a unique FT-IR spectrum. However, some compounds such as isomers have similar spectra. Compounds with similar spectra require additional analyses for identification. Additionally, some materials have the same chemical composition but differ in manufacturing process (e.g., weave, density); they

Figure 4.23   ATR FT-IR spectrometer.

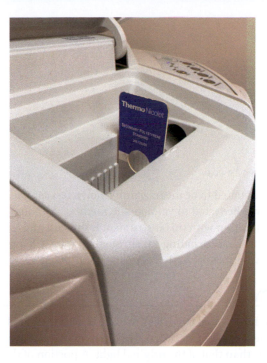

Figure 4.24   A polystyrene thin film standard in the sample holder in an infrared spectrophotometer.

will share the same IR spectra but may be differentiated using microspectrophotometry. Furthermore, not all bonds within a molecule will absorb IR radiation, even when the bond frequencies and energies match. While IR absorbances for pure or relatively pure molecules can be readily interpreted, trace materials of less than 5% of a sample will be undetectable. The IR spectrophotometer cannot be used to detect the concentration of an element in a substance. The identity of the components of complex mixtures may be determined by subtracting the IR spectra of pure compounds, if they are known. Water absorbs strongly in the IR region and while the instrument can subtract out water, dry samples with less than 0.01%–1% water are ideal. Standards and reference samples are used to create the

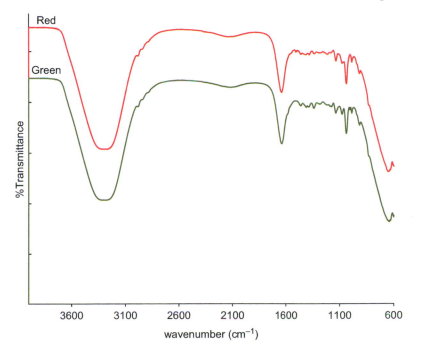

Figure 4.25   ATR FT-IR spectrum of red and green food dyes.

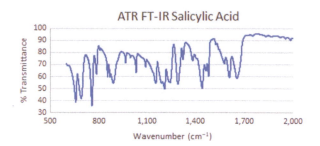

Figure 4.26 ATR FT-IR spectrum of salicylic acid.

library, which must be continually updated to be meaningful. Known test samples will be used to collect spectra and the library and search algorithm will be tested to validate the method. Although IR spectroscopy is routinely used for identification, performing quantitation with this instrument is difficult and less commonly attempted. IR instruments can be coupled to other instruments including gas chromatography (Chapter 6). Gas chromatography–Fourier transform infrared spectroscopy instruments can detect to less than 100 pg of an analyte.

RAMAN SPECTROSCOPY

In 1928, Sir C.V. Raman discovered that when particles are excited with incident light (UV/Vis/Near-IR) photons of less energy (longer wavelength), the light will be scattered in different directions and a small portion of the scattered light will acquire other wavelengths than that of the original light. A portion of the incoming photons' energy can be transferred to a molecule, giving its electrons a higher level of energy.

Raman scattering is the inelastic scattering of monochromatic light. Incident light energy is transferred to the molecule. This causes changes in the polarization of the molecule that are induced by molecular vibrations. Ground-state electrons are promoted to virtual excited states. They relax to a virtual ground state higher in energy than the ground state. The resulting energy and wavelength of incident and scattered light are no longer equal. Due to the lower energy released, a longer wavelength is observed. *Stokes* radiation is lower in energy than the incoming photon while anti-Stokes radiation is higher in energy. Other phenomena include anti-Stokes Raman scattering and Rayleigh scattering. *Anti-Stokes* transitions are characterized by the absorption of energy by electrons in virtual energy states that relax to the ground state thus releasing more energy. Stokes radiation is lower in energy (red side of spectrum) than the incoming photon while anti-Stokes is higher in energy (blue side of spectrum). Anti-Stokes Raman scattering is a weaker phenomenon owing to the fact that more electrons are in the ground state. In Rayleigh, or elastic scattering, the energy of the scattered light is equal to the energy of the incident light. Rayleigh scattering is approximately 10^7 times stronger than Raman scattering. The analyte ends in a different rotational and vibrational state than it was initially. The energy difference between the two states leads to a shift in the frequency of the emitted photon from the excitation wavelength. A Jablonski diagram for Raman scattering as compared to other forms of spectroscopy is shown in Figure 4.27.

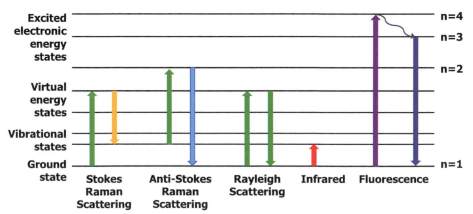

Figure 4.27 Jablonski diagram of Raman spectroscopy as compared to other spectroscopy types.

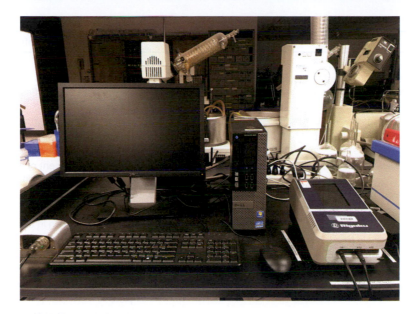

Figure 4.28 Raman spectrometer.

In a Raman spectrophotometer (Figure 4.28), a laser is used to project light onto a sample. The laser is directed toward a turning prism that bends the light 90°, and then toward a beam splitter, focusing lens, and a monochromator before it reaches the analyte where the energy of the radiation is absorbed to boost the electrons in the lowest rotational and vibrational energy level of the ground electronic state (or their excited counterparts) to virtual energy states. Filters guide the laser to the sample. Laser wavelengths of 785 nm (for nonfluorescent samples) and 1064 nm (for fluorescent samples) are commonly used. The emitted radiation, including Stokes and anti-Stokes Raman scattering, is collected by a focusing lens, focused through an entrance slit, dispersed by a diffraction grating into the various wavelengths to generate a spectrum and to discriminate Rayleigh scattered laser light, and transported to the detector by a fiber-optic cable.

In 1986, the CCD was used as a Raman detector. A photomultiplier can also be used as a detector. The photo-detector records the intensity of the inelastically scattered light in arbitrary units by wavelength.

The region typically employed for Raman spectroscopy is 200–2000 cm^{-1}. Benzonitrile is used to calibrate the instrument and perform instrument checks. Several spectra can be recorded and averaged to achieve a better signal-to-noise ratio and FT analysis can be performed. Dilute samples give weak signals that can be enhanced by tightly packing the sample and by applying signal-enhanced Raman spectroscopy (SERS) using and applying the samples to silver, gold, and copper metal nanoparticles to enhance the signal.

The collected spectrum can be considered a fingerprint for a given molecule and can be used to identify an analyte such as salicylic acid as shown in Figure 4.29. Raman spectroscopy can be used to generate information about molecular bonding as the polarizable electron density and bonds in molecules vibrate or rotate on exposure to the

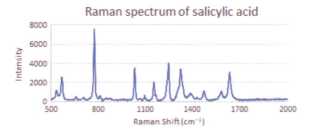

Figure 4.29 Raman spectrum of salicylic acid.

Table 4.5   Raman versus infrared spectroscopy: A contrasting view

Raman	Infrared
4000–200 cm^{-1}	4000–400 cm^{-1}
Detects emission of inelastically scattered monochromatic light (light scattered at a different frequency or color than the original frequency of light) by vibrating molecules	Detects absorption of infrared radiation at a given wavelength by vibrating molecules
Change in polarizable electron density and bonds	Change in net dipole moment of molecule
Does not require permanent dipole moment	Vibration of interest should have a change in the dipole moment due to that vibration
Can take spectra through glass or plastic container	Direct measurement of liquids and solids on diamond ATR or on salt plate or in thin film
Samples can be solids, liquids, or gases	Gases rarely sampled
Water can be used as a solvent	Water cannot be used as a solvent due to its strong absorption in this region
Rotational, vibrational, or electronic frequencies in bonds	Determination of functional groups
Gives an indication of the covalent character of the molecule	Gives an indication of the ionic character of the molecule
Simple measurement of bulk samples	Without ATR, press into disks, sample on salt plate
Quantification	No quantification of atoms in molecules or components in mixture
Biological molecules that fluoresce may mask scattering	
Scattering is weak	Absorptions can be strong to weak
Nondestructive	Nondestructive
Instrument cost can be high	Relatively inexpensive

radiation. IR spectroscopy gives complementary information and its spectra are often complementary to Raman spectra although they are different phenomena (Table 4.5). As in IR spectroscopy, Raman spectra can be analyzed using correlation charts (Table 4.6) to identify phenomena that give rise to the bands at various wavelengths. Spectra libraries can also be used for materials identification. In forensic chemistry, Raman is used to identify and differentiate drugs and controlled substances from inside their containers, art objects without direct contact, as well as fibers, pigments, inks, paints, textiles, and body fluids. Sampled items can be as small as a micron. It can be used for solids, liquids, and gases.

The advantages of Raman spectroscopy include how it can sample items through containers and without removing samples (much safer for controlled substance analysis) and thus is quick. Another advantage is that there is no water contribution to Raman spectra.

The disadvantages to this method include how Raman scattering can be very weak, the correlation charts are not as developed as those for interpreting IR spectra, and the instrument is more expensive than IR and less widely available in forensic labs.

MICROSPECTROPHOTOMETRY

Microspectrophotometers combine a microscope with a spectrophotometer. The most common types encountered in forensic laboratories couple a compound light microscope with a fluorimeter, UV-Vis-NIR, FT-IR, or Raman spectrophotometer. FT-IR microspectrophotometers have been available since the early 1980s and can detect materials as small as 10 microns. These tools are ideal for working with heterogeneous, altered, or contaminated microscopic materials or objects. The IR microspectrophotometer is also ideal for the analysis of plastics or fabrics that differ in weave but have the same chemical composition such as polyethylene terephthalate (PETE)/Dacron. The object is scanned on the microscope stage. Spectra can be recorded, as desired, from various locations of the object. There are drawbacks. The objects must be small enough to fit on the microscope stage and the spectra are only from the top layer, unless the top layer is removed to expose the underlying layers for scanning. Microspectrophotometers are used in the quality control of the thickness of films and for evaluating fibers, soil, paint, and ink.

Table 4.6 Correlation chart for Raman spectroscopy

Approximate wavenumber (cm⁻¹)	Group/bonding vibrations	Raman intensity
10–200	Lattice vibrations in crystals	Strong
150–450	Xmetal-O	Strong
250–400	C-C aliphatic chains	Strong
290–330	Se-Se	Strong
430–550	S-S	Strong
450–550	Si-O-Si	Strong
480–660	C-I	Strong
500–700	C-Br	Strong
550–800	C-Cl	Strong
600–1300	C-C acyclic aliphatic chain	Medium
630–790	C-S aliphatic	Strong
800–970	C-O-C	Medium
845–900	O-O	Strong
1000	C-C=C aromatic ring chain	Strong/Medium
1000–1250	C=S	Strong
1060–1150	C-O-C asymmetric	Weak
1080–1100	C-S aromatic	Strong
1340–1380	C-(NO_2)	Strong
1380	CH_3	Medium
1400–1470	CH_2, CH_3 asymmetric	Medium
1410–1440	N=N aromatic	Medium
1450, 1500	C-C=C aromatic ring chain	Medium
1500–1900	C=C	Strong
1530–1590	C-(NO_2) asymmetric	Medium
1550–1580	N=N aliphatic	Medium
1580, 1600	C-C=C aromatic ring chain	Strong
1610–1680	C=N	Strong
~1640	H_2O	Weak and Broad
1680–1820	C=O	Medium
2550–2600	-S-H	Strong
2800–3000	C-H	Strong
3000–3100	=(C-H)	Strong
3100–3650	O-H	Strong
3300–3500	N-H	Medium

Source: HORIBA, www.horiba.com/fileadmin/uploads/Scientific/Documents/Raman/bands.pdf.

QUESTIONS

1. Spectroscopy is used to investigate:

   a. The interaction of chemical materials with electromagnetic radiation as a function of wavelength

   b. The interaction of chemical materials with a polar stationary phase in the presence of a nonpolar mobile phase

   c. The interaction of chemical materials with a nonpolar stationary phase in the presence of a polar mobile phase

   d. The interaction of chemical materials with ions in a drift tube

2. Which of the following is not a spectroscopic method?

   a. Raman   b. IR

   c. UV-Vis   d. TLC

3. Which method(s) can be used to quantify a chromophore analyte?

   a. Raman spectroscopy   b. IR spectroscopy

   c. UV-Vis spectroscopy   d. TLC

4. Which method(s) can be used to determine the functional groups present in an analyte?
   a. Raman spectroscopy   b. IR spectroscopy
   c. UV-Vis spectroscopy   d. TLC

5. Which method(s) can be used to detect bonds in symmetrical compounds?
   a. Raman spectroscopy   b. IR spectroscopy
   c. UV-Vis spectroscopy   d. TLC

6. How does UV-Vis spectroscopy work? What is a use of UV-Vis spectroscopy in the forensic setting?

7. What is a use of fluorescence spectroscopy in the forensic setting?

8. How does IR spectroscopy work? Explain what types of molecules are IR active and inactive and why they vary in the amount of radiation they absorb.

9. How does Raman spectroscopy work? Explain what types of molecules are Raman active and inactive and why they vary in the amount of radiation they reflect.

10. Contrast Raman spectroscopy with IR spectroscopy.

Reference

Yan, J., J. Zhu, Q. Tan, L. Zhou, P. Yao, Y. Lu, J. Tan, et al. 2016. Development of a colorimetric and NIR fluorescent dual probe for carbon monoxide. *RSC Advances* 6:65373–65376.

Bibliography

Accardo, G., R. Cioffi, F. Colangelo, R. d'Angelo, L. De Stefano, and F. Paglietti. 2014. Diffuse reflectance infrared Fourier transform spectroscopy for the determination of asbestos species in bulk building materials. *Materials (Basel)* 7:457–470.

Ali, E.M., H.G. Edwards, M.D. Hargreaves, and I.J. Scowen.2008. In-situ detection of drugs-of-abuse on clothing using confocal Raman microscopy. *Anal. Chim. Acta.* 615(1):63–72.

Ali, E.M. and H.G. Edwards.2017. The detection of flunitrazepam in beverages using portable Raman spectroscopy. *Drug Test. Anal.* 9(2):256–259.

Andersson, P.O., C. Lejon, T. Mikaelsson, and L. Landström. 2017. Towards fingermark dating: A Raman spectroscopy proof-of-concept study. *Chemistry Open.* 6(6):706–709.

Bell, S. 2013. *Forensic Chemistry.* 2nd ed. Upper Saddle River, NJ: Pearson Prentice Hall.

Bianchi, F., N. Riboni, V. Trolla, G. Furlan, G. Avantaggiato, G. Iacobellis, and M. Careri. 2016. Differentiation of aged fibers by Raman spectroscopy and multivariate data analysis. *Talanta.* 154:467–473.

Bower, N.W., C.J. Blanchet, and M.S. Epstein. 2016. Nondestructive determination of the age of 20th-century oil-binder ink prints using attenuated total reflection Fourier transform infrared spectroscopy (ATR FT-IR): A case study with postage stamps from the Łódź Ghetto. *Appl. Spectrosc.* 70(1):162–173.

Brittain, H.G.2016. Attenuated total reflection Fourier transform infrared (ATR FT-IR) spectroscopy as a forensic method to determine the composition of inks used to print the United States one-cent blue Benjamin Franklin postage stamps of the 19th century. *Appl. Spectrosc.* 70(1):128–136.

Burnett, A.D., H.G. Edwards, M.D. Hargreaves, T. Munshi, and K. Page. 2011. A forensic case study: The detection of contraband drugs in carrier solutions by Raman spectroscopy. *Drug Test. Anal.* 3(9):539–543.

Causin, V., R. Casamassima, C. Marega, P. Maida, S. Schiavone, A. Marigo, and A. Villari. 2008. The discrimination potential of ultraviolet visible spectrophotometry, thin layer chromatography, and Fourier transform infrared spectroscopy for the forensic analysis of black and blue ballpoint ink. *J. Forensic Sci.* 53(6):1468–1473. doi:10.1111/j.1556-4029.2008.00867.x.

Collins, D. 2006. Experiment 5. In *Investigating Chemistry in the Laboratory*. 1st ed. New York: W.H. Freeman and Co.

Day, J.S., H.G. Edwards, S.A. Dobrowski, and A.M. Voice. 2004. The detection of drugs of abuse in fingerprints using Raman spectroscopy I: Latent fingerprints. *Spectrochim. Acta A Mol. Biomol. Spectrosc.* 60(3):563–568.

Day, J.S., H.G. Edwards, S.A. Dobrowski, and A.M. Voice. 2004. The detection of drugs of abuse in fingerprints using Raman spectroscopy II: Cyanoacrylate-fumed fingerprints. *Spectrochim. Acta A Mol. Biomol. Spectrosc.* 60(8–9):1725–1730.

Del Hoyo-Meléndez, J.M., K. Gondko, A. Mendys, M. Król, A. Klisińska-Kopacz, J. Sobczyk, and A. Jaworucka-Drath. 2016. A multi-technique approach for detecting and evaluating material inconsistencies in historical banknotes. *Forensic Sci. Int.* 266:329–337.

D'Elia, V., G. Montalvo, C.G. Ruiz, V.V. Ermolenkov, Y. Ahmed, and I.K. Lednev. 2018. Ultraviolet resonance Raman spectroscopy for the detection of cocaine in oral *fluid. Spectrochim.* Acta A Mol. Biomol. Spectrosc. 188:338–340.

de Souza Lins Borba, F., R.S. Honorato, and A. de Juan. 2015. Use of Raman spectroscopy and chemometrics to distinguish blue ballpoint pen inks. *Forensic Sci. Int.* 249:73–82.

Dick, R.M. 1970. A comparative analysis of dichroic filter viewing reflected infrared and infrared luminescence applied to ink differentiation problems. *J. Forensic Sci.* 15(3):357–363.

Do, T.T., F. Hadji-Minaglou, S. Antoniotti, and X. Fernandez. 2015. Authenticity of essential oils. Trends Analyt. Chem. 66:146–157. doi:10.1016/j.trac.2014.10.007.

Elkins, K.M. 2011. Rapid presumptive "fingerprinting" of body fluids by ATR FT-IR spectroscopy. *J. Forensic Sci.* 56:1580–1587.

Elkins, K.M. 2013. *Forensic DNA Biology: A Laboratory Manual*. Waltham, MA: Elsevier Academic Press.

Ewing, A.V. and S.G. Kazarian. 2017. Infrared spectroscopy and spectroscopic imaging in forensic science. *Analyst.* 142(2):257–272.

Gál L., M. Oravec, P. Gemeiner, and M. Čeppan. 2015. Principal component analysis for the forensic discrimination of black inkjet inks based on the Vis-NIR fibre optics reflection spectra. *Forensic Sci. Int.* 257:285–292.

Guirguis, A., S. Girotto, B. Berti, and J.L. Stair. 2017. Identification of new psychoactive substances (NPS) using handheld Raman spectroscopy employing both 785 and 1064nm laser sources. *Forensic Sci. Int.* 273:113–123.

HORIBA Jobin Yvon Inc. Raman Data Analysis. www.horiba.com/fileadmin/uploads/Scientific/Documents/Raman/bands.pdf (accessed January 25, 2018).

Joshi, B., K. Verma, and J. Singh. 2013. A comparison of red pigments in different lipsticks using thin layer chromatography (TLC). *J. Anal. Bioanal. Tech.* 4(1). doi:10.4172/2155-9872.1000157.

Kammrath, B.W., A. Koutrakos, J. Castillo, C. Langley, and D. Huck-Jones. 2017. Morphologically-directed Raman spectroscopy for forensic soil analysis. *Forensic Sci. Int.* 285:e25–e33. doi:10.1016/j.forsciint.2017.12.034.

Kenkel, J. 1994. *Analytical Chemistry for Technicians*. 2nd ed. Boca Raton: CRC Press.

King, N.M., K.M. Elkins, and D.J. Nelson. 1999. Reactivity of the invariant cysteine of silver hake parvalbumin (Isoform B) with dithionitrobenzoate (DTNB) and the effect of differing buffer species on reactivity. *Journal of Inorganic Biochemistry* 76: 175–185.

Kitson, R.E. and N.E. Griffith. 1952. Infrared absorption band due to nitrile stretching vibration. *Anal. Chem.* 24(2):334–337.

Kopainsky, B. 1989. Document examination: Applications of image processing systems. *Forensic Sci. Rev.* 1(2):85–101.

Lakowicz, J.R. 1983. *Principles of Fluorescence Spectroscopy*. New York: Plenum Press.

Lakowicz, J.R. 1999. *Principles of Fluorescence Spectroscopy*. 2nd ed. New York: Kluwer Academic/Plenum.

Lambert, D., C. Muehlethaler, P. Esseiva, and G. Massonnet. 2016. Combining spectroscopic data in the forensic analysis of paint: Application of a multiblock technique as chemometric tool. *Forensic Sci. Int.* 263:39–47.

Lanzarotta, A. 2016. Analysis of forensic casework utilizing infrared spectroscopic imaging. *Sensors (Basel).* 16(3):278.

Larkin, P.J. 2011. *IR and Raman Spectroscopy: Principles and Spectral Interpretation*. Waltham: Elsevier.

Lerner, K.L. and B.W. Lerner, ed. 2014. Newton's 1666 spectrum experiment. In *The Gale Encyclopedia of Science*. 5th ed. Farmington Hills, MI: Gale. *Science in Context*, http://link.galegroup.com/apps/doc/CV2210046435/SCIC?u=dc_demo&xid=1f246182. (accessed January 24, 2018).

Lin, I.C.P., J. Hemmings, V. Otieno-Alego, and C. Lennard. 2016. A comparison of conventional microspectrophotometry and hyperspectral imaging for the analysis of blue metallic paint samples. *J. Forensic Identification* 66(5):429–453.

L'vov, B.V. 2005. Fifty years of atomic absorption spectrometry. *J. Analyt. Chem.* 60(4):382–392.

McLauglin, G., K.C. Doty, and I.K. Lednev. 2014. Discrimination of human and animal blood traces via Raman spectrometry. *Forensic Sci. Int.* 238:91–95.

Media AB. 2014. Albert A. Michelson—Biographical. www.nobelprize.org/nobel_prizes/physics/laureates/1907/michelson-bio. html *(accessed January* 24, 2018).

Mitchell, M.B. 1993. Fundamentals and applications of diffuse reflectance infrared Fourier transform (DRIFT) spectroscopy. *Adv. Chem. Ser.* 236:351–375.

Mohamad Asri, M.N., W.N.S. MatDesa, and D. Ismail. 2018. Source determination of red gel pen inks using Raman spectroscopy and attenuated total reflectance Fourier transform infrared spectroscopy combined with Pearson's product moment correlation coefficients and principal component analysis. *J. Forensic Sci.* 63(1):285–291.

Murray, C.A. and S. B. Dierker. 1986. Use of an unintensified charge-coupled device detector for low-light-level Raman spectroscopy. *J. Opt. Soc. Am. A* 3(12):2151–2159.

Nam, Y.S., J.S. Park, N.K. Kim, Y. Lee, and K.B. Lee. 2014. Attenuated total reflectance Fourier transform infrared spectroscopy analysis of red seal inks on questioned document. *J. Forensic Sci.* 59(4):1153–1156.

Nam, Y.S., J.S. Park, Y. Lee, and K.B. Lee. 2014. Application of micro-attenuated total reflectance Fourier transform infrared spectroscopy to ink examination in signatures written with ballpoint pen on questioned documents. *J. Forensic Sci.* 59(3):800–805.

Pemberton, J.E., R.L. Sobocinski, and G.R. Sims. 1990. The effect of charge traps on Raman spectroscopy using a Thomson-CSF charge coupled device detector. *Appl. Spectrosc.* 44(2):328–330.

Pestaner, J.P., F.G. Mullick, and J.A. Centeno. 1996. Characterization of acetaminophen: Molecular microanalysis with Raman microprobe spectroscopy. *J. Forensic Sci.* 41(6):1060–1063.

Quinn, A.A. and K.M. Elkins. 2017. Analysis of ATR FT-IR spectra to differentiate menstrual and venous blood on various substrates. *J. Forensic Sci.* 62: 197–204.

Reedijk, J. and W.L. Groeneveld. 1968. Complexes with ligands containing nitrile groups: Part VII. Metal-ligand vibrations in methyl cyanide solvates. *Rec. Trav. Chim.* 87:552.

Risoluti, R., S. Materazzi, A. Gregori, and L. Ripani. 2016. Early detection of emerging street drugs by near infrared spectroscopy and chemometrics. *Talanta*. 153:407–413.

Saferstein, R. 2015. *Criminalistics: An Introduction to Forensic Science*. 11th ed. Boston: Pearson.

Senior, S., E. Hamed, M. Masoud, and E. Shehata. 2012. Characterization and dating of blue ballpoint pen inks using principal component analysis of UV-Vis absorption spectra, IR spectroscopy, and HPTLC. *J. Forensic Sci.* 57(4):1087–1093.

Socrates, G. 2001. *Infrared and Raman Characteristic Group Frequencies. Tables and Charts*. 3rd ed. West Sussex: John Wiley & Sons.

Sonnex, E., M.J. Almond, J.V. Baum, and J.W. Bond. 2014. Identification of forged Bank of England £20 banknotes using IR spectroscopy. *Spectrochim. Acta A Mol. Biomol. Spectrosc.* 118:1158–1163.

Tinoco, I., K. Sauer, and J.C. Wang. 1995. *Physical Chemistry: Principles and Applications in Biological Sciences*. 3rd ed. Upper Saddle River, NJ: Pearson Prentice Hall.

University of Wisconsin: Department of Chemistry. Simplified IR Correlation Chart. www.chem.wisc.edu/deptfiles/OrgLab/handouts/Simplified%20IR%20Correlation%20Chart.pdf (accessed January 25, 2018).

Wang, J., G. Luo, S. Sun, Z. Wang, and Y. Wang. 2001. Systematic analysis of bulk blue ballpoint pen ink by FTIR spectrometry. *J. Forensic Sci.* 46(5):1093–1097.

Wang, X.F., J. Yu, A.L. Zhang, D.W. Zhou, and M.X. Xie. 2012. Nondestructive identification for red ink entries of seals by Raman and Fourier transform infrared spectrometry. *Spectrochim. Acta A Mol. Biomol. Spectrosc.* 97:986–994.

Was-Gubala, J. and R. Starczak. 2015. Nondestructive identification of dye mixtures in polyester and cotton fibers using Raman spectroscopy and ultraviolet-visible (UV-Vis) microspectrophotometry. *Appl. Spectrosc.* 69(2):296–303.

Williamson, R., A. Raeva, and J.R. Almirall. 2016. Characterization of printing inks using DART-Q-TOF-MS and attenuated total reflectance (ATR) FTIR. *J. Forensic Sci.* 61(3):706–714.

Xu, B., P. Li, F. Ma, X. Wang, B. Matthäus, R. Chen, Q. Yang, et al. 2015. Detection of virgin coconut oil adulteration with animal fats using quantitative cholesterol by GC×GC–TOF/MS analysis. *Food Chem.* 178:128–135. doi:10.1016/j.foodchem.2015.01.035.

Zapata, F., M.A. Fernandez de la Ossa, and C. Garcia-Ruiz. 2015. Emerging spectroscopic techniques for the forensic analysis of body fluids. *Trends Analyt. Chem.* 64:53–63.

CHAPTER 5

Advanced spectroscopy

KEY WORDS: mass spectrometer, electron impact ionization, chemical ionization, molecular ion peak, base peak, fractionation ions, MALDI, DART, mass analyzer, resolution, resolving power, quadrupole, scan mode, single ion monitoring, internal standard, nuclear magnetic resonance spectroscopy, frequency, diamagnetic, internal standard, downfield, upfield, integration, shielded, multiplicity, spin-spin splitting, coupled, coupling constant, chemical shift

LEARNING OBJECTIVES

- To appreciate the historical contributions of scientists that led to the development of and advances in mass spectrometry (MS) and nuclear magnetic resonance (NMR) spectroscopy
- To identify and understand the functions of the components of a mass spectrometer
- To be able to explain electron impact ionization and chemical ionization
- To understand the key features of and how to interpret a mass spectrum
- To understand how an NMR spectrometer works
- To understand how to interpret a proton NMR spectrum
- To understand how to interpret a carbon NMR spectrum
- To be able to explain the use of mass spectra and NMR spectra in chemical analysis

DETECTION OF FENTANYL AND CARFENTANIL POISONING BY DETECTION OF ITS METABOLITES IN BODY FLUIDS

In May 2017, the Anne Arundel (Maryland) County Police reported that three people died from carfentanil overdoses. Nearby Baltimore has been coined the heroin capitol of Maryland. But the heroin, which is 50–100 times more potent than morphine, is being combined with fentanyl, a cheaper but more potent drug that is 50 times as potent as heroin, or with carfentanil, which is 100 times more potent than heroin—making it 10,000 times more potent than morphine. The Maryland Department of Health and Mental Hygiene reported that Maryland deaths due to heroin increased by 70% in the first nine months of 2016 as compared to the same period in 2015. Heroin deaths have increased by 540% since 2010. Fentanyl-related deaths increased from 31 to 738 deaths, or by 2280%, in the same period.

Fentanyl is a Drug Enforcement Agency Controlled Substance Schedule II drug with accepted medical use but high potential for abuse. Fentanyl may be ingested, inhaled, or delivered by intravenous administration. A medically therapeutic dose will deliver analgesic pain relief effects. Toxic doses can lead to decreased respiratory rate, lethargy, coma, and death. Even just touching or inhaling fentanyl can lead to a severe reaction. The estimated lethal dose for fentanyl is 2 mg. Carfentanil is used to tranquilize large zoo animals including elephants. The effects of carfentanil can be felt at only 1 mcg and a lethal dose for a human is only 20 mg according to the Canadian Border Services Agency and its dust in the air can be fatal. The effects of carfentanil on forensic caseloads is just beginning to be felt.

DETECTION OF FENTANYL AND CARFENTANIL POISONING BY DETECTION OF ITS METABOLITES IN BODY FLUIDS (continued)

Most chemical detection and identification methods used in the analysis of forensic casework test directly for the drug, poison, or chemical and not for the metabolites produced as they pass through the body. Metabolites are chemical breakdown products produced when the molecules are acted on by enzymes in the liver and other organs and are detectable in the urine. For example, acetyl fentanyl is metabolized by cytochrome P450 enzymes to acetyl norfentanyl. Forensic detection and identification of chemical compounds is most often performed using gas chromatography coupled to MS and infrared spectroscopy based on comparisons using search algorithms to related compounds in libraries. However, NMR spectroscopy and MS are two chemical instrumentation tools that can be used to elucidate the chemical structures of metabolites and metabolic pathways to understand fentanyl detoxification and excretion and analyze new fentanyl derivatives of forensic interest such as carfentanil, 4-fluoro isobutyryl fentanyl, or U-47700 without relying on a library or when library standards are not yet available.

NMR can be used to detect metabolites in blood without prior separation or pre-purification. Metabolomics studies typically employ 300–900 MHz NMR spectrometers, although stronger magnetic fields will yield a stronger NMR signal intensity. The resulting spectra is a biochemical fingerprint of the sample. Samples (approximately 600 µL) containing metabolites in low concentrations (1–10 µM) can be analyzed by NMR, although as little as 10 µg of analyte in 100 µL has been reported. Single dimensional (1D) spectra can be collected relatively quickly and the presence of metabolites can be detected by overlaying the spectra with library spectra of the metabolite of interest. For example, a ^{1}H NMR spectrum of complex overlapping peaks from a blood or urine sample can be overlaid with spectra of the pure metabolite, if the chemical structure is known and uploaded to the database, for confirmation. Using single and multidimensional methods, NMR can also be used to accurately elucidate the structures of pure, previously uncharacterized metabolites. This is invaluable in further elucidating metabolic pathways and understanding the metabolites produced.

In one method, the Agilent 1200 series quaternary liquid chromatography system interfaced with an API-4000 Q-Trap tandem mass spectrometer was used to identify acetyl norfentanyl and acetyl fentanyl in human urine. Calibration standards and quality control material were prepared from a common aqueous stock solution containing 10 µg/mL of acetyl fentanyl and acetyl norfentanyl. Using liquid chromatography–tandem mass spectrometry (LC–MS/MS), fentanyl could be quantitated at 1 ng/mL. The MS qualifying peaks used for the identification of acetyl fentanyl are 105, 188, and 323 atomic mass units (amu) and the MS qualifying peaks for acetyl norfentanyl identification are 94, 202, and 219 amu.

BIBLIOGRAPHY

Bartolotta, D. Drug 100 times more powerful than heroin responsible for multiple deaths in MD. May 6, 2017 http://baltimore.cbslocal.com/2017/05/06/drug-100-times-more-powerful-than-heroin-responsible-for-multiple-deaths-in-md/ (accessed July 25, 2017).

DEA issues carfentanil warning to police and public. www.dea.gov/divisions/hq/2016/hq092216.shtml (accessed January 27, 2018).

Dias, D.A., O.A.H. Jones, D.J. Beale, B.A. Boughton, D. Benheim, K.A. Kouremenos, J.-L. Wolfender, and D.S. Wishart. 2016. Current and future perspectives on the structural identification of small molecules in biological systems. *Metabolites* 6(4). doi:10.3390/metabo6040046.

Fentanyl. www.dea.gov/druginfo/fentanyl.shtml (accessed January 27, 2018).

New drug is 10,000 times stronger than morphine (and might be resistant to Narcan). www.fox19.com/story/32824614/new-drug-is-10000-times-stronger-than-morphine-and-might-be-resistant-to-narcan (accessed January 27, 2018).

Patton, A.L., K.A. Seely, S. Pulla, N.J. Rusch, C.L. Moran, W.E. Fantegrossi, L.D. Knight, J.M. Marraffa, P.D. Kennedy, L.P. James, G.W. Endres, and J.H. Moran. 2014. Quantitative measurement of acetyl fentanyl and acetyl norfentanyl in human urine by LC-MS/MS. *Anal. Chem.* 86(3):1760–1766.

Advanced spectroscopy tools include NMR spectroscopy and MS. These tools can be used in both screening and identification applications including total structure determination of highly purified small molecules (typically <1500 m/z) such as drugs, explosives, and drug metabolites, but can also be used for macromolecules such as protein toxins. Samples can be purified using liquid chromatography (LC) or gas chromatography (GC) or other chromatographic or electrophoretic techniques. These will be discussed in Chapter 6.

NMR and MS are categorized as Category A confirmatory methods by the Scientific Working Group for the Analysis of Seized Drugs (SWGDRUG). When a Category A method is used, only one other method is required to be used in an analytical scheme. Both NMR and MS have high specificity and high sensitivity. They can be used in qualitative and quantitative analysis and in the analysis of organic and inorganic materials. For *ab initio* chemical structure determinations, NMR and MS results are coupled with melting and boiling point data, Fourier transform-infrared (FT-IR) spectra and ultraviolet-visible spectra, polarimetry data, circular dichroism spectra, as well as any other known chemical or physical properties of the material.

MASS SPECTROMETRY

Mass spectrometers are commonplace and are workhorses in forensic laboratories. MS is a fast and reliable method for molecular identification. The mass spectrometer outputs a mass spectrum of the analyte that allows the forensic scientist to determine the mass-based chemical composition including the total mass, mass fragmentation pathways, and masses of the fragmenting groups. Photos of mass spectrometers coupled to GC and LC instruments are shown in Chapter 6. MS components include a sample input system to introduce the sample into the mass spectrometer, an ionization source, a mass analyzer (mass filter), a detector, vacuum pumps, and a computer for data acquisition and processing. The mass spectrometer is operated under a vacuum to ensure that analytes are vaporized and to avoid collisions between analytes.

Mass spectroscopy is an instrumental tool that has been developed over the past ca. 120 years. In 1898, Cavendish Professor of Physics J.J. Thomson at the University of Cambridge (also known as Cambridge University) experimentally measured the mass-to-charge ratio for electrons and then, in 1913, he was able to separate ^{20}Ne and ^{22}Ne isotopes with different mass-to-charge ratios. In 1919, Thomson's student, Francis Aston, also at Cavendish Laboratory in Cambridge, built the first fully functional mass spectrometer and determined its resolving power to be 130 amu, which he extended to 2000 amu in 1937. He identified at least 212 naturally occurring isotopes using his technique. Aston was awarded the 1922 Nobel Prize in Chemistry for his MS work as shown in Table 5.1. Roland Gohlke and Fred McLafferty coupled the mass spectrometer to a gas chromatograph for use as its detector in the 1950s. The gas chromatography–mass spectrometry (GC-MS) has been, and continues to be, widely used in forensic chemistry. John B. Fenn and Koichi Tanaka were awarded the Nobel Prize in 2002 for their work on MS. Their Nobel contributions are also shown in Table 5.1. Tanaka pioneered matrix-assisted laser desorption/ionization (MALDI) techniques and analysis for the study of biomacromolecules in 1983. Fenn pioneered electrospray ionization (ESI) techniques and analysis for biomacromolecules in 1984. Fourier transform mass spectrometry was inspired by Fourier transform developments for NMR spectroscopy.

After separation using gas or liquid chromatography, capillary electrophoresis, or another purification method, the analyte molecule is introduced to the chamber where it will be ionized in the mass spectrometer. Three commonly used *ionization methods* are electron impact ionization (EI), chemical ionization (CI), and desorption ionization. Other ionization methods include spark ionization, field desorption ionization (FI), photoionization (PI), plasma desorption (PD), thermospray ionization (TS) and MALDI. While EI and CI are widely used in GC-MS, ESI and MALDI are widely used in LC-MS as they can be used for nonvolatile and polar molecules. ESI does not denature proteins and MALDI is nondestructive. A newer innovation is the direct analysis in real-time (DART) instrument coupled with MS.

In EI, the analyte is bombarded by high-energy (70 eV) electrons produced from the ionization of argon. The electrons are contained in a beam by magnets on two sides and a material that repels the ions on the other side. An ionized molecule is one that has lost an electron. It is also referred to as a radical cation or molecular ion. On introduction into

Table 5.1 Nobel Prizes awarded for advances in mass spectrometry

Award recipient	Nobel Prize, year	Noted contribution
Francis W. Aston	Nobel Prize in Chemistry, 1922	For his discovery, by means of his mass spectrograph, of isotopes, in a large number of nonradioactive elements, and for his enunciation of the whole-number rule
John B. Fenn, USA Koichi Tanaka, Japan	Nobel Prize in Chemistry, 2002	For the development of methods for identification and structure analyses of biological macromolecules

Source: www.nobelprize.org/.

the electron beam, the analyte is ionized and fragments into characteristic pieces. The ionized molecule in the mass spectrum is termed the *molecular ion peak*. As the loss of an electron is negligible to the mass, the mass of the analyte is essentially the mass of the molecular ion. Additionally, the mass is the accurate mass of the isotopes of the atoms in the molecule—not a weighted average. The fragmentation at a given voltage is highly reproducible. The molecular ion peak and the molecule's ionized fragments form the mass fingerprint. The *mass spectrum* is reported as relative abundance versus mass divided by charge (m/z). The most abundant peak is termed the *base peak*; it is scaled to 100 relative abundance units. The characteristic peaks from commonly observed ion fragments (Table 5.2) resulting from the fragmented compound can be used to trace back to the structure of the analyte.

The analyte fragmentation pattern is important in determining its identity as multiple compounds may have the same mass such as acetone (2-propanone) and propionaldehyde (58 amu) as shown in Figure 5.1 and 1-propanol and 2-propanol (60 amu) as shown in Figure 5.2. Each set of compounds has the same molecular mass but exhibits different fragmentation patterns in its 70 eV EI ionization mass spectra. The tallest peaks of fragmenting groups in the acetone mass spectrum are 15, 31, and 58 amu, while the tallest peaks in the propionaldehyde spectrum are 28, 29, and 58 amu. The 15, 31, and 58 amu fragments in acetone are attributed to CH_3^+, $C=O^+$, and $HC=O^+$ atom groups respectively, while the 28, 29, and 58 amu fragment peaks in propionaldehyde are attributed to $C=O^+$, $HC=O^+$, and $CH_3\text{-}CH_2\text{-}CO+H^+$ atom groups respectively. The tallest peaks of fragmenting groups in the 1-propanol mass spectrum are 27, 29, and 31 amu, while the tallest peaks in the 2-propanol spectrum are 27, 43, and 45 amu. The 27, 29, and 31 amu fragments in 1-propanol are attributed to CH_3CH^+, COH^+, and CH_2OH^+ atom groups respectively, while the 27, 43, and 45 amu fragment peaks in 2-propanol are attributed to CH_3CH^+, $CH_3CHCH_3^+$, and CH_3CHOH^+ atom groups respectively.

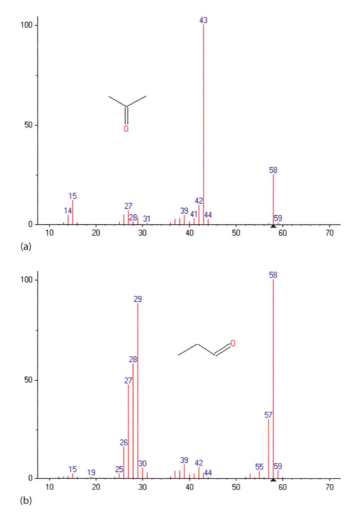

Figure 5.1 Mass spectra of (a) acetone and (b) propionaldehyde.

Table 5.2 Commonly observed fractionation ions in mass spectrometry

Ion	m/z
H_3C^+	15
O^+	16
OH^+	17
H_2O^+	18
F^+	19
$C{\equiv}N^+$	26
$C_2H_3^+$	27
$C{=}O^+$, N_2^+	28
$C_2H_5^+$, $HC{\equiv}O^+$	29
$CH_2NH_2^+$, NO^+	30
$H_2C{=}OH^+$, OCH_3^+	31
CH_3OH^+, O_2	32
SH^+	33
Cl^+	35, 37
HCl^+	36, 39
$C_3H_3^+$	39
$C_3H_5^+$	41
$C_3H_6^+$	42
$C_3H_7^+$, $CH_3C{=}O^+$	43
CO_2^+	44
$CH_3C(H)OH^+$	45
NO_2^+	46
H_2CSH^+, H_3CS+	47
CH_2Cl^+	49, 51
$C_4H_2^+$	50
$C_4H_3^+$	51
$C_4H_4^+$	52
$C_4H_5^+$	53
$C_4H_7^+$	55
$C_4H_8^+$	56
$C_4H_9^+$, $C_2H_5{-}C{=}O^+$	57
$CH_3C(O)CH_3^+$	58
$C_2H_5OCH_2^+$, $(CH_3O{-}C{=}O)^+$, $C_2H_5CHOH^+$, $CH_3O{-}CHCH_3^+$	59
$C_5H_3^+$	63
$C_5H_5^+$	65
$C_5H_7^+$	67
$C_5H_9^+$	69
$C_5H_{10}^+$	70
$C_5H_{11}^+$	71
$C_6H_5^+$	71
$C_6H_9^+$	81
$C_6H_{12}^+$	71
$C_6H_{13}^+$, $C_4H_9CO^+$	85
$C_6H_5.CH^+$	90
$C_7H_{13}^+$	97
$C_6H_5.CH_2O^+$	107

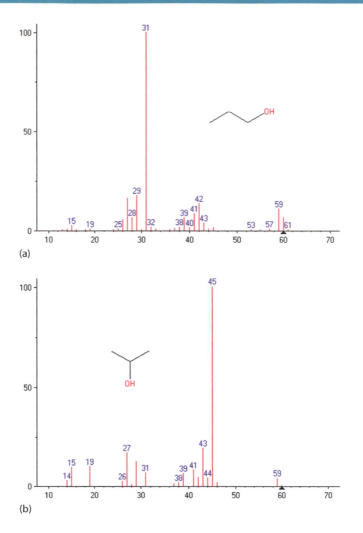

Figure 5.2 Mass spectra of (a) 1-propanol and (b) 2-propanol.

Some drawbacks of EI, a hard ionization method, include over-fragmentation of the molecule due to excessive voltage causing the molecular ion to be absent from the mass spectrum. Depending on the voltage used, the collected spectra may differ from literature spectra. Reducing the voltage can help to identify the molecular ion peak but the mass fingerprint may no longer be directly comparable to literature or library spectra.

An alternative to EI is *chemical ionization*. CI, developed in 1966, is a gentler ionization technique, or soft ionization method, that results in less fragmentation and intact molecular ions. Two modes are available: negative chemical ionization (NCI) and positive chemical ionization (PCI). In CI, a reagent gas (e.g., methane, isobutene, or ammonia) is introduced into the source for ionization (Figure 5.3). The ionized gas collides with the analyte molecules generating a [M+H]+ or [M+H]- ion (Figure 5.4). As most compounds will form positive ions, PCI is most commonly used. These ions are relatively stable. NCI is used for molecules that form stable negative ions such as analytes containing acidic groups or halogens (e.g., pesticides and herbicides). As EI and CI are different ionization methods, each will result in different fragmentation patterns, or mass fingerprints, for the analyte. Appropriate search libraries have to be used for analytes evaluated with each method.

Another type of ionization is *desorption/ionization*. Several different types are available including laser desorption/ionization and MALDI. Laser desorption/ionization employs a pulsed laser (e.g., IR laser: CO_2 laser; UV laser: Nd:YAG [yttrium aluminum garnet]) that is focused on the solid material. The ionization wavelength can be selected. A common desorption/ionization technique for use with LC instruments is MALDI. Franz Hillenkamp and Michael Karas pioneered the MALDI technique in 1985 when they ionized the amino acid alanine in a tryptophan matrix

PCI (simplified): $CH_4 + e^- \rightarrow CH_4^+ + 2e^-$

$CH_4 + CH_4^+ \rightarrow CH_3 + CH_5^+$

NCI: $CH_4 + e^- \rightarrow CH_4^-$

Figure 5.3 Chemical ionization of methane gas.

PCI: $CH_5^+ + M \rightarrow [M+H]^+ + CH_4$

NCI: $CH_4^- + M \rightarrow [M+H]^- + CH_4$

Figure 5.4 Molecular ion formation using ionized methane gas.

using a pulsed 266 nm laser. The tryptophan matrix absorbs some of the laser energy in the ultraviolet wavelength range. Koichi Tanaka of Shimadzu Corp. in Japan developed the "ultra fine metal plus liquid matrix method" that used a 337 nm nitrogen laser for the ionization of molecules in a glycerol matrix containing 30 nm cobalt particles. Tanaka ionized biomacromolecules such as the 34,472 Da protein carboxypeptidase-A. The analyte was co-crystallized with a matrix including 2,5-dihydroxybenzoic acid (DHB) and α-cyano-4-hydroxycinnamic acid (CHCA). The matrix disperses the large amounts of energy produced by the laser and minimizes the fragmentation of the molecules. MALDI can be used in the direct ionization of solids and is an excellent tool for analyzing biomacromolecules (e.g., proteins, nucleic acids, carbohydrates, and other biopolymers) with molecular weights over 10 kDa. The solid samples are placed on a support where they are bombarded with ions or photons. MALDI produces less multiply-charged ions than ESI. The matrix must be stable under vacuum and not react chemically but absorb strongly at a wavelength where the analyte absorbs weakly. For example, urea, alcohols, and carboxylic acids absorb strongly in the infrared region while salicylates (e.g., 5-chlorosalicylic acid) and 3-hydroxypicolinic acid absorb strongly in the UV region. A drawback is that there is significant signal suppression from low-mass ions (<500–700 Da) that interfere when organic matrices are used; this can be reduced using porous silicon chip matrices.

DART is being adopted by forensic laboratories for drug analysis at a rapid pace as it allows for direct ionization of analyte samples with minimal pre-treatment and with broad applicability and high accuracy. Advantages to DART include that it does not require elevated temperatures, laser irradiation, or a high-velocity gas stream. In DART, a gas is flowed through a chamber where the sample is ionized using an electrical discharge and the neutral molecules are allowed to flow through. The neutral molecules include excited state sample molecules. Molecular cations and anions are produced. DART has a constant resolving power of ~6000 full width at half maximum (FWHM).

After they have been ionized, analytes and their fragments are separated using a *mass filter*, or *mass analyzer*, according to their mass-to-charge (m/z) ratio in a vacuum chamber. The ions are transported to the mass filter through the use of ion-focusing electrodes, or focusing lenses. Focusing lenses use voltage power to attract or repel the ions and guide them to the mass filter. In scanning mass analyzers, such as the magnetic sector mass analyzer, only ions of an m/z of interest are allowed to pass through the analyzer at a given time. In simultaneous transmission mass analyzers, all of the ions are transmitted at the same time. Mass filters include the *time-of-flight (TOF), ion trap, Orbitrap, ion cyclotron resonance, and dispersive magnetic* mass analyzers. Tandem mass spectrometers have two or more mass analyzers in sequence. Common mass filters include ion traps and the quadrupole in GC-MS systems and the TOF mass filter in LC-MS systems.

Quadrupole mass analyzers are the most commonly used mass analyzers. They employ two pairs of cylindrical rods aligned in parallel. The opposite pairs of rods are connected to the opposite ends of a direct current (DC) source. The ions are introduced into the tunnel between the four rods and oscillate (corkscrew) through the quadrupole to the detector. A DC field is applied to two rods and a radio frequency (RF) to the other two. For a given DC-to-RF ratio at a fixed frequency, only the ions of a given m/z value will pass through the quadrupole electric field to the detector. Ions with different m/z values will be lost when they collide with the rods or travel outside the quadrupole. The analytes must be ionized and in the gas phase. Although it has a smaller m/z range and lower resolution than magnetic sector mass analyzers, the quadrupole mass analyzer is faster.

TOF mass analyzers yield high-resolution MS by employing a drift tube. Pulses of ions are accelerated into the evacuated drift tube. The velocity of the ions depends on their m/z (or mass only if they have the same charge). The lighter ions pass through the drift tube faster than the heavier ions, allowing the ions to be separated by their drift velocities (*v*).

Quadrupole and ion trap mass filters have constant resolution. *Resolution* is the difference in m/z values that can be separated from one another. In constant resolution, ions that differ by 1 m/z unit can be as easily separated at 100 and 101 amu as at 1000 and 1001 amu. High-resolution instruments do not have constant resolution but instead have constant resolving power. *Resolving power* (R) is defined as the ability to separate between closely related masses (M) (Equation 5.1).

$$R = M_n / (M_n - M_m)$$ (5.1)

Low-resolution instruments (e.g., quadrupole and ion trap) have a mass resolution of 1000–2000 Da while the TOF instruments have a mass resolution of 2500–40,000. The mass ranges of the quadrupole and ion trap mass analyzers are 50–6000 and 50–4000 Da, respectively. TOF instruments have a mass range of 20–500,000 Da and have constant resolving power; the resolution is highest at low m/z values. High-resolution Orbitrap MS instruments have a mass resolution in excess of 100,000 and a mass range of 40–4000 Da. TOF systems have mass errors of less than five parts per million (ppm) while Orbitrap, ion trap, and ion cyclotron resonance have mass errors of less than 2 ppm. The TOF mass analyzers are the fastest while the quadrupole and ion trap are medium speed and the Orbitrap is among the slowest. All except for the TOF mass analyzers can be used in MS/MS instruments.

Magnetic sector mass analyzers accelerate the ions in a field (sector) at a defined voltage (*V*). The energy of each ion is equal to the charge times the accelerating voltage, or *zV*, and is independent of mass. The sector can have any apex angle but 60° and 90° are common. Modern instruments often combine electric sector and magnetic sector for a double-focusing mass spectrometer. The electric sector acts as an energy filter. Ions with the right m/z reach the detector while others are lost when they collide with the sides of the instrument. The working m/z range is 1–1400 amu for single-focusing instruments and 5,000–10,000 amu for double-focusing instruments. Ions with small masses travel at a higher velocity than those with large masses.

Mass spectrometer detectors include array detectors used in TOF mass spectrometer instruments and electron multiplier (EM) detectors used mainly in GC-MS instruments. The EM detector is the most common detector. The continuous-dynode EM detector detects the charges that it receives and multiples the signal proportional to the quantity of analyte in the sample. The electrons hit the resistant conductive surface and are amplified in the detector funnel. It is highly sensitive and has a quick response time. The EM detectors have a limited lifetime that depends on instrument usage (e.g., number of molecules analyzed and the number of ions detected). Array detectors used in TOF instruments employ a focal plane camera that consists of an array of 31 Faraday Cups that serve to capture the ions and store the charge. The shape of the metal or plastic cup is designed to retain the electrons and reduce loss. Although it has a long response time, this detector type is the least expensive for detecting ions and up to 15 m/z values can be measured simultaneously. The array detector exhibits improved precision over the single-channel detector.

The mass spectrometer can be operated in two modes: *scan mode* (SCAN) and *single ion monitoring* (SIM) mode. In scan mode, the mass filter is set to allow a range of masses to pass through to the detector. The samples in Figures 5.1 and 5.2 were run in SCAN mode. SCAN mode is run to analyze new analytes and for interpretation using a library search. Isotope peak tree patterns will be observed. For example, if an analyte contains chlorine or bromine, multiple isotope patterns will be observed (see Table 5.2). To illustrate this, take the example of a molecule that contains four atoms of chlorine. Chlorine has two isotopes: 35-Cl and 37-Cl with relative abundances of 76% and 24% respectively. There will be five possible combinations of chlorine isotopes in the molecule. As three out of every four chloride atoms will be the 35-Cl isotope, the most intense peak in the mass spectrum will have an m/z corresponding to the fragment mass with three 35-Cl atoms and one 37-Cl atom. However, the m/z values will increase from four 35-Cl atoms, to three 35-Cl and one 37-Cl atom, two-35-Cl and two 37-Cl atoms, one 35-Cl and three 37-Cl atoms, and four 37-Cl atoms (most rare). SIM mode can be employed for monitoring specific ions, typically the two most intense fragment peaks, if they are known. In SIM mode, the mass filter passes only selected m/z ratio ions to the detector. Mass spectra in SIM mode will be much simpler than those run in SCAN mode. For the analytes shown in Figures 5.1 and 5.2, the tallest peaks could be selected for SIM mode and these would be the only ones detected and shown in the spectra. SCAN mode is less sensitive than SIM mode as most ions strike the quadrupole rods during the scan and never reach the detector. SIM mode increases sensitivity for detecting the desired ions and is used in quantitative analysis.

Perfluorotributylamine (PFTBA) is used as a calibration chemical to tune the MS instrument by detecting qualifying ions of masses (m/z) 69, 131, 219, and 502 amu at the appropriate relative peak heights. It is used to check the function of the detector over usage time.

GC-MS has several advantages over other separation and spectroscopic analysis methods. It is relatively inexpensive in terms of per sample cost as compared to LC-MS options. Separation is conducted using the GC prior to introducing the unknown to the MS. Separations are fast—run times may be as little as a few minutes. Electron ionization is standardized and reproducible.

Figures 5.5 and 5.6 show EI mass spectra of the positional isomers of stimulant drugs phentermine and methamphetamine respectively. Both have a mass of 149 amu. The tallest peaks in the mass spectrum for methamphetamine and phentermine are 58, 91, and 134 amu, although low relative intensity fragments at 67 amu are unique to the methamphetamine spectrum and a 56, 70, and 84 amu fragment is unique to the phentermine spectrum. An unrelated compound, salicylic acid, a starting material for and metabolite of aspirin, has a very different EI mass spectrum as shown in Figure 5.7, with its tallest or qualifier peaks at 92, 120, and 138 amu. There are extensive libraries of MS data including the National Institute of Standards and Technology (NIST) Mass Spectral Database, Wiley Spectra Laboratory, Golm Metabolome Database, and Agilent GC/MS Drug and Metabolomics libraries, as well as many other manufacturer and forensic-specific libraries that can be used to identify an unknown with its MS "fingerprint"

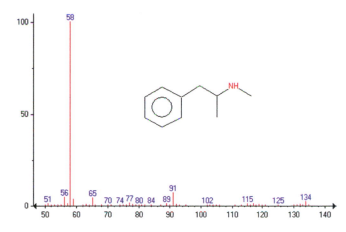

Figure 5.5 Mass spectrum of phentermine.

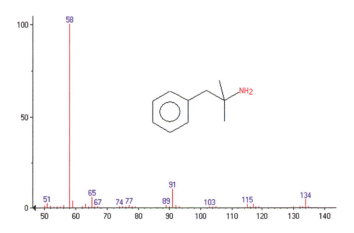

Figure 5.6 Mass spectrum of methamphetamine.

Figure 5.7 Mass spectrum of salicylic acid.

by matching it to the library using an algorithm. Figure 5.8 shows the collected MS spectrum of a caffeine standard as compared to the mass spectrum in the library. Caffeine is a legal stimulant found in coffee, tea, and energy drinks. GC can be used to separate the salicylic acid and caffeine as shown in Chapter 6.

Drawbacks of GC-MS systems include that the molecules must be volatile and the instrumental analysis is typically limited to small molecules of <500 Da. The molecular ion peak may be absent when using electron ionization if the voltage is too high. Higher boiling point compounds that are generally polar must be chemically derivatized first, prior to loading on GC-MS. Advantages of LC-MS systems include the separation and identification of nonvolatile and polar compounds without denaturation. These instruments are widely used in toxicology labs. The cost of LC-MS instruments have fallen to a price point that has made them accessible to more forensic labs. Databases and libraries for these systems include the Human Metabolome Database, METLIN Metabolomics database, and MassBank, among others.

By employing GC-MS-MS, scientists can record a mass spectrum of a mass spectrum. In 2D tandem MS-MS experiments, the first mass spectrometer is used to select the desired ion (specific m/z) for entry into the collision cell. The ion is energized by collisions in the collision cell. Some of the ions will fragment and are monitored by the second mass spectrometer. Typically, two different types of mass analyzers are used; for example, a quadrupole may be coupled with a TOF. By using different fragmentation approaches, the molecular connectivity and molecular formula can be determined.

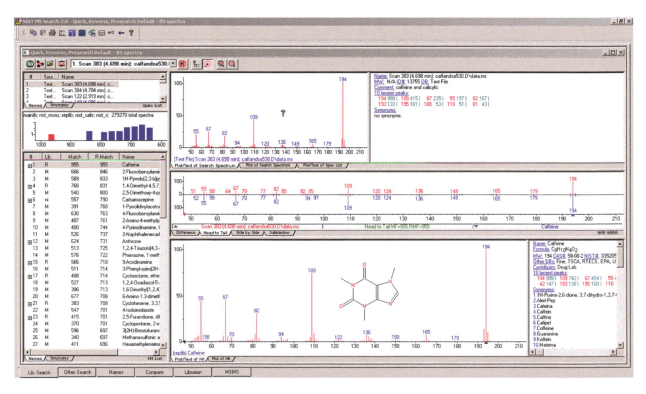

Figure 5.8 Collected mass spectrum of caffeine and library search result overlay (center).

NUCLEAR MAGNETIC RESONANCE SPECTROSCOPY

NMR spectroscopy is a nondestructive tool that can be used to determine the chemical structure of drugs, metabolites, small molecule and protein toxins, explosives, and other molecules *ab initio*. This is especially useful if a lab encounters a previously unseen compound that is not in its GC-MS or FT-IR chemical identification libraries. Owing to the cost of high field instruments, most crime laboratories do not have NMR spectrometers. Laboratories can outsource testing to private NMR testing labs or collaborate with academic labs with this instrument to determine the structures of their unknown molecules.

NMR spectroscopy is a tool that has been developed for over 80 years and several notable scientists have been awarded Nobel Prizes for their work, as shown in Table 5.3.

NMR spectroscopy uses radio waves to create a magnetic field; as shown in Chapter 4, radio waves have long wavelengths and low energy. Molecules can absorb the electromagnetic radiation, leading to the nuclear spins in the atoms of the molecule to change in the presence of the radio frequency waves. Recall that in quantum mechanics, spin is quantized as either $+1/2$ or $-1/2$. When charged particles such as protons in elements are exposed to a magnetic field, they spin on their axes and create a magnetic field. Normally, the protons in nuclei are randomly oriented but in the presence of the magnetic field, they will orient in one of two energy states: with or opposite to the magnetic field. Most nuclei orient with the applied magnetic field as it is a lower-energy state. When a magnetic pulse is applied that changes the magnetic field, nuclei that match the pulse energy with the difference between their two energy states will absorb the energy. The absorbed energy will cause the nucleus to change orientation or "spin flip" from a high to low or low to high energy state. The electrons circulating the protons in the nucleus will "shield" the protons from feeling the external magnetic field. The electrons absorb some of the electromagnetic radiation so a nucleus with more electrons will be more shielded than one with no electrons or fewer electrons. The molecules can reemit electromagnetic radiation when the nuclei relax to their lower-energy or original states; the energy released is detected by a detector and reported to a computer that records the spectrum.

The applied magnetic field is measured in tesla (T) and the *frequency* of the radiation is measured in megahertz ($1\,MHz = 10^6\,Hz$). NMR spectrometers are referred to by the frequency of the radiation they produce; low-power NMR spectrometers start at 60 MHz and high-powered NMR instruments exceeding 900 MHz are common today with very powerful magnets. A 400-MHz NMR spectrophotometer is shown in Figure 5.9. The magnet is cooled in liquid helium that allows it to conduct electricity with no resistance and reduces the thermal and electric noise observed in the spectra. In NMR spectrometers, the sample in a thin glass NMR sample tube (Figure 5.10) is inserted in a holder that is drawn into the center of the magnet and rotated in the magnetic field.

Diamagnetic nuclei including ^1H, ^{13}C, ^{15}N, ^{19}F, and ^{31}P can be probed using NMR. These nuclei have a $+1/2$ spin. *Diamagnetic* molecules have no unpaired electrons and are not attracted to magnetic fields. So, in an applied magnetic field, the protons in diamagnetic molecules are magnetized 180° to the applied magnetic field.

Table 5.3 Nobel Laureates in the field of NMR

Award recipient	Nobel prize, year	Noted contribution
Otto Stern, USA	Nobel Prize in Physics, 1943	For his contribution to the development of molecular ray method and his discovery of the magnetic moment of the proton
Isidor I. Rabi, USA	Nobel Prize in Physics, 1944	For his resonance method for recording the magnetic properties of atomic nuclei
Felix Bloch, USA, and Edward M. Purcell, USA	Nobel Prize in Physics, 1952	For their discovery of new methods for nuclear magnetic precision measurements and discoveries in connection therewith
Richard R. Ernst, Switzerland	Nobel Prize in Chemistry, 1991	For his contributions to the development of the methodology of high resolution nuclear magnetic resonance (NMR) spectroscopy
Kurt Wüthrich, Switzerland	Nobel Prize in Chemistry, 2002	For his development of nuclear magnetic resonance spectroscopy for determining the three-dimensional structure of biological macromolecules in solution
Paul C. Lauterbur, USA, and Peter Mansfield, UK	Nobel Prize in Physiology or Medicine, 2003	For their discoveries concerning magnetic resonance imaging

Source: www.nobelprize.org/.

Figure 5.9   400 MHz JEOL NMR spectrophotometer.

The most commonly probed nuclei are ^{13}C and ^1H due to their abundance in biological and organic molecules. ^1H NMR experiments are used to determine the number of hydrogens in a molecule and ^{13}C NMR experiments are used to determine the number of carbons in a molecule. ^1H NMR experiments are also used to determine how many, if any, hydrogen atoms are attached to adjacent atoms and what types of hydrogens those are. ^{13}C NMR experiments are used to determine the type and connectivities of the carbons in a molecule. Protons in different chemical and electronic environments will absorb electromagnetic radiation at slightly, but measurably different, frequencies.

NMR spectra are reported as plots of intensity versus chemical shift in ppm. Typically, absorption peaks are sharp in NMR spectra. An *internal standard*, such as tetramethylsilane (TMS), an inert compound, can be added to use its peak

Figure 5.10   An NMR sample tube.

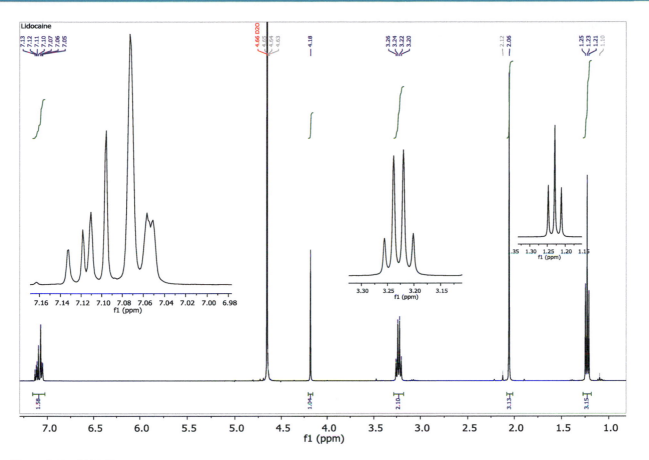

Figure 5.11 ¹H NMR spectrum of lidocaine in D_2O.

as a convenient reference. TMS is volatile and can easily be separated from samples afterward. TMS gives a single ¹H NMR peak. The chemical-shift scale is dimensionless but is expressed in ppm. The hydrogen protons in TMS absorb at a lower frequency than other compounds because the hydrogen atoms are bound to silicon which is less electronegative than carbon. The TMS peak is set as 0 ppm to calibrate the scale. Peaks at higher chemical shifts are said to be *downfield* of those with lower chemical shifts. ¹H NMR peaks range between 0 and 12 ppm (Table 5.4). The chemical shift in ppm can be computed by dividing the observed chemical shift (in Hz) downfield from TMS by the frequency of the NMR spectrometer in MHz. This allows data to be directly compared even if it is recorded on a lower or higher frequency power instrument. An alternative method to using an internal standard is to use an impurity in the solvent as a reference. For example, deuterated chloroform ($CDCl_3$) has a $CHCl_3$ peak at 7.24 ppm and dimethyl sulfoxide (DMSO) has a solvent peak at 2.50 ppm.

¹H spectra are the most quickly recorded NMR spectra. These spectra can reveal information about a compound's chemical structure. The recorded data can indicate the: (1) number of hydrogen atoms; (2) their environment(s); (3) how many are in each environment; and (4) how many other hydrogens (three bonds away) are bonded to an adjacent heteroatom.

Hydrogen atoms in the same environment give the same NMR signals stack to yield a higher-intensity peak while hydrogen atoms in different environments give rise to different NMR signals. For example, hydrogen atoms bonded to a carbon atom that is bound to another atom (or two atoms) via a single bond will have the same environment due to the free rotation around the single bond. So three methyl group substituents bonded to a carbon at one end of a molecule will, in sum, have nine hydrogens in the same environment. However, hydrogen atoms bonded to a carbon atom that is bound to the next atom with different substituent groups via a double bond will have a different environment because the double bond does not permit free rotation. In the case of a methyl group isolated by a carbonyl, the three hydrogens contribute to one peak that is three times the height the peak would be from a single-bound hydrogen, for example, from an alkyne. In ¹H NMR spectroscopy, the area under the peak can be used to

Table 5.4   Chemical shift for several proton types

Type of hydrogen proton	Chemical shift (ppm)
TMS	0[a]
Hydrogen bound to sp^3 hybridized carbon	0.9–0.2
Hydrogen bound to sp^3 hybridized carbon bound to one R group	~0.9
Hydrogen bound to sp^3 hybridized carbon bound to two R groups	~1.3
Hydrogen bound to sp^3 hybridized carbon bound to three R groups	~1.7
Hydrogen bound to sp^2 hybridized alkene carbon bound to a carbon bound to a C, N, or O atom	~2.5–1.5
Hydrogen bound to sp hybridized alkyne carbon	~2.5
Hydrogen bound to sp^3 hybridized carbon bound to N, O, or X	4–2.5
Hydrogen bound to oxygen in an alcohol bound to an R group	5–1
Hydrogen bound to nitrogen in an amine bound to an R group	5–1
Hydrogen bound to sp^2 hybridized alkene carbon	6–4.5
Hydrogen bound to a benzene ring	8–6.5
Hydrogen bound to carbon of aldehyde	10–9
Hydrogen bound to oxygen of carboxylic acid group	12–10

[a] Defined.

calculate the relative number of identical hydrogen atoms in a molecule contributing to a peak or group of peaks caused by spin-spin splitting using Beer's Law. The software can *integrate* the area under the peaks. The integral is plotted as a stepped curve with arbitrary units; the height of each step is proportional to the number of hydrogen protons. The integral steps are a ratio of the number of equivalent protons and may not be the absolute number. So, an integration with steps of 20:60 is equal to a ratio of 1:3 or 3:9 hydrogens in the two different environments.

In cases where the electrons surrounding the nucleus have *shielded* the protons by absorbing some of the electro-magnetic radiation, the resulting chemical shift will be *upfield*, or less ppm. Thus, protons in electron-rich environments require a lower frequency to come into resonance. Conversely, deshielded protons, including those near electronegative atoms that pull electrons away, require a higher frequency to come into resonance and experience a shift in the absorption downfield. For example, hydrogen atoms in chloroform (CH_3Cl) will absorb downfield of hydrogens in methane (CH_4). These and other trends are shown in the chemical-shift correlation chart in Table 5.4. The chemical shift of hydrogen atom protons increases with increasing R-group substituents. The protons in a benzene ring are deshielded and absorb downfield because the circulating pi electrons create a ring current, which induces a magnetic field near the protons. As compared to a double bond, electrons in a triple bond circulate, but the induced magnetic field opposes the applied magnetic field leading the proton to feel a weaker magnetic field and absorb less in the carbon–carbon triple bond than carbon–carbon double bond. Hydrogen protons bound to the carbon in an aldehyde or carboxylic acid functional group are the most deshielded and are shifted the most downfield.

Multiplicity is defined as the number of peaks in a signal from spin-spin splitting. *Spin-spin splitting* can yield doublet (2), triplet (3), quartet (4), quintet (5), sextet (6), septet (7), and other multiplet (>7) peak formations with equal spacing. The $n+1$ *rule* can be used to determine multiplicity. N is the number of equivalent protons bound to an adjacent carbon. The multiplicity can be used to determine how many other hydrogens within three bonds are bonded to adjacent heteroatoms. So, the observed splitting for methyl group hydrogens by adjacent ethyl group hydrogens is a triplet. Some observations of spin-spin splitting are as follows: (1) spin-spin splitting is not observed by equivalent protons; (2) spin-spin splitting only occurs between hydrogen protons on the same carbon or two adjacent carbons in different environments; and (3) a set of n nonequivalent protons bound to an atom will split the signal of a proton on an adjacent atom into $n+1$ peaks. Spin-spin splitting is not usually observed when protons are separated by more than three sigma (single) bonds. As the protons in different environments feel different magnetic fields, they absorb at different frequencies and can split a single-absorption peak based on how many fields are felt. Protons that split each other's signal are said to be *coupled*. Spin-spin coupling describes protons that are close enough to influence the spin of another nucleus. The difference in the split frequencies (peaks) is called the *coupling constant* (J). Spin-spin splitting can be used to determine how many other hydrogens (three bonds away) are bonded to an adjacent heteroatom. *Long-range*

Table 5.5 Chemical shift by carbon type

Type of carbon atom	Chemical shift (ppm)
sp³ hybridized carbon bound to hydrogen(s) and/or R groups	5–45
sp³ hybridized carbon bound to hydrogen(s) and/or R groups, one of which is a N, O, or X atom	30–80
Alkyne carbons (sp)	65–100
Alkene carbons (sp²)	100–140
Benzene or phenyl ring carbons	120–150
C=O (e.g., ketones, aldehydes, esters, carboxylic acids, anhydrides)	160–210

coupling is a small splitting that occurs when protons are separated by more than three bonds where one of the bonds is a double or triple bond.

A hydrogen (e.g., H_x) on a carbon adjacent to a carbon with two hydrogen atoms placed in a magnetic field can feel the magnetic field and when the adjacent protons, H_a and H_b, are aligned with or opposed to the applied magnetic field. Thus, H_x can feel three slightly different magnetic field strengths and absorbs at three different frequencies, which leads to a single absorption to be split into a triplet. Because the spin of H_a and H_b can be aligned with the field, or H_a can be aligned with the field and H_b can oppose the field, or H_a can oppose the field and H_b can align with the field, the ratio of the areas under the split peak in the triplet will be 1:2:1 or one opposed: two aligned: one opposed. The relative heights of the peaks in this triplet, and also other multiplets, follow Pascal's Triangle. (A doublet is 1:1, quartet is 1:3:3:1, and so on.) Hydrogen protons can also be split by hydrogen protons three bonds away on either side of themselves. This can lead to $n+1$ splitting on one side (to yield, for example, four peaks for a hydrogen proton adjacent to a methyl group) and $m+1$ splitting on the other side (to yield three peaks from a hydrogen proton adjacent to an ethyl group). To compute the total number of peaks expected, calculate $(n+1)$ times $(m+1)$ or $4 \times 3 = 12$ peaks, a quartet of triplets. Spin-spin splitting couplings are extremely useful in determining the connectivity of atoms used

ChemNMR ¹³C Estimation

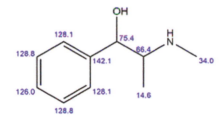

Estimation quality is indicated by color: good, medium, rough

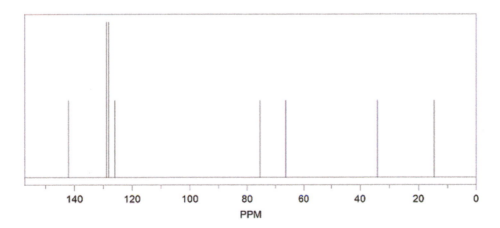

Figure 5.12 ¹³C NMR ephedrine in D_2O.

to solve a molecular structure. Hydrogens bonded *cis* to one another around a carbon–carbon double bond will have a smaller coupling constant (5–10 Hz) than *trans* hydrogens (11–18 Hz). Axial and equatorial hydrogens bound to cyclohexane will exhibit one peak for the average environment they feel as the cyclohexane ring interconverts very rapidly at room temperature. The six hydrogens bound to carbons in a benzene ring will be equivalent and exhibit a single ^{1}H NMR peak at 7.27 ppm. Due to magnetic anisotropy, the magnetic moment of the pi electrons or rings will align in the magnetic field so the protons will feel a larger effective magnetic field and will resonate at a higher frequency, thus leading to the downfield chemical shift.

The addition of a substituent group to the benzene ring will make the remaining five hydrogen atoms absorb differently and can create multiple irregularly split peaks at various chemical shifts. Under usual conditions, a hydrogen proton bound to oxygen in an alcohol (–OH) functional group will not split the NMR signal of adjacent protons, so it will appear as a singlet. The –OH group hydrogen is exchangeable with the solvent due to the weak acid properties of the group. For example, the ^{1}H NMR spectra of ethanol (CH_3CH_2OH) will consist of a triplet at ~1.2 ppm attributable to the –CH_3 hydrogen protons, a singlet between 1 and 5 ppm attributable to the –OH hydrogen proton, and a quartet at ~3.65 ppm attributable to the –CH_2 hydrogen protons.

^{13}C NMR spectroscopy is focused on determining the number of carbons based on the number of peaks and carbon types in a molecule based on their chemical shifts. The number of peaks in a ^{13}C NMR spectrum equals the number of unique carbon environments in the molecule. For example, $(CH_3)_3CCl$ will have two carbon environments and will present two ^{13}C NMR peaks. ^{13}C frequencies are not split due to the low natural abundance of ^{13}C. Peak splitting is only observed when two diatomic NMR-active nuclei are close in space. But because the natural abundance of ^{13}C atoms is only 1.1%, the chance that two ^{13}C atoms will be bonded to each other is very small (0.01%). The peak intensity is not proportional to the number of carbon atoms so ^{13}C NMR peak signals are not integrated. ^{13}C atoms can be split by neighboring ^{1}H atoms but this is usually eliminated using a radio-frequency pulse sequence that decouples

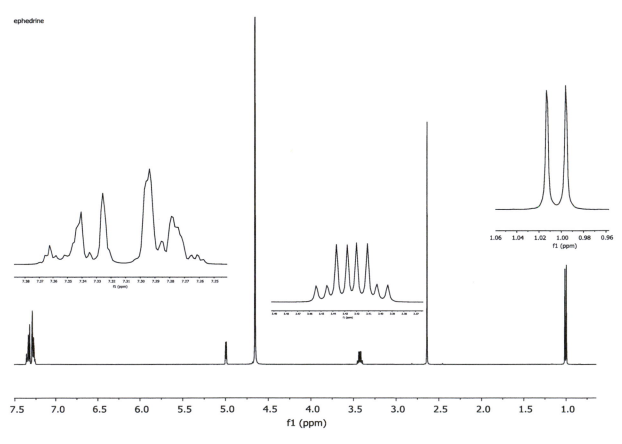

Figure 5.13 ^{1}H NMR ephedrine in D_2O.

the ^{13}C-^{1}H interactions so that the ^{13}C peaks are all singlets. Thus, ^{13}C NMR spectra are easier to interpret than ^{1}H NMR spectra.

Each carbon in a different environment will absorb at a different frequency. Unlike the ^{1}H NMR chemical shifts that are typically 1–12 ppm (and up to ~15 ppm), the ^{13}C chemical shifts cover a much broader range (0–250 ppm). This is a clear advantage of ^{13}C NMR over ^{1}H NMR as the potential for peak overlap is significantly reduced. Carbons with electron withdrawing groups (e.g., O, Cl, Br, F) attached will be highly deshielded and absorb downfield, whereas carbons attached to hydrogens only will be most shielded and absorb the farthest upfield. Common ^{13}C NMR chemical shifts are shown in Table 5.5.

NMR spectroscopy can be performed for samples in the solid or liquid phase. Solid-phase NMR requires a specialized NMR spectrometer. Most commonly, NMR samples are prepared in the liquid phase. Deuterated solvents that will not absorb the electromagnetic radiation are preferred. These include deuterated chloroform ($CDCl_3$) and deuterated water (D_2O). Samples of 1–10 mg/mL concentration are typical, although the high nanogram to milligram range is used. Metabolites of 1–10 µM in a total volume of only 100–600 µL can be detected with higher magnetic fields as signal intensity is proportional to the magnetic field intensity. As discussed previously, an internal standard, such as TMS, that does not absorb at the same frequencies as the analyte is added to the NMR tube with the sample. Modern NMR spectrometers are easy to use and have autoshim tools and preloaded pulse sequences that make running experiments easy. Control samples are run before running unknowns to determine that the NMR instrument is working properly. In addition to chemical structure determination, NMR can be applied to produce a "fingerprint" that can be used to screen samples quickly. ^{1}H NMR spectra for screening purposes can be collected in a few minutes. Figure 5.11 shows an 8-scan 1H NMR spectrum of the analgesic lidocaine in D_2O that took only a couple minutes to collect. The integration ratios and scale in ppm are shown and the insets enlarge the multiplet regions. The assignments (in ppm) for the lidocaine protons are 1.23(triplet, CH_3), 2.06(singlet, Aromatic-CH_3), 3.23(quartet, CH_2),

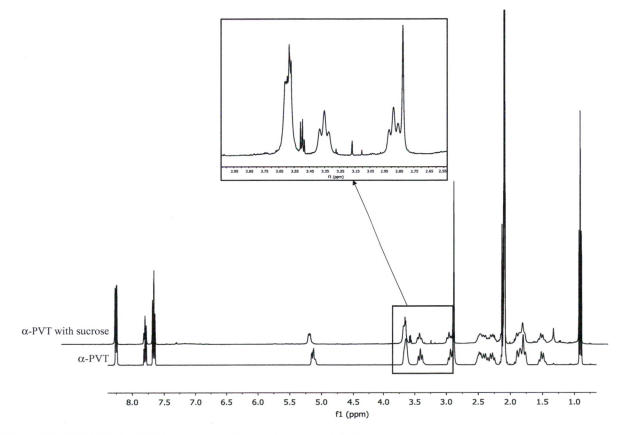

Figure 5.14 ^{1}H NMR of α-PVT alone and cut with powdered sugar in D_2O.

4.18(singlet, CH_2 closer to Aromatic ring), 4.65(singlet, NH_2), and 7.08(multiplet, Aromatic protons). Alternatively, peak-pattern matching can be used to identify molecules in libraries. A multifunctional software tool for analyzing NMR data is MNova™ by Mestrelab Research.

NMR can also be used to evaluate the properties of the bulk of the sample, much like infrared spectroscopy. However, cutting agents added to drugs and the drug itself will both absorb the electromagnetic radiation and give rise to peaks at the appropriate frequencies based on their chemical structure. When the components of the mixture are similar in proportion, the peaks are equally observed. The peaks can be used to determine that the sample is a mixture and can be deconvoluted to determine the components of the mixture using a chemical library. Software is also available to predict NMR spectra; for example, ChemDraw has a spectral prediction function and was used to predict the ^{13}C spectrum of ephedrine (Figure 5.12). The ^1H spectrum of ephedrine is shown in Figure 5.13. ^1H spectra of the cathinone α-PVT alone and cut with powdered sugar (sucrose) are shown in Figure 5.14. The cathinone and sugar peaks are all present and distinguishable; the characteristic sugar peaks are at ~3.5 ppm.

Multidimensional NMR spectroscopy aids in the structure elucidation of complicated and larger molecules. Two frequencies are collected instead of one. 2D NMR experiments that are often employed include correlation spectroscopy (COSY), nuclear Overhauser effect spectroscopy and experiments (NOESY), total correlation spectroscopy (TOCSY), rotating frame nuclear Overhauser effect spectroscopy (ROESY), heteronuclear single quantum coherence (HSQC), and heteronuclear multiple bond correlation (HMBC). For example, COSY spectra are used to determine ^1H to ^1H coupling. HMBC experiments are used to determine coupling between ^{13}C and ^1H nuclei two, three, or even four bonds apart. Pulse sequences have been published for each of these experiments and are uploaded to modern NMR instruments for selection.

Several software packages including ACD Lab's Structure Elucidator, SENECA, CAST (Canonical-Representation of Stereochemistry)/CNMR Structure Elucidator, Logic for Structure Elucidation, and Automics are available to help forensic scientists to elucidate chemical structures. The identification of metabolites is aided by the use of Bayesil software and the Metabolomics Consortium Database. In addition to NMR spectroscopy, magnetic resonance imaging has also been used in studying metabolites and in metabolomics research.

QUESTIONS

1. The technique of _____ exposes molecules to a beam of high-energy electrons in order to fragment them.
   a. HPLC
   b. MS
   c. NMR spectroscopy
   d. IR spectroscopy

2. To determine the presence of cocaine by its mass and fingerprint, _____ should be used.
   a. TLC
   b. GC
   c. HPLC
   d. MS

3. The observation that no two substances produce the same fragmentation patterns under carefully controlled conditions is a unique feature of:
   a. Electromagnetic radiation
   b. Ultraviolet spectrophotometry
   c. Thin-layer chromatography
   d. MS

4. NMR can be used to probe _____ molecules.
   a. Paramagnetic
   b. Diamagnetic

5. The tallest peak in a mass spectrum is termed the:
   a. Base peak
   b. Molecular ion peak
   c. Fragment peak
   d. Terminal peak

6. Diagram and label the parts of a mass spectrometer.

7. Why can both positive and negative ions be produced with CI but not EI?

8. Explain how NMR works.

9. Why is peak splitting often observed with proton NMR but not with carbon NMR?

10. Explain chemical shielding and coupling in NMR.

Bibliography

Al-Meshal, I.A., K.A. Al-Rashood, M. Nasir, and F.S. El-Faraly. 1987. (-)-Cathinone: Improved synthesis and carbon-13 NMR assignments. *J. Nat. Prod.* 50(6):1138–1140.

Balayssac, S., E. Retailleau, G. Bertrand, M.-P. Escot, R. Martino, M. Malet-Martino, and V. Gilard. 2014. Characterization of heroin samples by 1H NMR and 2D DOSY 1H NMR. *Forensic Sci. Int.* 234:29–38.

Black, C., O.P. Chevallier, and C.T. Elliott. 2016. The current and potential applications of ambient mass spectrometry in detecting food fraud. *Trends Anal. Chem.* 82:268–278.

Bodnar, W.M., V. McGuffin, and R. Smith. 2017. Statistical comparison of mass spectra for identification of amphetamine-type stimulants. *Forensic Sci. Int.* 270:111–120.

Brand, W.A. 2004. Chapter 38: Mass spectrometer hardware for analyzing stable isotope ratios. In *Handbook of Stable Isotope Analytical Techniques*, Volume I, ed. P.A. deGroot. 835–856. Amsterdam: Elsevier B. V.

Collins, G.P. 1991. Nobel chemistry prize recognizes the importance of Ernst's NMR work. *Phys. Today* 44(12):19.

Cui, Q., I.A. Lewis, A.D. Hegeman, M.E. Anderson, J. Li, C.F. Schulte, W.M. Westler, H.R. Eghbalnia, M.R. Sussman, and J.L. Markley. 2008. Metabolite identification via the Madison Metabolomics Consortium Database. *Nat. Biotechnol.* 26:162–164.

Dias, D.A., O.A.H. Jones, D.J. Beale, B.A. Boughton, D. Benheim, K.A. Kouremenos, J.-L. Wolfender, and D.S. Wishart. 2016. Current and future perspectives on the structural identification of small molecules in biological systems. *Metabolites* 6(4):46. doi:10.3390/metabo6040046.

Draffan, G.H., R.A. Clare, and F.M. Williams. 1973. Determination of barbiturates and their metabolites in small plasma samples by gas chromatography-mass spectrometry. Amylorbarbitone and 3'-hydroxyamylobarbitone. *J. Chromatogr.* 75(1):45–53.

Houck, M.M., ed. 2016. *Materials Analysis in Forensic Science*. Cambridge, MA: Academic Press.

Koichi, S., M. Arisaka, H. Koshino, and H. Satoh. 2014. Chemical structure elucidation from C-13 NMR chemical shifts: Efficient data processing using bipartite matching and maximal clique algorithms. *J. Chem. Inf. Modeling* 54(4):1027–1035. doi:10.1021/ci400601c.

Krummel, J.N., L.N. Russell, D.N. Haase, S.M. Schelble, E. Tsai, K.M. Elkins. 2015. Application of high-field NMR spectroscopy for differentiating cathinones for forensic identification. *Colonial Acad. Alliance Undergraduate Res. J.* 5. http://publish.wm.edu/caaurj/vol5/iss1/1/.

LaBelle, M.J., C. Savard, B.A. Dawson, D.B. Black, L.K. Katyal, F. Zrcek, and A.W. By. 1995. Chiral identification and determination of ephedrine, pseudoephedrine, methamphetamine and methcathinone by gas chromatography and nuclear magnetic resonance. *Forensic Sci. Int.* 71:215–223.

Lesiak, A.D. and J.R. Shepard. 2014. Recent advances in forensic drug analysis by DART-MS. *Bioanalysis* 6(6):819–842.

Lewis, J.K., Wei, J., and Siuzdak, G. 2000. Matrix-assisted laser desorption/ionization mass spectrometry in peptide and protein analysis. In *Encyclopedia of Analytical Chemistry*, ed. R.A. Meyers, pp. 5880–5894. Chichester: Wiley.

Maheux, C.R., C.R. Copeland, and M.M. Pollard. 2010. Characterization of three methcathinone analogs: 4-methylmethcathinone, methylone, and bk-MBDB. *Microgram J.* 7(2): 42–49.

McMurry, J.E. 2015. *Organic Chemistry*, 9th edn. Pacific Grove: Brooks-Cole Publishing.

Moigradean, D., M.A. Poiana, L.M. Alda, and I. Gogoasa. 2013. Quantitative identification of fatty acids from walnut and coconut oils using GC-MS method. *J. Agroalimentary Processes Technol.* 19(4):459–463.

Munson M.S.B. and F.H. Field. 1966. Chemical ionization mass spectrometry. I. General Introduction. *J. Am. Chem. Soc.* 88(12):2621–2630.

Nier, K.A., A.L. Yergey, and P. Jane Gale, eds. 2015. *The Encyclopedia of Mass Spectrometry: Volume 9: Historical Perspectives Part B: Notable People in Mass Spectrometry*. Waltham, MA: Elsevier.

Poetzsch, M., A.E. Steuer, C.M. Hysek, M.E. Liechti, and T. Kraemer. 2016. Development of a high-speed MALDI-triple quadrupole mass spectrometric method for the determination of 3,4-methylenedioxymethamphetamine (MDMA) in oral fluid. *Drug Test Anal.* 8(2):235–40.

Power, J.D., P. McGlynn, K. Clarke, S.D. McDermott, P. Kavanagh, and J. O'Brien. 2011. The analysis of substituted cathinones. Part 1: Chemical analysis of 2-,3- and 4-methylmethcathinone. *Forensic Sci. Int.* 212(1–3): 6–12.

Reitzel, L.A., P.W. Dalsgaard, I.B. Müller, and C. Cornett. 2012. Identification of ten new designer drugs by GC-MS, UPLC-QTOF-MS, and NMR as part of a police investigation of a Danish internet company. *Drug Test Anal.* 4(5):342–354.

Roepstorff, P. 2014. Franz Hillenkamp (1936–2014). *Angew. Chem.* 53(47):12673 doi:10.1002/anie.201409504.

Rosano, T.G., M. Wood, K. Ihenetu, and T.A. Swift. 2013. Drug screening in medical examiner casework by high-resolution mass spectrometry (UPLC-MSE-TOF). *J. Anal. Toxicol.* 37(8):580–593.

Schelble, S.M., E. Tsai, A. Drotar, D. McElwee, M.J. Wieder, and N. Shoup. 2014. *On-Line Organic I Laboratory Manual*. Dubuque, IA: Great Rivers Technology Publishing. ISBN: 9781615496075.

Schelble, S.M., E. Tsai, A. Drotar, D. McElwee, M.J. Wieder, and R. Barrows. 2014. *On-Line Organic II Laboratory Manual*. Dubuque, IA: Great Rivers Technology Publishing. ISBN: 9781615496303.

Schelble, S.M., K.M. Elkins, E. Tsai, M. Wieder, and R.D. Walker. 2016. Creating scholarship opportunities for undergraduate students through use of high field NMR. In *ACS Symposium Series 1225: NMR Spectroscopy in the Undergraduate Curriculum*, Volume 3, eds. D. Soulsby, T. Wallner, and L. Anna, pp. 183–205. Washington, DC: American Chemical Society.

SWGDRUG. 2016. Scientific Working Group for the Analysis of Seized Drugs Recommendations Version 7.1. www.swgdrug.org/Documents/SWGDRUG%20Recommendations%20Version%207-1.pdf (accessed October 10, 2017).

The Nobel Prize in Chemistry. 1922. www.nobelprize.org/nobel_prizes/chemistry/laureates/1922/ (accessed January 27, 2018).

The Nobel Prize in Chemistry. 1991. www.nobelprize.org/nobel_prizes/chemistry/laureates/1991/ (accessed January 27, 2018).

The Nobel Prize in Chemistry. 2002. www.nobelprize.org/nobel_prizes/chemistry/laureates/2002/ (accessed January 27, 2018).

The Nobel Prize in Physics. 1943. www.nobelprize.org/nobel_prizes/physics/laureates/1943/ (accessed January 27, 2018).

The Nobel Prize in Physics. 1944. www.nobelprize.org/nobel_prizes/physics/laureates/1944/ (accessed January 27, 2018).

The Nobel Prize in Physics. 1952. www.nobelprize.org/nobel_prizes/physics/laureates/1952/ (accessed January 27, 2018).

The Nobel Prize in Physiology or Medicine. 2003. www.nobelprize.org/nobel_prizes/medicine/laureates/2003/ (accessed January 27, 2018).

Vigli, G., A. Philippidis, A. Spyros, and P. Dais. 2003. Classification of edible oils by employing 31P and 1H NMR spectroscopy in combination with multivariate statistical analysis. A proposal for the detection of seed oil adulteration in virgin olive oils. *J. Agric. Food Chem.* 51(19):5715–5722.

Westphal, F., T. Junge, U. Girreser, W. Greibl, and C. Doering. 2012. Mass, NMR and IR spectroscopic characterization of pentedrone and pentylone and identification of their isocathinone by-products. *Forensic Sci. Int.* 217(1–3):157–167.

Wilkinson, D.J. 2018. Historical and contemporary stable isotope tracer approaches to studying mammalian protein metabolism. *Mass Spectrom. Rev.* 37:57–580.

Xu, B., P. Li, F. Ma, X. Wang, B. Matthäus, R. Chen, Q. Yang, W. Zhang, and Q. Zhang. 2015. Analytical methods: Detection of virgin coconut oil adulteration with animal fats using quantitative cholesterol by GC×GC–TOF/MS analysis. *Food Chem.* 178:128–135.

CHAPTER 6

Chromatography

KEY WORDS: chromatography, thin-layer chromatography, stationary phase, mobile phase, paper chromatography, column chromatography, column, cation exchange, anion exchange, chelating column, normal phase, reverse phase, fast-performance liquid chromatography, high-performance liquid chromatography, ultra-performance liquid chromatography, gas chromatography, boiling point, detector, capillary electrophoresis

LEARNING OBJECTIVES

- To explain how to use thin layer, paper, column, liquid, and gas chromatography in forensic separations and analysis
- To identify and explain the difference between the mobile and stationary phases in a chromatographic system
- To understand the difference between normal and reverse-phase chromatography
- To understand the difference between cation-exchange and anion-exchange chromatography
- To identify and understand the functions of the parts of a gas chromatograph
- To understand the basis of separations using a gas chromatograph
- To be able to explain why molecules elute in a particular order based on polarity, charge, structure, molecular weight, and boiling point using chromatography
- To define the terms retention factor and retention time and their relevance in forensic analysis

JWH-018 IMPLICATED IN DEATH OF COLLEGE BASKETBALL PLAYER

On October 4, 2011, Anderson University basketball player Lamar Jack collapsed during a preseason workout. Before collapsing, the 19-year-old freshman had complained of vision problems and had cramping.

When he arrived at the emergency room, Jack was found to have an extremely high body temperature due to an increased heart rate and blood pressure. Toxicological lab testing revealed the presence of JWH-018, a synthetic cannabinoid substance. Its pharmacological effects are similar to that of tetrahydrocannabinol (THC), the primary psychoactive compound in marijuana.

JWH-018 (1-pentyl-3-(1-naphthoyl)indole) is sprayed on herbs or other plant material, which is referred to as "fake pot." Around this time, it was sold in drug and convenience stores in the region as herbal incense and marketed as "K2" or "Spice." It was touted as a "legal high" alternative to marijuana. It was also sold at head shops, gas stations, and on the internet. The plant material can be ingested or smoked like marijuana.

JWH-018 was developed in basic research to study cannabinoid receptors but does not share a chemical backbone with THC or other naturally occurring cannabinoids from marijuana. It was found to bind to cannabinoid receptor 1 (CB1) with increased affinity as compared to THC with similar agonist effects.

JWH-018 can be detected and identified using gas or liquid chromatography–mass spectrometry (LC-MS). In 2010, Dr. Craig Banks and colleagues at the Manchester Metropolitan University in Manchester, England, reported developing a GC-MS urinalysis assay to quickly detect, identify, and quantify JWH compounds in

JWH-018 IMPLICATED IN DEATH OF COLLEGE BASKETBALL PLAYER (continued)

herbal incense. In 2011, a team from the Arkansas Department of Public Health and Arkansas State Crime Laboratory reported the use of liquid chromatography–tandem mass spectrometry (LC-MS/MS) to quantitate JWH-018 metabolites, 5-(3-(1-naphthoyl)-1H-indol-1-yl)-pentanoic acid, and, (1-(5-hydroxypentyl)-1H-indol-3-yl) (naphthalene-1-yl)-methanone, excreted in urine. Another group from the University of Verona in Italy reported a direct screening test using matrix-assisted laser desorption ionization–time of flight mass spectrometry (MALDI-TOF MS) in the range of 150–550 m/z to test for synthetic cannabinoids in 2012.

Jack died four days later. Anderson County (South Carolina) Coroner Jack Shore reported that the cause of death was acute drug toxicity that led to Jack's multiple organ failure. The manner of death was reported as an accident.

JWH-018 was identified in 3264 forensic cases in 2011; synthetic marijuana accounted for 6959 calls to poison centers in 2011. JWH-018 and "its salts, isomers, and salts of isomers" were temporarily controlled by the DEA and placed on the Controlled Substance Act Schedule I on March 1, 2011; the control status was made permanent by the Synthetic Drug Prevention Act of 2012 on July 9, 2012. In 2012, JWH-018 was identified in 982 forensic cases. The National Collegiate Athletic Association (NCAA) banned the drug in 2011 and began testing college athletes for it at its championships in 2013.

BIBLIOGRAPHY

Chinese chemical supplier pleads guilty to conspiracy and importation of synthetic drugs, controlled substances, March 13, 2015. www.justice.gov/usao-mdfl/pr/chinese-chemical-supplier-pleads-guilty-conspiracy-and-importation-synthetic-drugs (accessed January 27, 2018).

D'Andrea, N. JWH "synthetic marijuana" drug tests coming soon. *Phoenix New Times*, June 16, 2010. www.phoenixnewtimes.com/arts/jwh-synthetic-marijuana-drug-tests-coming-soon-6576652 (accessed January 27, 2018).

Engdahl, K. Children leave messages of support for neighborhood cop. www.wyff4.com/article/children-leave-messages-of-support-for-neighborhood-cop/10259892 (accessed January 27, 2018).

Gottardo, R., A. Chiarini, I. Chiarini, C. Dal Prà, C. Seri, G. Rimondo, U. Armato Serpelloni, and F. Tagliaro. 2012. Direct screening of herbal blends for new synthetic cannabinoids by MALDI-TOF MS. *J. Mass Spectrom.* 47(1):141–146.

JWH-018. www.deadiversion.usdoj.gov/drug_chem_info/spice/spice_jwh018.pdf (accessed January 27, 2018).

Moran, C.L., V.-H. Le, K.C. Chimalakonda, A.L. Smedley, F.D. Lackey, S.N. Owen, P.D. Kennedy, G.W. Endres, F.L. Ciske, J.B. Kramer, et al. 2011. Quantitative measurement of JWH-018 and JWH-073 metabolites excreted in human urine. *Anal. Chem.* 83(11): 4228–4236. http://pubs.acs.org/doi/abs/10.1021/ac2005636?src=recsys.

Scarbinsky, K. The spread of Spice: Colleges, NCAA deal with the problem of synthetic marijuana, July 29, 2012. www.al.com/sports/index.ssf/2012/07/the_spread_of_spice_colleges_n.html (accessed January 27, 2018).

Smith, J.P., O.B. Sutcliffe, and C.E. Banks. 2015. An overview of recent developments in the analytical detection of new psychoactive substances (NPSs). *Analyst* 140:4932–4948.

The Associated Press. SC coroner: Synthetic pot killed college athlete. October 16, 2011, *USA Today*. https://usatoday30.usatoday.com/sports/college/mensbasketball/2011-10-16-1691565843_x.htm (accessed January 27, 2018).

Chromatography is a technique used to separate components of a mixture by their physical properties including polarity, boiling point, solubility, size, charge, and structure. In chromatography, there are two phases: the mobile and the stationary phase. Mixture components are separated by exploiting the differences in their strengths of interaction with the two phases such as adsorption or binding to the stationary phase and solubility or gas state in the mobile phase. There are several types of chromatography including paper and thin-layer chromatography (TLC) that are named for their stationary phases. Other types, including gas chromatography (GC) and high-performance liquid chromatography (HPLC), have names that reflect their mobile phases. Thin layer, paper, column, and liquid chromatography are all types of liquid chromatography. Column chromatography methods include GC, gravity column chromatography, fast protein liquid chromatography (FPLC), HPLC, and ultra-performance liquid chromatography (UPLC).

THIN-LAYER CHROMATOGRAPHY

TLC is a simple and inexpensive method for separating sample mixtures. TLC typically employs silica $[SiO_2]_n$ or alumina (Al_2O_3) as the *stationary phase*. A thin veneer of the silica or alumina is spread on glass or plastic plates (inert supports) that can be cut to the desired size and dimensions. The stationary phase is more polar than the mobile phase. Silica is used in the separation of polar compounds and the alumina is used to separate less polar or nonpolar compounds. The *mobile phase* is an organic solvent or mixture of organic solvents in desired proportions. The mobile phase (approximately half a centimeter) is added to a beaker, jar, or other developing chamber that is covered with a glass lid or watch glass. The covered mobile phase is equilibrated for 10 minutes in the fume hood prior to adding the plate with the standard(s) and evidence samples. TLC is a relative technique; standards or reference compounds must be run on the same plate as the samples for analysis and direct comparison. Typically, the plates are cut into small rectangles that can accommodate three to four samples at a time. However, larger plates can be used to process a dozen or more samples simultaneously.

Solids and liquids can be analyzed using TLC. Solids are dissolved in a small amount (0.1–1 mL) of volatile organic solvent, usually in a ceramic spot plate. The solvent is allowed to partially evaporate. A line parallel with the bottom of the plate is drawn gently using a ruler and a pencil, approximately a centimeter from the bottom of the plate. The samples are then added to the plate using a glass microcapillary tube along the pencil line. Each sample is applied three times to the same spot using the microcapillary once the solvent has evaporated from the spot. Appropriate TLC final sample sizes are 0.1–50 mg. Care must be taken not to enlarge the spot with each application. The final spot should be only 1–2 mm in diameter. Samples and standards are spaced approximately a centimeter apart; two to three samples and a standard are routinely run side-by-side on small plates. The standards can be mixed if it is known that they will be well separated using the mobile phase. After all samples and standards have been spotted, the plate is carefully inserted into the equilibrated mobile-phase separation chamber as shown in Figure 6.1.

The components of a mixture are separated by means of polarity using a selected mobile phase. The mobile phase serves to move the components over the stationary phase sieve. The solvent is drawn up the plate by capillary action taking the most nonpolar components with it most readily. The more polar components are drawn to the polar plate by intermolecular attractive forces and remain at or near the origin. Components that are less polar may remain mostly adsorbed to the plate but move slightly. For example, butyl amine is more polar than cyclohexane and will migrate more slowly up the TLC plate. Using a polar solvent, the most polar compounds can be carried up the plate.

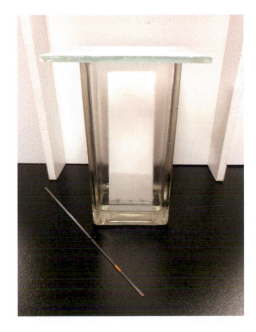

Figure 6.1   A silica TLC plate in a developing chamber.

Returning to our previous example, butyl amine will migrate further in a polar solvent than a nonpolar solvent but will still not migrate as far as cyclohexane as it is also attracted to the stationary phase.

The plate should be removed from the system before the mobile phase elutes from the plate as it may take the nonpolar mixture components with it. A run may take 3 to 60 minutes depending on the compounds being separated, the mobile phase selected, and the length of the plate. The final location of the mobile-phase solvent front should be marked with a pencil line. Increasing the development time will increase the distance traveled by the solvent front.

TLC can be used to qualitatively determine if an evidence sample has the same component(s) as a reference or standard sample or if two evidence samples contain the same components. It can also be used to determine which mixture is most complex or which sample has the most polar (or nonpolar) component. TLC can also be used in quantitative analysis if the analytes are compared to quantitative standards applied to the same plate.

TLC is used to analyze evidence containing drugs, inks, and pigments. Figure 6.2 shows a TLC plate used to separate the components of five different inks. From the separation, some of the inks can be determined to be chemically different based on their differing chemical makeup. As the ink dyes are visible, the compounds do not need to be developed for visualization. However, under an ultraviolet visible (UV-Vis) wand or alternate light source, some dyes are observed to be fluorescent, further aiding in their differentiation. TLC is routinely used in some forensic labs to determine if THC is present in a sample suspected to be marijuana. While the colored ink and THC samples are visible to the naked eye, white drug samples will be difficult to differentiate from the plate without employing a visualization method. Visualization or development methods render TLC spots temporarily or permanently visible and include ultraviolet light, iodine fuming, wet chemical reagents that react with the spots to render them a visible color (e.g., Ninhydrin, Marquis test), and other reagents that will selectively stain only desired spots while the others remain colorless. Some of the visualization methods will only temporarily render the spots visible so they should be circled in pencil to ensure their locations will be known after the method concludes, such as when UV light is removed or a chemical applied (e.g., iodine) sublimes.

The TLC plate analysis is completed by computing the *retention factor* (Rf) value. The Rf value is the ratio of the distance traveled by the component from the origin to its stopping point over the total distance traveled by the solvent front (from the pencil line at the origin to the pencil line on removal from the developing chamber). The Rf value is generally reported as a decimal with two significant figures. Rf values can be compared only if the same mobile and stationary phase is used. TLC is nondestructive; the separated components can be removed from the plate with the silica using a razor blade and recovered by dissolving in a solvent.

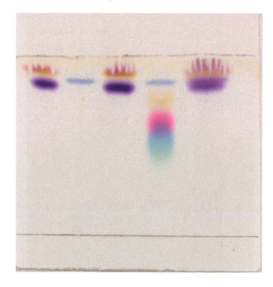

Figure 6.2   A silica TLC plate used to separate inks.

PAPER CHROMATOGRAPHY

In *paper chromatography*, the stationary phase is filter paper of the desired ash and strength. It is performed in the same manner as TLC with an organic or aqueous liquid mobile phase. Polarity is used to separate the components of the mixture. Paper chromatography is inexpensive and simple to perform and can be used instead of TLC to differentiate analytes such as components of inks as shown in Figure 6.3, although TLC usually leads to sharper bands and better separation.

COLUMN CHROMATOGRAPHY

Column chromatography is a type of liquid chromatography (LC). In column chromatography, the stationary phase is contained in a solid support such as a plastic or glass tube fitted with a filter or fiber (e.g., cotton, wool, or glass) plug and stopcock at one end. The tube may be approximately 6 inches to 6 feet in length. It is held vertical to the floor or benchtop with a ring stand or metal frame and clamps. The stationary phase suspended in buffer or solvent is called a slurry and is applied to the column containing only buffer or solvent and filled to approximately one-third. As the applied stationary phase settles, more slurry is added. Once a base of stationary phase is established, the buffer is pipetted off the top or released very slowly by opening the stopcock. The column continues to be packed with stationary phase until the desired height is reached or until there is only approximately 1–5 cm of space remaining in the tube. After the column is completed, two to three volumes of the column of mobile phase are passed through. Buffer or solvent mobile phase must remain on top of the stationary phase to keep it moist. In lieu of pouring columns, they can also be purchased pre-poured.

To prepare the column to receive the sample, the mobile phase above the stationary phase is carefully removed using a pipette. The sample mixture is carefully applied inside the solid support containing the stationary phase without disturbing the column bed, is allowed to migrate into the stationary phase (by gravity or high pressure due to pump action), and then the mobile phase is added to carry the components through the column while they are sieved by the stationary phase material. Column chromatography is most commonly used in preparative work with a large quantity of sample mixture to separate. The separated components are collected in test tubes manually or using a fraction collector. Fractions containing the analyte(s) are determined by visual inspection (if colored) or using a detector such as UV-Vis spectroscopy. Column chromatography is not widely used in forensic laboratories except when performing or evaluating synthetic pathways in drug production used by perpetrators.

There are several types of column chromatography named after the stationary phase separation method used, including size exclusion, anion exchange, cation exchange, chelating, normal phase, and reverse phase. In *size-exclusion chromatography*, cellulose-based stationary phase beads that vary in pore size are used as the stationary phase. These beads restrict entry to molecules of a maximum size. Large particles cannot enter the gel and are excluded and thus elute faster. Small particles that can enter the pores of the beads are retarded and elute later. Examples of size-exclusion stationary phases include G-25, G-50, G-75, G-150, and so on; the smaller number such as G-25 has smaller pore sizes

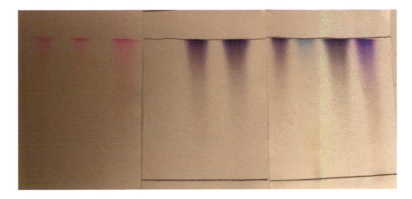

Figure 6.3 Paper chromatography used to separate inks.

and is useful for separating buffer from a protein toxin analyte while a larger number such as G-150 can be used to separate isoforms of protein toxins.

In a *cation-exchange* column, stationary phases such as carboxymethyl may be used in which silica is substituted with anionic residues that interact strongly with cationic species. The positively charged species are eluted with acid. The most positively charged species elute last. Conversely, in *anion-exchange* columns, silica is substituted with cationic residues that interact strongly with anionic species; diethyl amino ethyl cellulose (DEAE) is an example of an anion-exchange column stationary phase. The most negatively charged species elute last.

In a *chelating column*, a resin that binds strongly to a tag or analyte of interest is used in the separation. These columns tend to be short due to the binding affinity and separation properties of the column. An example of a chelating column stationary phase is the nickel–nitrilotriacetic acid resin that is used to bind proteins tagged with six terminal histidine residues in protein purification.

In *normal-phase* column chromatography, the material used to pack the column is very polar. The polar stationary phase will attract and adsorb or bind polar compounds using hydrogen bonding or dipole–dipole interactions. In contrast, nonpolar compounds will pass through the material quickly and elute first. Eluting polar compounds may require a change in the mobile phase that increases its affinity for the mobile phase or reduces the stationary phase's ability to bind and retain the compound. A gradient may be used to elute polar compounds in order of increasing polarity. In contrast, *reverse-polarity* stationary phases are composed of silica or alumina functionalized with long-chain hydrocarbons that bind to hydrophobic analytes and elute polar species first. Polar solvents such as methanol, water, and acetonitrile are used as the mobile phase in reverse-phase column chromatography.

Fast-performance (or fast protein) liquid chromatography (FPLC) is a variation of column chromatography in which a pump is used to increase the separation speed of the column (Figure 6.4). As a high-pressure pump is used to move the components through for separation, the column packing material will compress if it is not already tightly packed. As the columns can be packed tighter, shorter columns will provide equivalent separation as longer gravity columns. Due to the increased pressure from using the pump, FPLC columns may have strong plastic walls or steel jackets. FPLCs may be connected to fraction collectors such as the one shown in Figure 6.4 and use a UV-Vis detector. FPLC can be used for the same preparative uses as gravity columns. These include separating proteins, toxins, drugs, and other analytes. The proteins shown in the separation in Figure 6.5 eluted at approximately 170 minutes and were detected using a UV-Vis detector.

Figure 6.4   FPLC instrument with fraction collector.

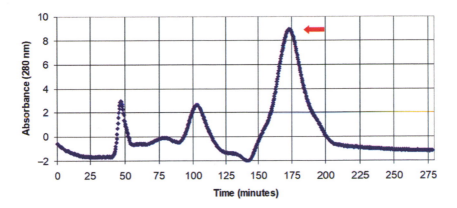

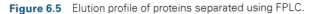

Figure 6.5 Elution profile of proteins separated using FPLC.

HIGH-PERFORMANCE LIQUID CHROMATOGRAPHY

In *HPLC*, the sample mixture is separated via a steel-jacketed chromatography column that can withstand high pressures and is packed with one of several available stationary phases (Figure 6.6). HPLC instruments have containers for one or more mobile phases, a degasser, a mixing vessel, a high-pressure pump, an autosampler, pressure gauges, a pre-column, a separation or analytical column, column thermostats to control the column temperature, a detector, a waste container or fraction collector, and a computer. An HPLC instrument is shown in Figure 6.7. Samples may be injected as gases or liquids. Solids can be dissolved in a solvent and injected but a solvent peak will be seen. The components are eluted with a solvent at high pressure, typically 4000 psi or 130 atm. The solvent is usually a mixture of organic and aqueous liquids including acetonitrile and water. Mixtures are separated by the polarity or size of their components. The most common detectors are UV-Vis detectors due to their ability to detect a large range of analytes and their relatively low cost.

The stationary phases include normal polarity, ion exchange, size exclusion, and reverse polarity. The normal-polarity stationary phase consists of silica or alumina. As these materials are highly polar, they interact with the polar components of the mixture. The least-polar compounds will elute from the column first. In a cation-exchange column, silica is substituted with anionic residues that interact strongly with cationic species such as sodium ions (Na^+). The positively charged species are eluted with acid. The most positively charged species elute last. Conversely, in anion-exchange columns, silica is substituted with cationic residues that interact strongly with anionic species such as chloride ions (Cl^-). The most negatively charged species elute last. In size-exclusion chromatography, mixtures are separated by the size of the compounds using gel beads. The beads are often cellulose-based and characteristically have pores of a set size. Only compounds with sizes smaller than the pore can enter so small molecules tend to interact with the beads in the pores while larger molecules pass by and are eluted. Due to the high pressure and bead pore size, larger molecules elute first and smaller molecules elute last. Finally, reverse polarity stationary phases are composed of silica or alumina functionalized with long-chain (e.g., 18 carbon) hydrocarbon group caps. These columns are commonly referred to as C-18 columns. These nonpolar caps interact strongly with nonpolar compounds. Thus, the most nonpolar compounds will elute last while the least nonpolar or more polar compounds will elute first. Polar solvents such as methanol and water are used as the mobile phase in reverse-phase column chromatography. HPLC instruments can be used to separate

Figure 6.6 Steel-jacketed HPLC C18 column.

Figure 6.7 HPLC instrument.

and detect drugs and toxins in body fluids such as blood or urine. An HPLC chromatogram with gradient elution (light blue line) is shown in Figure 6.8. The target protein (marked with the yellow arrow) was separated from several other proteins and eluted at 74 minutes and collected in fraction 9. Due to the high pressure used, HPLC is not used for explosives.

ULTRA-PERFORMANCE LIQUID CHROMATOGRAPHY

A UPLC instrument is shown in Figure 6.9. Waters launched the first UPLC in 2004. Like HPLC and FPLC instruments, the UPLC uses columns and pumps in the separation. The ultra performance is due to the ultra-high pressures the instrument uses. While HPLC separates particles sized between 2.5 and 5 microns, UPLC is used to separate particles less than 2 microns in size. The pressures used exceed 6000 psi and may be as high as 15,000 psi. The use of such high pressures allows the UPLC to achieve analyte separation resolution equal to or better than HPLC instruments. UPLC instruments are used to separate and detect performance enhancing drugs used by athletes as well as drugs and drug combinations taken by medical patients and drug users, among others. Methods or standard operating procedures need to be developed for each analyte or class of drugs. As in other chromatography instruments, internal standards are used to compare chromatograms and identify substances based on their retention times. When coupled to MS instruments, the fragmentation pattern and qualifying ions can be used to identify a substance.

GAS CHROMATOGRAPHY

GC is a physical method of separation used to separate volatile materials based on their boiling points and polarity and their interaction with the mobile and stationary phases of the instrument. The mobile phase is an inert gas such as argon, helium, or nitrogen used to move the sample through the chromatography column. Gas pressure gauges, flow meters, and regulators are used to control and quantify the flow of the gas. The stationary phase is a column consisting of fine tubing containing an adsorbing liquid such as a polymer or wax with a high boiling point that is immobilized on an inert solid such as diatomaceous earth. This technique is also called gas–liquid chromatography because a gaseous mobile phase and liquid immobilized on the surface of an inert solid stationary phase are used. A schematic of a GC instrument coupled to a mass spectrometer is shown in Figure 6.10. A GC-MS instrument is shown in Figure 6.11.

The columns can be packed or open tubular and vary in length from 1–100 meters. Up to 1000 mg can be loaded onto packed columns while 10–1000 ng can be loaded onto open tubular columns. The open tubular, or capillary, columns

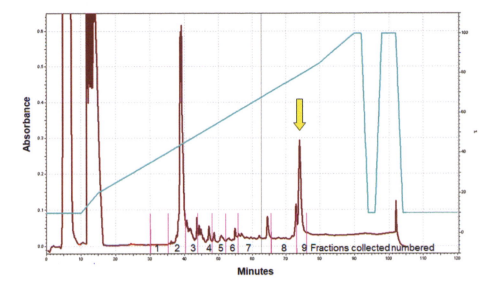

Figure 6.8 HPLC elution profile of protein separation.

may be wall-coated or support-coated with films and have lengths of 10–100 meters. The film thickness will affect the retentive character of the column; thick films are used for highly volatile solutes to retain them longer and improve separation. Film thickness ranges from 0.1 to 5 µm, although thicker films can be used in wide-bore columns. They can support relatively low pressure. Wall-coated open tubular columns are capillary tubes (0.25–0.75 mm diameter) coated with a thin layer of stationary phase. Support-coated open tubular columns (0.5 mm diameter) hold

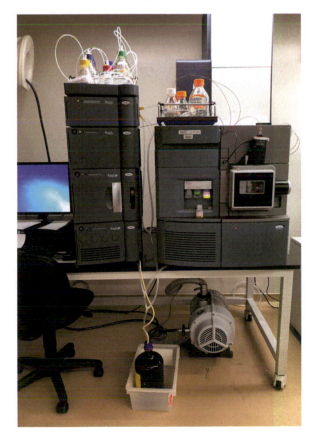

Figure 6.9 UPLC instrument.

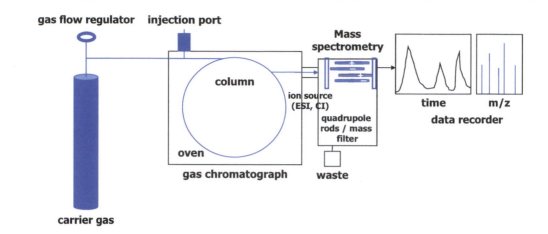

Figure 6.10   Schematic of GC-MS.

more stationary phase than the wall-coated columns and have a greater sample capacity. Support-coated columns consist of capillaries lined with a thin film (approximately 30 μm) of stationary phase such as diatomaceous earth. Diatomaceous earth is a naturally occurring material composed of hollow, spherical to oval-shaped diatom skeletons with a large surface area. These single-celled plants took in nutrients and disposed of wastes via molecular diffusion through pores. They are excellent support material for GC as the technique is based on the same kind of molecular diffusion. Packed columns can withstand high pressure as they are constructed of tubes of metal such as stainless steel, copper or aluminum, glass, or Teflon in lengths of 1–6 m and internal diameters of 2–4 mm. The tubes are densely packed with a uniform fine packing material or solid support and are coated with a thin layer of liquid stationary phase. As the particle diameter increases, the efficiency increases. The pressure difference required to maintain a flow rate of carrier gas is inversely related to the square of the particle diameter. The column is usually coiled from 10–30 cm in diameter to fit into an oven. The temperature of the column must be precisely controlled with the thermostat to a few tenths of a degree to perform precise analysis. The open tubular columns are more chemically inert than the packed columns. The columns are routinely exposed to temperatures of 300°C or more, but temperatures will vary depending on the boiling point of the sample to be separated.

The liquid and solid in the column must be chemically inert and be able to withstand high temperatures without decomposing. The liquid stationary phase must also have a boiling point of 100°C higher than the maximum operating temperature of the column and the ability to resolve solutes within a reasonable range. The retention time of an

Figure 6.11   GC-MS instrument.

Table 6.1 Gas chromatography stationary phases and their applications

Stationary phase	Applications	Maximum temperature (°C)
Polydimethyl siloxane	General purpose, semi-volatile, drugs, amines, halogenated compounds	340
5% Phenyl-polydimethyl siloxane	Alkaloids, drugs, halogenated compounds	325
50% Phenyl-polydimethyl siloxane	Drugs, steroids, pesticides, herbicides	320
Polyethylene glycol	Free acids, alcohols, glycols, fragrances, essential oils	260
50% Cyanopropyl-polydimethyl siloxane	Free acids, alcohols, ethers, essential oils, glycols	250
50% Trifluoropropyl-polydimethyl siloxane	Chlorinated aromatics, nitroaromatics, alkyl substituted benzenes	240

Source: Agilent, www.hichrom.com/assets/ProductPDFs/GC/GC_Agilent.pdf.

analyte will depend on its distribution constant, which is based on the chemical nature of the stationary phase. To be retained for some time in the column, the analyte must have some degree of chemical complementarity (e.g., "like dissolves like") to the stationary phase liquid. Polar stationary phases with functional groups such as cyano, carboxyl, carboxylic acid, and hydroxyl groups or polyesters will interact with polar analytes whereas nonpolar analytes will interact with the nonpolar stationary phase components including hydrocarbons and dialkyl siloxanes. Table 6.1 lists several types of chromatography stationary phases used in columns and their applications. Polydimethylsiloxane columns are commonly used to separate hydrocarbons, steroids, polychlorinated biphenyls, and polynuclear aromatics and can be run up to 350°C. Ethers, ketones, and aldehydes are found in medium-polarity stationary phases. Polyethylene glycol can be used to separate more polar compounds such as free acids, alcohols, glycols, ethers, and essential oils.

GC can be used for dilute samples in the liquid and gas phase. Solids should be dissolved in an appropriate solvent and filtered prior to injection. Some samples may need to be derivatized to increase their volatility prior to injection. The samples are placed in milliliter-sized glass vials with self-sealing Teflon-sealed rubber caps in their lids (Figure 6.12). Inserts can be used that reduce the volume needed to fill the vial (Figure 6.12). A very small amount, or plug, of sample is introduced to the instrument via an injector and flash vaporized in a port at the head of the column. A microsyringe is most commonly used for the injection. The microsyringe punctures the disks in the vials to remove the sample and introduce it to the head of the column via a rubber septum. This improves resolution while slow injection and/or oversized samples will lead to band spreading and poor resolution. The sample is carried by the gaseous mobile phase from the vaporization chamber into and through the column. The column temperature is controlled by a thermostat. The injection temperature is set about 50°C above the highest boiling point of the component in the sample. The instrument operator defines the temperature to start the chromatography, the time spent at each temperature, and the final temperature.

Figure 6.12 GC-MS vials with insert (left) and without (right). (Courtesy of Tim Phillips.)

Table 6.2 Boiling point and order of elution of hydrocarbons in fuels on a GC equipped with a polydimethylsiloxane column

Fuel	Chemical formula	Boiling point (°C)	Order of elution
Butane	$CH_3(CH_2)_2CH_3$	−1	1[a]
Pentane	$CH_3(CH_2)_3CH_3$	36.1	2[a]
Hexane	$CH_3(CH_2)_4CH_3$	69	3
Heptane	$CH_3(CH_2)_5CH_3$	98	4
Octane	$CH_3(CH_2)_6CH_3$	126	5
Nonane	$CH_3(CH_2)_7CH_3$	151	6
Decane	$CH_3(CH_2)_8CH_3$	174	7
Undecane	$CH_3(CH_2)_9CH_3$	196	8
Dodecane	$CH_3(CH_2)_{10}CH_3$	216	9
Tetradecane	$CH_3(CH_2)_{12}CH_3$	254	10
Hexadecane	$CH_3(CH_2)_{14}CH_3$	287	11
Octadecane	$CH_3(CH_2)_{16}CH_3$	316	12

Source: PubChem.
[a] These compounds often elute in the solvent delay or prior to detection and are undetectable.

Separation and retention on the column is affected by the volatility or boiling point of the analyte, polarity of the analyte and its interaction with the column stationary phase, and column temperature. The more volatile and least-polar compounds elute first and the less volatile elute later. Table 6.2 shows the order of elution for hydrocarbons in fuels and ignitable liquids. If the boiling point of the target is known, the oven temperature can be set equal to, or slightly above, the boiling point. This will result in a short elution time (2–10 minutes). If the components of the sample are unknown or if the components are known to have a broad range of boiling points, the column temperature is gradually increased or increased in steps to facilitate the separation resulting in a method run of around 30 minutes or more. GC is used to separate drugs, including alkaloids and steroids, and accelerants, for example. As GC is a relative technique, an internal standard is used to calibrate the retention time between sample runs. The Automated Mass Spectral Deconvolution and Identification System (AMDIS) software can be used to deconvolute GC peaks. Changes in pressure or flow of the carrier gas will change GC elution times. GC is destructive and the sample is not recovered. Figure 6.13 shows the GC separation of two drugs, salicylic acid and caffeine, using a polydimethylsiloxane column.

GC detectors (Table 6.3) include the mass spectrometer and flame ionization detector (FID) that have been widely used for forensic applications and for many others. The detector should have adequate sensitivity, good stability and reproducibility, be reliable and easy to use, have a linear range that extends over several orders of magnitude, and a temperature range from room temperature to at least 400°C. It should also have a short response time independent of the flow rate, respond well to all analytes or be highly selective to one or more classes of analytes, and be nondestructive to the sample.

Although it is destructive, for many years, the *FID* was the most widely used GC detector for forensic applications due to its ability to detect accelerant and other organic hydrocarbons. As the compounds elute from the column, they are mixed with hydrogen and air and ignited in a flame. The hydrocarbons produce ions and electrons that conduct electricity through the flame. The current (approximately 10^{-12} A) is measured. The sensitivity is 1 pg/s with low noise and the linear range extends over seven orders of magnitude.

Thermal conductivity detectors (TCD) were early universal detectors for the GC. They are very simple in construction, have a dynamic range of five orders of magnitude, and are nondestructive. The temperature of an element heated with

Figure 6.13 Salicylic acid and caffeine are separated by GC. Their boiling points are 211°C and 178°C, respectively.

Table 6.3 Gas chromatography detectors and their applications

Detector	Application
Flame ionization (FID)	Hydrocarbons, accelerants
Mass spectrometer (MS)	Tunable for any compound or species
Thermal conductivity	Universal detector
Electron capture (ECD)	Halogenated compounds, environmental samples
Thermionic	Nitrogen and phosphorous compounds
Electrolytic conductivity	Halogen-, nitrogen-, or sulfur-containing compounds
Photoionization (PID)	Compounds ionized by UV radiation
Fourier transform infrared spectroscopy (FTIR)	Organic compounds

constant electrical power is detected as it varies with the thermal conductivity of the organic or inorganic analyte in the gas. The heating element may be constructed of platinum, gold, or tungsten wire, or a semiconducting thermistor. However, its sensitivity is relatively low at only 500 pg solute/mL carrier gas.

Electron capture detectors (ECD) are very sensitive (5 fg/s) and widely used to analyze environmental forensic samples as they detect halogen-containing compounds including pesticides, herbicides, and polychlorinated biphenyls. They can also detect peroxides and quinones but not amines, alcohols, and hydrocarbons. The detector consists of a beta emitter, usually Ni-63. As the eluent passes the detector, the carrier gas is ionized by an electron from the emitter causing a burst of electrons and a pair of electrodes registers a current. However, the current is disrupted and markedly decreased by atoms, such as halogens, which tend to capture electrons.

Most GC units at forensic labs are now outfitted with *mass spectrometry (MS) detectors* (covered in Chapter 5) as they are tunable to detect any compound or species. A schematic of the parts of the GC-MS are shown in Figure 6.10 and the instrument is shown in Figure 6.11. The analytes are bombarded with high-energy electrons (in EI) and a molecular ion (radical cation) and other fragments are generated from the compound resulting in a unique fragmentation-pattern fingerprint for each compound. The related *GC-MS-MS* enables the forensic scientist to record a mass spectrum of selected ions from the first mass spectrum. GC-MS-MS is used for quantitative analysis and is a highly sensitive and specific technique. It can be used to determine structures of unknown compounds and map fragmentation pathways.

GC can also be used quantitatively to determine the concentration of analytes including or such as blood alcohol content (Figure 6.14). A series of standards of known concentration can be prepared and the same quantity of each injected on the column. The samples are injected from milliliter-sized glass vials with Teflon-sealed rubber caps in 0.2 μL to microliter volumes. From the chromatogram, the area under the peak can be determined for each standard. The area is proportional to concentration. By plotting the area under the peak versus the concentration of the standards (e.g., alcohol) and determining the slope of the line, the concentration of an unknown can be determined (Figure 6.14).

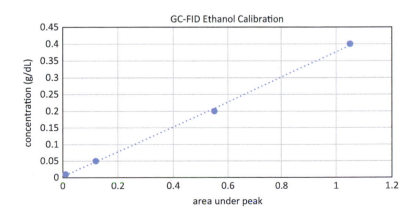

Figure 6.14 GC-FID used in determination of blood alcohol content.

Figure 6.15   A capillary electrophoresis instrument.

OTHER SEPARATION METHODS

Other types of separation methods include *capillary electrophoresis* (CE). In CE, a viscous polymer pumped into a thin glass capillary is used to separate ions based on size based upon the elution time as compared to an internal standard. CE instruments can be coupled to a MS or fluorescence detector. CE instruments used in DNA fragment sizing and sequencing applications employ fluorescence detectors (Figure 6.15).

QUESTIONS

1. Which of the following has a gas as its mobile phase?
   a. GC
   b. HPLC
   c. CE
   d. TLC

2. How are samples differentiated using GC?
   a. Retention time
   b. Fluorescence
   c. Absorption
   d. Angle scattering

3. Which of the following is a technique that can be used to separate a complex mixture of barbiturates?
   a. GC
   b. UV
   c. IR
   d. Microcrystalline tests   e. Fluorescence

4. One advantage to LC methods over GC methods is that _____.
   a. The samples are heated
   b. Samples do not need derivitization
   c. Mass spectrometry can be readily employed   d. Runs are typically shorter

5. Which type of material is run with instruments such as GC and CE with the sample to calibrate the run for comparison to other runs?
   a. Internal standard
   b. Evidence sample
   c. Elimination sample
   d. Reference sample
   e. All of the above

6. How does TLC work? Draw a schematic of its implementation and explain what is compared between an analyte and a standard.

7. Define and explain the difference between normal and reverse-phase HPLC.

8. How does GC work? Draw a schematic of the instrument and label the key features.

9. How can GC be used to quantitate an unknown sample? What samples are run, what is measured, and how is this used to calculate the unknown quantity?

10. What are the main differences between HPLC and UPLC?

Bibliography

Abdel-Hay, K.M., J. DeRuiter, and C.R. Clark. 2015. GC-MS and IR studies on the six ring regioisomeric dimethoxyphenylpiperazines (DOMePPs). *J. Forensic Sci.* 60(2):285–294.

Barker, J., R. Ramotowski, and J. Nwokoye. 2016. The effect of solvent grade on thin layer chromatographic analysis of writing inks. *Forensic Sci. Int.* 266:139–147.

Belal, T., T. Awad, J. DeRuiter, and C.C. Randall. 2008. GC–MS studies on acylated derivatives of 3-methoxy-4-methyl- and 4-methoxy-3-methyl-phenethylamines: Regioisomers related to 3,4-MDMA. *Forensic Sci. Int.* 178(1):61–82.

Bell, S. 2013. *Forensic Chemistry*, 2nd edn. Upper Saddle River, NJ: Pearson Prentice Hall.

Berg, T., B. Jørgenrud, and D.H. Strand. 2013. Determination of buprenorphine, fentanyl and LSD in whole blood by UPLC-MS-MS. *J. Anal. Toxicol.* 37(3):159–165.

Betancourt, J. and S. Gottlieb. 2017. Liquid chromatography. https://chem.libretexts.org/Core/Analytical_Chemistry/Instrumental_Analysis/Chromatography/Liquid_Chromatography(accessed January 25, 2018).

Bogusz, M.J., ed. 2007. *Handbook of Analytical Separations Vol. 6: Forensic Science*, 2nd edn. New York: Elsevier.

Causin, V., R. Casamassima, C. Marega, P. Maida, S. Schiavone, A. Marigo, and A. Villari. 2008. The discrimination potential of ultraviolet-visible spectrophotometry, thin layer chromatography, and Fourier transform infrared spectroscopy for the forensic analysis of black and blue ballpoint inks. *J. Forensic Sci.* 53(6):1468–1473.

Djozan, D., T. Baheri, G. Karimian, and M. Shahidi. 2008. Forensic discrimination of blue ballpoint pen inks based on thin layer chromatography and image analysis. *Forensic Sci. Int.* 179(2–3):199–205.

Dołowy, M., K. Kulpińska-Kucia, and A. Pyka. 2014. Validation of a thin-layer chromatography for the determination of hydrocortisone acetate and lidocaine in a pharmaceutical preparation. *Sci. World J.* 2014 http://dx.doi.org/10.1155/2014/107879.

Elkins, K.M. 2013. *Forensic DNA Biology: A Laboratory Manual*. Waltham, MA: Elsevier Academic Press.

Harris, D.C. 2006. *Quantitative Chemical Analysis*, 7th edn. New York: W.H. Freeman.

Holler, J.M., S.P. Vorce, J.L. Knittel, B. Malik-Wolf, B. Levine, and T.Z. Bosy. 2014. Evaluation of designer amphetamine interference in GC-MS amine confirmation procedures. *J. Anal. Toxicol.* 38(5):295–303.

King, N.M., K.M. Elkins, and D.J. Nelson. 1999. Reactivity of the invariant cysteine of silver hake parvalbumin (Isoform B) with dithionitrobenzoate (DTNB) and the effect of differing buffer species on reactivity. *J. Inorg. Biochem.* 76:175–185.

Lesiak, A.D. and J.R. Shepard. 2014. Recent advances in forensic drug analysis by DART-MS. *Bioanalysis.* 6(6):819–842.

Lillsunde, P. and T. Korte. 1991. Comprehensive drug screening in urine using solid-phase extraction and combined TLC and GC/MS identification. *J. Anal. Toxicol.* 15(2):71–81.

Lin, D.L., S.M. Wang, C.H. Wu, B.G. Chen, and R.H. Liu. 2016. Chemical derivatization for forensic drug analysis by GC- and LC-MS. *Forensic Sci. Rev.* 28(1):17–35.

Liu, C., Z. Hua, and Y. Bai. 2015. Classification of illicit heroin by UPLC-Q-TOF analysis of acidic and neutral manufacturing impurities. *Forensic Sci. Int.* 257:196–202.

Moore, J.M. 1990. The application of chemical derivatization in forensic drug chemistry for gas and high-performance liquid chromatographic methods of analysis. *Forensic Sci. Rev.* 2(2):79–124.

Philipp, A.A., M.R. Meyer, D.K. Wissenbach, A.A. Weber, S.W. Zoerntlein, P.G. Zweipfenning, and H.H. Maurer. 2011. Monitoring of kratom or Krypton intake in urine using GC-MS in clinical and forensic toxicology. *Anal. Bioanal. Chem.* 400(1):127–135.

Poetzsch, M., A.E. Steuer, C.M. Hysek, M.E. Liechti, and T. Kraemer. 2016. Development of a high-speed MALDI-triple quadrupole mass spectrometric method for the determination of 3,4-methylenedioxymethamphetamine (MDMA) in oral fluid. *Drug Test Anal.* 8(2):235–240.

Reitzel, L.A., P.W. Dalsgaard, I.B. Müller, and C. Cornett. 2012. Identification of ten new designer drugs by GC-MS, UPLC-QTOF-MS, and NMR as part of a police investigation of a Danish internet company. *Drug Test Anal.* 4(5):342–54.

Rosano, T.G., M. Wood, K. Ihenetu, and T.A. Swift. 2013. Drug screening in medical examiner casework by high-resolution mass spectrometry (UPLC-MSE-TOF). *J. Anal. Toxicol.* 37(8):580–593.

Saferstein, R. 2007. Exercise 10. In *Basic Laboratory Exercises for Forensic Science*. Upper Saddle River, NJ: Pearson Prentice Hall.

Saferstein, R. 2015. *Criminalistics: An Introduction to Forensic Science*, 11th edn. Boston, MA: Pearson.

Skoog, D.A., F.J. Holler, and S.R. Crouch. 2006. *Principles of Instrumental Analysis*, 6th edn. Boston, MA: Brooks Cole.

Smith, F.T., J. DeRuiter, K. Abdel-Hay, and C.R. Clark. 2014. GC-MS and FTIR evaluation of the six benzoyl-substituted-1-pentylindoles: Isomeric synthetic cannabinoids. *Talanta* 129:171–182.

Stout, P.R., J.M. Gehlhausen, C.K. Horm, and K.L. Klette. 2002. Evaluation of a solid-phase extraction method for benzoylecgonine urine analysis in a high-throughput forensic urine drug-testing laboratory. *J. Anal. Toxicol.* 26:401–405.

Tebbett, I.R. 1991. Chromatographic analysis of inks for forensic science applications. *Forensic Sci. Rev.* 3(2):71–82.

Thaxton, A., T.S. Belal, F. Smith, J. DeRuiter, K.M. Abdel-Hay, and C.R. Clark. 2015. GC-MS studies on the six naphthoyl-substituted 1-n-pentyl-indoles: JWH-018 and five regioisomeric equivalents. *Forensic Sci. Int.* 252:107–113.

Thet, K. and N. Woo. 2015. Gas chromatography. https://chem.libretexts.org/Core/Analytical_Chemistry/Instrumental_Analysis/Chromatography/Gas_Chromatography(accessed January 25, 2018).

Warfield, R.W. and R.P. Maickel. 1983. A generalized extraction-TLC procedure for identification of drugs. *J. Appl. Toxicol.* 3(1):51–57.

Waters Corporation. 2005. Ultra performance LC™ separation science redefined. www.waters.com/webassets/cms/library/docs/720001136en.pdf (accessed January 25, 2018).

Zhang, L., Z.H. Wang, H. Li, Y. Liu, M. Zhao, Y. Jiang, and W.S. Zhao. 2014. Simultaneous determination of 12 illicit drugs in whole blood and urine by solid phase extraction and UPLC-MS/MS. *J. Chromatogr. B Analyt. Technol. Biomed. Life Sci.* 955–956:10–19.

Zivanovica, L., S. Agatonovic-Kustrina, M. Vasiljevica, and I. Nemcovab. 1996. Comparison of high-performance and thin-layer chromatographic methods for the assay of lidocaine. *J. Pharm. Biomed. Anal.* 14(8–10):1229–1232.

CHAPTER 7

Inorganic poisons and contaminants

KEY WORDS: inorganic, trace elements, photons, ground state, excited, excited state, flame test, continuous spectrum, line spectrum, emission spectrograph, thin-layer chromatography, ultraviolet-visible spectroscopy, infrared spectroscopy, Raman spectroscopy, x-ray fluorescence, photoelectric effect, atomic absorption spectroscopy, inductively coupled plasma-mass spectrometry, x-ray diffraction, diffraction pattern, neutron activation analysis, scanning electron microscopy

LEARNING OBJECTIVES

- To understand the value of identifying trace elements in materials in forensic investigations
- To understand how an emission spectrograph can be used to differentiate elements
- To understand the difference between a continuous spectrum and a line spectrum
- To explain the uses and limitations of ultraviolet-visible, fluorescence, infrared, and Raman spectroscopy in inorganic analysis
- To explain the uses and limitations of x-ray fluorescence, atomic absorption spectroscopy, inductively coupled plasma-mass spectrometry, x-ray diffraction, neutron activation analysis, and scanning electron microscopy in inorganic analysis

HOW COULD AN ELEMENT SO LUMINOUS BE SO HARMFUL? THE STORY OF THE "RADIUM GIRLS"

Radium was discovered in 1898 by scientist Marie Sklodowska Curie in her lab work for her Ph.D. thesis. Curie isolated the radium from pitchblende ore from which uranium had been extracted. Curie was awarded the 1903 Nobel Prize in Physics for her work jointly with her husband, Pierre Curie, who served as her assistant, and Henri Becquerel, for his work on uranium. Radium was found to have numerous military, industrial, and medical uses, including cancer treatment which Curie promoted in her later years. However, radium was also used for its luminescent properties in watch dials. During World War I, women were recruited to work in factories while men were at war. Beginning in 1917 to serve World War I, women were hired by factories in Orange, New Jersey, Ottawa, Illinois, and Waterbury, Connecticut, to paint the numbers on watch and clock dials using paint containing radium salts. The watches, whose radium dials luminesced, or emitted their own light, were sent to the front to be used by the military. To give the paintbrushes the fine tip they needed to paint the numbers clearly, the women placed them between their lips and in their mouths. Pointing the paintbrushes using rags or water was found to take too much time and led to the loss of too much paint. Some women also used the radium paint to paint their fingernails, face, and teeth. In using their lips to give the brushes their fine tips, they knowingly ingested radium which they were told was harmless. At the time, this was not a concern as radium was also an ingredient in women's cosmetics including powder, cream, soap, and toothpaste, among others.

HOW COULD AN ELEMENT SO LUMINOUS BE SO HARMFUL? THE STORY OF THE "RADIUM GIRLS" (continued)

Radium-226 is a fissionable element that gives off alpha rays when it decomposes to radon-222 gas (which is also radioactive and produces high-energy gamma particles). The radium produced a spectral line observed with a grating spectrograph at a wavelength of 381.48 nm that had never before been documented. Alpha rays are not considered as harmful as higher-energy beta or gamma rays as they are heavier and they are stopped from penetrating human skin by clothes, gloves, or even a piece of paper. Alpha rays are harmful, however, if ingested or inhaled. With its half-life of 1600 years, and the fissionable products it produces down chain, radium is extremely toxic. Radium circulates in the bloodstream and deposits itself into bones and makes them brittle and the gamma radiation from the radon kills cells and causes gene mutations that lead to cancer.

The female factory workers, or "radium girls" as they were known, exhibited ulcers and sores, anemia, and bone cancer and were diagnosed with radium poisoning. Five of the women from the New Jersey factory, Grace Fryer, Edna Hussman, Katherine Schaub, Quinta McDonald, and Albina Larice, sued their employer under the state's occupational injuries law. When the case went to court in January 1928, two of the plaintiffs were bedridden and the others could not even raise their arms to take the oath; the case was settled out of court. The Illinois company, Radium Dial Company, was established in 1922. In 1926–1927, women began exhibiting signs of radium poisoning. Five of the women from the Illinois factory also sued their employer; they won their case in 1938. The cases led to several new lab laws and the implementation of new industrial safety precautions including protective gear. Radium paint was used to paint clock and compass dials and airplane and military instruments until the 1960s.

BIBLIOGRAPHY

Blum, D. 2011. *The Poisoner's Handbook: Murder and the Birth of Forensic Medicine in Jazz Age New York*. New York, NY: Penguin Press.

Gunderman, R.B. and R.S. Gonda. 2015. Radium girls. *Radiology* 274:314–318.

NIST Atomic Spectra Database Lines Form. http://physics.nist.gov/PhysRefData/ASD/lines_form.html (accessed January 25, 2018).

Rasmussen, E. 1933. Serien im Funkenspektrum des Radiums. Ra II. *Z. Phys.* 86:24–32.

The Nobel Prize in Physics 1903. www.nobelprize.org/nobel_prizes/physics/laureates/1903/ (accessed January 25, 2018).

Inorganic matter and materials are widely encountered in criminalistics. Inorganics contain metal ions and non-carbon elements. As 75% of the earth's crust contains silicon and oxygen, and the third and fourth most abundant elements are aluminum (7.9%) and iron (4.5%), inorganics are much more common than carbon or carbon-containing molecules as they make up approximately 99% of the atoms on earth. Inorganic evidence includes tools, weapons, ammunition, explosives, poisons, trace materials in paints and dyes, coins, components of soil, glass, and some fibers. Metalloorganics, such as the dye sodium nitroprusside that contains iron bound by cyano and nitro ligands, combine metals in carbon-containing molecules.

For differentiation, criminalists rely on the detection of *trace elements*, or those elements present in concentrations of less than 1% (and typically parts per million or parts per billion) in naturally occurring and manufactured materials. Although the incorporation of trace elements may have been unintended and they may not impact the appearance or usefulness of products, they serve as "invisible markers" and useful points for comparison or attribution. For example, copper, silver, and bismuth are trace elements that are found in bullet lead at 5–15 ppm concentrations and antimony is a hardening agent found at 20–1200 ppm in bullet lead. The corrosion of metal pipes can lead to the detection of lead, copper, zinc, aluminum, and iron in water supplies. Trace metals including barium, cadmium, chromium, nickel, and silver are also often found in groundwater and tap water and are detected in environmental forensic analysis.

The interaction of inorganic elements or compounds with heat and light is used to identify and differentiate inorganic materials. Energy from light (electromagnetic radiation) is quantized into packets called *photons*. Atoms contain electrons in discrete energy levels. Electrons in their lowest energy shells and orbitals are said to be in the *ground state*. Electrons can absorb photons and be *excited*, or promoted, to a higher energy level shell (further from the nucleus) consistent with the energy of the photon. An electron in a higher energy level due to excitation is said to be in its *excited state*. If an excited electron descends to its ground state lower-energy level, it will emit a photon of light equal to the energy difference between the two orbitals. Different energy photons correspond to different colors with higher-energy photons corresponding to violet and lower-energy photons corresponding to red. On exposure to high-energy photons, electrons can be removed from atoms in a process termed *ionization*.

Several methods are appropriate for inorganic analysis and will be the focus of this chapter. These include the flame test, emission spectrograph, thin-layer chromatography (TLC), ultraviolet-visible (UV-Vis) spectroscopy, infrared (IR) spectroscopy, Raman spectroscopy, x-ray fluorescence spectroscopy (XRF), atomic absorption (AA), x-ray diffraction and x-ray crystallography, inductively coupled plasma-mass spectrometry (ICP-MS), neutron activation analysis (NAA), and scanning electron microscopy (SEM). As several of these have been covered in previous chapters, they will be only briefly covered here.

FLAME TEST

A simple and inexpensive test for inorganics and an early one used in toxicology is the *flame test*. The analyte metal salt is heated to a high temperature and the electrons in its atoms are boosted to an excited state from the heat energy. When the electrons relax to their ground state, light is emitted as shown in Figure 7.1. In the flame test, a small amount of the evidence is placed in a flame and the color (from the specific wavelengths of light energy) emitted is noted. The light spectrum produced is a *continuous spectrum* in which the colors blend or merge together. Sunlight is an example of a continuous spectrum. Each salt will produce a unique color (Table 7.1), although mixtures' colors will be additive. For example, strontium sulfate emits red light, potassium chlorate emits lilac light, barium chlorate emits green light, and a mixture would present a hue of these three colors. Copper (I) chloride and sodium chloride emit blue and yellow-orange light respectively. A drawback to the flame test as an investigative tool is that it is a destructive technique and the color is a composite that is produced by all of the metal ions.

EMISSION SPECTROGRAPH

An *emission spectrograph* extends the flame test to produce a line spectrum of the analyte. Again, heat is used to achieve an excited state and light is emitted following the electrons' return to the ground state (Figure 7.1). The analyte is placed between two carbon or graphite electrodes. A direct current is used to vaporize the analyte and a tungsten incandescent bulb or neon light is used to excite the atoms. The light is passed through a focusing lens. A prism is used to separate the light into its component frequencies to produce a light spectrum that is recorded on a photographic plate. Each element can be identified by its characteristic line frequencies so the emission spectrum is considered a fingerprint of the element. A spectroscope (Figure 7.2) can also be used to view the line frequencies of elements. The specific frequency of light absorbed or emitted can be determined by Equation 7.1 where E is the energy difference between two orbitals, h is Planck's constant (6.63×10^{-34} Js), and υ is frequency.

$$E = h\upsilon \tag{7.1}$$

In 1885, J. Balmer, a Swiss schoolteacher, derived the mathematical formula (Equation 7.2) used to calculate the wavelengths of the colors of the emission lines that are now equated with the quantized energy levels of nuclei orbitals of atoms. The energy emitted is the difference in the energy states of the higher and lower orbital. Evidence samples can be compared to standards, a chart, or web tool by matching the colored lines and their calculated frequencies. At the turn of the twentieth century, Dr. Alexander Gettler used emission spectroscopy to identify poisons including

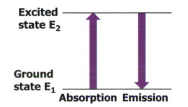

Figure 7.1   Principle of emission spectroscopy.

Table 7.1 Flame colors of metals

Metal	Symbol	Compound	Flame color
Barium	Ba	$BaCl_2$	Green
Boron	B	H_3BO_3	Green
Calcium	Ca	$CaCl_2$	Red-Orange
Copper	Cu	$CuCl_2$	Green
Lithium	Li	LiCl	Red
Potassium	K	KCl	Violet
Sodium	Na	NaCl	Yellow
Strontium	Sr	$SrCl_2$	Red-Orange
Zinc	Zn	$ZnCl_2$	White-Green

arsenic, lead, chromium, and thallium in his New York City forensic lab. Emission spectroscopy can also be used to identify trace elements in glass, metals, and soil. The method is reliable, rapid, and easy to perform but its drawbacks include the interpretation of many emission lines (Figure 7.3) such as those recorded for neon, argon, and nitrogen.

$$v = C\left(\frac{1}{2^2} - \frac{1}{n^2}\right) \quad \text{where } n = 3,4,5,\text{etc...}$$

$$C = \text{constant}$$

(7.2)

THIN-LAYER CHROMATOGRAPHY

TLC can be used to separate and differentiate materials containing inorganic pigments. As an example, lipstick contains inorganic pigments and organic dyes as colorants in a wax and oil matrix. The wax is soluble in toluene so this can be used as the mobile phase to run the plate. In a recent study by Joshi et al. (2013), lipsticks were separated into up to six spots, depending on brand and composition, using a toluene/benzene/diethyl ether (4:12:2) mobile phase. More details on running and analyzing TLC plates are included in Chapter 6.

UV-VIS SPECTROSCOPY

Examples of uses of *UV-Vis spectroscopy* in inorganic chemistry include the detection and quantitation of metals or poisons in blood, urine, or water samples and absorption features of colorants in paints, inks, and pigments. In

Figure 7.2 Student spectroscope.

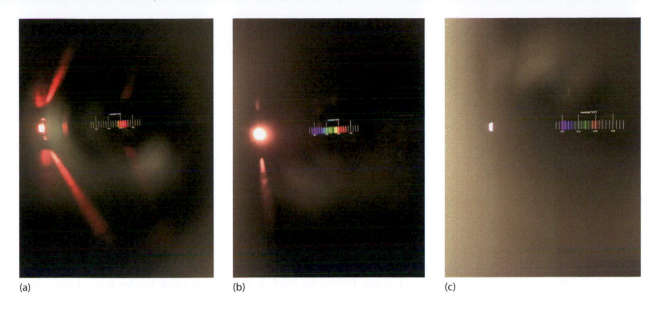

(a)   (b)   (c)

Figure 7.3   Line spectra of elements (a) neon, (b) argon, and (c) nitrogen.

addition to the π-bonded electrons that are largely responsible for the absorbance of UV light by small organic molecules, d and f electrons found in inorganics and organometallics contribute to the UV-Vis absorption in those molecules. UV-Vis spectroscopy can be employed to determine the presence and quantity of iron in toxicology samples such as hemoglobin in blood or urine or environmental samples such as iron in drinking water. The EPA standards for drinking water allow for no more than 0.3 mg/L of iron. Based on data using salicylic acid and iron standards prepared in the lab, the quantity of iron of unknown solutions can be tested by the formation of a salicylic acid-iron complex. Salicylic acid, iron, and the salicylic acid-iron complex all absorb light in the UV-Vis range. UV-Vis spectrophotometers have a range largely in the visible portion of the spectrum, from 300 nm (or 325 nm) to 800 nm (or 1100 nm), depending on the model. Iron absorbs strongly from 325 to 600 nm and has maxima at approximately 325 and 590 nm. Salicylic acid also absorbs strongly from 325 to 600 nm and has maxima at approximately 325 nm (300/310/313 nm) and 1100 nm. Salicylates react with ferric salts (Fe(III)) to produce a violet color, which is proportional to the concentration of the iron in the solution (Figure 7.4). For the toxicological detection of iron (or conversely, salicylates), a wavelength of 540 nm is used to measure the maximum absorption intensity of the complex. Because of the simplicity of this salicylate procedure, the assumption of iron poisoning may be verified or disputed within minutes, especially as severe poisoning produces a very strong and easily seen violet color with the color reagent. By adding varying amounts of iron to salicylate and observing the purple colors using a UV-Vis spectrophotometer for several known standard solutions, this data can be used to prepare a Beer's Law linear calibration plot. The calibration plot can be used to analyze unknowns including iron in water samples. Iron is soluble in an acidic solution. Nitric acid (HNO_3) is used to acidify the solution as it does not absorb in the UV-Vis region. The basis of UV-Vis spectroscopy is covered in detail in Chapter 4.

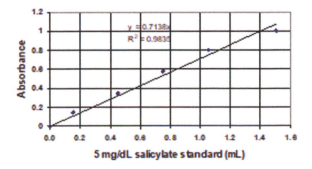

Figure 7.4   Quantitation of iron using UV-Vis spectroscopy and Beer's Law on complexation with salicylic acid.

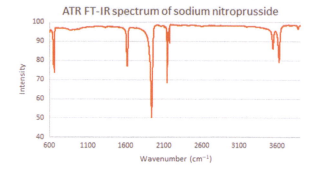

Figure 7.5   ATR FT-IR spectrum of the inorganic pigment sodium nitroprusside.

IR SPECTROSCOPY

IR spectroscopy can be used to differentiate and identify functional groups and bonding in inorganic pigments in paints, inks, fibers, poisons, pollutants, and drugs. IR can be used to distinguish paint binder formulations and inorganic explosives by the stretches observed in the spectra. Characteristic stretches for the inorganic pigment sodium nitroprusside ($Na_2[Fe(CN)_5NO]$) using attenuated total reflectance Fourier transform-infrared (ATR FT-IR) spectroscopy are shown in Figure 7.5. The 2140 cm^{-1} peak is attributed to the nitrile-iron metal ion stretch. IR spectroscopy is covered in detail in Chapter 4.

RAMAN SPECTROSCOPY

Raman spectroscopy can be used to differentiate and identify inorganic materials. For example, Raman spectroscopy can be used to detect metal binding by the chelating agent ethylene diamine tetra acetic acid (EDTA) or the biological molecule and citric acid cycle intermediate oxalic acid. Raman spectroscopy has been used to characterize metal-oxygen bonds to aluminum, gallium, and indium in inorganic materials and DNA bonds with divalent metal cations barium, cadmium, calcium, cobalt, copper, magnesium, manganese, nickel, palladium, and strontium (Duguid et al., 1995).

In forensics, this instrument has also been shown to differentiate fibers, inks, pigments, paints, art objects, and drugs from samples as small as a micron and can be used for solid, liquid, and gaseous materials without direct contact. A Raman spectrum for the inorganic pigment sodium nitroprusside is shown in Figure 7.6. Raman spectroscopy is ideal for the analysis of art objects such as paintings and porcelain calling cards as the objects can be analyzed and rotated on the instrument base without removing any paint or disturbing the object in any way. Porcelain calling cards were status symbols of the bourgeoisie and were used like calling cards, business cards, or party invitations. Raman spectroscopy has been used to determine the colorants that were used in these cards including Prussian blue ($Fe_4[Fe(CN)_6]_3$), ultramarine ($Na_{8..10}Al_6Si_6O_{24}S_{2..4}$), red vermillion (HgS), chrome yellow ($PbCrO_4$), and white lead paint (2 $PbCO_3$ · $Pb(OH)_2$). For example, Prussian blue exhibits a characteristic stretch at 2150 cm^{-1} due to the cyano moiety. Sodium nitroprusside exhibits a characteristic stretch at 408 cm^{-1} due to the C≡N-Fe^{3+} and N≡O-Fe^{3+}

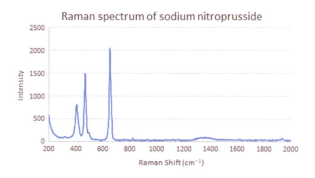

Figure 7.6   Raman spectrum of the inorganic pigment sodium nitroprusside.

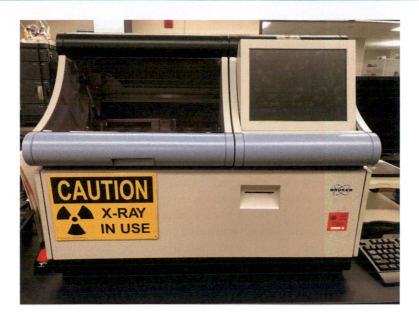

Figure 7.7 XRF instrument.

interactions. Some fibers, including aramid, cotton, and polypropylene, are Raman active. These fibers exhibit a shift in Raman frequency with applied stress.

Raman spectroscopy was used to analyze the authenticity of the Vineland Map, which purported to predate Columbus' voyage in age by 50 years. On analysis, the map was found to have carbon ink overlaid with a yellow line determined to be anatase (Brown and Clark, 2002). As anatase has only been found in post-1923 iron gallotannate/carbon inks and never on medieval artifacts, it indicated that the document was hundreds of years newer than claimed. In addition, the black ink had bled at the border due to iron leeching. The ink did not match medieval-type iron-based reference inks including chromite ($FeCr_2O_4$) and limenite ($FeTiO_3$). Additionally, a map that is a few hundred years old would be expected to exhibit parchment loss and this map had none. More detail about Raman spectroscopy is included in Chapter 4.

X-RAY FLUORESCENCE

X-rays were discovered by Wilhelm Conrad Roentgen in 1895. X-rays are short-wave radiation of much higher energy than visible light. *XRF* is another technique that can be used to fingerprint elements by detecting the fluorescence emitted by atoms on their absorption of x-ray radiation. X-ray radiation is absorbed and transferred to an electron in an inner orbital or shell, a phenomenon known as the *photoelectric effect*, causing the electron to be ejected from the inner shell and a vacancy to exist. This resulting vacancy causes the atom to be unstable. As a result, an electron from an outer shell is transferred to the inner shell orbital, giving off a characteristic fluorescence of x-ray energy corresponding to the difference between the binding energies of the corresponding shells. The emission of characteristic x-rays is called XRF. Spectra reflect multiple such events and contain multiple peaks of different intensities depending on the energy differences between the shells. As each element has a unique set of filled energy levels, the x-ray fluorescence is unique and the set of energies provides a fingerprint for that element.

Sometimes, as the electron in the analyte atom relaxes to its ground state, the energy is transferred to an outer electron, instead of being emitted, causing it to be ejected from the atom. The ejected atom is termed an "Auger" electron and the process is a competing one to XRF.

XRF is a rapid and nondestructive technique that is sensitive to parts per million. Samples that can be analyzed include solids, liquids, powders, thin films, and thick films. It can be used to analyze all elements and their isotopes from boron to uranium and detect both their presence and relative abundance in a sample. As a result, this method finds uses in analyzing gunshot residue (e.g., Ba, Sb, Pb, Sr), paint chips, explosives, soil, glass, rocks, and even body fluids including blood and semen in which metal poisons may be present. The Mars Exploration Rover Spirit was equipped with an XRF instrument to evaluate the geologic history, elemental composition, and abundance on Mars. An XRF instrument is shown in Figure 7.7 and an XRF spectrum is shown in Figure 7.8.

ATOMIC ABSORPTION

AA is an atomic spectroscopy technique that can determine the identity and concentration of inorganic elements by their absorption of light. Liquid samples have to be converted to the gas phase. In AA, the analyte is heated enough to vaporize the atoms but also leave a substantial number of the atoms in the ground state. Light from a hollow-cathode lamp discharge tube source for a specific element is directed at the vaporized atoms. If that element is present, it will absorb the radiation in the ultraviolet and visible range (selective absorption), boosting electrons to their excited states. As the electrons return to their ground states, energy will be released in the form of light emission. The emission wavelengths are determined by the monochromator detector and amplified by the photomultiplier. The concentration of the analyte is directly proportional to the quantity of the light absorbed. The concentration is determined from a calibration curve (absorption versus concentration) produced by running standards under the same conditions and with the same operating procedure as the analyte. An AA spectrometer is shown in Figure 7.9. AA can be used to detect metals in drinking water; rice; infant cereal and formula; gunshot residue; and lead in paint, hair, toys, or shoes.

AA was first demonstrated in 1955 using an air-acetylene flame. Newer AAs employ a heated graphite furnace or tantalum metal strip for the flame as these lead to a more efficient volatilization and sensitivity to parts per trillion. Graphite furnace AA has been used to determine the quantity of lead in milk for infants. For safety, a fume hood is placed above the flame. AA can be employed to probe 60 elements, and is more sensitive than emission spectroscopy, simple to operate, and fast to process several samples. Despite its sensitivity, there are a few disadvantages to the technique. Elements must be probed specifically and individually with separate light sources. Standards must be prepared and processed separately with the appropriate light sources. Using the known concentrations of the standards and the experimental determination of the metal concentrations, calibration curves are constructed to quantitate unknown samples. Thus, if a criminalist wants to determine the concentration of three possible inorganic contaminants in a drinking water sample, she or he will need to probe them sequentially. Additionally, the technique is destructive.

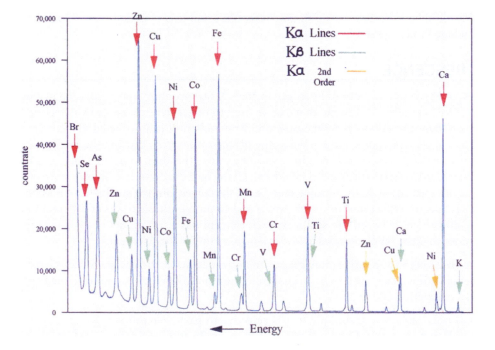

Figure 7.8 XRF spectrum. (From LinguisticDemographer at English Wikipedia, https://commons.wikimedia.org/w/index.php?curid=6229332.)

Figure 7.9 Atomic absorption spectrometer.

DETECTING AUTHENTIC AND COUNTERFEIT FOOD PRODUCTS

Cumin is a seed that is ground to a powder. Cumin is used as a spice in traditional Indian cooking to flavor curry powder and is used in the garam masala spice mixture. It is also used in Mexican food and to flavor some cheeses. Food safety standards require accurate labeling and consumer information for food products, including spices.

Counterfeit food products are a billion-dollar problem and the penalties if convicted are trivial—a maximum of 1 to 2 years imprisonment—compared with drug trafficking. Herbs and spices are particularly easy targets; counterfeiters add industrial dyes to cheaper products to brighten and alter the colors to make them more attractive. But the dyes are often toxic. Thus, methods are being developed to authenticate food origins. In a recent study, 24 cumin samples from China, India, Syria, and Turkey were differentiated by a Towson University research team led by Dr. Ellen Hondrogiannis using laser ablation-inductively coupled plasma-time of flight-mass spectrometry (LA-ICP-TOF-MS) using five elements: calcium, magnesium, manganese, sodium, and zinc. LA-ICP-TOF-MS was used to collect the elemental concentration of 11 elements in the samples. The samples were differentiated by a statistical model using discriminant functional analysis computed in SPSS. The model was composed of three discriminant functions and was able to accurately determine the origins of each of the cumin samples. The model could be used as a method to differentiate food samples of unknown origin and other food products.

BIBLIOGRAPHY

Geographical origin determination of cumin using laser ablation-inductively coupled plasma-time of flight-mass spectrometry (LA-ICP-TOF-MS) and discriminant analysis. www.towson.edu/fcsm/departments/chemistry/facultystaff/documents/hondrogiannis-poster-1.pdf (accessed January 25, 2018).

The big cash in counterfeit food: Why you might not be eating what you think you're eating. www.cbc.ca/radio/day6/episode-279-playing-ball-on-grass-vs-turf-taytweets-big-fail-narco-subs-fake-food-and-more-1.3514966/the-big-cash-in-counterfeit-food-why-you-might-not-be-eating-what-you-think-you-re-eating-1.3515053 (accessed January 25, 2018).

INDUCTIVELY COUPLED PLASMA-MASS SPECTROMETRY

In inductively coupled plasma (ICP) emission spectroscopy, a hot plasma torch is used to produce the emission spectra instead of the carbon arc used in the emission spectrograph. An ICP-MS instrument is shown in Figure 7.10. The plasma torch consists of three concentric quartz tubes wrapped with a radio-frequency coil (radio-frequency

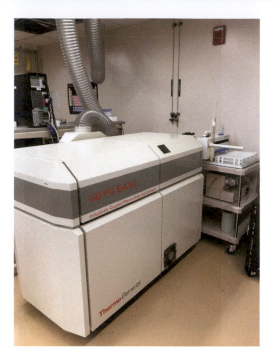

Figure 7.10   ICP-MS instrument.

generator). Argon flows through the tubes and the coil carries a current creating an intense magnetic field. When a spark is generated, electrons are removed from the argon. The charged atoms collide with other argon atoms, creating more argon ions and electrons that form an inductively coupled plasma discharge which, at 7,000–10,000°C, acts as a flame. The aerosolized sample is introduced to this flame of argon ions. Charged particles are generated. These charged particles emit light of characteristic wavelengths corresponding to the identity of the elements present. The ICP can be coupled to an MS. ICP has been applied to the identification and characterization of mutilated bullets and glass fragments recovered from a crime scene.

A deformed .22-caliber bullet was recovered from the victim's head. An Eley LR .22 caliber pistol and a box of bullets was seized from the suspect. These were compared to nine brands (e.g., Remington Target, Remington Yellow Jacket, Remington large velocity, Dynamit Nobel RWS, CCI Green Tag, Mauser KK80, Fiocchi rifle V320, and Winchester pistol 0.22) of .22 caliber bullets from the lab collection including ammunition for a .22 caliber Eley LR pistol. The $^{203}Pb/^{206}Pb$ isotope ratio was plotted against the $^{207}Pb/^{206}Pb$ isotope ratio. The isotope ratios of the bullet from the victim's head clustered tightly with the Remington ammunition whereas the Eley pistol ammunition clustered in an entirely different region of the graph (Ulrich et al., 2004).

The advantages of ICP-MS are that it can be used to analyze for, and differentiate between, all elements, and that it is a fast and sensitive technique that requires no sample preparation. The analysis results in an elemental profile from which compositions can be compared.

Of the many metals that may be present in the object, one can be assigned as an internal standard and elemental ratios of the other metals can be determined. Even though there is no database for the comparisons, the New Jersey Supreme Court ruled in 1999 that these comparisons are meaningful. However, disadvantages include that the determination of some elements is made difficult by a common matrix and some molecular species (e.g., doubly charged or molecular ionic species) that can interfere with their determination and quantification.

X-RAY CRYSTALLOGRAPHY AND X-RAY DIFFRACTION

X-ray diffraction analysis employs an x-ray beam that penetrates mounted solid, crystalline materials or powders in its path. As the atoms in crystals are composed of a series of parallel planes, the x-ray beam is reflected by the atomic planes. As the x-rays are reflected from the crystal's atomic planes, they combine to form a pattern of light and dark bands called a *diffraction pattern*. The diffraction pattern is recorded on a photographic film or digitized to a computer for analysis. The intensity of the bands will vary. The intensities are proportional to the square of the structure-factor amplitude. Each crystalline substance exhibits a unique diffraction pattern. X-ray diffraction can be used to understand the relationships between the atoms in a molecule and the repeating units, or symmetry, of the molecules in the crystal unit cell and thus enable the determination of structures of molecules. When it is used to determine molecule structure, it is referred to as *x-ray crystallography*. Bragg's Law is solved to determine the angles of beams reflected off the atomic planes.

In forensic science, it can be used to differentiate and identify components of explosives including potassium chlorate and potassium nitrate salts, but also various drugs such as naproxen sodium. It can also be used in "fingerprinting" inorganic materials as all compounds will produce unique diffraction patterns. The drawback to this technique is that it is not very sensitive to trace materials in a sample; it cannot detect components of a mixture present in proportions of less than 5%.

NEUTRON ACTIVATION ANALYSIS

NAA is a form of atomic spectroscopy. In NAA, a nuclear reactor is used to probe the nucleus of the atoms. The method is highly sensitive and can simultaneously identify and quantitate 20–30 trace elements including those contained in hair, paint, soil, or gunshot residue. In the nuclear reactor, neutrons are produced and used to bombard the analyte. The analyte will capture some of the neutrons and form *isotopes*, atoms with the same number of protons but a different number of neutrons, some of which may be radioactive. Radioactive isotopes emit radiation including beta particles and gamma rays. The isotopes can be detected and quantified by the energy of the radiation they emit. From the gamma radiation emitted, the concentration of the analyte and element identity can be evaluated by the intensity of the radiation and wavelength. The technique is nondestructive with sensitivity to the picogram or part-per-trillion range. Another advantage is that it can be used to simultaneously analyze many elements to provide the elemental composition of the sample, and it can be used with any physical state (solid, liquid, or gas) or chemical form (compound, ion) of samples. Unfortunately, access to this technique is limited as it requires the nuclear reactor, but it has been used in high-profile cases such as the 2004 analysis of the bullets used to assassinate President John F. Kennedy, and to solve mysteries in the death of Napoleon Bonaparte whose death can be at least partially attributed to arsenic poisoning detected by sampling his hair (Weider and Fournier, 1999). Thus, NAA is especially useful in cases where there is an extremely limited sample and a nondestructive method is required.

SCANNING ELECTRON MICROSCOPY

SEM is used to identify inorganic constituents of gunshot primer residue encountered in firearms analysis (Chapter 13) including metals such as barium, antimony, and lead by their characteristic shapes and sizes. SEM is covered in greater detail in Chapter 3.

QUESTIONS

1. Inorganics can be detected by _____.
   - a. AA spectrometry
   - b. SEM
   - c. NAA
   - d. ICP-MS
   - e. All of the above
2. Which of the following methods can detect only one metal at a time?
   - a. AA spectrometry
   - b. SEM
   - c. NAA
   - d. ICP-MS
   - e. All of the above
3. Which of the following is a destructive method useful for determining the presence of lead in drinking water and its concentration?
   - a. AA
   - b. HPLC
   - c. XRF
   - d. GC-MS
   - e. ICP-MS

4. Which of the following methods is a nondestructive technique that employs a laser that beams neutrons at a sample and records emitted gamma radiation?

   a. NAA   b. XRF

   c. NMR   d. X-ray diffraction   e. All of the above

5. X-ray diffraction patterns are obtained from ____ materials.

   a. Crystalline   b. Amorphous

6. Name an instrument that can be used to identify isolated poisons by their spectral lines.

7. Name an instrument that can be used to detect and quantitate several metals in a single analysis based on emitted gamma rays.

8. Name an instrument that can be used to detect and identify gunshot residue based on its shape and morphology.

9. Name an instrument that can be used to compare and differentiate metalloorganic ink pigments used in art objects.

10. Name a method that is applied to solid, crystalline materials where the pattern fingerprint is compared for inorganic compounds such as potassium chlorate and potassium nitrate.

References

Brown, K.L. and R.J. Clark. 2002. Analysis of pigmentary materials on the Vinland Map and tartar relation by Raman microprobe spectroscopy. *Anal. Chem.* 74(15):3658–3661. www.webexhibits.org/vinland/paper-clark02.html (accessed January 25, 2018).

Duguid, J.G., V.A. Bloomfield, J.M Benevides, and G.J. Thomas, Jr. 1995. Raman spectroscopy of DNA-metal complexes. II. The thermal denaturation of DNA in the presence of Sr^{2+}, Ba^{2+}, Mg^{2+}, Ca^{2+}, Mn^{2+}, Co^{2+}, Ni^{2+}, and Cd^{2+}. *Biophys. J.* 69(6):2623–2641.

Ulrich, A., C. Moor, H. Vonmont, H.-R. Jordi, and M. Lory. 2004. ICP–MS trace-element analysis as a forensic tool. *Anal. Bioanal. Chem.* 378(4):1059–1068.

Weider, B. and J.H. Fournier. 1999. Activation analyses of authenticated hairs of Napoleon Bonaparte confirm arsenic -poisoning. *Am. J. Forensic Med. Pathol.* 20(4):378–382.

Bibliography

Abel, J.E. and P.J. Kemmey. 1975. *Identification of Explosives by X-ray Diffraction: Technical Report 4766.* Dover, NJ: Picatinny Arsenal. www.dtic.mil/dtic/tr/fulltext/u2/a013378.pdf (accessed January 25, 2018).

Bartelink, E.J., S.B. Sholts, C.F. Milligan, T.L. VanDeest, and S.K. Wärmländer. 2015. A case of contested cremains analyzed through metric and chemical comparison. *J. Forensic Sci.* 60(4):1068–1073.

Berendes, A., D. Neimke, R. Schumacher, and M. Barth. 2006. A versatile technique for the investigation of gunshot residue patterns on fabrics and other surfaces: m-XRF. *J. Forensic Sci.* 51(5):1085–1090.

Blake, C. and B. Bourqui. 1998. Determination of lead and cadmium in food products by graphite furnace atomic absorption spectroscopy. *At. Spectrosc.* 19(6):207–213. http://citeseerx.ist.psu.edu/viewdoc/download?doi=10.1.1.463.543&rep=rep1&type=pdf (accessed January 25, 2018).

Blum, D. 2011. *The Poisoner's Handbook: Murder and the Birth of Forensic Medicine in Jazz Age New York.* New York, NY: Penguin Press.

Charles, S., B. Nys, and N. Geusens. 2011. Primer composition and memory effect of weapons—Some trends from a systematic approach in casework. *Forensic Sci. Int.* 212 (1–3): 22–26.

Crespy, C., P. Duvauchelle, V. Kaftandjian, F. Soulez, and P. Ponard. 2010. Energy dispersive x-ray diffraction to identify explosive substances: Spectra analysis procedure optimization. *Nucl. Instrum. Methods Phys. Res. Sec. A: Accelerators, Spectrometers, Detectors and Associated Equipment* 623(3):1050–1060.

Dias, D., J. Bessa, S. Guimarães, M.E. Soares, M. de L. Bastos, and H.M. Teixeira. 2016. Inorganic mercury intoxication: A case report. *Forensic Sci. Int.* 259:e20–e4.

Environmental Protection Agency (EPA). 2017. National primary drinking water regulations. www.epa.gov/ground-water-and-drinking-water/national-primary-drinking-water-regulations#Inorganic (accessed January 25, 2018).

Girard, J.E. 2007. *Criminalistics: Forensic Science and Crime*, 1st edn. Sudbury, MA: Jones & Bartlett Learning.

Goullé, J.P., E. Saussereau, L. Mahieu, and M. Guerbet. 2014. Current role of ICP-MS in clinical toxicology and forensic toxicology: A metallic profile. *Bioanalysis* 6(17):2245–2259.

Guinn, V.P. 1979. JFK assassination: Bullet analyses. *Anal. Chem.* 51(4):484–493.

Hua, L., M. Nishida, A. Fujiwara, M. Yashiki, M. Nagao, and A. Namera. 2009. Preliminary screening method for the determination of inorganic arsenic in urine. *Leg. Med. (Tokyo)* 11(2):80–82.

Joshi, B., K. Verma, and J. Singh. 2013. A comparison of red pigments in different lipsticks using thin layer chromatography (TLC). *J. Anal. Bioanal. Tech.* 4(1). doi:10.4172/2155-9872.1000157.

Kuras, M.J. and M.J. Wachowicz. 2011. Cannabis profiling based on its elemental composition—Is it possible? *J. Forensic Sci.* 56(5):1250–1255.

L'vov, B.V. 2005. Fifty years of atomic absorption spectrometry. *J. Anal. Chem.* 60(4):382–392.

National Institute of Standards and Technology (NIST). NIST Atomic Spectra Database Lines Form. http://physics.nist.gov/PhysRefData/ASD/lines_form.html (accessed January 25, 2018).

Nobel Media AB. 2014. The Nobel Prize in Physics 1903. www.nobelprize.org/nobel_prizes/physics/laureates/1903/ (accessed January 25, 2018).

O'Neill, E., D. Harrington, and J. Allison. 2009. Interpretation of laser desorption mass spectra of unexpected inorganic species found in a cosmetic sample of forensic interest: Fingernail polish. *Anal. Bioanal. Chem.* 394(8):2029–2038.

Orfila, M. 1832. Forensic chemistry. On the detection of mixed poisons. *Lancet.* 18(468):613–615.

Pollock, J. 1999. Supreme Court of New Jersey. STATE of New Jersey, Plaintiff-Appellant, v.JudelNOEL, Defendant-Respondent. Decided: February 10, 1999.http://caselaw.findlaw.com/nj-supreme-court/1340159.html (accessed January 25, 2018).

Rasmussen, E. 1933. Serienim Funkenspektrumdes Radiums. Ra II. *Zeitschrift für Physik* 86(1–2):24–32.

Saferstein, R. 2015. *Criminalistics: An Introduction to Forensic Science*, 11th edn. Boston, MA: Pearson.

Spectroscopy. 2011. X-ray analysis goes to mars. *Spectroscopy* 26(7). www.spectroscopyonline.com/x-ray-analysis-goes-mars (accessed January 25, 2018).

Taudte, R.V., A. Beavis, L. Blanes, N. Cole, P. Doble, and C. Roux. 2014. Detection of gunshot residues using mass spectrometry. *BioMed Res. Int.* http://dx.doi.org/10.1155/2014/965403.

University of Rhode Island. 2004. Neutron activation analyses proves Oswald acted alone in JFK assassination. *Sci. Daily.* www.sciencedaily.com/releases/2004/10/041025131255.htm (accessed January 25, 2018).

Vandenabeele, P. and L. Moens. 2000. The application of Raman spectroscopy for the non-destructive analysis of art objects. *15th World Conference on Nondestructive Testing Roma (Italy).* www.ndt.net/article/wcndt00/papers/idn163/idn163.htm (accessed January 25, 2018).

von Aderkas, E.L., M.M. Barsan, D.F. Gilson, and I.S. Butler. 2010. Application of photoacoustic infrared spectroscopy in the forensic analysis of artists' inorganic pigments. *Spectrochim. Acta A Mol. Biomol. Spectrosc.* 77(5):954–959.

Vuki, M., K.K. Shiu, M. Galik, A.M. O'Mahony, and J. Wang. 2012. Simultaneous electrochemical measurement of metal and organic propellant constituents of gunshot residues. *Analyst* 137(14):3265–3270.

Waters, H. 2011. The first x-ray, 1895. *The Scientist.* www.the-scientist.com/?articles.view/articleNo/30693/title/The-First-X-ray--1895/ (accessed January 25, 2018).

Zeichner, A., S. Ehrlich, E. Shoshani, and L. Halicz. 2006. Application of lead isotope analysis in shooting incident investigations. *Forensic Sci. Int.* 158(1):52–64.

CHAPTER 8

Controlled substances

KEY WORDS: controlled substances, Controlled Substances Act, stimulants, amphetamine, methamphetamine, cathinone, BZP, cocaine, caffeine, depressants, antianxiety drugs, benzodiazepines, gamma hydroxybutyrate, flunitrazepam, barbiturate, hallucinogens, marijuana, THC, LSD, mescaline, phencyclidine, psilocybin, MDMA, DMT, ibogaine, ketamine, *Salvia divinorum*, narcotics, morphine, codeine, heroin, anabolic androgenic steroids, performance-enhancing drugs, new psychoactive substances, synthetic cannabinoids, kratom

LEARNING OBJECTIVES

- To understand the role of the Controlled Substances Act and the meaning of the five schedules
- To describe the differences between classes of drugs including stimulants, depressants, hallucinogens, club drugs, and performance-enhancing drugs and be able to give examples of each class
- To understand how drugs function biochemically in the human body
- To identify clandestine crime labs by the chemicals and drug paraphernalia seized at the scene
- To understand the term *new psychoactive substances* and be able to give examples of drugs in this class
- To understand how drugs can be differentiated and identified by their chemical and spectroscopic properties

PRINCE: CAUSE OF DEATH

Music and performance artist superstar Prince Rogers Nelson, known to the world as "Prince," died on April 21, 2016; he was found unresponsive in an elevator in his home in Chanhassen, Minnesota. An investigation was conducted by the Midwest Medical Examiner's office in Ramsey, Minnesota. The report was released on June 2, 2016, via Twitter. The cause of death based on the toxicology tests was toxicity due to fentanyl.

Fentanyl, a synthetic opioid, is a Drug Enforcement Agency Controlled Substance Schedule II substance. Fentanyl is a prescription medication used in surgery and cancer treatment. According to the Drug Enforcement Agency (DEA), fentanyl is 25–50 times more potent than heroin and 50–100 times more potent than morphine. According to the Centers for Disease Control, 91 Americans die from an opioid overdose—including prescription opiates, such as morphine, oxycodone, hydrocodone, and methadone, and illegal opiates, such as heroin—every day. Opioid deaths account for over 60% of drug overdose deaths. The number of deaths due to opioids from 2000 to 2015 has quadrupled with over half a million people dying from drug overdoses. This is in direct proportion to the nearly quadruple number of prescription opioids sold in the United States since 1999. Most of the fentanyl overdose and death cases are due to illegally made fentanyl, much of it produced in China. In illegal markets, fentanyl is cheaper than heroin and is often mixed with heroin. Taking a dose of a drug expected to be heroin but laced with fentanyl unknowingly causes users to overdose.

In the report by Dr. A. Quinn Strobl, Chief Medical Examiner, Prince is indicated to have self-administered the drug and Strobl ruled his death an accident. Prince did not have any active prescriptions for opiates. Prince is known to have been addicted to Percocet, a pain medication that contains oxycodone and acetaminophen. His plane made

PRINCE: CAUSE OF DEATH (continued)

an emergency landing in Illinois 6 days before his death so that he could be treated at a hospital for a potential overdose of pain medication. A day before his death, his team sought help for him from an eminent opioid specialist with the goal of evaluating his health and encouraging him to seek treatment for pain management and potential addiction. A criminal investigation led investigators to uncover pills in unlabeled bottles throughout his residence. His body guard and friend Kirk Johnson arranged for a doctor, Dr. Michael Schulenberg, to prescribe medications including clonidine, hydroxyzine pamoate, diazepam, and oxycodone under his name to protect Prince's privacy.

The DEA and federal prosecutors investigated the sources of the prescription medication and how Prince obtained them and no one was criminally charged. As of March 1, 2017, China has banned the manufacture and sale of four derivatives of fentanyl including carfentanil, valeryl fentanyl, acryl fentanyl, and furanyl fentanyl—the DEA has called this a "potential game-changer." Carfentanil, a legal tranquilizer used by zookeepers for elephants and other large animals, is 100 times more potent than fentanyl.

BIBLIOGRAPHY

China's fentanyl ban a "game-changer" for opioid epidemic, DEA officials say. www.cnn.com/2017/02/16/health/fentanyl-china-ban-opioids/ (accessed January 29, 2018).

Coscarelli, J. and S. M. Eldred. 2018. Prince's overdose death results in no criminal charges. *The New York Times*, April 19, 2018, https://www.nytimes.com/2018/04/19/arts/music/prince-death-investigation.html (accessed July 10, 2018).

Drug overdose deaths in the United States continue to increase in 2016. www.cdc.gov/drugoverdose/epidemic/ (accessed January 29, 2018).

Investigation into Prince's death reveals pills were hidden throughout Paisley Park. www.npr.org/sections/therecord/2017/04/17/524398523/investigation-into-prince-s-death-reveals-pills-were-hidden-throughout-paisley-p (accessed January 29, 2018).

Prince died of accidental overdose of opioid fentanyl, medical examiner says. www.cnn.com/2016/06/02/health/prince-death-opioid-overdose/ (accessed January 29, 2018).

The controlled substances most commonly encountered by crime laboratories are cocaine, heroin (and now fentanyl), marijuana, and new psychoactive substances (including synthetic cannabinoids and synthetic cathinones). Worldwide, controlled substances vary due to differences in state, regional, and federal laws. With the popularity and abuse of controlled and "legal" substances increasing, forensic scientists' ability to differentiate and characterize seized drug material is crucial (Figure 8.1). Forensic scientists may be called to identify prescription pills, illicit powders, and plant material packaged in blocks, baggies, and vials, among others, as well as suspected starting materials or chemical intermediates seized from a suspected clandestine drug lab.

CONTROL AND SCHEDULING

In the United States, the DEA controls access to, and works to prevent the abuse of, drugs through the 1970 Comprehensive Drug Abuse and Prevention and Control Act federal law known as the *Controlled Substances Act* (CSA). Controlled substances are classified into one of five schedules based on the medical value and potential for physical and psychological dependence and abuse of the substances (Table 8.1). Schedule I drugs have the highest potential of abuse and no acceptable medical use (e.g., heroin, methaqualone [Quaalude], and lysergic acid diethylamide [LSD]) while Schedule V drugs have a low potential for abuse and dependence and are accepted for medical use (e.g., low-dose codeine with Tylenol). Accordingly, storage and recording requirements for Schedule I and II drugs is more stringent than drugs that are otherwise scheduled. The differences in schedule also result in varying criminal penalties; the legal penalties are often higher for higher-scheduled drugs. The CSA was amended in 1984 to allow the DEA administrator to place a substance temporarily in Schedule I upon imminent hazard to public safety.

CLASSES OF DRUGS

Controlled substances are grouped by their main or primary physical effect or original plant source into one of several groups including opiates, stimulants, hallucinogens, depressants, antianxiety drugs, anabolic androgenic steroids, other drugs abused in sport, and new psychoactive substances. The drugs gamma hydroxybutyrate (GHB), 3,4-methylene-dioxymethamphetamine (MDMA), ketamine, and rohypnol are placed in one of these several groups but are also

Figure 8.1 Drug seizure of cannabis, cocaine, and MDMA. (From West Midlands Police, https://www.flickr.com/photos/westmidlandspolice/11172420845/in/photolist-i2gztB-XNq4JU-5EsSQG-9T9yMb-oYsM2U-WAekn-9TbaoW-9KgshS-8gkmEa-8gkmEk-9KdA7V-5EsX6C-9KdzAe-9KdAH2-5Mx3ef-9Kgstj-9KdAR8-5LAGKx-9KdCia-9KdB6X-9KdBx8-5Mx3fb-9Kgqt9-9KdFo8-9Kgq97-5Mx3ey-9KgpGm-5LAGLe-9Kgv6w-9KdE6R-gnB89D-oYsM2y-9KdDVT-9KdEgT-9KgtiG-9KgsEj-9KgtBh-9KgpcC-9KdBka-Wx5e5w-5LAGKM-9Kgt7L-9Kgufm-9KgoYq-9KgpC9-5LAGKR-5Mx3es-9Kdz34-5LAGKF-5Mx3eU.)

considered to be "club drugs." Selected drugs in each of these groups will be described in the sections that follow. Many drugs are alkaloids, meaning that they are alkaline and contain one or more nitrogen atoms. The chemical structures of amphetamine, heroin, fentanyl, delta-9-tetrahydrocannabinol (THC), phenobarbital, JWH-018, and testosterone controlled substances are shown in Figure 8.2.

STIMULANTS

The two main controlled stimulants are amphetamines and cocaine. However, other controlled stimulant drugs include cathinones and BZP (N-benzylpiperazine). In contrast, caffeine and ephedrine are commonly used legal stimulants. Caffeine is mixed with illegal drugs as an enhancing or cutting agent. The adulterants lidocaine, levamisole, phenacetin, diltiazem, and hydroxyzine have also been found in cocaine. Ephedrine is found in asthma medicines and produced in the ma-huang (*Ephedra vulgaris*) plant. *Stimulants* stimulate or upregulate processes in the central nervous system, leading to an increased heart rate and increased alertness or activity in users, and dilate the bronchial tubes in the lung, which is followed by a decrease in fatigue and a loss of appetite.

Amphetamine and its derivatives, including *methamphetamine*, belong to the phenylethylamines class; the variety of derivatized compounds in this class is extensive due to R-group substituent possibilities for positions R1–R9. Amphetamine was first synthesized in Germany in 1887 by L. Edeleano; in the 1930s, it was investigated for its therapeutic properties. Amphetamine and methamphetamine are structurally similar to the dopamine and norepinephrine neurotransmitters in the brain that increase attention and pleasure and improve mood.

Table 8.1 Controlled Substance Act drug schedule

Schedule I	High potential for abuse, no medical use
Schedule II	High potential for abuse, some accepted medical use, severe dependence
Schedule III	Some potential for abuse, accepted medical use, ranging dependence
Schedule IV	Low potential for abuse, accepted medical use, limited dependence
Schedule V	Low potential for abuse, accepted medical use, very low dependence, non-narcotic ingredients

Source: DEA, www.deadiversion.usdoj.gov/schedules/.

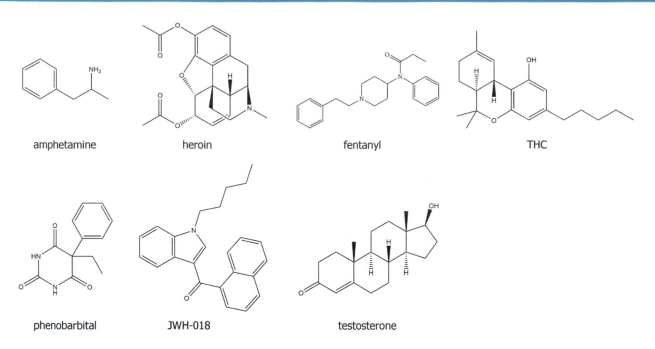

Figure 8.2 Chemical structures of controlled substances amphetamine, heroin, fentanyl, THC, phenobarbital, JWH-018, and testosterone. (Structure prepared by Ashley Cowan.)

Like other stimulants, amphetamines, known as "speed," improve physical alertness and energy levels. Amphetamines are used medically as mental-focusing and cognitive-enhancer therapeutics and to treat attention deficit hyperactivity disorder, narcolepsy, and obesity. They have been used on the battlefield by soldiers in World War II to overcome fatigue. Pharmaceutical preparations of amphetamines include amphetamine–dextroamphetamine (Adderall®), dextroamphetamine (Dexedrine®), methylphenidate (Ritalin®), and sibutramine (Meridia®). Methylphenidate blocks the reuptake of dopamine. Amphetamines can produce euphoria due to their ability to elevate blood pressure, heart rate, and respiratory rate. Amphetamine use can also lead to agitation, memory loss, mood swings, tremors, delusions, violent behavior, psychotic episodes, cardiac arrest, stroke, and seizures. Amphetamine users may exhibit poor hygiene, sweating, weight loss, discolored and rotten teeth, and skin sores.

Crime scene investigators may be called to investigate illicit drug labs. Over 150 procedures to synthesize amphetamines have been published in scientific journals and they are widely produced in clandestine labs. Knowledge of the starting materials and key intermediates for extractions and syntheses used in preparing controlled substances is instrumental in identifying the type of clandestine lab and potential safety risks to the investigators.

Clandestine labs that synthesize methamphetamine utilize several toxic chemicals and dangerous chemistry without the protection of modern safety equipment including chemical fume hoods and ventilation. The chemicals can adsorb and be splashed or spilled onto wall board, ceiling tiles, and floor coverings. Methamphetamine synthesis may begin by grinding ephedrine (or pseudoephedrine) tablets (restricted by the US Patriot Act in 2006); in home labs, a blender is often found for this purpose. The starting materials are extracted using denatured alcohol or another obtainable solvent and the solid impurities are filtered out using a coffee or other filter such as cheesecloth or bedsheets. Red phosphorus is collected from match strike pads—a staggering 10,000 strike pads are needed to recover an ounce of red phosphorous. Iodine crystals are obtained and acidified to produce hydroiodic acid. The ephedrine is then reacted with the red phosphorous and hydroiodic acid in a flask or glass pot using a heating mantle or stovetop at a slow boil for 5–8 hours. In a strongly exothermic process, lye or sodium hydroxide is added to remove the excess phosphorous. Nonpolar or less-polar solvents including toluene, methanol, acetone, mineral spirits, isopropyl alcohol, ether, and charcoal lighting fluid may be used to extract the free base. Methamphetamine sulfate is not water soluble so it is converted to methamphetamine hydrochloride by bubbling hydrogen gas through it. Hydrogen gas can be produced by reacting table salt with sulfuric acid. Glassware including separatory funnels, connecting adaptors, condensers, and three-necked flasks may be found in methamphetamine labs. A lithium catalyst may be prepared by dissolving lithium batteries in anhydrous ammonia. The presence of one or more of these starting materials, solvents, or paraphernalia can be used to identify a methamphetamine lab in crime scene investigation. In addition to the above, kitty litter is often found in these labs as it is used to absorb the odors.

Pure amphetamines are white powders while impure preparations lead to dark yellow or brown, sticky or solid substances. "Ice" and "crank" are terms used to describe pure (>90%) methamphetamine. There are two enantiomers of amphetamine: dextroamphetamine and levoamphetamine. The dextro form is two to three times more potent than the mixture of the stereoisomers (depending on the ratio) and five times more potent than the levorotatory form. Amphetamine and methamphetamines are often cut with other white powders including methyl sulfonyl methane, caffeine, sugar, procaine, and baking soda.

Amphetamines function as trace amine-associated receptor 1 (TAAR1) and G_s-coupled and G_q-coupled G protein-coupled receptor (GPCR) agonists that inhibit monoamine transporters which regulate monoamines in brain catecholamine neurons. This leads to an increase in dopamine and norepinephrine. Dopamine is a neurotransmitter with roles in pleasure and reward. Norepinephrine is a neurotransmitter that controls attention, impulse control, and mood. In addition, at high doses, amphetamine inhibits the monoamine oxidase metabolizing enzyme that reduces dopamine and phenethylamine metabolism. Amphetamine can also bind to the serotonin or 5-hydroxytryptamine (5-HT) 1A receptor.

Due to the differences in activity of stimulant drugs, the compounds vary in their scheduling. *Cathinone* and *BZP* are CSA Schedule I agents while amphetamine, methamphetamine, and cocaine are CSA Schedule II agents. In the United States, the Comprehensive Methamphetamine Control Act of 1996 placed pseudoephedrine under the same restrictions as ephedrine and controlled key materials used to synthesize methamphetamine including iodine and phosphorus. While methamphetamine is a Schedule II controlled substance due to its strong psychological dependency, a positional isomer of methamphetamine, phentermine, is a Schedule IV controlled substance with a low potential for abuse and little to no dependency.

Another stimulant is *cocaine*. Cocaine has been used by Native Indian tribes in South America for social and religious occasions for over 2500 years. It grows naturally in the Andean Mountains. Approximately 80% of cocaine is grown in Colombia. It is grown legally in Bolivia and Peru. Cocaine was first extracted from the leaf of the *Erythroxylum coca* plant in 1859 by Albert Nieman. It is also found in the *Erythroxylum novagranatense* plant. The leaves of the plant contain 1%–2% cocaine. Cocaine was first used medically as an anesthetic in eye surgery by Dr. Carl Koller in 1884; it is still used medically as a topical anesthetic to induce localized numbing and a vasoconstrictor to decrease bleeding in nasal surgery. Cocaine acts to block dopamine, serotonin, and norepinephrine neurotransmitter reuptake by binding to the dopamine and other transporters in transmitting neurons so the cocaine continually stimulates receptors in receiving neurons. Continued use leads to a decrease in dopamine receptors and thus a higher dose of the drug is needed for the same euphoric effect. In the 1890s, it was given to construction crews in the United States to increase their energy levels and it was found in the original Coca-Cola recipe and in wines in the early 1900s. Withdrawal symptoms include fatigue, lassitude, and irritability. In the United States, its use was restricted by the Pure Food and Drug Act of 1906 and it was (falsely) labeled and controlled as a narcotic by the Harrison Narcotic Act in 1914.

In field operations, *Erythroxylum coca* leaves are harvested and pulverized using a weed trimmer. Then the leaves are degraded using cement as an alkali. To extract the alkaloids, gasoline is added and the mixture is stirred for 2 to 3 hours. Sulfuric acid is added to precipitate the cocaine in the form of cocaine sulfate or cocaine "base," which is isolated by filtering using a cloth. At this stage, the purity is approximately 40%. The cocaine sulfate is washed in fresh sulfuric acid with potassium permanganate to remove the two main impurities: cis- and trans-cinnamoyl-cocaine. The material has a yellow coloration and is extracted again in sulfuric acid to achieve a purity of approximately 80%. A base is added to precipitate the cocaine base and the solid is dried and shipped to a lab where the drug is converted to cocaine hydrochloride using acetone and hydrochloric acid leading to a purity of 86%–90%. The cocaine hydrochloride is dried and pressed and sealed in plastic wrap. It is often found cut with vitamin B, mannitol, lactose, inositol, lidocaine, benzocaine, amphetamine, caffeine, corn starch, dextrose, ephedrine, flour, phenylpropanolamine, procaine, sucrose, talc, and tetracaine. Crack cocaine is produced by adding baking soda to the cocaine hydrochloride to "free" the base; ether is used to precipitate the drug. Black cocaine, cocaine negra, is formulated using charcoal, iron dust, cobalt, and ferric chloride so that it resembles soil or organic fertilizer to avoid detection. Cocaine can also be synthesized using nitrone and methyl 3-hexanoate as starting materials.

Users smoke freebase or crack forms of cocaine while they may dissolve cocaine hydrochloride in water and squirt it into their nasal passages or inject it into a vein. Users who snort cocaine may develop a hole in their septum. Cocaine is metabolized to benzoylecgonine, which can be found in urine for up to 3 days following the use of cocaine.

Legal stimulants such as *caffeine* are ubiquitous in modern society and found in coffee, tea, and energy drinks. Caffeine can also be fatal if too much is consumed in one dose—the consumption of multiple caffeinated energy drinks has been associated with death.

TEEN DIED FROM "HUFFING" AIR DUSTER

On March 18, 2013, Aria Doherty, 13, was found dead in her bedroom by a sibling. As an honors student, her parents and teachers did not suspect she would use drugs.

Propellants are legal and are used in numerous consumer and household products. The products are sold in dollar and discount stores; they are cheap and easy to obtain. Air duster, compressed air used to clean computer keyboards, contains difluoroethane or tetrafluorethane as propellants. These propellants and other products are used as inhalant drugs by young people. Other inhalants include common household products such as nail polish remover, hair spray, strong glues, nail polish, spray paint, air fresheners, lighter fluid, gasoline, propane, and butane.

According to the National Inhalant Prevention Coalition, by the time they reach the eighth grade, one in five children in the United States has used an inhalant to get high. For many, it will be the first drug they try. More than 100 teens die from "huffing" each year; some die from a heart attack from their first huff, a condition called Sudden Sniffing Death Syndrome. Approximately 2.6 million pre-teens and teens ages 12–17 use inhalants each year. Inhalants are taken anywhere—in bedrooms, the backyard, cars, at school or at the mall, or even walking home from school.

Inhalants are transferred from the lungs to the bloodstream where they travel to the brain. The nonpolar inhalant molecules dissolve the protective covering around brain cells, leading to a reduction in the messaging of chemical signals in the brain. Children and teens, with their developing brains and mucosa, are more susceptible to the effects of inhalants than adults. Inhalant use can cause permanent damage to speech and memory, dementia, and other brain damage as well as damage to the heart, lungs, and liver.

Doherty was found in bed with her nostrils taped up and the straw from the inhalant container still hanging from her mouth. On investigation, the cause of death was determined to be cardiac arrest. Her parents think it may have been her first time "huffing."

BIBLIOGRAPHY

DiConsiglio, J. 2011. Death by huffing. *Scholastic.* http://choices.scholastic.com/story/death-huffing (accessed July 4, 2017).

Doherty, C. Aria Doherty (age 13). www.consumered.org/personal-stories/aria-doherty-age-13 (accessed July 4, 2017).

Kuruvilla, C. 2013. L.A. student, 14, dies in 'huffing' incident. *The New York Daily News.* www.nydailynews.com/news/national/student-14-dies-huffing-incident-article-1.1295298 (accessed January 29, 2018).

DEPRESSANTS AND ANTIANXIETY DRUGS

Depressants include alcohol, barbiturates, antianxiety and antipsychotic drugs, tranquilizers, methaqualone, and "huffing" or "glue sniffing." Depressant drugs depress the central nervous system, calm irritability and anxiety, reduce muscle coordination, impair judgment, and may induce sleep. Depressants function by inducing changes in membrane fluidity, thus affecting neural activity. The most widely used depressant is ethanol (ethyl alcohol), a legal drug commonly known as alcohol. Ethanol will be examined in greater detail in Chapter 9.

Antianxiety drugs are those used to treat anxiety and panic disorders and acute stress reactions without inducing sleep or impairing thinking. They are typically *benzodiazepines* that have sedative and hypnotic activity. Benzodiazepines can cause irritability, confusion headaches, depression, and memory impairment when taken routinely at high doses. These include alprazolam (Xanax®), chlordiazepoxide hydrochloride (Librium®), clonazepam (Klonopin®), diazepam (Valium®), and lorazepam (Ativan®). Chlorpromazine (CPZ, Thorazine®) and reserpine (Raudixin®, Serpalan®,

Serpasil®) are two antipsychotics. Methaqualone (Quaalude®) is a sedative-hypnotic drug used as a sedative and a muscle relaxant and meprobamate (Miltown®, Equanil®) is used as a tranquilizer.

Depressants may be inhaled, "sniffed," or "huffed." Such activities are common with teenage users. Such substances include toluene, methylethylketone, paint thinner, propane, butane, gasoline, trichloroethylene, lighter fluid, nail polish, nail polish remover, airplane glue, model cement, and aerosol gas propellants such as freon. Other popular inhalants include compressed air used to clean computer keyboards; in addition to the air, they contain difluoro- ethane or tetrafluorethane propellants. Inhalant users experience immediate effects ranging from immediate but fleeting exhilaration and euphoria to dizziness, headache, slurred speech, impaired judgment, double vision, spasms, drowsiness, hallucinations, and stupor. "Huffing" and "sniffing" can lead to heart, liver, and brain damage and has caused heart failure and sudden deaths in teens.

Gamma hydroxybutyrate and rohypnol are "club drugs" that may be slipped into drinks (often alcoholic ones) of unknowing users so they are more accurately termed "date rape drugs." GHB was synthesized in 1874 by Alexander Zaytsv and is a white crystalline solid that is water soluble and tastes salty. GHB is an endogenous neurotransmitter that is a precursor to gamma-aminobutyric acid (GABA) and glutamate. At low doses, GHB has stimulant and aph- rodisiac effects. GHB binds the GHB receptor and stimulates dopamine release to act as a stimulant and, at higher doses, is a $GABA_B$ receptor agonist which inhibits dopamine release, acts as a central nervous system depressant, and elicits a calming effect. As it is metabolized by the alcohol and aldehyde dehydrogenases (described in Chapter 9), it slows the metabolism of alcohol and is extremely dangerous when taken with alcohol. GHB is easily synthesized from gamma-butyrolactone and sodium hydroxide. Alone, GHB is a Schedule I drug, but formulated in an FDA-approved drug, sodium oxybate, it is Schedule III. GHB use leads to dizziness, sedation, headache, nausea, euphoria, disinhibi- tion, and increased libido and unconsciousness and death at high doses. It can be detected in urine within four hours of ingestion.

Flunitrazepam or rohypnol ("roofies") is a depressant and sleep aid whose effects include relaxation and sedation, and at higher doses, impairment of speech and balance, respiratory depression, loss of consciousness, and inability to remember what has happened for several hours, including sexual assault or rape. Thus, this substance is also used by perpetrators as a "date rape" drug and is slipped into the drink of an unknowing victim. Rohypnol is an odorless, colorless, and tasteless benzodiazepine. It was first synthesized by Roche and a patent was filed in 1962. It is used medically to treat insomnia in other countries and is a Schedule IV drug. It is metabolized by CYP3A4 in the liver. The metabolite 7-aminoflunitrazepam is used to detect ingestion. Rohypnol may be used by attendees of all-night dance parties, raves, and electronic dance music festivals.

The first *barbiturate* was barbituric acid, which was synthesized by Adolph von Bayer, the namesake of the Bayer Chemical Company, in 1864. Barbiturates, or "downers," include amobarbital, phenobarbital, secobarbital, and pentobarbital. Barbiturates are sedatives that can be taken orally or intravenously. They are used to treat anxiety and insomnia. Barbiturates enhance GABA activity like GHB and are also extremely dangerous when taken with alcohol.

HALLUCINOGENS

Hallucinogens alter normal perception, moods, and thought processes. The most commonly abused hal- lucinogen is marijuana, the common name for the *Cannabis sativa* plant from which the drug is derived. Marijuana has been used in China since 2700 BC. The active drug substance in marijuana is the cannabinoid *delta-9-tetrahydrocannabinol*, which was first isolated in 1964, although marijuana is known to contain at least 400 chemicals including 66 cannabinoids. Plant-derived cannabinoids such as THC bind to the cannabinoid (CB) receptors 1 and 2 on presynaptic neurons that are targeted by endogenous cannabinoids including anan- damide and 2-arachidonoylglycerol.

The 1937 Marijuana Tax Act prohibited the use and possession of marijuana in the 46 US states. Marijuana is cur- rently listed as a CSA Schedule I substance. However, in the United States, 30 states and the District of Columbia have legalized marijuana for medical and/or recreational use as of this writing. Medicinal uses of the marijuana plant include reducing pressure in the eye due to glaucoma, as an anti-nausea medication for patients undergoing cancer treatment, and as a muscle relaxant. Synthetic THC is a CSA Schedule II agent that is prescribed for medicinal use as

Dronabinol (marketed as Marinol). It is used by patients undergoing chemotherapy to reduce nausea. As of 2017, the recreational use of marijuana has been legalized by several US states including Alaska, Colorado, California, Maine, Massachusetts, Nevada, Oregon, Washington, and the District of Columbia.

The *Cannabis sativa* plant has several unique characteristics among plants that are typically encountered in cases of drug crimes. These are used in macroscopic evaluations of seized plant material. The plant sprouts into four-leaved plants with fluted stems consisting of two serrated and two rounded slender leaves. They grow to contain 3, 5, 7, 9, or 11 leaflets per palmate leaf and reach 5–8 ft or more in height. They have a red thread in the bud. Microscopically, the leaves can be seen to contain bear-shaped trichome cystolithic hairs that act like miniature thorns on the top surface with small calcium carbonate crystals at the base of the bear claw. A photo of marijuana is shown in Figure 8.3.

The marijuana plant can be separated into many parts that vary widely in their THC content. The resin and the flowers of the female plants contain the most THC followed by the leaves, stem, seed, and roots in decreasing order. Overall, controlled substance and medical marijuana may have a THC content at or above 25%–30%. In contrast, the strain of the *Cannabis sativa* plant that has been bred to grow very tall and produce very little THC (e.g., maximum of 0.3% in Canada) is known as hemp. Hemp is used legally to produce fiber to be used in rope. Related plants in the same genus are *Cannabis indica* and *Cannabis ruderalis*. *Cannabis sativa* also has limited uses as an oilseed crop.

The highest concentration of THC is in the resin (2%–8%), which is processed to make hashish that can have a final THC content of 50% or higher. To make hashish, the resin can be removed by rubbing the leaves using the hands, and by shaking or using a sieve to separate the resin from the leaves. The resin may then be extracted using ice water and separated using a coffee filter. The resulting resin is light brown, tan, or green in color when fresh but it oxidizes to a darker color over time. The recovered material can be formed into patties or balls using the fingers or a press.

Another product from the *Cannabis sativa* plant is hash oil. Hash oil can be extracted from finely ground or chopped leaves using a nonpolar solvent such as isopropyl alcohol, ether, acetone, butane, ethanol, benzene, methylene chloride, turpentine, or camping fuel. A metal or silk screen or sieve can be used to separate the solid plant material from the liquid extract. The hash oil can be isolated by evaporating the solvent. The hash oil ranges from a light honey to red, green, brown, or black oil color with a final THC content of 8%–20%.

Marijuana is used as a drug by smoking the leaves or the resin. Hash oil may be processed into gelatin capsules, candies, or brownies for ingestion or smoked by users in a glass tube or bulb (e.g., light bulb).

Other hallucinogens include 4-methoxyamphetamine (PMA), 5-methoxy-3,4-methylenedioxyamphetamine (MMDA), 3,4-methylenedioxyamphetamine (MDA, Sassafrass, the "hug drug"), 3,4-methylenedioxymethamphetamine (MDMA, also known as Molly and Ecstasy), dimethyltryptamine (DMT), ibogaine, LSD, mescaline, peyote, psilocybin, and psilocin. All are Schedule I controlled substances.

Figure 8.3 Seized marijuana plant material submitted to the Maryland State Police crime lab for analysis. (Courtesy of Ashley Cowan.)

LSD is a semisynthetic drug that causes delirium and vision and sound distortions and was first synthesized in 1938 by Albert Hofmann of Sandoz Company of Switzerland. It is synthesized from lysergic acid extracted from the fungus *Claviceps purpurea* that grows on grains including rye, wheat, and barley or ergotamine tartrate from morning glory, baby Hawaiian woodrose, or *Stipa robusta* seeds. It was sold in 25 μg tablets and 100 μg ampules under the brand name Delysid; production was discontinued in 1966. In popular culture, it was the drug taken in the 1960s by Dr. Timothy Leary, the Harvard psychologist who encouraged people to "turn on—tune in—drop out." LSD is white or almost colorless unless exposed to light or heat, which causes it to turn black, so LSD liquid is often stored in darkened bottles (such as those for food coloring or extracts used in baking). LSD is often dosed on art blotter paper, on sugar cubes, or on candy dots in 20–80 μg doses. Users dissolve the substrate on the tongue or ingest the drug. It can also be used as eye drops. LSD, like atypical antipsychotics, binds to the 5-HT-2A and 2C receptors.

Mescaline (3,4,5-trimethoxyphenethylamine) was used historically in Mexico and Native American culture in religious rites and medicine to relieve fatigue and hunger. Its use was outlawed in the United States by the Indian Religious Crime Code in 1883 but it is allowed for use by members of the Native American Church for religious purposes. The source of mescaline is a small (3 inches in diameter and 0.5 inches tall after a decade of growth), spineless cactus (or "button") called peyote, *Lopophora williamsii* (Lamaire), that is endogenous to North Central Mexico and Central Texas where up to 60,000 can be found in one acre as multiple cacti can share a root. Mescaline is one of the thirty alkaloids of the peyote cactus. Each button contains approximately 25 mg of mescaline. The San Pedro cactus also contains mescaline (about a third as much as the peyote cactus) but is legal as an ornamental. The cactus can be dried and chewed, boiled as a tea, ground and loaded into a gelatin capsule, or extracted with solvents including lye, benzene, sulfuric acid, and water to yield mescaline. The extracted mescaline in the sulfate form is characterized by translucent white needles. Pure mescaline hydrochloride is a white powder.

Phencyclidine (1-1-phenylcyclohexyl piperidine hydrochloride, PCP) was synthesized in 1957 by the Parke-Davis Company and was tested as a general anesthetic under the brand name Sernyl. It never made it to market for humans as it causes post-operation delirium, delusions, and psychotic behavior. PCP is a CSA Schedule II drug permitted for use by veterinarians for anesthetizing large animals; the drug name is Sernylan. Its use as an illicit drug, the PeaCe pill, began in 1967 at a music festival in San Francisco. In powder form, it is referred to by users as "angel dust," although it can be found in tablet, capsule, and liquid form or laced into cigarettes and smoked, injected, or ingested as a pill. Its simple structure is easy to synthesize in clandestine labs and a dose is 1–6 mg. The synthesis of PCP may begin with piperidine (distribution is controlled), sodium or potassium cyanide, and distilled water. The solution is cooled on ice, then cyclohexanone is added in basic solution with potassium hydroxide to yield is 1-piperidinocyclo-hexanecarbonitrile (PCC). Bromobenzene is prepared with a Grignard reagent (magnesium turnings in anhydrous ether) to produce a phenyl magnesium bromide Grignard reagent that is reacted with the PCC over ether to form PCP. Pure PCP is a metallic or bitter tasting, odorless, white powder or clear liquid. Impurities lead it to be a tan to brown solid with a strong odor or a yellow liquid. The use of pyrrolidine as a starting material instead of piperidine leads to the formation of 1-(1-phenylcyclohexyl)pyrrolidine (PHP). PCP is a dissociative anesthetic that can lead users to ignore severe burns and even pull their own teeth. PCP binds to the N-methyl-D-aspartic acid channel in the central nervous system to block the flow of calcium ions and thus prevent neuronal activation.

Psilocybin (phosphorylated 4-hydroxydimethyltryptamine) is a natural product indole drug produced by the hallucinogenic "magic" mushrooms *Psilocybe cubensis* and *Psilocybe cyanescens*, although it is also produced by genera *Conocybe, Panaeolus, Inocybe, Pluteus, Copelandia*, and *Gymnopilus*. The first recorded use of the mushrooms was in 1502 during the coronation feast of Montezuma; it is thought to have been used since 1500 B.C. in religious ceremonies by Indian cultures of Mexico and Central America. The mushrooms were rediscovered in Central Mexico by R. Gordon Wasson in 1953 and psilocybin was isolated by Albert Hofmann of Sandoz Labs in Switzerland and synthesized for evaluation as a drug for psychiatric research. Psilocybin can be synthesized from the amino acid tryptophan in five steps. Psilocybin is metabolized to psilocin (4-hydroxydimethyltryptamine) by dephosphorylation. Metabolites from psilocin include psilocin O-glucuronide, 4-hydroxyindole-3-acetaldehyde, 4-hydroxytryptophol, and 4-hydroxindole-3-acetic acid. Psilocybin and psilocin are both CSA Schedule I drugs, although the spores are legal in all US states except California. Structurally, the two compounds are similar to serotonin—they bind to the 5-HT receptors and inhibit the activity of serotonin in the hippocampus, corpus striatum, and cerebral cortex portions of the brain. A street test used to tentatively identify magic mushrooms is to look for a bluish color on bruising or breakage attributable to oxidized indoles or to place the cap gill side down on a piece of white paper for several hours; most psilocybin-containing spores are purplish-black in color. Users may take the mushrooms in gelatin capsules to mask the taste, brew them as a tea or as an ingredient in food, or smoke them. A dose is 20–60 mg.

MDMA was synthesized in 1914 and patented by Merck and was tested by the US Army in 1953. It is commonly known as Molly or Ecstasy. It has hallucinogenic and amphetamine-like effects. MDMA enhances self-awareness and decreases inhibitions in users; however, seizures, muscle breakdown, stroke, kidney failure, and cardiovascular system failure often accompany chronic abuse. It is synthesized in clandestine labs around the world starting from sassafras oil. The sassafras oil is distilled to yield 8%–90% safrole. There are multiple routes to synthesis. In one, glacial acetic acid is added and hydrogen bromide gas is bubbled through to produce bromosafrole, which has a burgundy color. Stirring for 24–48 hours under temperature control using a water bath yields 3,4-methylenedioxphenyl-2-propane. To remove the bromosafrole, the mixture is extracted with toluene, shaken, and separated using a separatory funnel. The bromosafrole is separated from the toluene by fractional distillation and converted to MDMA by adding methylamine and isopropyl alcohol and heating at 265°F for 3–4 hours using a metal "bomb" or pressure cooker. The MDMA is purified in ether and converted to the water-soluble hydrochloride form by bubbling in hydrochloric acid gas. It became popular in the 1980s as a "club drug." The drug is typically sold in 25–75 mg capsules or tablets; a dose is 75–125 mg. MDMA users snort, inject, smoke, or orally ingest the drug. It is often cut with powdered sugar, ephedrine, phenylpropylamine, caffeine, dextromethorphan, atropine, and glyceryl guaiacolate (guaifenesin). Biochemically, MDMA causes axons to release a large amount of serotonin in the brain and blocks its reuptake. The serotonin is degraded by monoamine oxidase; this leads to lowered serotonin levels and may cause neurotoxic damage. Closely related drugs include MMDA and MDA.

N,N-Dimethyltryptamine (DMT) was synthesized in 1931 by Richard Helmuth Fredrick Manske. DMT is a powerful psychedelic drug used to produce euphoria and hallucinations with a short duration of action. A dose is usually 20–70 mg. When inhaled or injected, effects last only 5–15 minutes. At low doses, it can be mood elevating and raise blood pressure, heart rate, and temperature, and increase pupil diameter. As it is metabolized by deamination, it can be taken orally with effects lasting up to 3 hours when taken with a monoamine oxidase inhibitor. It is used by South American shamans. It is structurally related to serotonin and other tryptamine drugs including 5-OH-DMT (bufotenin), MeO-DMT, 5-HO-DMT, as well as psilocin and psilocybin.

Ibogaine is a tryptamine indole alkaloid produced by the plant *Tabernanthe iboga* as well as the *Tabernanthe manii*, *Voacanga Africana*, and *Tabernaemontana undulata* plants. The root bark is ground and ingested by the Bwiti in religious ceremonies. It can also be synthesized from the plant alkaloid voacangine. The total synthesis was published in 1956. Ibogaine is metabolized to noribogaine by cytochrome P450 2D6. Noribogaine acts as a serotonin reuptake inhibitor, moderate κ-opioid receptor agonist, and weak μ-opioid receptor agonist. Users struggle with muscle coordination, and may experience dry mouth, nausea, and vomiting.

Ketamine is a CSA Schedule III drug with a legal use as an animal anesthetic by veterinarians. It was synthesized in 1962 by Calvin Stevens of Wayne State University. When used in high doses by humans, it causes euphoria, trance, and hallucinations as well as slurred speech, sedation, and memory loss. It also causes increased blood pressure, muscle tremors, and breathing problems and causes agitation and confusion as it wears off. It is very short acting—its effect starts fast and the hallucinogenic effects wear off fast. It functions as a NDMA receptor antagonist.

Salvia divinorum, Diviner's sage, is a psychoactive plant that produces the diterpenoid salvinorin A at 0.18% of the dry plant weight. It is native to Oaxaca, Mexico; Mazatec shamans used Salvia for healing rituals. It was first recorded in writing by Jean Basset Johnson in 1939. It is a kappa opioid and D_2 receptor agonist. It is very potent—a dose is 200 μg. It is chewed, taken sublingually as a tincture, or smoked. When chewed, it is held in the mouth and absorbed via the oral mucosa. Smoking requires a high temperature to volatilize the drug and causes an immediate effect that lasts about 8 minutes, leading to alterations of visual reality and time, temporary loss of speaking ability, and loss of muscle coordination. Salvia is not currently controlled under the Controlled Substances Act but many US states and other countries have regulated its use.

OPIATES/OPIOIDS

The *opiates* refer to natural product, semisynthetic, and synthetic drugs related in chemical structure and properties to plant-derived drugs from the opium poppy, *Papaver somniferum*. Opiates are also referred to as narcotics, a word derived from the Greek word *narkotikos*, meaning a state of lethargy or sluggishness. *Narcotics* have analgesic, or pain relief, effects and cause euphoria when used at high concentrations. The opium poppy has a long history of over 6000 years of medical use and produces both *morphine* (approximately 10% of the total weight of the dried opium) and *codeine* as natural products. Morphine was first isolated from the opium poppy in 1803. By the late 1800s in the United States, over 170 opium products were offered with applications in treating severe pain, diarrhea, and insomnia, among others. In addition to codeine and morphine, opiates include the semisynthetic compound *heroin*, which is synthesized from morphine. Felix Hoffman performed the first heroin synthesis in 1897. By the turn of the nineteenth century in the United States, over 250,000 people (4.6 per 1000) were addicted to opiates. The Harrison Narcotic Act of 1914 regulated the use of opium, morphine, and heroin. The Heroin Control Act of 1924 made the manufacture and possession of heroin illegal in the United States. The United States Poppy Control Act of 1942 restricted the cultivation of the opium poppy in the United States. Other narcotics include oxycodone (Oxycontin®, Percocet®), which was introduced in 1995, and hydrocodone (Vicodin®), meperidine (Demerol®), and propoxyphene (Darvon®), as well as fentanyl and fentanyl derivatives including carfentanil and acetyl fentanyl. The DEA has temporarily scheduled many fentanyl-related substances in Schedule I. Methadone (Dolophine®), naloxone (Narcan™), buprenorphine, and naltrexone (Revia®) are opiate antagonists that are used to help break addiction and help to stop physical dependence on the drugs.

Most opiates are extremely dangerous due to their potency and the fact that they are highly addictive—they may be addictive after only one use in people with certain genetic traits. For example, exposure to only $250\,\mu g$ (1/4 of a milligram) of elephant tranquilizer carfentanil can be deadly to an adult male. Opiate use, both legal and illegal, resulted in 33,091 of the over 50,000 drug-related deaths in 2016 in the United States according to the Centers for Disease Control. Opiate overdoses have quadrupled since 1999. Heroin is scheduled as a Schedule I drug with a high potential for abuse and physical dependence and no accepted medical use. Other opiates including morphine, methadone, oxycodone, fentanyl, and carfentanil are Schedule II drugs that have accepted medical uses but also have a risk of severe physical dependence and a high potential for abuse. They are used in hospital and veterinary medicine as surgical painkillers and prescription narcotics prescribed for severe pain. Codeine is a Schedule IV drug with current medical uses in Tylenol with codeine products and has a low potential for abuse and dependence. It is used in cold and flu medicines as a cough suppressant.

The opium poppy plant grows up to four feet in height and produces flowers of various colors including red, pink, white, blue, and purple; the flowers are legal. Users may grind the flowers for use in preparing a tea containing opium products. The unripe pod of the poppy contains a white gummy, milky substance that flows out on incision and blackens on oxidation. It contains morphine and codeine and is collected and used to synthesize heroin.

Opium from the opium poppy is extracted using hot water followed by calcium hydroxide to yield opium alkaloids and calcium morphenate. Morphine is precipitated from calcium morphenate using ammonium chloride at a basic pH. Morphine is combined with acetic anhydride and sodium carbonate or acetyl chloride and sodium chloride to form heroin base, which is washed with acetone and precipitated using hydrochloric acid to produce diacetylmorphine (heroin). Pure heroin is a white crystalline solid that is soluble in water. Cutting agents include quinine, starch, lactose, procaine (Novocaine), and mannitol, or brown sugar, cocoa mix, chocolate milk powder, and coffee.

The DEA has determined that two routes are being used in the clandestine synthesis of fentanyl: the Janssen and Siegfried synthesis routes. The Janssen route is the 1965 route patented by Janssen Pharmaceutical and uses the starting material n-benzyl-4-piperidone; the benzylfentanyl (322 amu) impurity indicates the use of this method. An easier and most commonly used method to illicitly manufacture fentanyl is the use of the Siegfried method (since the 1980s). The Siegfried method uses 4-anilino-N-phenethyl-4-piperidine (ANPP, 280 amu) as its starting material; its use is detected if an ANPP impurity is present without the of benzylfentanyl impurity. Recent seizures of fentanyl are of such a high purity that the determination of the method is more difficult. Both ANPP and benzylfentanyl precursors are Schedule I substances.

Opiates act as opiate receptor agonists and activate the central nervous system by binding to the endogenous opiate receptors mu, kappa, and delta, and the noriceptin receptor opiate-like receptor 1 (OLR-1) of the sensory neurons in the brain, spinal cord, and digestive tract. Opiate receptors are inhibitory GPCRs folded to seven transmembrane helices that are embedded in the receptor membrane. These receptors are naturally stimulated by endogenous opioids including endomorphins, enkephalins, endorphins, nociception, and dynorphins in the body. Opiate GPCRs are activated by phosphorylation by the G protein receptor kinase, a process that takes less than 20 seconds. Heroin is metabolized by the liver carboxyesterase 1 that rapidly hydrolyzes ester linkages to morphine and 6-acetyl morphine. Oxycodone is metabolized to oxymorphone and morphine is metabolized to morphine-3-glucuronide (M3G) and morphine-6-glucuronide (M6G). Morphine binds the opiate receptors leading to the release of GABA from synaptic vesicles in the nerve terminals. The GABA binds to GABA receptors that stimulate the release of dopamine in the interneuron, which binds to the dopamine receptor in the postsynaptic membrane. The GPCR is recycled by dephosphorylation within a minute and resensitized. However, in opiate abuse, the receptor is turned off by beta-arrestin-2 binding, which leads to receptor endocytosis that takes an hour or more to recycle, thus leading to less receptors available for the drug to stimulate.

ANABOLIC STEROIDS

Anabolic androgenic steroids are Schedule III drugs that are chemically related to the male sex hormone testosterone and synthetic compounds used for muscle building. Anabolic steroids are *performance-enhancing drugs* that were controlled by the DEA in 1991. (Stimulants, narcotics, diuretics, and oxygen-delivery promoters may also be used as performance-enhancing drugs.) Although they are not as addictive as Schedule I and II drugs, they are still addictive. Anabolic androgenic steroids are used medically to treat hormonal issues including delayed puberty and to promote the replacement of muscle lost in diseases such as AIDS. Other steroids such as corticosteroids (e.g., cortisone, hydrocortisone, prednisone, and prednisolone) are used to treat asthma and allergic reactions by opening breathing pathways and reducing swelling, redness, and inflammation in rashes. Although the use of prednisone and prednisolone is allowed in some sports, they and glucocorticosteroids are banned for use by the World Anti-Doping Agency (WADA). Table 8.2 lists the exogenous anabolic androgenic steroids banned by WADA. The endogenous anabolic androgenic steroids androstenediol, androstenedione, dihydrotestosterone, prasterone (dehydroepiandrosterone, DHEA), and testosterone are also banned when administered as a drug.

Anabolic androgenic steroids are used illegally to accelerate muscle growth in athletes including body builders to improve performance or improve their physical appearance. Anabolic androgenic steroids may be used orally, by injection, or by application to the skin via a cream, gel, or patch. Negative side effects or "roid rage" include mood and personality changes such as extreme jealousy and irritability, aggression, depression, delusions, and impaired judgment. Steroid use can also lead to health problems including liver cancer and damage, enlarged heart, high blood pressure, kidney problems or failure, and halting bone growth in teens. In males, excess testosterone can lead to infertility, baldness, breast development, and increased risk for prostate cancer while in females it can lead to facial

Table 8.2 Banned exogenous anabolic androgenic steroids as of January 1, 2012. Related drugs with the same parent structure and biological activity are also banned

1-Androstenediol	Gestrinone	Norboletone
1-Androstenedione	4-Hydroxytestosterone	Norclostebol
Bolandiol	Mestanolone	Norethandrolone
Bolasterone	Mesterolone	Oxabolone
Boldenone	Metenolone	Oxandrolone
Boldione	Methandienone	Oxymesterone
Calusterone	Methandriol	Oxymetholone
Clostebol	Methasterone	Prostanozol
Danazol	Methyldienolone	Quinbolone
Dehydrochlormethyltestosterone	Methyl-1-testosterone	Stanozolol
Desoxymethyltestosterone	Methylnortestosterone	Stenbolone
Drostanolone	Methyltestosterone	1-Testosterone
Ethylestrenol	Metribolone	Tetrahydrogestrinone
Fluoxymesterone	Mibolerone	Trenbolone
Formebolone	Nandrolone	
Furazabol	19-Norandrostenedione	

Source: WADA, www.wada-ama.org/sites/default/files/resources/files/WADA_Prohibited_List_2012_EN.pdf.

hair growth, deepened voice, and enlarged clitoris, and can stop the menstrual cycle. Withdrawal symptoms may include mood swings, fatigue, problems sleeping, decreased sex drive, and loss of appetite.

OTHER DRUGS ABUSED IN SPORTS

Other drugs that may be used by athletes to enhance their performance and recovery but are controlled for use in sport include human growth hormone (HGH), erythropoietin (EPO), diuretics, creatine, and stimulants. HGH is a 191-amino acid polypeptide used medically to stimulate growth in children and adults with growth deficiencies. Recombinant HGH is sold as Somatotropin or Genotropin®. EPO is a 34 kDa glycoprotein that stimulates red blood cell production. It is used in the treatment of kidney disease and anemia. Creatine is an endogenous molecule that functions in recycling adenosine triphosphate (ATP) as phosphocreatine in brain and muscle tissue. It also serves as a pH buffer.

NEW PSYCHOACTIVE SUBSTANCES

New psychoactive substances (NPS) are recreational designer drugs sold as "legal highs." They may be labeled as "bath salts," "spice," incense, or plant food "not intended for human consumption." They include herbal psychoactive substances or plant-derived NPS, synthetic cathinones, and synthetic cannabinoids. Many of these products remain uncontrolled although several have been controlled in the past several years owing to their significant impacts on human health, including death.

Synthetic cathinones are structurally related to the cathinone drug produced by the *Catha edulis* plant that is widely used in Middle Eastern countries ("Khat") where users chew the plant and hold it in their oral mucosa. Synthetic cathinones have stimulant and hallucinogenic properties. They include 3,4-methylenedioxypyrovalerone (MDPV), butylone, pentylone, ethylone, methedrone, methylone, dimethylcathinone, ethcathinone, 3- and 4-flouromethcathinone, 4-fluoromethcathinone, pyrovalerone, 4-methyl-N-ethylcathinone, 4-methyl-alpha-pyrrolidinopropiophenone, alpha-pyrrolidinopentiophenone (alpha-PVP), and alpha-pyrrolidinobutiophenone (alpha-PBP). All are Schedule I drugs.

Synthetic cannabinoids have a hallucinogenic activity similar to that of THC and are cannabinoid receptor agonists. Some, like HU-210 (1,1-Dimethylheptyl-11-hydroxy-tetrahydrocannabinol), are structurally related to THC, while others have a different chemical backbone. HU-210 shares THC's analgesic activity, although it is 800 times more potent as a hallucinogen than THC. HU-210 was first synthesized in 1988 by Dr. Raphael Mechoulam's group at Hebrew University (the source of the HU in the compound identifiers)—the same group that isolated and synthesized THC. According to the US Customs and Border Protection, HU-210 was identified in Spice Gold incense in January 2009 in seized product. HU-210 is not federally controlled by the DEA in the United States but several states have passed laws controlling this substance.

Several other synthetic aminoalkyl indole (AAI) cannabinoids were designed and synthesized in Dr. John W. Huffman's lab at Clemson University to study CB1 and CB2 receptor binding and bear his initials in their names. They include JWH-018 (1-pentyl-3-(1-naphthoyl)indole), JWH-019 (1-hexyl-3-(1-naphthoyl)indole), JWH-073 (1-butyl-3-(1-naphthoyl)indole), JWH-081 (1-pentyl-3-[1-(4-methoxynaphthoyl)]indole), JWH-122 (1-pentyl-3-(4-methyl-1-naphthoyl)indole), JWH-200 (1-[2-(4-morpholinyl)ethyl]-3-(1-naphthoyl)indole), JWH-203 (1-pentyl-3-(2-chlorophenylacetyl)indole), JWH-210, JWH-250 (1-pentyl-3-(2-methoxyphenylacetyl)indole), and JWH-398 (1-pentyl-3-(4-chloro-1-naphthoyl)indole). They are marketed as "K2," "spice," or "synthetic marijuana." All except JWH-210 are CSA Schedule I agents.

The AM synthetic cannabinoids, such as AM-694 (1-(5-fluoropentyl)-3-(2-iodobenzoyl)indole), AM-678 (same as JWH-018), and AM-2201 (1-(5-fluoropentyl)-3-(1-naphthoyl)indole), were produced by Dr. Alexandros Makriyannis' group at Northeastern University. N-dealkylation of AM-2201 releases the fluoropentane metabolite (as compared to pentane for JWH-018). CP-47 497 (5-(1,1-dimethylheptyl)-2-[(1R,3S)-3-hydroxycyclohexyl]-phenol) and CP-47 497 C8 (5-(1,1-dimethyloctyl)-2-[(1R,3S)-3-hydroxycyclohexyl]-phenol or cannabicyclohexanol) were developed at Pfizer. Other synthetic cannabinoids include RCS-4 (1-pentyl-3-[(4-methoxy)-benzoyl]indole) and RCS-8 (1-cyclohexylethyl-3-(2-methoxyphenylacetyl)indole). As of this writing, all of these drugs are CSA Schedule I drugs.

Figure 8.4   Kratom capsule and contents.

NPS derived from plants include d-lysergic acid amide (LSA) from *Ipomoea purpurea* (morning glory), ergine alkaloid or LSA from *Merremia tuberosa* (Hawaiian woodrose), scopolamine and atropine from *Datura stramonium* (jimson weed), and mitragynine from *Mitragyna speciosa* (kratom). Capsules of ground kratom are shown in Figure 8.4. All of these are legal plants not regulated by federal control in the United States. Plant seeds may be ingested whole while the leaves are typically prepared for use in teas. Consuming too much of the material or using these in combination with other substances has led to deaths in users. A multiplex polymerase chain reaction high-resolution melt assay was developed to detect and differentiate these NPS plants by their DNA: marijuana, Hawaiian woodrose, jimson weed, and morning glory could simultaneously be detected and identified by their different melt temperatures using the assay. Another similar assay can be used to detect marijuana and kratom DNA in a duplex assay by their different melt temperatures. The chemical substances in these plants can be separated and identified by thin-layer chromatography, gas chromatography–mass spectrometry (GC-MS), or high-pressure liquid chromatography using published protocols.

CHEMICAL ANALYSIS: IDENTIFICATION AND QUANTITATION

According to the United Nations Office of Drugs and Crime, a three-pronged approach should be implemented for the chemical analysis of drugs. For example, a color test, attenuated total reflectance Fourier transform-infrared spectroscopy (ATR FT-IR), and GC-MS can be used. These tests would exceed the Scientific Working Group for the Analysis of Seized Drugs (SWGDRUG) minimum testing criteria as well. Similarities in structural characteristics of drugs and their derivatives can cause problems for analysts and make drug analysis difficult and time consuming. On-site screening of controlled substances may be performed using the presumptive chemical color tests described in Chapter 2, the macroscopic and microscopic screening as described in Chapter 3, or by using handheld versions of instrumental spectroscopic methods including Fourier transform-infrared spectroscopy (FT-IR) and Raman spectroscopy as described in Chapter 4. Immunoassays may also be used in screening. Several commonly seized drugs, their chemical features and positive results for presumptive color tests, microcrystalline tests, and spectroscopic analyses are shown in Table 8.3.

Some drugs are acids while others are bases (Figure 8.5). Knowing the chemical form is useful for determining what solvent should be used to extract the drug from body fluids. For example, aspirin, ibuprofen, and cocaine hydrochloride are acidic drugs with ionizable protons. Caffeine, nicotine, and crack cocaine (hence the term "free base") are all examples of basic drugs with hydrogen acceptors. Aspirin is a legal nonsteroidal anti-inflammatory drug (NSAID) that is prepared using salicylic acid, a component of willow bark. It was first synthesized by Felix Hoffman in 1897. By binding to its cyclooxygenase target, it inhibits the synthesis of prostaglandins, thus reducing fever, inflammation, and blood clotting. Ibuprofen is another NSAID. As ionic compounds tend to be soluble in aqueous solution, cocaine hydrochloride that readily ionizes is soluble in water, which explains its use as a drug through moist regions

Table 8.3 Drugs and their chemical and spectroscopic features

Drug name	Class	Schedule	Molecular mass (g/mol)	Color test result(s)	Microcrystalline reagent(s)	MS qualifying ions (m/z)	IR (cm⁻¹)	UV/Vis λ_{max} (nm)
Heroin	Narcotic	I	369.417	Marquis (dark violet) Mecke (dark green) Froehde (purple)	Mercuric iodide Platinic chloride Mercuric chloride	162, 268, 310, 327, 369	1738, 1447, 1367, 1230, 1192, 1037, 885, 775 (FT-ATR) 1762, 1740, 1621, 1447, 1367, 1232 (FT-IR)	279
Fentanyl	Narcotic	II	336.479	Marquis (orange)	–	146, 189, 202, 336	1724, 1577, 1221, 702, 584 (FT-ATR)	–
Cocaine base	Stimulant	II	303.358	Acidified cobalt thiocyanate (blue, flaky precipitate) Scott's (blue ppt, then pink, then pink over blue) Household bleach (base floats and separates from HCl)	Gold chloride Platinic chloride	82, 182, 198, 303	2946, 1738, 1710, 1452, 1276, 713 (FT-IR: Base) 1732, 1713, 1267, 1109, 731 (FT-IR: HCl)	233, 275
Codeine	Narcotic	II	299.37	Marquis (dark violet) Mecke (green blue) Froehde (blue green)	Potassium cadmium iodide Potassium tri-iodide	124, 162, 229, 299	3422, 2917, 2838, 1534, 1452, 1277, 1254, 1203, 1157, 1120, 1100, 1054, 788 (FT-IR: Base)	–
Carfentanil	Narcotic	II	394.515	–	–	42, 105, 187, 303	1725, 1634, 1592, 1352, 1225, 1179, 1099, 761, 703, 613, 422 (ATR-FT-IR)	–
Marijuana (THC)	Hallucinogen	I	314.469	Duquenois-levine (violet-purple color in chloroform layer)	–	231, 243, 258, 271, 299, 314	2929, 1624, 1578, 1425, 1184, 1039 (FT-IR)	276, 283
Morphine	Narcotic	II	285.343	Marquis (purple violet) Mecke (dark green) Froehde (purple-grey)	Potassium cadmium iodide Potassium tri-iodide	124, 162, 215, 268, 285	3210, 2937, 1473, 1446, 1247, 1120, 1607, 978, 943, 834, 832, 759 (FT-IR: Base)	285, 298
Mescaline	Hallucinogen	I	211.261	Marquis (orange) Liebermann (black) Froehde (brown; colorless) Mandelin (green; violet; grey) Mecke (greenish brown; brown) Vitali Morin (dull red; purple)	Wagenaar Picric acid Mercuric chloride Gold chloride Potassium bismuth iodide	139, 151, 167, 182, 211	3351, 2930, 1684, 1590, 1513, 1422, 1322, 1242, 1106, 1047, 995, 829, 779, 530 (ATR FT-IR: Base)	268

(Continued)

Table 8.3 (Continued) Drugs and their chemical and spectroscopic features

Drug name	Class	Schedule	Molecular mass (g/mol)	Color test result(s)	Microcrystalline reagent(s)	MS qualifying ions (m/z)	IR (cm⁻¹)	UV/Vis λ_{max} (nm)
Psilocybin	Hallucinogen	I	284.252	Marquis (dull orange) Froehde (greenish blue; yellow)	–	58, 171, 180, 185, 204	3180, 2325, 1504, 1442, 1349, 1230, 1101, 1042, 921, 855, 786, 747, 497 (ATR FT-IR)	268
Oxycodone	Narcotic	II	315.369	Marquis (yellow; brown; violet) Liebermann (strong scarlet) Froehde (strong brown-yellow)	Platinic bromide Platinum bromide Iodine-potassium iodide	70, 140, 230, 258, 315	2936, 1728, 1604, 1500, 1259, 1038, 936, 814 (FT-IR: Base)	280
Ketamine	Hallucinogen	III	237.727	Janovsky (HCl: purple ppt; Base: weak purple)	Platinic iodide	152, 180, 182, 209, 237	3352, 2944, 1700, 1429, 1130, 1035, 752, 718, 429 (FT-IR: Base)	269
Lysergic acid diethylamide (LSD)	Hallucinogen	I	323.44	p-DMAB (violet)	–	181, 207, 221, 323	1626, 1446, 777, 748, 627 (FT-IR: Base)	311
Methamphetamine	Stimulant	II	149.237	Marquis (orange-brown) Sodium nitroprusside (blue)	Platinic chloride Gold chloride Bismuth iodide	58, 91, 134, 149	2966, 1493, 1453, 741, 700 (FT-IR: Base)	257
Phencyclidine (PCP)	Hallucinogen	II	243.394	p-DMAB (red)	Potassium permanganate	84, 91, 200, 243	1440, 1100, 755, 702, 585, 524 (FT-IR: Base)	258
Amphetamine	Stimulant	II	135.21	Marquis (orange; brown) Mandelin (dark green) Liebermann (red/orange)	Gold chloride Platinic chloride Gold bromide	44, 65, 91, 120, 135	1478, 1446, 743, 700, 503 (FT-IR: Base)	257
MDMA (Ecstasy)	Hallucinogen	I	193.246	Marquis (purple; black)	Gold chloride Platinic chloride	58, 77, 135, 193	2964, 2787, 1487, 1440, 1371, 1336, 1244, 1188, 1036, 930, 862, 802, 636, 604 (ATR FT-IR: Base)	–
Methaqualone	Depressant	III	250.301	Acidified cobalt thiocyanate (blue) Fischer-Morris (pink)	Potassium permanganate Sodium carbonate	91, 132, 235, 250	1676, 1608, 779, 760, 658 (FT-IR: Base)	234

Source: DEA, www.dea.gov/pr/multimedia-library/publications/drug_of_abuse.pdf.

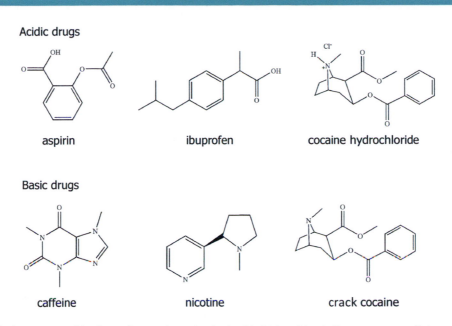

Figure 8.5   Acidic (top row: aspirin, ibuprofen, and cocaine hydrochloride) and basic (bottom row: caffeine, nicotine, and crack cocaine) drugs. (Structure prepared by Ashley Cowan.)

such as the nose or mouth. In contrast, crack cocaine is insoluble in water and is smoked. The smoke is absorbed by nonpolar membranes including those in the brain. ATR FT-IR spectroscopy can be used to differentiate the acid and base forms of cocaine.

Seized methamphetamine, or ice, is often a crystalline white powder. In a recent case, a shipment was seized by the Australian Federal Police Forensics Officer in Melbourne, Australia. A total of 903 kg (worth 900 million Australian dollars or $911 million US dollars) of methamphetamine smuggled from China was concealed in hollow wooden floorboards. The methamphetamine testing procedure exemplified the three-pronged approach. Colorimetric testing of methamphetamine using the Marquis test yielded an orange to orange-brown color and a light tan color with each of the Simon and sulfuric acid tests. The color and morphology were documented using a stereoscope. Confirmatory testing for methamphetamine is often conducted using GC-MS and FT-IR for dual confirmation at crime laboratories, although labs vary in their capabilities and some labs will instead utilize liquid chromatography–mass spectrometry.

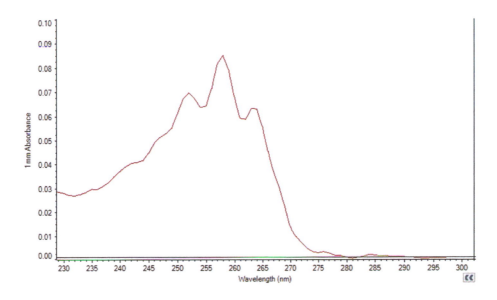

Figure 8.6   UV-Vis spectrum for methamphetamine collected using a NanoDrop UV-Vis spectrophotometer.

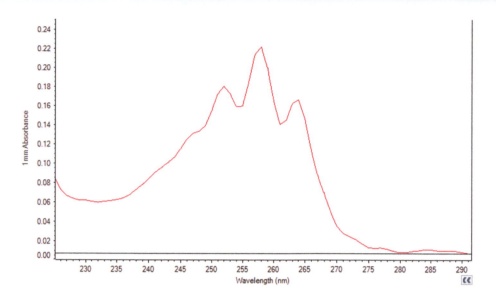

Figure 8.7   UV-Vis spectrum for phentermine collected using a NanoDrop UV-Vis spectrophotometer.

A lab may also need to differentiate very similar drugs including positional isomers, such as those of methamphetamine. Positional isomers are difficult to differentiate using MS. IR spectroscopy or GC-NMR can be used to differentiate and identify positional and cis/trans isomers. Phentermine, a positional isomer of methamphetamine with the same molecular formula and mass of 185.7 g/mol, has presented difficulty in methamphetamine identification when the drugs are comingled. Methamphetamine contains a secondary amine while phentermine has a primary amine due to differences in the connection of the functional group to the parent compound. The two drugs have essentially identical ultraviolet-visible (UV-Vis) spectra (Figures 8.6 and 8.7) and similar electron impact ionization mass spectra (Figures 8.8 and 8.9). Both drugs exhibit a base peak at m/z 58 amu in the mass spectrum. While UV-Vis can be coupled to LC instruments for identification of the eluting drug with the retention time from the chromatography, UV-Vis alone is not unique enough to differentiate and identify methamphetamine from phentermine as both compounds exhibit three strong absorbance peaks at 250, 257, and 265 nm, as documented in the IDDA reference library. ATR FT-IR spectra (Figure 8.10) *can* be used to differentiate the two drugs if the seized material is pure:

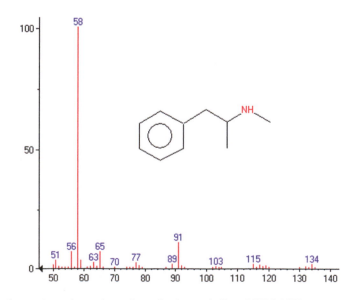

Figure 8.8   Mass spectrum for methamphetamine collected using an Agilent 5975C MSD.

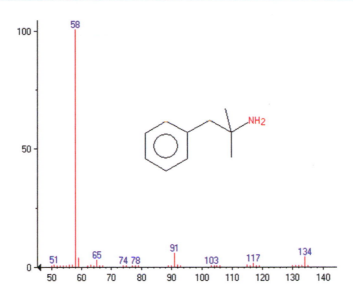

Figure 8.9 Mass spectrum for phentermine collected using an Agilent 5975C MSD.

methamphetamine is characterized by strong peaks at 1491, 1456, 752, and 702 cm⁻¹ and phentermine is character-ized by strong peaks at 1393, 1372, 730, and 702 cm⁻¹.

Gas chromatography (GC) can resolve a unique retention time for the drugs depending on the method used. The retention time for methamphetamine using the Agilent Technologies 7890A GC is very similar to phen-termine but can be resolved at 6.658 min using a ramp of 5°C/min from 100 to 270°C equipped with a DB-5 (30 m × 25 µm × 0.250 µm) column while phentermine elutes at 7.122 min. A search using a validated library can be used for comparison. Although the isomers can be distinguished based on retention time, SWGDRUG states that this method alone is not sufficient for identification. As a mass spectrometer is typically coupled to the GC in the crime laboratory, using electron impact ionization, methamphetamine is characterized by its fragment qualifying ions of 56, 58, and 91 atomic mass units (amu). Phentermine can be resolved from methamphetamine by its mass spectrum qualifying ions 58, 91, and 134 amu. Peak height ratios may also be used. Together with a third method, such as ATR FT-IR spectroscopy, SWGDRUG recommendations can be met.

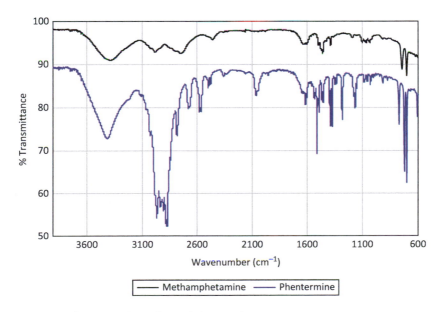

Figure 8.10 ATR FT-IR spectra of methamphetamine and phentermine.

QUESTIONS

1. The Controlled Substances Act was amended in 1984 to allow the DEA administrator to ____.
   a. Place a substance temporarily in Schedule I upon imminent hazard to public safety
   b. Place a substance permanently in Schedule I upon imminent hazard to public safety
   c. Place a substance temporarily in Schedule V upon imminent hazard to public safety
   d. Place a substance permanently in Schedule V upon imminent hazard to public safety

2. Drugs that cause marked alterations in mood, attitude, thought processes, and perceptions are called ___.
   a. Stimulants   b. Depressants
   c. Analgesics   d. Hallucinogens   e. None of the above

3. The principal drug found in coca leaves is known as ____.
   a. Morphine   b. Caffeine
   c. Cocaine   d. Codeine   e. Cannabis

4. The principal narcotic drug found in opium is known as ____.
   a. Morphine   b. Caffeine
   c. Oxycontin   d. Codeine   e. Demerol

5. Which of the following is a confirmatory test that can differentiate close compounds such as amphetamines and methamphetamines?
   a. Fluorescence   b. NIK kits
   c. FT-IR   d. Light microscopy

6. List the classes of drugs and give an example of each.

7. Compare and contrast different types of controlled substances cases.

8. What are new (or novel) psychoactive substances and why are forensic chemists concerned with them?

9. Which features of seized substances do forensic chemists in a crime lab use in their analysis?

10. You are asked to investigate "street" drug labs. From the description of chemicals seized, what drug is most likely being produced?
    a. Brown milky pods, lime, acetic anhydride, hydrochloric acid, sodium carbonate, ethyl ether: _____
    b. Ephedrine tablets, denatured alcohol, red phosphorous, iodine crystals, sodium hydroxide, mineral spirits, sulfuric acid: _____
    c. Piperidine, potassium cyanide, cyclohexanone, bromobenzene, magnesium metal shavings: _____
    d. Finely chopped plant material, ice water, butane: _____
    e. Sassafras oil, glacial acetic acid, hydrogen bromide gas, toluene, isopropyl alcohol, methylamine: _____
    f. Finely chopped plant material, acetone, gasoline, lime, sulfuric acid, potassium permanganate, ethyl acetate, hydrochloric acid: _____

Bibliography

Arbo, M.D., M.L. Bastos, and H.F. Carmo. 2012. Piperazine compounds as drugs of abuse. *Drug Alcohol Depend.* 122(3):174–185.

Arunotayanum, W. and S. Gibbons. 2012. Natural product 'legal highs'. *Nat. Prod. Rep.* 29(11):1304–1316.

Ashton, J.C. 2012. Synthetic cannabinoids as drugs of abuse. *Curr. Drug Abuse Rev.* 5(2):158–168.

Bell, S. 2013. *Forensic Chemistry*, 2nd edn. Upper Saddle River, NJ: Pearson Prentice Hall.

Büchi, G., D.L. Coffen, K. Kocsis, P.E. Sonnet, and F.E. Ziegler. 1966. The total synthesis of iboga alkaloids. *J. Am. Chem. Soc.* 88(13):3099–3109.

Bodnar, W.M., V. McGuffin, and R. Smith. 2017. Statistical comparison of mass spectra for identification of amphetamine-type stimulants. *Forensic Sci. Int.* 270:111–120.

Brandt, S.D., S. Freeman, H.R. Sumnall, F. Measham, and J. Cole. 2011. Analysis of NRG 'legal highs' in the UK: Identification and formation of novel cathinones. *Drug Test. Anal.* 3(9):569–575.

Brennan, R. and M.C. Van Hout. 2014. Gamma-hydroxybutyrate (GHB): A scoping review of pharmacology, toxicology, motives for use, and user groups. *J. Psychoactive Drugs* 46(3):243–251.

Center for Substance Abuse Research UMD. 2013. Amphetamines. www.cesar.umd.edu/cesar/drugs/amphetamines.pdf (accessed January 29, 2018).

Clarkson, E.D., D. Lesser, and B.D. Paul. 1998. Effective GC-MS procedure for detecting iso-LSD in urine after base-catalyzed conversion to LSD. *Clin. Chem.* 44(2):287–292.

Cottencin, O., B. Rolland, and L. Karila. 2014. New designer drugs (synthetic cannabinoids and synthetic cathinones): Review of literature. *Curr. Pharm. Des.* 20(25):4106–4111.

Cowan, A.F. and K.M. Elkins. 2017. Detection and identification of *Psilocybe cubensis* DNA using a real-time polymerase chain reaction high resolution melt (PCR-HRM) assay. *J. Forensic Sci.* 2017 Dec 1. doi:10.1111/1556-4029.13714.

Dagne, E., Y. Adugna, E. Kebede, and Y. Atilaw. 2010. Determination of levels of cathine in khat (*Catha edulis*) leaves and its detection in urine of khat chewers: A preliminary report. *Ethiopian e-Journal for Research and Innovation Foresight* 2(1):7–22.

DEA. Control of a chemical precursor used in the illicit manufacture of fentanyl as a List I chemical.https://www.deadiversion.usdoj.gov/fed_regs/rules/2007/fr0423.htm (accessed May 25, 2018).

DEA. Schedules of controlled substances: Temporary placement of fentanyl-related substances in Schedule I. https://www.deadiversion.usdoj.gov/fed_regs/rules/2018/fr0206_4.htm (accessed May 25, 2018).

DEA. 2016. N,N-Dimethyltryptamine (DMT) www.deadiversion.usdoj.gov/drug_chem_info/dmt.pdf (accessed January 29, 2018).

Draffan, G.H., R.A. Clare, and F.M. Williams. 1973. Determination of barbiturates and their metabolites in small plasma samples by gas chromatography-mass spectrometry. Amylorbarbitone and 3′-hydroxyamylobarbitone. *J. Chromatogr.* 75(1):45–53.

Drug Enforcement Administration, Department of Justice. 2017. Schedules of controlled substances: Placement of 10 synthetic cathinones into Schedule I. Final rule. *Fed. Regist.* 82(39):12171–12177.

Elkins, K.M., A.C.U. Perez, and A.A. Quinn. 2017. Simultaneous identification of four "legal high" plant species in a multiplex PCR high resolution melt assay. *J. Forensic Sci.* 62:593–601.

Elkins, K.M., A.F. Cowan, T.H. Boise. 2018. Detection and identification of kratom using chemical tests: A review. *Drug Test. Anal.*, in preparation.

Elkins, K.M. 2018. Identification of "legal-high" plant species in forensic labs. *Expert Witness J.*, in press.

EMCDDA. 2015. Methylenedioxymethamphetamine (MDMA or 'Ecstasy') drug profile. www.emcdda.europa.eu/publications/drug-profiles/mdma (accessed January 29, 2018).

Gautam, L., A. Shanmuganathan, and M.D. Cole. 2013. Forensic analysis of cathinones. *Forensic Sci. Rev.* 25:47–64.

German, C.L., A.E. Fleckenstein, and G.R. Hanson. 2014. Bath salts and synthetic cathinones: An emerging designer drug phenomenon. *Life Sci.* 97(1):2–8.

Gil, D., P. Adamowicz, A. Skulska, B. Tokarczyk, and R. Stanaszek. 2013. Analysis of 4-MEC in biological and non-biological material—Three case reports. *Forensic Sci. Int.* 228:e11–e15.

Girard, J.E. 2007. *Criminalistics: Forensic Science and Crime.* Burlington, MA: Jones and Bartlett Learning.

Goerig, M., D. Bacon, and A. van Zundert. 2012. Carl Koller, cocaine, and local anesthesia: Some less known and forgotten facts. *Reg. Anesth. Pain Med.* 37(3):318–324.

Governing. 2018. State Marijuana Laws in 2018 Map. www.governing.com/gov-data/state-marijuana-laws-map-medical-recreational.html (accessed January 29, 2018).

Hepner, F., E. Cszasar, E. Roitinger, and G. Lubec. 2005. Mass spectrometrical analysis of recombinant human growth hormone (Genotropin®) reveals amino acid substitutions in 2% of the expressed protein. *Proteome Sci.* 3:1.

Hilderbrand, R.L. 2011. High-performance sport, marijuana, and cannabimimetics. *J. Anal. Toxicol.* 35(9):624–637.

Huffman, J.W. 2005. CB2 receptor ligands. *Mini Rev. Med. Chem.* 5(7):641–649.

Jakabova, S., L. Vincze, A. Farkas, F. Kilar, B. Boros, and A. Felinger. 2012. Determination of tropane alkaloids atropine and scopolamine by liquid chromatography-mass spectrometry in plant organs of Datura species. *J. Chromatogr. A* 1232:295–301.

Johnson, J.B. 1939. The elements of Mazatec witchcraft. *Ethnol. Stud.* 9:119–149.

Johnson, L.A., R.L. Johnson, and R.-B. Portier. 2013. Current "Legal Highs". *J. Emerg. Med.* 44(6):1108–1115.

Juszczak, G.R. and A.H. Swiergiel. 2013. Recreational use of D-lysergamide from the seeds of Argyreia nervosa, *Ipomoea tricolor, Ipomoea violacea*, and *Ipomoea purpurea* in Poland. *J. Psychoact. Drugs* 45(1):79–93.

Kavanagh, P.J., J.F. O'Brien, C. O'Donnell, R. Christie, J.D. Power, and S.D. McDermott. 2012. The analysis of substituted cathinones. Part 3. Synthesis and characterisation of 2,3-methylenedioxy substituted cathinones. *Forensic Sci. Int.* 216:19–28.

Lanthorn, T.H., D.A. Dimaggio, P.C. Contreras, J.B. Monahan, L.M. Pullan, G. Handelmann, N.M. Gray, and T.L. O'Donohue. 1988. Alpha- and beta-endopsychosins: Physiological actions and interactions with excitatory amino acids. In *Neurobiology of Amino Acids, Peptides and Trophic Factors*, eds. J.A. Ferrendelli, R.C. Collins, and E.M. Johnson, 181–197. Boston, MA: Springer.

Laussmann, T., I. Grzesiak, A. Krest, K. Stirnat, S. Meier-Giebing, U. Ruschewitz, and A. Klein. 2015. Copper thiocyanato complexes and cocaine—A case of "black cocaine." *Drug Test. Anal.* 7(1):56–64.

Leffler, A.M., P.B. Smith, A. de Armas, and F.L. Dorman. 2014. The analytical investigation of synthetic street drugs containing cathinone analogs. *Forensic Sci. Int.* 234:50–56.

Li, L. and P.E. Vlisides. 2016. Ketamine: 50 years of modulating the mind. *Front. Hum. Neurosci.* 10:612.

Makino, Y., S. Tanaka, S. Kurobane, M. Nakauchi, T. Terasaki, and S. Ohta. 2003. Profiling of illegal amphetamine-type stimulant tablets in Japan. *J. Health Sci.* 49(2): 29–137.

Mansnerus, L. 1996. Timothy Leary, pied piper of psychedelic 60's, dies at 75. www.nytimes.com/1996/06/01/us/timothy-leary-pied-piper-of-psychedelic-60-s-dies-at-75.html (accessed January 29, 2018).

Mariotti, K.C., L.G. Rossato, P.E. Fröehlich, and R.P. Limberger. 2013. Amphetamine-type medicines: A review of pharmacokinetics, pharmacodynamics, and toxicological aspects. *Curr. Clin. Pharmacol.* 8(4):350–357.

Marnell, T., ed. 1999. *Drug Identification Bible*, 4th edn. Grand Junction, CO: Amera-Chem.

Metzner, R., ed. 2005. *Sacred Mushroom of Visions: Teonanácatl: A Sourcebook on the Psilocybin Mushroom*, 2nd edn. Paris, ME: Park Street Press.

Millard, J.T. 2006. *Adventures in Chemistry*. Boston, MA: Cengage Learning.

Mills, B., A. Yepes, and K. Nugent. 2015. Synthetic cannabinoids. *Am. J. Med. Sci.* 350(1):59–62.

Miotto, K., J. Striebel, A.K. Cho, and C. Wang. 2013. Clinical and pharmacological aspects of bath salt use: A review of the literature and case reports. *Drug Alcohol Depend.* 132(1–2):1–12.

Momaya, A., M. Fawal, and R. Estes. 2015. Performance-enhancing substances in sports: A review of the literature. *Sports Med.* 45(4):517–531.

Musselman, M.E. and J.P. Hampton 2014. "Not for human consumption": A review of emerging designer drugs. *Pharmacotherapy* 34(7):745–757.

NPS Expert Review Panel. 2014. New psychoactive substances review. www.gov.uk/government/uploads/system/uploads/attachment_data/file/368583/NPSexpertReviewPanelReport.pdf (accessed December 2, 2017).

Pawlik, E., H. Mahler, B. Hartung, G. Plässer, and T. Daldrup. Drug-related death: adulterants from cocaine preparations in lung tissue and blood. *Forensic Sci Int.* 2015;249:294–303. doi:10.1016/j.forsciint.2015.02.006.

PubChem. https://pubchem.ncbi.nlm.nih.gov/ (accessed January 31, 2018).

Rodriquez, R., H. Bergkvist, W. Boonchuay, R. Dahlenburg, A. Kemmenoe, T. Kishi, W. Krawczyk, I. Lurie, Y. Makino, T. McKibben, et al. 2006. Recommended methods for the identification and analysis of amphetamine, methamphetamine and their ring-substituted analogues in seized material. *United Nations Office on Drugs and Crime*, pp. 1–88.

Rudd, R.A. and P. Seth. 2016. Increases in drug and opioid-involved overdose deaths—United States, 2010–2015. *MMWR* 65(50–51):1445–1452. www.cdc.gov/mmwr/volumes/65/wr/mm655051e1.htm (accessed January 29, 2018).

Saferstein, R. 2015. *Criminalistics: An Introduction to Forensic Science*, 11th edn. Boston, MA: Pearson.

Santali, E.Y., A.-K. Codogan, N.N. Daeid, K.A. Savage, and O.B. Sutcliffe. 2011. Synthesis, full chemical characterisation and development of validated methods for the quantification of (±)-4'-methylmethcathinone (mephedrone): A new "legal high." *J. Pharmaceut. Biomed.* 56:246–255.

Seely, K.A., A.L. Patton, C.L. Moran, M.L. Womack, P.L. Prather, W.E. Fantegrossi, A. Radominska-Pandya, G.W. Endres, K.B. Channell, N.H. Smith, et al. 2013. Forensic investigation of K2, spice, and "bath salt" commercial preparations: A three-year study of new designer drug products containing synthetic cannabinoid, stimulant, and hallucinogenic compounds. *Forensic Sci. Int.* 233:416–422.

Shulgin, A.T. and D.E. Mac Lean. 1976. Illicit synthesis of phencyclidine (PCP) and several of its analogs. *Clin. Toxicol.* 9(4):553–560.

Sosa, N. and K.M. Elkins. 2017. Methodology for the separation and differentiation of methamphetamine and phentermine through gas chromatography-mass spectrometry (GC/MS), unpublished data.

Stout, S.M. and N.M. Cimino. 2014. Exogenous cannabinoids as substrates, inhibitors, and inducers of human drug metabolizing enzymes: A systematic review. *Drug Metab. Rev.* 46(1):86–95. doi:10.3109/03602532.2013.849268.

Stuart, J.H., J.J. Nordby, and S. Bell. 2014. *Forensic Science: An Introduction to Scientific and Investigative Techniques*, 4th edn. London: Taylor & Francis.

Sutter, M.E., J. Chenoweth, and T.E. Albertson. 2014. Alternative drugs of abuse. *Clin. Rev. Allerg. Immunol.* 46(1):3–18.

SWGDRUG. 2016. Scientific Working Group for the Analysis of Seized Drugs Recommendations Version 7.1. www.swgdrug.org/Documents/SWGDRUG%20Recommendations%20Version%207-1.pdf (accessed October 10, 2017).

SWGDRUG. 2017. SWGDRUG history. www.swgdrug.org/history.htm (accessed January 23, 2018).

The Associated Press. April 5, 2017. "Australian police make 'largest ever' crystal meth seizure." https://www.thestar.com/news/world/2017/04/05/australian-police-make-largest-ever-crystal-meth-seizure.html (accessed December 2, 2017).

The Centre for Social Justice. www.centreforsocialjustice.org.uk/ (accessed December 2, 2017).

United Nations Office on Drugs and Crime. 2014. World drug report 2014. New York, NY: United Nations. www.unodc.org/documents/data-and-analysis/WDR2014/World_Drug_Report_2014_web.pdf (accessed December 2, 2017).

Valente, M.J., P.G. Pinho, M.L. Bastos, F. Carvalho, and M. Carvalho. 2013. Khat and synthetic cathinones: A review. *Arch. Toxicol.* 88:15–45.

Weib, J., K. Kadkhodaei, and M. Schmid. 2017. Indirect chiral separation of novel amphetamine derivatives as potential new psychoactive compounds by GC-MS and HPLC. *Sci. Justice* 57:6–12.

World Anti-Doping Agency (WADA). 2011. The 2012 prohibited list international standard. www.wada-ama.org/sites/default/files/resources/files/WADA_Prohibited_List_2012_EN.pdf (accessed January 29, 2018).

Zaitsu, K., H. Nakayama, M. Yamanaka, K. Hisatsune, K. Taki, T. Asano, T. Kamata, M. Katagai, Y. Hayashi, M. Kusano, et al. 2015. High-resolution mass spectrometric determination of the synthetic cannabinoids MAM-2201, AM-2201, AM-2232, and their metabolites in postmortem plasma and urine by LC/Q-TOFMS. *Int. J. Legal Med.* 129(6):1233–1245.

Zawilska, J.B. and J. Wojcieszak. 2013. Designer cathinones—An emerging class of novel recreational drugs. *Forensic Sci. Int.* 231:42–53.

CHAPTER 9

Toxicology

KEY WORDS: forensic toxicology, toxicologist, alcohol, ethanol, ingestion, absorption, absorption rate, post-absorption, metabolism, blood alcohol concentration, alcohol deterrent drugs, excretion, Henry's Law, Breathalyzer, Intoxilyzer, fuel cell breath test, gas chromatography, metabolites, retention time, analytical scheme, poisons

LEARNING OBJECTIVES

- To understand the role of the forensic toxicologist in forensic testing
- To explain the process by which alcohol is consumed, absorbed, metabolized, and excreted
- To explain how field breath tests to detect alcohol work
- To list the factors that influence blood alcohol concentration
- To explain the use of gas chromatography–mass spectrometry (GC-MS) in determining blood alcohol concentration
- To identify secondary metabolites for commonly used drugs
- To understand that the retention time in urine varies for different drugs
- To describe how toxicologists use instrumental methods to detect and quantitate poisons

BLOOD ALCOHOL CONCENTRATION EXTRAPOLATION

Ms. Floyd was stopped while driving and, on administration of a blood alcohol concentration (BAC) test, registered a blood alcohol concentration of 0.069% at 10:30 p.m. Floyd was charged with aggravated driving under the influence (DUI) of alcohol.

In the United States, the legal drinking age for alcohol is 21 years old. Noticeable cognitive changes from alcohol consumption are present at a blood alcohol level of 0.02%–0.03%. In the United States, the legal limit of alcohol in the blood is 0.08% for persons operating a motor vehicle and persons operating a motor vehicle on a public highway are subject to the 1973 "implied consent" law. The law mandates that drivers must comply with blood alcohol testing if requested by a member of law enforcement or be subject to the loss of their driver's license. Aggravated DUI charges can result when the defendant is caught driving on a suspended or revoked license.

Drinking alcohol, CH_3CH_2OH, is a colorless liquid that is often mixed with other beverages for consumption. After ingestion, alcohol is absorbed in the stomach and small intestine over a period of 30 minutes to an hour after the alcohol is consumed. Factors that influence alcohol absorption rates include the sex and weight of the person consuming the alcohol, tolerance for alcoholic beverages, food consumed before or with the alcohol, and the number of drinks consumed. After absorption, the alcohol is metabolized in the body by alcohol dehydrogenase in the liver and excreted through the urine. Owing to its volatility, some alcohol is also released through the skin and breath. If the blood for the blood alcohol determination is sampled immediately after the defendant consumed the alcohol or prior to full absorption, the BAC would not have reached its peak concentration value. Likewise, if the blood alcohol is sampled in the post-absorption phase, samples taken earlier would be expected to reflect a

BLOOD ALCOHOL CONCENTRATION EXTRAPOLATION (continued)

higher BAC. Samples should be taken at arrest or soon thereafter to accurately reflect the driver's blood alcohol content at the time of the incident. In the laboratory, BAC determinations are typically performed using GC-MS spectroscopy.

The scientific expert witness in the case performed a retrograde time extrapolation of the BAC data. The extrapolation indicated that Floyd's BAC was between 0.082% and 0.095%, or above 0.08%, at the time she was driving. Based on the witness testimony, Floyd was convicted. However, the conviction was reversed in appellate court on appeal.

The scientific expert witness based the calculations on the elimination rate of alcohol of 0.01 g and 0.02 g per deciliter of blood per hour. However, for this calculation to be valid, a person must be known to metabolize alcohol at the normal rate and be in the post-absorption phase when the breath test is administered. Alcohol absorption is influenced by the type of food a person has eaten, the type of alcohol a person has consumed, the length of time during which the alcohol was consumed, and the isozyme of acetaldehyde dehydrogenase a person has. However, the expert witness had no data on what food or alcohol Floyd had consumed or how long she had been drinking. The expert witness assumed she had entered the elimination phase at 9:10 p.m. based on statements that she had her last drink at 7:30 p.m.

The appellate court reversed and stated "A retrograde extrapolation calculation based on a single breath test, and when many of the factors necessary to determine whether the defendant was in the elimination phase are unknown, is insufficient to provide a reliable calculation and invites the jury to determine guilt on an improper basis. Based on the specific circumstances presented in this case, we believe that the prejudicial effect of the retrograde extrapolation calculation substantially outweighed its probative value and that the trial court abused its discretion in admitting it."

BIBLIOGRAPHY

A court looks at BAC assumptions that aren't based on scientific investigation. https://csidds.com/2017/06/14/a-court-looks-at-bac-assumptions-that-arent-based-on-scientific-investigation/ (accessed January 25, 2018).

People V.F. 2014. IL App (2d) 120507, 1-2, 11 N.E.3d 335, 336. http://caselaw.findlaw.com/il-court-of-appeals/1662026.html (accessed January 25, 2018).

Forensic Toxicology is focused on the identification and quantification of legal and illegal substances to answer questions pertaining to violations of criminal law. Forensic toxicology assists with determinations of cause and manner of death through the testing of samples from body fluids, tissues, and organs. A major focus of forensic toxicology is the quantification of alcohol in the blood. Toxicologists also determine the presence of poisons including cyanide, carbon monoxide, and arsenic, and quantify the presence of drugs, including legal and illegal substances, contributing to unusual or erratic behavior, illness, or death. Toxicologists can determine if the quantities measured constitute alcohol use above the legal limit for operating a motor vehicle, and the quantities of substances in a drug overdose, poisoning, and drug interaction case.

Toxicologists are employed by the medical examiner or coroner offices, hospital laboratories, forensic laboratories, or private testing laboratories. Toxicologists evaluate submitted body fluid samples including blood, urine, vitreous humor, breastmilk, feces, meconium, hair, and tissue samples. Toxicologists often work with death investigators, medical examiners, and pathologists to determine the cause of death for a deceased.

The US Department of Transportation's National Highway Traffic Safety Administration reported that 32,719 people died in traffic crashes in 2013. The agency reports that 31% of those people (10,076) died in alcohol-related traffic incidents in 2013. National Institutes of Health (NIH) research shows that alcohol dependence is indicated in over 70% of people involved in alcohol-related crashes, but most of these people have never been arrested or received treatment to curb alcohol abuse. According to the US NIH, while the number of traffic-related deaths involving alcohol in the mid-1970s was over 60% and comprised two thirds of traffic deaths of young people aged 16–20, today the number of alcohol-related traffic deaths has been reduced to 37% for those of young people aged 16–20, representing a significant improvement in highway safety.

Drinking alcohol is *ethanol* or *ethyl alcohol* (CH_3CH_2OH). Its chemical structure is shown in Figure 9.1. In its pure form, it is a colorless liquid. Alcohol is usually mixed with ice, water, or other liquids for consumption as a beverage. In the United States, the minimum age to legally purchase and consume alcohol is 21 years, although in most countries the drinking age is 18 or 19 years old. In some areas of India, the drinking age is as high as 25 and 30 years old

Figure 9.1   Structure of ethanol. (Structure prepared by Ashley Cowan.)

while in some countries there is no minimum age required to buy or consume alcohol and in several Middle Eastern countries alcohol consumption is banned.

The digestive and circulatory systems are linked and alcohol is ingested and subsequently transported throughout the body. On *ingestion* of an alcoholic drink via the mouth, and its passage down the esophagus to the stomach and intestines, the blood concentration of alcohol slowly rises through *absorption* from the stomach where approximately 20% of the alcohol is absorbed and the small intestine where the rest is absorbed. In the blood, the alcohol in arteries travels away from the heart, in veins toward the heart, and in capillaries, or tiny blood vessels, exchanged via their walls to tissues. On reaching the lungs via the pulmonary artery, alveoli, small sacs in the lungs, allow for the exchange of alcohol from the blood to the lungs to be transported by the trachea.

The maximum concentration of alcohol in the blood, or *BAC*, registers as quickly as 30 minutes or as long as hours after the alcohol has been completely absorbed, depending on how many drinks were consumed and if there was food in the stomach to slow the absorption (Figure 9.2). The *absorption rate* is affected by the time taken to consume the alcoholic drink (or drinks), its alcohol content, the number of ounces consumed, the body weight and tolerance of the individual, the sex of the individual, and the presence or absence of food in the stomach. Following absorption, the *post-absorption* period begins and the BAC begins to decline.

Alcohol *metabolism* begins immediately on the alcohol reaching the liver via the blood, primarily through the alcohol dehydrogenase-acetaldehyde dehydrogenase pathway (Figure 9.3). Alcohol is detoxified by one or more of the many alcohol dehydrogenase (ADH) enzymes present in humans. The product of the reaction is acetaldehyde, which is toxic. Interestingly, the ADH enzyme that is more common in Asians has a higher catalytic activity than the isoform possessed by Europeans and their descendants. A higher catalytic activity leads to a more rapid conversion of alcohol to acetaldehyde. In chronic alcohol users, the microsomal ethanol oxidizing system special cytochrome P450 monooxygenase is upregulated and metabolizes alcohol independently of the ADH pathway. While the ADH pathway is coupled to the respiratory chain in the mitochondria, the microsomal ethanol oxidizing pathway is coupled directly to molecular oxygen and the energy is discarded as heat which leads to flushing in users. Acetaldehyde dehydrogenases (ALDHs) metabolize the acetaldehyde to yield acetic acid. As with ADH, there are multiple isozymes, or forms, of this enzyme including ALDH1 (in the cytosol) and the much higher activity ALDH2 (in the mitochondria) (Figure 9.4). Some people, especially Asians, display a higher sensitivity to alcohol as compared to Europeans and Africans due to a mutation in their ALDH2 enzyme that renders it catalytically inactive. The rapid ADH and mutation in ALDH2 are the reason many Asians avoid alcohol. The acetic acid produced in the ADH-ALDH pathway enters the Krebs cycle (also known as the citric acid cycle or tricarboxylic acid cycle) where it is used to generate energy for the cell and is broken down to carbon dioxide and water. The detoxification rate, or elimination rate, is approximately 0.015% w/v (0.015 g/100 mL) per hour, on average, although this can vary by as much as 30% between individuals due to the varying levels of ADH and ALDH enzymes and liver function.

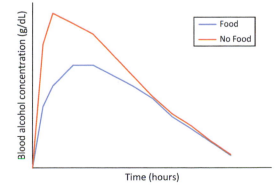

Figure 9.2   BAC over time in hours with and without food intake.

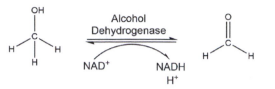

Figure 9.3   ALDH enzymatic reaction to convert ethanol to formaldehyde.

Individuals with low ALDH levels or ALDH2 accumulate acetaldehyde, which has a powerful vasodilating effect, leading to unpleasant side effects including flushing (red color) of the skin and hangover-like symptoms such as headache, ill feeling, and trembling. *Alcohol deterrent drugs* are focused on blocking the activity of the functional ALDH2 enzyme by serving as inhibitors. Two such examples are Antabuse® (disulfuram) and naltrexone. Figure 9.5 shows Antabuse® binding to ALDH2.

Alcohol and its detoxification products are *excreted* through the skin, breath, and urine. The carbon dioxide gas is transported to the lungs and transported by the trachea to the breath for excretion where some alcohol is also excreted as volatile oxygen vapors. A small amount of ingested alcohol is excreted through the skin due to perspiration. Alcohol and its metabolites can also be excreted in the urine.

In sum, alcohol is a drug that affects all organs and systems, but especially the central nervous system and particularly the brain. Through research, the BAC has been determined to be directly proportional to the brain-alcohol concentration and the effect of alcohol is directly proportional to its concentration in nerve cells. In the brain, alcohol

```
CLUSTAL 2.1 multiple sequence alignment

ALDH1A1__HUMAN     ----------------MSSSGTPDLPVLLTDLKIQYTKIFINNEWHDSVSGKKFPVFNPA
ALDH2_HUMAN        MLRAAAARFGPRLGRRLLSAAATQAVPAPNQQPEVFCNQIFINNEWHDAVSRKTFPTVNPS
                              :*::.*   :*.    : ::   .:*********:** *.**..**:

ALDH1A1__HUMAN     TEEELCQVEEGDKEDVDKAVKAARQAFQIGSPWRTMDASERGRLLYKLADLIERDRLLLA
ALDH2_HUMAN        TGEVICQVAEGDKEDVDKAVKAARAAFQLGSPWRRMDASHRGRLLNRLADLIERDRTYLA
                   * * :*** *************** ***:***** ****.***** :********* **

ALDH1A1__HUMAN     TMESMNGGKLYSNAYLNDLAGCIKTLRYCAGWADKIQGRTIPIDGNFFTYTRHEPIGVCG
ALDH2_HUMAN        ALETLDNGKPYVISYLVDLDMVLKCLRYYAGWADKYHGKTIPIDGDFFSYTRHEPVGVCG
                   ::*:::.** *  :** **   :* *** ****** :*:******.**:******:****

ALDH1A1__HUMAN     QIIPWNFPLVMLIWKIGPALSCGNTVVVKPAEQTPLTALHVASLIKEAGFPPGVVNIVPG
ALDH2_HUMAN        QIIPWNFPLLMQAWKLGPALATGNVVVVMKVAEQTPLTALYVANLIKEAGFPPGVVNIVPG
                   *********:*   **:****: **.**:* *********:**.****************

ALDH1A1__HUMAN     YGPTAGAAISSHMDIDKVAFTGSTEVGKLIKEAAGKSNLKRVTLELGGKSPCIVLADADL
ALDH2_HUMAN        FGPTAGAAIASHEDVDKVAFTGSTEIGRVIQVAAGSSNLKRVTLELGGKSPNIIMSDADM
                   :********:** *:*********:::*: *** ***************** *:::***:

ALDH1A1__HUMAN     DNAVEFAHHGVFYHQGQCCIAASRIFVEESIYDEFVRRSVERAKKYILGNPLTPGVTQGP
ALDH2_HUMAN        DWAVEQAHFALFFNQGQCCCAGSRTFVQEDIYDEFVERSVARAKSRVVGNPFDSKTEQGP
                   * *** **..:*::***** *.** **:*.******.*** ***. ::***: . . ***

ALDH1A1__HUMAN     QIDKEQYDKILDLIESGKKEGAKLECGGGPWGNKGYFVQPTVFSNVTDEMRIAKEEIFGP
ALDH2_HUMAN        QVDETQFKKILGYINTGKQEGAKLLCGGGIAADRGYFIQPTVFGDVQDGMTIAKEEIFGP
                   *:*: *:.***. *:*:**:***** **** .::::***:*****.:* * * ********

ALDH1A1__HUMAN     VQQIMKFKSLDDVIKRANNTFYGLSAGVFTKDIDKAITISSALQAGTVWVNCYGVVSAQC
ALDH2_HUMAN        VMQILKFKTIEEVVGRANNSTYGLAAAVFTKDLDKANYLSQALQAGTVWVNCYDVFGAQS
                   * **:***:::*: ****: ****: ***:*.*****:*** :*.************.*..**.

ALDH1A1__HUMAN     PFGGFKMSGNGRELGEYGFHEYTEVKTVTVKISQKNS
ALDH2_HUMAN        PFGGYKMSGSGRELGEYGLQAYTKVKTVTVKVPQKNS
                   ****:****.*********:: **:*******:.****
```

Figure 9.4   ClustalW 2.1 ALDH1A1 and ALDH2 protein sequence alignment with Glu504Lys mutation in yellow.

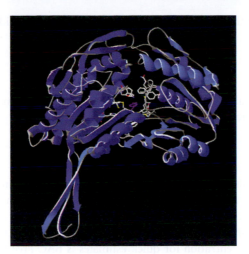

Figure 9.5 Crystal structure of human ALDH2 (3sz9.pdb) bound to 1-(4-ethylbenzene)prop-2-en-1-one inhibitor.

affects the forebrain, involved in thought, reasoning, behavior, and memory; and then the rear brain, responsible for speech, emotion, and understanding abstract concepts; and finally the medulla, which regulates respiration and heart activity. As previously discussed, alcohol causes flushing of the skin and sweating and can cause bruising. In the circulatory system, it can cause changes in the red blood cells in the blood and high blood pressure, irregular pulse, and an enlarged heart. In the musculoskeletal system, it causes weakness and loss of muscle tissue. It is linked to an increased risk of infection, including tuberculosis. In the digestive system, it can cause the lining of the stomach to become inflamed, bleeding and ulcers in the stomach, and an inflamed lining and ulcers in the intestines. It can cause impotence in males due to shrinking testicles and damaged and lowered sperm counts. Alcohol can cause inflammation in the pancreas resulting in pain. Finally, it can cause liver swelling, pain, loss of metabolic function, and diseases including hepatitis, cirrhosis, and cancer.

The use of alcohol increases the risk of accidents when operating a motor vehicle or heavy equipment. The rate of accidents increases exponentially due to alcohol's effects (Figure 9.6). A BAC (w/v) of as low as 0.02%–0.03% will cause noticeable cognitive changes. At 0.10%–0.15%, a person will appear obviously intoxicated and may exhibit signs of delirium. BACs of 0.24%–0.36% typically lead to loss of consciousness and a concentration of 0.48% has been linked to severe coma and even death. While alcohol consumption is legal in the United States by persons 21 years of age or older, the current BAC legal limit for a person operating a motor vehicle is 0.08% (w/v). Legal limits are lower in other countries including France, Germany, Ireland, Japan (0.05%), and especially Sweden (0.02%).

The concentration of alcohol in the blood can be determined directly by sampling the blood or indirectly by sampling the breath. Research has shown the concentration of alcohol exhaled by the breath to be directly proportional to the BAC. According to *Henry's Law* (Equation 9.1), when a volatile chemical is dissolved in a liquid—such as blood—and

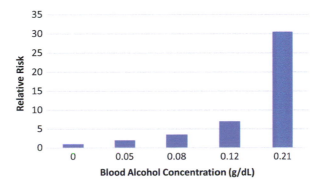

Figure 9.6 Relative risk increases as blood alcohol concentration increases.

brought to equilibrium with air in the alveoli, the pressure of the gas (p_{gas}) divided by the concentration of the gas (c_{aq}) in the liquid is a constant (k).

$$k = p_{gas}/c_{aq}$$
(9.1)

BAC can be estimated mathematically (Equation 9.2) or by using a smartphone app that makes the computation (Figure 9.7). Alternatively, a nomagram can be used.

$$C_p = D/\big((V_d)(W)\big)$$
(9.2)

In Equation 9.2, C_p is the BAC in units of grams per liter or percent (w/v), D is the dose of alcohol in grams, V_d is the volume of distribution (0.70 for men and 0.60 for women), and W is the weight in kilograms. In a case in which a 150-pound female consumes 3.5 1.5-ounce drinks consisting of 80 proof alcohol, the computed BAC will be 0.1126% w/v. Inserting the same values, but using the volume of distribution for a male, leads to a computation of 0.1026% w/v. One app that computes blood alcohol concentration for quick estimates is BAC Free (Apple Marketplace). BAC Free estimates blood alcohol concentration based on submitted values including number of drinks, type of drink, sex, weight, and elapsed time. For example, entering the values 150 pounds, female, and 3.5 1.5-ounce drinks leads to a BAC calculation of 0.109% w/v (Figure 9.7). Using the same values for a male, the BAC computed by the app is 0.099% w/v. Therefore, the app conservatively estimates (overestimates) blood alcohol (due to assuming that the person consumed the alcohol on an empty stomach, had a low tolerance, or other reason). Using a nomogram for each situation such as the one in Saferstein's *Criminalistics* (2015) textbook, full or empty stomach, the estimated BAC is determined to be 0.062% w/v or 0.105% w/v respectively. The BAC app output values correlate most strongly with the values in Equation 9.2 using the volume of distribution for a male and nomogram estimate for a full stomach.

The concentration of alcohol in the blood can be measured indirectly using breath tests and directly using blood tests. There are several breath tests. The first alcohol breath test for car drivers was the "Drunkometer" first demonstrated on December 31, 1938. A person being tested breathed into a balloon and the breath was transferred to be tested in a chemical reaction with potassium permanganate and sulfuric acid. Breath tests are rapid, convenient, performed on

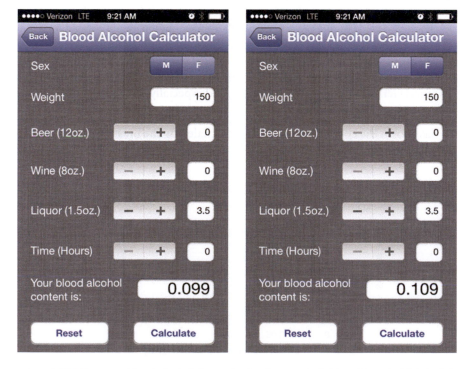

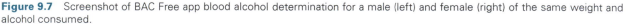

Figure 9.7   Screenshot of BAC Free app blood alcohol determination for a male (left) and female (right) of the same weight and alcohol consumed.

site, clearly report the BAC, require no invasive sampling measures, and are reliable. Three breath tests include the Breathalyzer, an infrared test, and a fuel cell test.

The Breathalyzer is a tool that was developed in 1954. Based on a reaction of known stoichiometry, the alcohol (ethanol) is reacted with the potassium dichromate reagent under acidic conditions and using a silver nitrate catalyst; the chromium sulfate product is detected using spectrometry at a fixed wavelength. The Breathalyzer collection cylinder collects 56.5 mL of breath, but some of the sample remains in the delivery tube and does not reach the test sample of 52.5 mL. The breath originating from the mouth is 34°C; it is heated to 50°C in the cylinder. Experimentally, it has been determined that 2100 mL of alveolar breath contains the same quantity of alcohol as 1 mL of blood. Thus, the test volume, 52.5 mL, is equal to 1/40th of the alcohol in alveolar breath. The instrument uses the absorbance intensity to calculate the BAC, which is shown on the face of the measurement tool as a meter.

$$2K_2Cr_2O_7 + 3C_2H_5OH + 8H_2SO_4 \rightarrow 2Cr_2(SO_4)_3 + 2K_2SO_4 + 3CH_3COOH + 11H_2O$$

Newer breath testers, such as the Intoxilyzer, determine the blood alcohol level by evaluating the absorbance of infrared light by the captured breath as determined by a detector. One advantage of the infrared (IR) method is that it does not rely on chemical reagents and a chemical reaction. The instrument also has two beams so that a negative control sample can be tested simultaneously to check for interferences within the instrument. The quantity of alcohol in the blood is evaluated indirectly using the breath by determining the decrease in light transmitted in the presence of the alcohol in comparison to that transmitted in an alcohol-free breath sample. Figure 9.8 shows attenuated total reflectance Fourier transform-infrared spectroscopy (ATR FT-IR) spectra of a sample of ethanol. The IR spectrum of ethanol contains peaks at 3391, 2981, and 1055 cm^{-1}, correlated to the O–H stretch, C–H stretch, and C–O stretch respectively.

Another breath test employs a fuel cell. The fuel cell breath test determines the quantity of alcohol in the blood using the alcohol as a fuel that is ignited and burned in the presence of oxygen, creating energy that is converted to an electrical current.

$$2C_2H_5OH + O_2 \rightarrow 2CH_3COOH + electricity + 2H_2$$

To directly determine the BAC, blood needs to be drawn. This is performed by a health professional or certified phlebotomist—the person that is suspected of driving with a BAC above the legally accepted limit is transported to a facility that can draw the blood under medically accepted conditions; the sampling is not performed on site. The time elapsed until the blood is drawn in relation to the time at which the alcohol was consumed can greatly influence

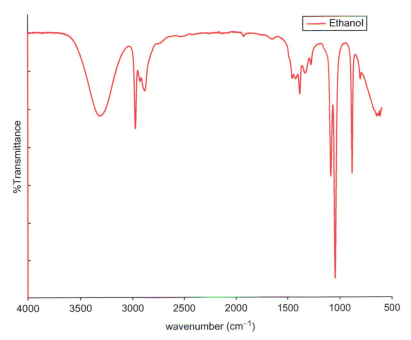

Figure 9.8   ATR FT-IR spectrum of ethanol.

the results of the test. Initially after consumption and up to approximately 30 minutes thereafter, the BAC is 41 times higher in arterial blood than venous blood. After this time, the alcohol is absorbed and the arterial and venous blood concentration is equal. Immediately on absorption, the liver begins to break down the alcohol enzymatically. Thus, the BAC will depend greatly on the time and circumstances surrounding the draw: depending on the elapsed time and food ingested prior to consumption and sampling, absorption and breakdown will influence the BAC. In drawing blood for a blood alcohol test, care must be taken to use only a non-alcohol disinfectant to clean the skin. The blood must be stored in a sealed container (purple-capped tubes) in the presence of an anticoagulant (e.g., EDTA) and a preservative (e.g., NaF) at 4°C (refrigeration). Typically, a sample of 10 mL is collected for blood alcohol testing; sufficient blood is drawn to enable duplicate or triplicate testing of the samples.

GC-MS is used to determine the BAC as discussed in Chapter 6. GC can be used to separate similar alcohols including methanol, ethanol, isopropanol, and N-propanol, and other small volatiles including acetone. Mass spectrometry is used to confirm the identity of the drug. The quantity of alcohol is determined using the area under the alcohol peak recorded on the GC chromatogram by comparing the peak area of the unknown to the peak areas of alcohol standards of known concentration. The line equation using linear regression developed using the standards is used to determine the concentration of the sample.

Alcohol is not the only drug that toxicologists are tasked with quantitating or identifying. In fact, toxicologists must consider a multitude of drugs and poisons and their metabolites as discussed in Chapter 8. Cases include drug overdose, drug interaction, poisoning, product counterfeiting, and product tampering cases, to name a few. For example, in the United States, spending on prescription drugs for pain reduction accounted for $18.2 billion (16.2%) of spending on prescription drugs in 2012. The Substance Abuse and Mental Health Services Administration reported an uptick in the number of emergency room visits related to prescription sleep aids, or sleeping pills, from 21,824 in 2005–2006 to 42,274 in 2009–2010. In 2010, approximately one third (20,793 of the 64,175) of emergency department visits involving zolpidem were related to overdoses of sleeping pills including Ambien®, Ambien CR®, Edluar®, and Zolpimist® that contain the active ingredient zolpidem (The DAWN Report). One high-profile case of product tampering was a 1982 Chicago case in which Tylenol capsules intentionally poisoned with potassium cyanide (KCN) were rebottled, repackaged, and replaced on store shelves. The KCN concentration was 10,000 times the quantity needed to kill a human. Seven people died in the case; it remains unsolved.

Pills and powders or empty pill containers may be found near the deceased body or in the possession of the individual. Potential drugs and poisons include thousands of prescription or illegal drugs. These can be used to tentatively identify what drugs and secondary metabolites to look for using colorimetric chemical tests. Apps including Prescription Pill Identifier (Apple Marketplace) and books including *The Drug Bible* are useful in determining the identities of legal and illegal drugs in the form of pills and powders. Books including the *Instrumental Data for Drug Analysis* (IDDA) and computerized libraries including the NIST and Japanese structure databases serve as a source of comparison for spectra. The NIST library has over 140,000 compounds that can be searched for matches.

Rarely are drugs excreted in the same chemical form as they are taken into the body; often, like alcohol, they are metabolized by liver enzymes and other pathways to a different chemical or *metabolite* (Table 9.1). For example, as shown in Table 9.1 and Figure 9.9, heroin is metabolized to morphine. Heroin and cocaine are both metabolized by human liver

Table 9.1 Drugs and secondary metabolites

Substance	Metabolite(s)
Ethanol	Acetaldehyde → Acetic acid → Carbon dioxide and water
Heroin	6-Acetyl morphine → Morphine
Codeine	Hydrocodone → Hydromorphone of Dihydrocodeine, Morphine, Norhydrocodone
S-Cocaine	Homatropoine → Naloxone methiodide
THC	11-nor-Δ9-tetrahydrocannabinol-9-carboxylic acid (Δ9-THC-COOH)
Amphetamine	Norephedrine, Phenylacetone
Morphine	Morphine-3-gucuronide (M3G), Morphine-6-glucuronide (M6G), Hydromorphone
Oxycodone	Oxymorphone

Source: Bencharit, S., C.L. Morton, Y. Xue, P.M. Potter, and M.R. Redinbo. 2003. Structural basis of heroin and cocaine metabolism by a promiscuous human drug-processing enzyme. *Nat. Struct. Mol. Biol.* 10:349–356.

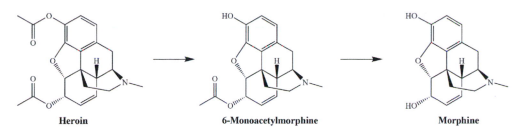

Heroin   **6-Monoacetylmorphine**   **Morphine**

Figure 9.9 Heroin metabolism reaction. (Structure prepared by Ashley Cowan.)

carboxylesterase 1 (Figure 9.10), which rapidly hydrolyzes ester linkages. (The enzyme also detoxifies chemical weapon nerve agents including sarin, soman, and tabun, which are covered in Chapter 15.) The mechanism was determined using x-ray crystallography: human carboxylesterase 1 (hCE1) was complexed to naloxone methiodide, a stable heroin analogue. Although the prescribed or ingested dose of a drug can be estimated from the concentration of the drug detected in body fluids, the actual dose will depend upon if a user followed the prescribed dose or if they were a tolerant user.

The *retention time* or length of time a drug may be found in a body fluid depends on the type of body fluid, dose, drug, individual metabolism, and body hydration. For example, some drugs such as alcohol may take only a couple of hours to be excreted in urine while others including daily use of cannabinoids, long-acting barbiturates, phencyclidine, and benzodiazepines may take weeks to show up in urine (Table 9.2). The condition of a body found postmortem will also influence whether a drug can be found: tissue or body fluid samples may be colonized by microorganisms that can produce alcohol as a metabolic product.

Based on the information available, the toxicologist must devise an *analytical scheme* to successfully detect, isolate, and specifically identify poisons, drug substances, and their metabolites. Oftentimes, the substances or metabolites are present only in nanogram to microgram quantities, so the analytical methods used must be very sensitive. The procedure used depends on the chemical properties of the substances being tested. To be analyzed, samples may need to be separated from the body fluid or may be analyzed *in situ* depending on the method. For example, methods may initially separate acidic and basic drugs. Acidic drugs have an ionizable proton (e.g., barbiturates, aspirin, cocaine HCl) while basic drugs will have a proton acceptor (e.g., phencyclidine, methadone, amphetamines, crack cocaine). These structures are shown in Chapter 8.

Screening and confirmatory methods may be employed; colorimetric, microscopic, chromatographic, and spectroscopic methods may be used according to the Scientific Working Group for the Analysis of Seized Drugs (SWGDRUG). The first test often performed is a color test or immunoassay. Both are susceptible to false positives so the result must

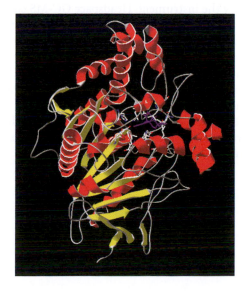

Figure 9.10 Human carboxylesterase 1 (hCE1) bound to naloxone methiodide, a heroin analogue (1MX9.pdb).

Table 9.2   Retention time in urine for several drugs

Drug	Retention time in urine
Ethanol	2–14 hours
Propoxyphene	6–48 hours
Cocaine	12–72 hours
Barbiturates—Secobarbital, short-acting	24 hours
Amphetamines	1–3 days
Cocaine metabolite	2–4 days
Opiates	1–3 days
Methadone	3 days
Benzodiazepines	3 days–4 weeks
Cannabinoids (occasional)	1–7 days
Cannabinoids (4×week)	5 days
Cannabinoids (daily)	10 days–4 weeks
Methaqualone	8 days
Barbiturates—Phenobarbital, long-acting	2–3 weeks
Phencyclidine (PCP)	8–30 days

Source: www.eapcareinc.com/resource-room/drug-retention-times/.

be confirmed with a confirmatory test. Confirmatory tests or SWGDRUG Category A methods include mass spectrometry (MS), FT-IR, and nuclear magnetic resonance spectroscopy (NMR) methods. These tests are highly reliable and are rarely subject to interferences. Tandem instrumental methods including GC-MS and high-pressure liquid chromatography-ultraviolet (HPLC-UV) or high-pressure liquid chromatography–mass spectrometry (HPLC-MS) are widely used to detect, identify, and quantitate drugs and poisons in body fluids including blood and urine.

The identification of inorganic diatomic poisons (e.g., CO, CN, Cl_2) (Figure 9.11) is often accomplished by methods including colorimetric tests and ultraviolet-visible (UV-Vis) spectroscopy. For example, UV-Vis spectroscopy can be used to detect carboxyhemoglobin (COHb) and oxyhemoglobin as discussed in Chapter 4. Hemoglobin binds both oxygen and carbon monoxide ligands in its ligand-binding site (Figure 9.12) but it binds carbon monoxide 200 times more strongly than oxygen. The ratio of the two species can be used to determine if a person suffered from carbon monoxide poisoning. Symptoms of carbon monoxide poisoning include headache, confusion, fatigue, high blood pressure, and rapid pulse. Fatal levels of carbon dioxide can vary based on age, fitness level, and general health. GC-MS can also be used to detect poisons such as carbon dioxide: a reagent can be used to liberate the molecule so that it can be detected and identified by the instrument. Headspace GC-MS can be used to quantitate carbon monoxide using the area under the peak and the signal response using total ion chromatogram integration. Treatment for blood poisons such as carbon monoxide, cyanide, and others that bind to hemoglobin and reduce its efficiency in delivering oxygen to the tissues is administration of 100% oxygen or hyperbaric oxygen therapy. A carbon monoxide ligand concentration of 50%–60% in hemoglobin is considered fatal for a middle-aged individual whereas with a BAC of 0.20% (w/v), a ligand concentration of greater than 35%–40% is fatal. The detection of carbon monoxide and carbon dioxide can be integral to understanding the circumstances of a case. A high concentration of these gases in the victim's blood would indicate that the victim breathed combustion products or exhaust fumes and was alive when a fire began or an automobile was running inside a garage. Carbon monoxide poisoning may indicate arson or automobile exhaust poisoning that was used as a cover-up for a murder.

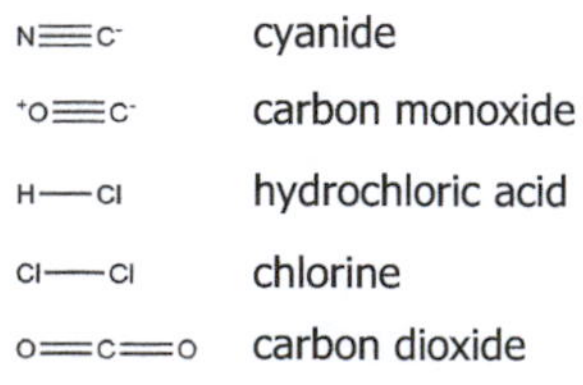

N≡C⁻	cyanide
⁺O≡C⁻	carbon monoxide
H—Cl	hydrochloric acid
Cl—Cl	chlorine
O≡C≡O	carbon dioxide

Figure 9.11   Chemical structures of poisons including CO_2, CO, CN, HCl, Cl_2.

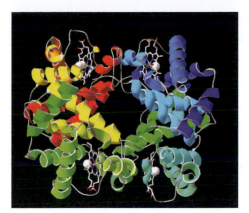

Figure 9.12 Structure of human hemoglobin (1gzx.pdb), a hetero-tetramer consisting of two alpha and two beta chains. The ribbons, which represent the alpha helices, are colored by the order of the secondary structure elements and trace the backbone structure of the protein chains. The oxygens are shown as four white diatomic balls, each in one quadrant of the structure, and are tethered by the heme moieties shown in stick form. The proximal and distal histidine residues that are essential for oxygen binding are shown in purple. Poisons including diatomic gases CO, CN, HCl, and Cl_2 can bind at the oxygen binding site in the protein. Carbon dioxide (CO_2) binds at the allosteric site to induce a conformational change in the protein to promote oxygen release into the cell.

Heavy metal poisons include arsenic (As), bismuth (Bi), lead (Pb), cadmium (Cd), mercury (Hg), antimony (Sb), and thallium (Tl). A presumptive or tentative test for heavy metals is the Reinsch Test. If a piece of tissue homogenate or body fluid is placed in hydrochloric acid and a copper strip is placed in the sample, a silvery or dark coating on the copper indicates the presence of a heavy metal. The identity of the metals can be determined using methods including emission spectroscopy, atomic absorption spectroscopy, x-ray fluorescence spectroscopy, or inductively coupled plasma-mass spectrometry as discussed in Chapter 7. Quantitation of the metal poisons can be determined using UV-Vis or atomic absorption spectroscopy.

QUESTIONS

1. Who evaluates body fluids for drugs and other compounds using chemical tests?
   a. Entomologist
   b. Toxicologist
   c. Pathologist
   d. Histologist
   e. Odontologist
2. Breath tests determine the amount of alcohol in the _____.
   a. Alveolar air
   b. Blood
   c. Heart
   d. Lungs
3. In which test is alcohol burned with oxygen in the air to determine the quantity of alcohol?
   a. Fuel cell test
   b. Infrared breath test
   c. Sobriety test
   d. Breathalyzer
4. _____ is metabolized on entering the body into morphine.
   a. Cocaine
   b. LSD
   c. Heroin
   d. MSMA
5. Which of the following body fluids or materials should be assayed to test for chronic rather than acute drug exposure?
   a. Blood
   b. Urine
   c. Hair
   d. Gastric contents
   e. Vitreous humor
6. Explain the process by which alcohol is absorbed.
7. Explain the process by which alcohol is metabolized.

8. List four factors that influence BAC.

9. Explain how GC-MS is used to detect and quantitate BAC.

10. Why does the retention time in the human body vary for different drugs?

References

Saferstein, R. 2015. *Criminalistics: An Introduction to Forensic Science*, 11th edn. Boston: Pearson.

The DAWN Report. Emergency department visits attributed to overmedication that involved the insomnia medication zolpidem. https://www.samhsa.gov/data/sites/default/files/DAWN-SR150-Zolpidem-2014/DAWN-SR150-Zolpidem-2014.pdf (accessed July 10, 2018).

Bibliography

Alston, T.A. 2014. Why does methylene blue reduce methemoglobin in benzocaine poisoning but beneficially oxidize hemoglobin in cyanide poisoning? *J. Clin. Anesth.* 26(8):702–703.

Bartholow, M. 2013. Top 200 drugs of 2012. *Pharm. Times* 79 (7):42–44.

Bencharit, S., C.L. Morton, Y. Xue, P.M. Potter, and M.R. Redinbo. 2003. Structural basis of heroin and cocaine metabolism by a promiscuous human drug-processing enzyme. *Nat. Struct. Mol. Biol.* 10:349–356.

Burch, H.J., E.J. Clarke, A.M. Hubbard, and M. Scott-Ham. 2013. Concentrations of drugs determined in blood samples collected from suspected drugged drivers in England and Wales. *J. Forensic Leg. Med.* 20(4):278–289.

Chen, C.H., J.C. Ferreira, E.R. Gross, and D. Mochly-Rosen. 2014. Targeting aldehyde dehydrogenase 2: New therapeutic opportunities. *Physiol. Rev.* 94(1):1–34.

Crabb, D.W., H.J. Edenberg, W.F. Bosron, and T.K. Li. 1989. Genotypes for aldehyde dehydrogenase deficiency and alcohol sensitivity. The inactive ALDH2(2) allele is dominant. *J. Clin. Invest.* 83(1):314–316.

Doyle, M.L., E. Di Cera, C.H. Robert, and S.J. Gill. 1987. Carbon dioxide and oxygen linkage in human hemoglobin tetramers. *J. Mol. Biol.* 196(4):927–934.

Drummer, O.H., I. Kourtis, J. Beyer, P. Tayler, M. Boorman, and D. Gerostamoulos. 2012. The prevalence of drugs in injured drivers. *Forensic Sci. Int.* 215(1–3):14–17.

Eichhorst, J.C., M.L. Etter, P.L. Hall, and D.C. Lehotay. 2012. Opiate screening and quantitation in urine/blood matrices using LC-MS/MS techniques. *Methods Mol. Biol.* 902:53–64.

Elkins, K.M., S.E. Gray, and Z.M. Krohn. 2015. Evaluation of technology in crime scene investigation. *CS Eye*. www.cseye.com/content/2015/april/research/evaluation-of-technology (accessed January 25, 2018).

Fenton, J.J. 2002. *Toxicology: A Case Approach*. Boca Raton: CRC Press.

Guzman, J.A. 2012. Carbon monoxide poisoning. *Crit. Care Clin.* 28(4):537–548.

Hampson, N.B. and S.L. Dunn. 2015. Carbon monoxide poisoning from portable electrical generators. *J. Emerg. Med.* 49(2):125–129.

Hao, H., H. Zhou, X. Liu, Z. Zhang, and Z. Yu. 2013. An accurate method for microanalysis of carbon monoxide in putrid postmortem blood by head-space gas chromatography-mass spectrometry (HS/GC/MS). *Forensic Sci. Int.* 229(1–3):116–121. doi:10.1016/j.forsciint.2013.03.052.

Holt, S., M.J. Stewart, R.D. Adam, and R.C. Heading. 1980. Alcohol absorption, gastric emptying and a breathalyser. *Br. J. Clin. Pharmacol.* 9(2):205–208.

Hughes, M.M., R.S. Atayee, B.M. Best, and A.J. Pesce. 2012. Observations on the metabolism of morphine to hydromorphone in pain patients. *J. Anal. Toxicol.* 36(4):250–256.

Jensen, F.B. 2004. Red blood cell pH, the Bohr effect, and other oxygenation-linked phenomena in blood O2 and CO2 transport. *Acta Physiol. Scand.* 182(3):215–227.

Jones, A.W., A. Holmgren, and J. Ahlner. 2012. Concentrations of free-morphine in peripheral blood after recent use of heroin in overdose deaths and in apprehended drivers. *Forensic Sci. Int.* 215(1–3):18–24.

Kamendulis, L.M., M.R. Brzezinski, E.V. Pindel, W.F. Bosronand, and R.A. Dean. 1996. Metabolism of cocaine and heroin is catalyzed by the same human liver carboxylesterases. *J. Pharm. Exp. Ther.* 279(2):713–717.

Kenrick, E.B. 1902. Reinsch's test for arsenic. *J. Am. Chem. Soc.* 24(3):276–276.

Khanna, M., C.H. Chen, A. Kimble-Hill, B. Parajuli, S. Perez-Miller, S. Baskaran, J. Kim, K. Dria, V. Vasiliou, D. Mochly-Rosen, and T.D. Hurley. 2011. Discovery of a novel class of covalent inhibitor for aldehyde dehydrogenases. *J. Biol. Chem.* 286:43486–43494.

Li, Y., D. Zhang, W. Jin, C. Shao, P. Yan, C. Xu, H. Sheng, Y. Liu, J. Yu, Y. Xie, et al. 2006. Mitochondrial aldehyde dehydrogenase-2 (ALDH2) Glu504Lys polymorphism contributes to the variation in efficacy of sublingual nitroglycerin. *J. Clin. Invest.* 116(2):506–511.

Mitchell, M.C., Jr, E.L. Teigen, and V.A. Ramchandani. 2014. Absorption and peak blood alcohol concentration after drinking beer, wine, or spirits. *Alcohol Clin. Exp. Res.* 38(5):1200–1204.

Montgomery, M.R. and M.J. Reasor. 1992. Retrograde extrapolation of blood alcohol data: An applied approach. *J. Toxicol. Environ. Health* 36(4):281–292.

National Center for Statistics and Analysis. 2014. *Traffic Safety Facts—2013 Motor Vehicle Crashes: Overview*. Washington, D.C.: National Highway Traffic Safety Administration. https://crashstats.nhtsa.dot.gov/Api/Public/ViewPublication/812101 (accessed January 17, 2018).

Nelson, D.L. and M.M. Cox. 2004. *Lehninger Principles of Biochemistry*, 4th edn. New York: W.H. Freeman and Co.

Oshima, T., K. Yonemitsu, A. Sasao, M. Ohtani, and S. Mimasaka. 2015. Detection of carbon monoxide poisoning that occurred before a house fire in three cases. *Leg. Med. (Tokyo)* 17(5):371–375.

Paoli, M., R. Liddington, J. Tame, A. Wilkinson, and G. Dodson. 1996. Crystal structure of T state haemoglobin with oxygen bound at all four haems. *J. Mol. Biol.* 256:775–792.

Roberts, C. and S.P. Robinson. 2007. Alcohol concentration and carbonation of drinks: The effect on blood alcohol levels. *J. Forensic Leg. Med.* 14(7):398–405.

Royal Society of Chemistry. 2017. On this day—December 31: The drunkometer was invented on this day in 1938. http://rsc.org/learn-chemistry/resource/rdc00001231/on-this-day-dec-31-drunkometer-was-invented (accessed January 25, 2018).

Smith, R.P. and H. Kruszyna. 1974. Nitroprusside produces cyanide poisoning via reaction with hemoglobin. *J. Pharmacol. Exp. Ther.* 191(3):557–563.

Stuart, J.H., J.J. Nordby, and S. Bell. 2014. *Forensic Science: An Introduction to Scientific and Investigative Techniques*, 4th edn. London: Taylor & Francis.

Suddendorf, R.F. 1989. Research on alcohol metabolism among Asians and its implications for understanding causes of alcoholism. *Public Health Rep.* 104(6):615–620.

US Department of Health and Human Services. 2013. Results from the 2012 National Survey on Drug Use and Health: Summary of National Findings. www.samhsa.gov/data/sites/default/files/NSDUHresults2012/NSDUHresults2012.pdf (accessed January 17, 2018).

Yeoh, M.J. and G. Braitberg. 2004. Carbon monoxide and cyanide poisoning in fire related deaths in Victoria, Australia. *J. Toxicol. Clin. Toxicol.* 42(6):855–863.

Yoshida, A., I.Y. Huang, and M. Ikawa. 1984. Molecular abnormality of an inactive aldehyde dehydrogenase variant commonly found in Orientals. *Proc. Natl. Acad. Sci. USA.* 81(1):258–261.

Zouaoui, K., S. Dulaurent, J.M. Gaulier, C. Moesch, and G. Lachâtre. 2013. Determination of glyphosate and AMPA in blood and urine from humans: About 13 cases of acute intoxication. *Forensic Sci. Int.* 226(1–3):e20–e25. https://icsw.nhtsa.gov/people/injury/research/drug_impaired.html#ptor (accessed May 24, 2018).

CHAPTER 10

Trace evidence

KEY WORDS: trace evidence, microanalysis, physical properties, chemical properties, immersion method, match point, Locard's exchange principle, density, glass, radial fracture, concentric fracture, conchoidal lines, Wallner lines, refractive index, Becke line, soil, paint, electrocoat primer, primer surface, basecoat, clearcoat, stain, varnish, enamel, polymers, plastic, hair, follicle, root, tip, shaft, cuticle, cortex, medulla, anagen, catagen, telogen, follicular tag, fiber, natural fiber, synthetic fiber, duct tape, backing, scrim, adhesive

LEARNING OBJECTIVES

- To list examples of trace evidence
- To differentiate between physical and chemical characteristics of trace evidence
- To explain how to detect differences in trace evidence using microscopy and instrumental methods
- To explain how hair is formed and diagram the physical characteristics of hair
- To list the phases of hair growth and describe features of each
- To describe differences between human and animal hairs
- To describe differences between natural and synthetic fibers
- To list the parts of duct tape and how these can be evaluated by physical and chemical analysis

DETECTING AND LINKING FIBERS HELPS SOLVE RELATED CASES COMMITTED BY SERIAL KILLER

Sofia Silvia, 16, was abducted from her front porch in Spotsylvania, Pennsylvania, on September 9, 1996. Eight months later, Kati and Kristin Lisk, ages 12 and 15 respectively, also went missing in Spotsylvania. Silvia's body was found in a creek and the Lisk girls' bodies were found in a river. There were no witnesses in the cases. The only evidence recovered were a few hairs and fibers from the girls' clothing.

Federal Bureau of Investigation (FBI) trace evidence expert Doug Deedrick led the trace evidence analysis in the case. Hairs and acrylic, cotton, polyester, and nylon fibers are routinely differentiated and identified by species or as synthetic and by type in forensic labs using polarized light microscopy. Forensic labs also utilize ultraviolet-visible (UV-Vis), infrared (IR), and fluorescence microspectrophotometry in their fiber examinations. Deedrick recovered pink fibers from Kati Lisk that reminded him of a bath rug. He found dark blue acrylic fibers on all three girls; several of these blue fibers were recovered from the inside of Silvia's shirt. In total, Deedrick and his team at the FBI recovered three hairs and 190 fibers from the girls' bodies and clothing. Based on his findings and the similarities in the cases, he suspected a serial killer was the perpetrator in the three cases. Even after the police chased over 12,000 leads, compared the evidence data with that from over 45,000 other unsolved cases, and compared DNA samples to 1.2 million genetic profiles, the girls' cases remained cold cases.

DETECTING AND LINKING FIBERS HELPS SOLVE RELATED CASES COMMITTED BY SERIAL KILLER (continued)

In June 2002, a 15-year-old girl was abducted from Columbia, S.C. She was transported in a container in a car and was handcuffed and raped repeatedly by the perpetrator. When the rapist fell asleep, she escaped and sought help from the police. The perpetrator had left the scene but a suspect was caught by the police 2 days later in a chase in Sarasota, Florida. He was found to have previously lived in Spotsylvania and he had a 1997 newspaper article about the Lisk girls' abduction in his South Carolina apartment.

Deedrick traveled to South Carolina for the cases and in searching the apartment, found a pink bath rug in the bottom of a box in the linen closet. He also found "furry" handcuffs with fibers that looked black but were actually dark blue in the light. In addition to the fiber evidence, fingerprint evidence from the lid of the trunk from one of his cars matched those of Kristin Lisk and the DNA from the three recovered hairs including one from Kati Lisk's sock matched the suspect. From the judicial proceedings and using the evidence, the suspect, Richard Marc Evonitz, 38, was determined to be the perpetrator and serial killer.

Research on dyes and fibers continues to be an active area of research. Recently in a study funded by the US National Institutes of Justice, time-of-flight secondary ion mass spectrometry (TOF-SIMS) was demonstrated for use in the analysis and differentiation of dye colors used in nylon carpets or clothing. The TOF-SIMS method developed by researchers from Dr. David Hinks' group at North Carolina State University can help determine the chemical fingerprint of the dye, fiber, and trace chemicals or contaminants on the fiber's surface. This can be used in fiber attribution to determine a facility or process used in industrial production. A small piece of a cross section of a nylon fiber sample was used to minimize destruction. The cross-section surface is cleaned using a beam of ionized C60 fullerenes prior to recording the TOF-SIMS spectra. Dye colorants can be differentiated and identified using other methods including thin layer chromatography but the evidence needs to be dissolved in a solvent to release the dye.

BIBLIOGRAPHY

Deedrick, D.W. and S.F. Koch. 2004. Microscopy of hair part I: A practical guide and manual for human hairs. *Forensic Sci. Commun.* 6 (1). https://archives.fbi.gov/archives/about-us/lab/forensic-science-communications/fsc/jan2004/research/2004_01_research01b.htm (accessed January 24, 2018).

Deedrick, D.W. and S.F. Koch. 2004. Microscopy of hair part II: A practical guide and manual for human hairs. *Forensic Sci. Commun.* 6 (3). https://archives.fbi.gov/archives/about-us/lab/forensic-science-communications/fsc/july2004/research/2004_03_research02.htm (accessed January 24, 2018).

Everts, S. 2012. Identifying material evidence from crime scene carpets. *Chem. and Eng. News.* http://cen.acs.org/articles/90/web/2012/11/Identifying-Material-Evidence-Crime-Scene.html (accessed July 6, 2017).

Morgan, S.L., A.A. Nieuwland, C.R. Mubarak, J.E. Hendrix, E.M. Enlow, B.J. Vasser, and E.G. Bartick. 2004. Forensic discrimination of dyed textile fibers using UV-VIS and fluorescence microspectrophotometry. *Proceedings of the European Fibres Group (Annual Meeting, Prague, Czechoslovakia).* www.sjsu.edu/people/steven.lee/courses/JS111FLUOR/s12/Bartick_et_al_EuropeanFibresGroup_2004.pdf (accessed July 6, 2017).

Tiny fibers helped FBI nab serial killer. http://abcnews.go.com/GMA/story?id=125152 (accessed July 6, 2017).

Zhou, C., M. Li, R. Garcia, A. Crawford, K. Beck, D. Hinks, and D.P. Griffis. 2012. Time-of-flight-secondary ion mass spectrometry method development for high-sensitivity analysis of acid dyes in nylon fibers. *Anal. Chem.* 84(22): 10085–10090.

Trace evidence has been, and continues to be, integral to solving forensic cases. Due to the nature of crimes, small quantities or broken items may be left at the crime scene or may be transferred on a victim or suspect from place to place. *Trace evidence* includes a wide range of materials such as polymers, plastics, paints, paper, hair, fiber, soil, tape, and gunshot residue. The evidence is evaluated both qualitatively and quantitatively with the goal of identifying the minor or ultraminor components of the sample. This is also referred to as *microanalysis*. The sample may include the entire submitted item or a quantity of a subsample of the evidence. As with all evidence items, care should be taken to minimize the potential for loss, cross-transfer, and contamination. As some evidence samples may contain only milligram or microgram sizes, the microscope is an invaluable tool for screening, partitioning, and evaluating trace evidence. The evidence may be difficult to partition so it is especially important for criminalists to perform destructive or invasive techniques last, if at all. Examinations should be sequenced to maximize the potential value of the submitted evidence.

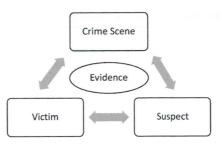

Figure 10.1 Locard's exchange principle.

Physical and chemical properties of trace evidence can be used to differentiate and identify the samples. *Physical properties* include density, magnetism, size and diameter, morphology, fracture edge matching, and refractive index. *Chemical properties* include percent composition, pH, spectroscopic properties, and elemental composition.

The refractive index of a substance can be determined by the *immersion method*. In the immersion method, pieces of glass, fibers, or soil minerals are immersed into liquids or oils of a known refractive index. The *Becke line* is a halo-like glow that appears around the object immersed in the liquid. When the refractive index of the sample is equal to the liquid, the *match point* has been reached. Several oils can be tested to narrow down the refractive index range or a hot stage can be used to make a more precise determination.

Instrumental tools including pyrolysis gas chromatography, infrared spectroscopy, microspectrophotometry, and scanning electron microscopy are most commonly applied to trace analysis. The methods used are chosen based on which will provide the greatest level of discrimination for the materials under examination. A minimum of two confirmatory analytical techniques must be used to make a positive identification.

Edmond Locard was a forensic geologist who was most interested in dust that was transferred from the crime scene to the perpetrator. He is most famous for his Principe de l'echange (known widely as *Locard's exchange principle*) that "Il est impossible au malfaiteur d'agir avec l'intensit que suppose l'action criminelle sans laisser des traces de son passage" (1923) translated as "It is impossible for a criminal to act, especially considering the intensity of a crime, without leaving traces of this presence." A longer writing of Locard's was translated (by Roux et al., 2015) as "The truth is that none can act with the intensity induced by criminal activities without leaving multiple traces of his path. [...] The clues I want to speak of here are two kinds: Sometimes the perpetrator leaves traces at a scene by their actions; sometimes, alternatively, he/she picked up on their clothes or their body traces of their location or presence" or is described as "whenever two objects come into contact, a transfer of material will occur. Trace evidence that is transferred can be used to associate objects, individuals, or locations" (The Scientific Working Group on Materials Analysis, 1999) (Figure 10.1). Or put more simply, "every contact leaves a trace."

GLASS

One common type of trace evidence that may be submitted to the laboratory for examination is *glass*. Glass is a hard, amorphous, brittle, usually transparent solid. Glass may originate from broken windows and other items. Glass exhibits a conchoidal or a smooth, rounded fracture. Glass analysis can reveal whether the breakage was due to a low or high-velocity impact. Glass can be differentiated by physical features such as color, thickness, texture, uniformity, curvature, and fitting the pieces together as well as its reflective, refractive, fluorescence, and density properties.

Table 10.1 Chemical composition of six common glass types

Glass type	SiO_2	B_2O_3	Na_2O (soda)	CaO (lime)	MgO	Al_2O_3	PbO	P_2O_5
Fused silica	100%							
Ninety-six percent silica	96%							
Borosilicate	70%–81%	>5%	4%–8%			2%–7%		
Soda-lime silica	60%–75%		12%–18%	5%–12%	4%	1%		
Aluminosilicate	62%	5%	1%	8%	7%	17%		
Lead	54%–65%		13%–15%			2%	>20%	
Phosphate								>50%

Table 10.2   Densities of various types of glass

Type of glass	Density (g/mL)
Fused silica (96%)	2.18
Pyrex	2.23
Porcelain	2.3–2.5
Borosilicate	2.4
Window glass	2.4.2.8
Bottle glass	2.4–2.8
Headlight glass	2.47–2.63
Lead crystal	3.1
Flint (densest)	7.2

Source: Chemistry LibreTexts, https://chem.libre-texts.org/Exemplars_and_Case_Studies/Exemplars/Forensics/Glass_Density_Evidence.

The chemical properties of glass such as elemental composition can also be determined. A major component of glass is sand or silicon dioxide (SiO_2). Other elemental oxides impart varying colors and properties to glass. Common types of glass are soda-lime, soda-lead, borosilicate, silica, aluminosilicate, lead glass, tempered, and laminated glass (Table 10.1). Soda-lime glass is used in plate and window glass as well as glass containers and is the most common and least expensive of the glass types. Soda-lead glass is used in fine tableware and is found in art objects. Borosilicate glass is used for heat-resistant glass applications including lightbulbs, Pyrex bakeware, laboratory glassware, and automobile sealed-beam headlights. Lead glass is used in art glass and thermometer tubing. Aluminosilicate glass is used in resistors. Silica glass is used in chemical glassware. Tempered glass is used in the side windows of cars. Unlike ordinary glass, tempered glass does not shatter into pieces when broken but rather fragments into cubes. Laminated glass is the type of glass used for the windshield of most cars. A plastic layer is sandwiched between two layers of glass.

The density of glass varies by composition and thermal history. This physical property can quickly eliminate from further analysis two samples of glass that did not come from the same source: two samples of glass with different densities could not have originated from the same source. Glass density ranges for some glass types are shown in

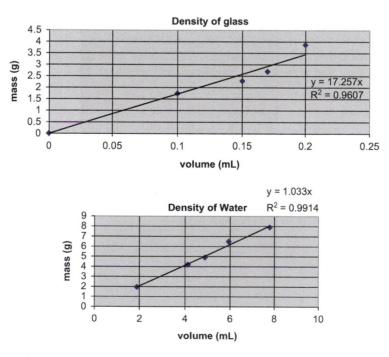

Figure 10.2   Determination of glass density.

Table 10.2. A determination of glass density can be performed using the immersion method as shown in Figure 10.2 by submerging fragments of the glass evidence in a solvent that is less dense than the glass. This technique is nondestructive and the glass fragments can be recovered for further analysis.

Glass fracture examination can identify the direction of impact, heat fractures, origin of a projectile, and the sequence of bullets or other projectiles or impacts. Initial impacts in non-tempered glass result in the formation of radial, or radiating, fractures that emanate from the break point. After the formation of *radial fractures*, concentric fractures may follow. *Concentric fractures* are characterized by circular lines that spear to form a circle around the point of impact or between the radial lines. The sequence of high-velocity projectiles, including bullets, can be determined by analyzing the fractures: the cracks that radiate furthest and are not stopped originated first and the cracks that are terminated by other fractures originated later. The radiating lines form a right angle at the reverse side of the force. The concentric fractures originate after the radial fractures and start at the same side as the impact. The origin of the incoming high-velocity projectile can also be determined: a wider hole is left at the exit side of glass. *Conchoidal lines*, or Hackle marks, on the edge of broken glass can be identified under a microscope. Conchoidal lines are arch-shaped stress markings that are perpendicular to one glass surface and curved parallel to the opposite surface. The perpendicular surface can be used to detect the side where the crack originated as it faces the side of the origin. Other stress markings are called *Wallner lines*.

Fragments of glass recovered from a crime scene and suspect (or the suspect's location or belongings) can also be individualized if the discrete fragments are found to "match" or fit together like puzzle pieces. Patterns that match like this are considered to be unique.

To determine the *refractive index* of glass, a microscope equipped with a hot stage (variable temperature stage) is used. The glass is placed in a high-boiling liquid such as silicone oil. The refractive index changes with temperature. As the temperature is increased, the analyst watches for the disappearance of the *Becke line*. If the Becke line is noted on the inside of the glass fragment, the glass has a higher refractive index than the liquid immersion medium. Conversely, if the Becke line is observed on the outside of the piece of glass, the refractive index of the glass is lower than the medium. The temperature is noted and the refractive index is determined using a calibration chart for that liquid; the refractive indices for several liquids and oils are shown in Table 10.3. The GRIM3 software can be used in glass refractive index measurements; it automates the measurements of the match temperature and refractive index for glass fragments. Other methods for determining the refractive index of glass are dispersion staining color and oblique illumination. The refractive indices of some glass types are shown in Table 10.4. Headlight and bottle glass have relatively low densities as compared to leaded glass.

Additional instrumental methods for glass analysis include scanning electron microscopy (SEM), x-ray fluorescence (XRF), SEM coupled with energy dispersive x-ray spectrometry (SEM-EDS), and laser ablation inductively coupled plasma-mass spectrometry (LA-ICP-MS) for elemental analysis.

Table 10.3 Refractive indices for liquids and oils

Liquid	Refractive index
Water	1.333
Ethyl acetate	1.373
Olive oil	1.467
Glycerin	1.473
Castor oil	1.482
Clove oil	1.543
Bromobenzene	1.560
Bromoform	1.597
Cinnamon oil	1.619

Source: Forensic geology. The legal application of earth and soil science, by Todd Hocking, www.cargille.com/refractivestandards.shtmlwww.gemsociety.org/article/refractive-index-list-of-common-household-liquids/.

Table 10.4   Refractive indices for several types of glass

Glass	Refractive index
Headlight	1.47–1.49
Window	1.51–1.52
Bottle	1.51–1.52
Quartz	1.544–1.553
Lead	1.56–1.61

Source: Girard, J.E. 2007. *Criminalistics: Forensic Science and Crime.* Burlington, MA: Jones and Bartlett Learning.

SOIL

Another type of trace evidence that may be presented for forensic examination is soil. By definition, soil is the upper layer composed of naturally deposited materials, including minerals and organic matter, that covers the earth's surface and is capable of supporting plant growth. Soil may be analyzed to determine its constituent minerals and trace pollen and biological material that may be present. Often, using these features of soil, criminalists can determine the origin of a soil sample. As the earth's surface is vast in size, the evidential value of soil can be excellent. In terms of statistical probability, the likelihood of a given sample having the same properties as another is very small. Soil varies widely in composition but consists of varying proportions of sand, silt, clay, and organic matter. The probative value can be increased by finding rare or unusual minerals in the sample, fossils, rocks, pollen and spores, and manufactured particles. More than 2000 mineral components have been identified in soil. Of these, approximately twenty are common in soil but most soils contain only three to five minerals. Minerals commonly found in soil include sodium, manganese, iron, boron, copper, zinc, molybdenum, and chlorine. Minerals can be identified by their size, color, luster, density, fracture, streak, and magnetism. Rocks can be natural (e.g., granite) or man-made (e.g., concrete). Fossils are the remains of plants and animals. They can be used to determine the age of rocks. As some are scarce, they can be used to identify specific regions or locations. Pollen and spores can also be characteristic to certain areas.

The subfield of forensic science that deals with soil may be referred to as *forensic geology*, the legal application of earth and soil science. Forensic geologists employ stereoscopes, SEM, microchemical analysis, DNA analysis, and x-ray diffraction in their work. The origin of this science can be traced back to Sir Arthur Conan Doyle who wrote, in 1887–1893, about techniques for solving crimes, including soil and its composition, and techniques that had never actually been used. In 1893, an Austrian criminal investigator, Hans Gross, observed that "dirt on shoes can often tell us more about where the wearer of those shoes had last been than toilsome inquiries" and wrote that criminalists should study "dust, dirt on shoes and spots on cloth." In 1904, Georg Popp, a German forensic scientist, analyzed the first criminal case in which the analysis of soil was presented as evidence in a criminal case, the strangulation of Eva Disch. Locard's interest was also in studying dust as he published in three papers in 1930. A more recent case involved the analysis of soil in a South Dakota case. The body of a nine-year-old was found in a wooded area along a river. Soil collected from where the body was found was compared to evidence samples obtained from the fenders of a suspect's truck. Gahnite was found in both soils. As this rare blue mineral has never before been reported in South Dakota, the evidential value of this finding was high.

The class characteristics of soil include percentage of components, color, density, and so on. Individual characteristics of soil include unusual components such as pollen, seeds, spores, vegetation, fossils, or other fragments. Sand is characterized by natural particles including clay ranging from less than 0.002 mm to very coarse sand of up to 2.00 mm and rocks and gravel that exceed these sizes (Table 10.5). Its color is dependent on the parent rock and surrounding plant and animal life trace material that may be contained in the soil. Sand can also be sorted by its type using microscopy. There are four types of sand. *Continental sand* is formed from weathered continental rock such as granite. *Ocean floor sand* is typically formed from basalt volcanic material. *Carbonate sand* contains calcium carbonates. *Tufa sand* is a type of sand formed by the reaction of calcium ions from underground springs with carbonate ions present in the salt water of salt lakes. Chemical analysis can be used to identify the calcium carbonates.

Using a stereomicroscope, botanists can classify and identify pollens, seeds, and spores. Trained criminalists can also identify pollens and spores. Additionally, DNA analysis may be employed to aid in the identification, especially when a botanist or trained criminalist is unavailable.

Table 10.5   Soil particles by size

Soil particle	Size (diameter, mm)
Clay	>0.002
Silt	0.002–0.05
Very fine sand	0.05–0.10
Fine sand	0.10–0.25
Medium sand	0.25–0.50
Coarse sand	0.50–1.00
Very coarse sand	1.00–2.00
Gravel	2–75

Source: USDA, www.nrcs.usda.gov/Internet/FSE_DOCUMENTS/
stelprdb1044818.pdf.

PAINT

Paint forensic evidence is often submitted in burglary and hit-and-run cases but forensic analysis of paint may also include art. Paint is used for protecting and decorating surfaces as well as providing warnings (e.g., reflective paint, fluorescent paint). One of the most common types of evidence submitted to the laboratory is paint originating from vehicles. In hit-and-run cases, a small chip of paint may be all that is available for analysis. The paint "chip" is a small piece of the paint layer or layers that has detached from its surface. Typically, all layers detach together. Care must be taken not to damage the paint chip or smooth out any impressions in the paint. Paint smears may also be encountered. The smear may derive from a wet paint top layer or loose oxidized layer that is transferred, or smeared, onto another surface. Unlike paint chips, paint smears will not contain all of the layers. Reference samples are collected for comparison to evidence samples when possible. In collecting reference samples from the suspect's car, a clean knife or scalpel is used to carefully remove a paint chip with all of the layers (down to the metal) near the point of contact. When the submitted evidence samples are sufficiently large to allow the criminalist to perform physical matching, paint will have individualizing characteristics.

The chemistry of paint is very complex. Paint is often a homogenous mixture applied with a volatile solvent. Paint is a suspension of a pigment and additives in a film former (or vehicle). The film former is usually an organic polymer whose purpose is to allow the application of the pigments and protect the surface. When the paint dries, the pigment(s) and additives remain suspended in the thin polymer film that serves as a binder to the surface. When paint dries, the solvent evaporates, and is heat set (thermoset) or oxidizes. Solvent evaporation is characteristic of rust-proof paints while thermosetting is used for automobile paints. Oxidation is characteristic of artistic paints and linseed oil. The pigment may include inorganic metal salts and/or organic chromophores used to impart color. The pigments may differ in chemical properties; attenuated total reflectance Fourier transform-infrared spectroscopy (ATR FT-IR) can be used to determine if two paints are from the sample manufacturer or source (Figure 10.3).

Automobile manufacturers typically apply four coatings to the exterior of new automobiles. The coatings include the electrocoat primer, primer surfacer, basecoat, and clearcoat. The *electrocoat primer* layer consists of an epoxy-based resin applied to reduce corrosion. It is black to gray in color and uniformly applied for an even thickness and appearance. The *primer surfacer layer* also serves in corrosion control and smooths out seams or imperfections. It is composed of an epoxy-modified polyester or urethane material and is highly pigmented to minimize the contrast between the electrocoat primer and the top clearcoat layers. The *basecoat* consists of an acrylic-based polymer that contains pigments to impart the color of the vehicle. The pigments may include organic colorants and inorganic features such as pearl to provide luster, mica to generate interference colors, or aluminum flakes to impart a metallic look. Toxic substances and heavy metals (e.g., lead, chromium) are no longer used due to environmental and human health concerns. In addition to providing the color layer, the basecoat must also resist weathering due to acid rain, ultraviolet radiation, and heat. Finally, the *clearcoat* is the top layer applied to the automobile. It is an acrylic or polyurethane that imparts gloss, and improves durability and appearance.

Paint evidence is compared to color charts provided (annually) by automobile manufacturers or a database such as the Paint Data Query (PDQ). This allows the criminalist to determine the possible make and model of the vehicle. The

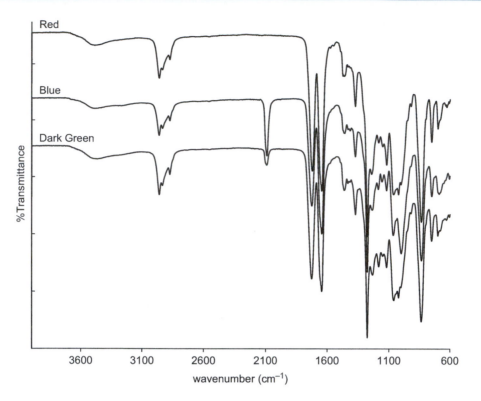

Figure 10.3   ATR FT-IR spectra of nail polish paint of different colors.

evidential value of the color differentiation depends on how many vehicles were made with that color. Additionally, although the color of the basecoat is strictly controlled, there is a wide variation in binder formulations that criminalists can use to differentiate samples. Figure 10.4 shows the ATR FT-IR spectra of the clearcoat layer from a 1975 sky blue VW Beetle, a red Kia Spectra, and a red Toyota Celica. The IR fingerprints of the clearcoat are different for the three automobiles. IR spectroscopy also can be used to determine the type of film former.

Paint evidence is evaluated using a stereomicroscope to determine the number of layers present in the sample, identify the color layer sequence, view the surface texture, perform color matching, and compare the evidence with a known sample. Due to the many manufacturers and aftermarket paints that may be applied, the color of vehicles is diverse. Vehicles that have been repainted or that have additional layers added to the original coating are of greater forensic significance. Careful cross-sectioning with a scalpel can reveal the color layer structure and allow an accurate layer count. The presence of many layers adds to the individualization of the sample. In a hit-and-run case, a paint examiner testified that an eleven-layer paint chip of various colors recovered from the victim matched the paint color layer sequence observed in a reference sample taken from the suspect's car. There is currently no minimum for a match. It has been estimated that the odds a paint chip from a crime scene originated from a randomly chosen vehicle are 33,000:1.

Sectioning cannot be done on paint smears but other methods can be used. ATR FT-IR spectroscopy is used to compare the chemical features of the paint sample. UV-Vis, fluorescence and Raman spectroscopy, electron microscopy, x-ray diffraction, x-ray spectroscopy, and neutron activation analysis can also be employed to analyze and differentiate paint samples. In order to use UV-Vis spectroscopy, the pigment must be separated from the film former and dissolved in a liquid. The liquid used depends on the solubility of the pigments but solvents such as acetone, dichloromethane, and pyridine are commonly tested. For example, acrylic layers dissolve in acetone. Inorganic pigments in single or top layers and paint smears can be analyzed by electron microscopy. Pyrolysis gas chromatography, a destructive method, can also be used to analyze and distinguish paint layers and film formers with only a 20 μg sample. However, this is a bulk analysis technique as all layers are analyzed together. If sufficient material is available, reactivity in chemical reactions or via emission spectroscopy can also be used to evaluate samples.

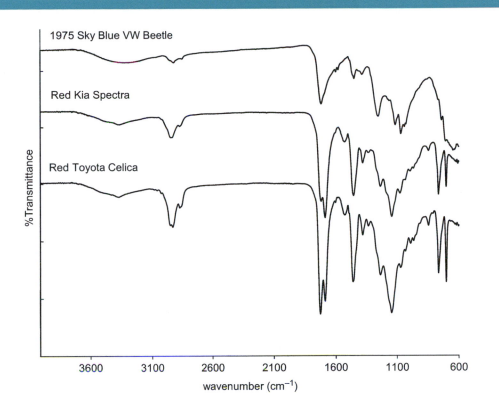

Figure 10.4 ATR FT-IR spectra of clearcoat from three automobiles.

Other coatings that may be encountered include varnish, stain, and enamel. *Stain* is composed of a mixture of organic dyes dissolved in a solvent used to impart a color to wood. Although it penetrates the wood, it does not serve as a protectant. *Varnish* is a film former used to protect wood, typically a polyurethane dissolved in an organic solvent that does not contain a color or stain. *Enamel* was originally a term used to refer to a glossy, thermosetting paint but now includes any paint that dries glossy.

POLYMERS

Polymers are also common items that can link objects, people, and places in cases. Plastics are polymers used to make products including water and soda bottles, cups, pipes, foam insulation, and food containers. Table 10.6 lists common plastic polymers and examples of their uses and recycling codes. Figure 10.5 shows the chemical structures of some plastic polymers shown in Table 10.6. Figure 10.6 shows the ATR FT-IR spectra of selected materials from the six recycling codes; the materials with the same chemical composition such as LDPE and HDPE (recycling codes #2 and #4) share an IR fingerprint. Plastics with a different chemical composition can be differentiated using this method. Many fibers share the same chemical composition as plastics and are also polymers; these are the topic of the section on fibers.

Table 10.6 Plastic polymers by uses and recycling code

Plastic	Examples of uses	Recycling code
Polyethylene terephthalate (PETE)	Water bottles, polyester polar fleece	1
High-density polyethylene (HDPE)	Milk jugs, cups	2
Polyvinyl chloride (PVC)	Pipes, flooring, toys, lawn chairs	3
Low-density polyethylene (LDPE)	Plastic bags, soda six-pack rings	4
Polypropylene (PP)	Food containers, auto parts	5
Polystyrene (PS)	Coffee cup lids, Styrofoam, plastic utensils	6
Other	Acrylic, nylon, safety glasses, headlight covers	7

Source: Polychem-USA.com.

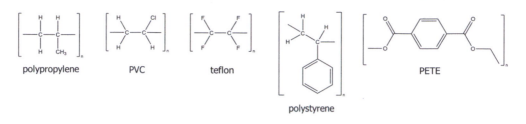

Figure 10.5   Chemical structures of plastic polymers. (Structure prepared by Ashley Cowan.)

Pyrolysis gas chromatography can also be applied to the analysis of polymer evidence. The applied heat energy will fragment bonds in the polymer chain of the fibers. It may allow the criminalist to identify the unknown fiber or identify to which subclass it belongs.

HAIR

Hairs are commonly found at crime scenes at which there was a physical struggle, including rape and kidnapping, and are submitted for evaluation by the crime laboratory. In these cases, most of the time the hair has been forcibly shed or pulled from the scalp or pubic areas of the body. Pubic hairs are short in length and exhibit a characteristic kink or twist. A hair analysis may seek to answer the question: Is the hair recovered from the crime scene consistent with hair from the suspect? The value of the evidence depends on the degree of probability an examiner can assign to it. Studies have been conducted to determine the mathematical probability of a match for hair examinations. The United States FBI Laboratory neither uses the mathematical calculations of other researchers, nor supports the feasibility of establishing a numerical probability of a hair match. In a 2002 study (Houck and Budowle) of hair examination by FBI examiners, 11% of the hairs determined to be a positive match by microscopy were later determined to differ in source by DNA analysis. This does not mean, however, that hair evidence has no value. Non-matching hairs can be used to exclude an

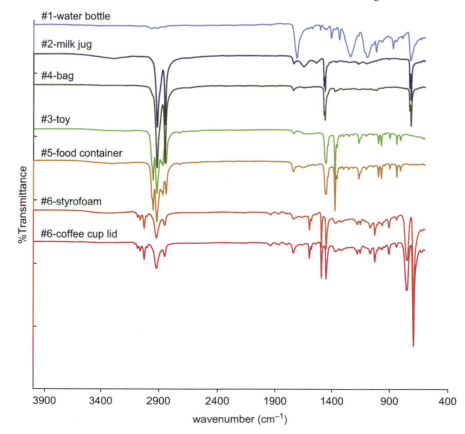

Figure 10.6   ATR FT-IR spectra of selected materials from the six recycling codes.

individual or suspect or to rule out proposed scenarios. Hair evidence can also be used to corroborate (support) other physical evidence if it is consistent with the rest of the evidence. Hair is excellent for examination as forensic evidence as it resists decomposition, is abundant and easily transferred, and can provide valuable information.

Hair is composed of the proteins collagen and keratin. Keratin is also the primary component of finger and toe nails. Hair color is mostly the result of pigments—chemical compounds that reflect certain wavelengths of visible light. There are two main pigments found in human hair: eumelanin, which gives color to brown or black hair, and pheomelanin, which produces the color in blond or red hair. Hair color may also be influenced by the optical effects of light reflecting and bouncing off the surfaces of the different hair layers. Hair shape (round or oval cross section) and texture (curly or straight) is influenced heavily by genes. However, nutritional status and intentional alteration (heat curling, "perms") can affect the physical appearance of hair.

Hair is an outgrowth of the skin and grows from a structure called the *follicle*. Humans develop hair follicles during fetal development, and no new follicles are produced after birth. The hair *root* (composed of melanocyte cells) may contain external adhering cells from the external root wreath of the follicle. The texture and diameter of hairs may vary significantly from the root to the *tip* (far end opposite the root) of the hair. From the hair root comes the hair *shaft*. The hair shaft can be dissected into three regions. Figure 10.7 is a schematic and cross section of a hair. The outermost layer that covers and protects the hair shaft is the *cuticle*. The cuticle can appear as overlapping shingles on a roof and varies in its appearance under the microscope from imbricate (looks like skin cells) to petals (spinous) to stacked cups (coronal). Other cuticle scale patterns include mosaic, irregular petal, pectinate, chevron, and diamond petal. The scales of the cuticle may vary in how many there are per unit of measure, how much they overlap, their overall shape, and how much they protrude from the surface. The thickness of the cuticle may vary as well, and the cuticles of some species' hairs may contain pigment. Characteristics of the cuticle are not useful in distinguishing between different people as human hairs are all imbricate. If hair is dyed with temporary hair dye or henna, the dye sits on the cuticle. The *cortex* is the second layer from the surface. This protein-rich cellular region contains the hair pigment. The cortex varies in thickness, texture, and color and distribution of pigments. Permanent hair dye opens the cortex via hydrogen peroxide. Bleach reacts with the melanin in the cortex to remove the color by oxidation. The yellow color that remains is that of the keratin protein of the cortex. The innermost portion of the hair shaft is the *medulla*. It may or may not be present in the hair, but, if present, it runs down the center of the hair shaft. The medulla may vary in thickness, continuity (one continuous structure or broken into pieces), and opacity (how much light is able to pass through it). The medulla may be continuous, segmented, interrupted (regular interruption) or fragmented (irregular interruption) and amorphous, fine or wide lattice, globular, aeriform lattice, or multi-serial ladder. Like the cuticle, the medulla does not lend much important information to the differentiation between hairs from different people. The structure of the hair shaft has been compared to that of a pencil with the medulla being the lead, the cortex being the wood, and the cuticle being the paint on the outside.

There are three growth phases of hair. They include the anagen, catagen, and telogen phases. The *anagen* is the longest-lasting phase. It is the growth phase and may last up to 6 years. The root is attached to the follicle for continued growth and hair pulled at this stage will have a flame-shaped root bulb. The *catagen* phase lasts 2 to 3 weeks. Hair continues to grow in this phase but at a decreasing rate. The root at this stage appears elongated and the root bulb shrinks in size as the hair is pushed out of the follicle. Hanging follicular material (the *follicular tag*) bound to the root of the hair will indicate that a hair was in an active growth phase and has been forcibly pulled. Hair is shed naturally over a 2- to 6-month period when it is in the *telogen* phase when growth ends. The telogen root will appear club-shaped.

Human hair is often separated into three classes: Caucasoid, Mongoloid, and Negroid (Table 10.7). Even so, as many individuals have multiple contributions to their racial background, analysts must be careful when making race determinations from hair microscopic analysis. Negroid hair exhibits an oval to flat cross section. Caucasoid hair exhibits the least pigment and an oval to round cross section. Mongoloid hair exhibits the coarsest or widest diameter, a round cross section, and the densest pigmentation. It also usually has a continuous medulla. In coarse hair, the cuticle: cortex ratio may be 10% of the diameter of the hair shaft. For fine hair, the cortex will be reduced so the cuticle: cortex ratio will be 20%–40%. Artificial coloring, bleaching, and dying may be encountered in hairs from all races. This treatment may serve as an additional tool for differentiation. The dye color may be distributed in the cortex and medulla. As hair grows, the natural color pigments the new, untreated portion of the hair. Based on a growth rate of 1 cm per month, the time since coloring or dyeing can be estimated. Signs of damage (e.g., burned hair, split ends, and razor cuts), disease, and infection can also be used in hair examinations.

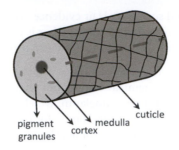

Figure 10.7 Schematic and cross section of hair.

Hairs are examined using the compound light microscope. Hair from any part of the body exhibits a wide range of characteristics such as color, length, and diameter. Even hair from different parts of the same area will differ slightly. It is necessary for the forensic examiner to have multiple reference samples (several dozen) from the same individual and the same area (head, pubic, etc.). The parts of the hair that can easily be seen by the microscope under magnification are the medulla, cortex, and cuticle (Figure 10.8). Many animal hairs are distinguished from human hairs by the size and shape of their medullae and the patterns of their cuticle or scale structure. Microscopic examination can also reveal the condition and shape of the root and tip. Typically, hairs with a medulla that fills only one third or less of the central portion of the hair are human. In addition to the physical features of the hair that can be analyzed under the microscope, cellular material on the hair root can be used in autosomal DNA typing analysis, cellular material of the hair shaft can be used in mitochondrial DNA typing, and adhering material and material within the hair shaft can be used in drug and poison analysis.

FIBERS

Fiber trace evidence is often carried or transferred in criminal cases. Fibers recovered as forensic evidence may be naturally occurring, man-made, or mineral. There are over 1000 different types of fibers. A list of manufactured fibers and their uses is shown in Table 10.8. A study of fibers by Cantrell, Roux, Maynard, and Robinson in 2001 evaluated fibers found on seats in a Sydney movie theater in the winter; in total, they classified 3025 fibers from 16 seats. As expected, cotton was the most common type (84%). Rayon accounted for 15% of the fibers. As fiber evidence is just as easily lost as transferred or carried, it is important to collect fiber evidence as soon as possible. It has been demonstrated that, in 4 hours, 80% of fibers can be expected to be lost. Only 5%–10% of fibers are expected to remain after 24 hours. Six of these types are the most commonly encountered of the 21 classifications of the US Federal Trade Commission. *Natural fibers* include cotton, linen, jute, sisal, hemp, kapok, wool, alpaca, mohair, cashmere, silk, and flax and also animal hairs from rabbit, fox, goat, and camel used in fur gloves, linings of coats and boots, and stoles. *Man-made* or *synthetic fibers* include nylon, acetate, rayon, acrylic, polyester, and Orlon. Asbestos is a *mineral fiber*, and the only natural fiber found in this category. Hemp, jute, and sisal are used in ropes, cords, and bags.

Fibers can be invaluable in criminal analysis, especially if the fiber that is transferred is rare. Natural fibers comprise approximately 25% of textile fibers produced in the United States. Naturally occurring vegetable fibers such as cotton are very common due to their use in clothing. In 1979, 24% of all US textile fiber was cotton. Thus, undyed cotton

Table 10.7 Characteristics of human head hairs

Hair type	Cross section	Mean diameter (μm)	Cortex pigment	Cuticle	Medulla
Caucasian	Oval	80 (finest)	Sparse to moderately dense, even distribution	Imbricate	May be present or absent
Negroid	Flattened	>80 (medium)	Dense distribution, clumps	Imbricate	May be present or absent
Mongoloid	Round	>80 (coarsest)	Dense distribution, arranged in streaks	Imbricate	Prominent, broad, continuous

Source: Deedrick, D.W. and S.L. Koch. 2004. *Forensic Science Communications*, Microscopy of hair part I: A practical guide and manual for human hairs, https://archives.fbi.gov/archives/about-us/lab/forensic-science-communications/fsc/jan2004/research/2004_01_research01b.htm.

Figure 10.8   Images of human Caucasian (left) and Negroid (right) hairs under a comparison microscope.

fibers have little to no forensic value. In fact, cotton is found in samples of household dust. Woolen fibers make up less than 1% of all textile fibers. Even less common than wool is silk. Its use has decreased significantly since the development of synthetic fibers. As silk does not shed easily, its recovery as evidence is rare. Similarly, goat (cashmere, mohair), llama (alpaca, vicuna, guanaco), and camel hair are not frequently encountered and can be quite valuable as forensic evidence. Cattle and rabbit hairs may be found in felts, material produced when fibers settle out from water and are pressed together.

Like human hairs, animal hairs will exhibit a cuticle, cortex, and medulla as well as damage and forceful removal features. Standard samples and/or a comparison database can be used for the identification of the 35 most commonly encountered animal hairs. Animal hairs have some class characteristics. The cuticle, or exterior surface, of wool, alpaca, cashmere, silk, and linen is imbricate. Cat hair is fine in diameter. Animal fibers have a medullary index (diameter of the hair's medulla divided by the diameter of the hair) of 0.5 or greater. The medulla of cat hair resembles square beads on a string. Deer hair exhibits a wide lattice medulla. Rabbit hair exhibits a multi-serial ladder medulla. The mouse hair medulla appears as parallel rows of beads. Animal hair roots may also vary widely. The dog hair root is shaped like an arrow while the cat hair root is frayed. Deer hair exhibits a wineglass-shaped root and the medulla extends into the cattle hair root. Fur types will vary in hairs from different areas of the animal and between animals. Images of a few animal hairs are shown in Figures 10.9 and 10.10. Figure 10.9 shows a micrograph of red fox and deer hairs and Figure 10.10 shows horse and rabbit hairs.

Synthetic fibers represent approximately 75% of textile fiber production in the United States. Synthetic fibers have no medulla or scale pattern and are readily distinguishable from animal hair. Textile fibers are the smallest part of a textile, the yarns and fibers that are used to produce it. The chemical process used to manufacture the textile will impact the appearance and physical features of the fiber. For example, in the mercerizing process of cotton, the fibers are treated with alkali that causes them to swell up and become less twisted and more rounded. This process is performed to improve the texture and feel of the fiber.

Asbestos mineral fibers are used in siding and fireproof materials. However, as they are rarely used in clothing or household objects, these fibers are not commonly seen in forensic evidence. Asbestos minerals are crystalline materials that fracture or cleave, like glass, into long, thin rods. The rods continue to break until microscopic particles are produced that are easily airborne, transferred from one place to another, and may be drawn into the lungs on breathing. When collected, asbestos fibers can be valuable forensic evidence. Forensic scientists group the asbestos fibers by type: cysotile, a form of the mineral serpentine, and crocidolite, a form of amphibole.

Fibers, like other trace evidence, are first evaluated using a stereomicroscope. Under the stereomicroscope, it can be determined if the fiber is synthetic, naturally occurring, or inorganic. Physical features such as length, color, crimp, relative diameter, luster, cross-section features, damage, and adhering debris can be noted. A point-by-point and side-by-side

Table 10.8 Fibers and their uses

Fiber	Example uses
Acetate	Blouses, clothing linings, wedding and party attire, home furnishings, draperies, cigarette filters
Acrylic	Sweaters, socks, fleece, activewear, blankets, area rugs, upholstery, luggage, outdoor furniture, awnings, yarn
Anidex	Blouses, athletic wear, dresses, hosiery, lingerie, jackets, linings, rainwear, shirts, suits, sweaters, upholstery
Aramid	Flame-resistant clothing, protective vests and helmets, hot air filtration fabrics, ropes and cables, sail cloth, sporting goods
Azlon	Fine clothing; blended with wool, cotton, or synthetics for weaving as it takes color well
Carbon	Bikes, running shoes, hockey sticks, tennis rackets, golf clubs, robotic equipment, planes, missiles, helmets
Glass	Fiberglass insulation
Lyocell	Coats, jeans, dresses, slacks
Melamine	Fire blocking fabrics for aircraft seating, upholstered furniture, firefighters' gear, heat-resistant gloves
Metallic	Decorative fibers in apparel, upholstery, braids, military uniform decorations, ribbons, draperies, laces, table linens
Modacrylic	Coats, trims and linings, fleece, simulated fur, wigs and hair pieces, blankets, curtains, paint rollers, stuffed toys
Nylon	Nets, hoses, carpets, blouses, shirts, dresses, hosiery, ski and cycling apparel, windbreakers, swimsuits, lingerie, raincoats, upholstery, seat belts, air bags, parachutes, racket strings, ropes, sleeping bags, tents, thread, dental floss
Nytril	Sweaters, fake fur, wool blends
Olefin (polypropylene or polyethylene)	Activewear, sportswear; socks, thermal underwear, linings, truck liners, carpets; carpet backing, upholstery, ropes, filters
PBI (polybenzimidazole)	Space suits, firemen's turnout coats
PEEK (polyether ether ketone)	Bone and cartilage replacement
PEN (polyethylene naphthalate)	Cordage, tire cord, high-performance sailcloth for racing
PLA (polylactic acid)	Sports, performance and outdoor apparel
Polyacrylate	Adhesives, coating SPME fibers
Polyimide	Spacecraft insulation, semiconductor chips
Polyester	Every type and form of clothing, carpets, curtains, draperies, sheets, upholstery, ropes, nets, thread, sails, fill
PSA (polysulfone)	Permeable membranes
Rayon	Blouses, dresses, jackets, lingerie, linings, slacks, sportswear, suits, ties, work clothes, drapes, tablecloths, sheets, curtains, bedspreads, blankets, slipcovers, upholstery, tire cord, feminine hygiene products
Saran	Deck chairs, garden furniture
Spandex	Swimsuits, hosiery, aerobic/exercise wear, ski pants, golf jackets, disposable diapers, waist bands, bra straps, support hose, cycling shorts
Sulfar	Papermakers' felts, electrical insulation, filter fabrics for liquid and gas filtration, gaskets/packings, electrolysis membranes
Triexta (PTT, polytrimethylene terephthalate)	Carpets, rugs, upholstery, automotive mats
Ultra-high-molecular-weight (UMHW) polyethylene	Tough plastic sheets, rods, tape
Vinal (PVA, Polyvinylalcohol)	Papermaking, coatings
Vinyon (PVC, polyvinyl chloride)	Bonding agent for nonwoven fabrics in industrial applications

Source: American Fiber Manufacturer's Association, Inc., www.fibersource.com/fiber-products/.

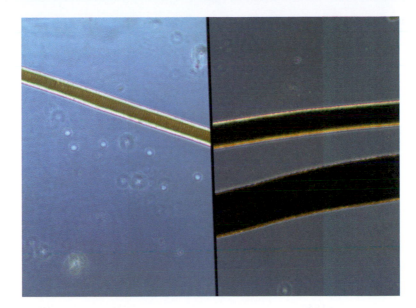

Figure 10.9 Micrograph of red fox (left) and deer (right) hairs.

microscopic comparison provides a discriminating method of determining if two or more fibers are consistent with originating from the same source. Fibers that can be differentiated using a comparison stereomicroscope do not need to undergo further microscopic analysis but fibers that exhibit the same physical features will next be examined using the compound light microscope. Compound light microscopes are used to quickly and accurately examine fibers including their physical characteristics such as color and dye content, shape, direction and numbers of twists, surface features, cross section, diameter, fabric weave (if present), length, and interaction with plane-polarized light in a nondestructive manner. Foreign material adhering to or embedded in fibers can also be identified under the microscope.

Cotton has a flattened helical shape and silk exhibits a slight twist while polyester and other synthetic fibers are straight. The cross section of cotton is hoof-shaped while wool is round and synthetic fibers are irregular but oval-like. Synthetic fibers can be microfiber, bicomponent, and high performance. *Microfiber* is characterized by a fine composition and feel. *Bicomponent fibers* are prepared when two polymers of different chemical composition and/or physical features are extruded from the same spinneret leading to both fibers in the same filament. Bicomponent

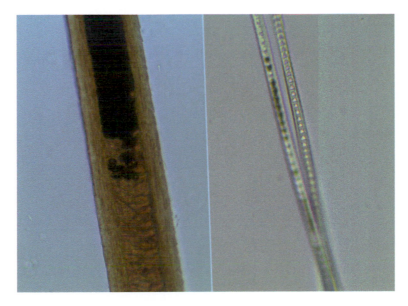

Figure 10.10 Micrograph of horse (left) and rabbit (right) hairs.

Table 10.9 Fiber characteristics

Fiber	RI	Birefringence	Sign of elongation	Pleochroism	Extinction	Interference colors
Acetate	1.480 1.476	.004	Positive	Yes	Yes	Orange, blue
Acrilan	1.520 1.524	−.004	Negative	Yes	Yes	Orange, blue
Acrylic	1.520 1.524	−.004	Negative	Yes	Yes	Orange, blue
Dacron	>1.640 1.54			No	Yes	Yellow
Kodel	1.632 1.532	.10	Positive	No	Yes	Bright yellow
Modacrylic	1.532 1.520	.012	Positive	Yes	Yes	Orange, blue
Nylon	1.560 1.512	.050	Positive	No	Yes	Rainbow
Orlon	1.508 1.512	−.004	Negative	Yes	Yes	Blue, orange
Polyester (PET)	>1.640 1.532			No	Yes	Yellow
Polypropylene	1.520 1.492	.028	Positive	No	Yes	Yellow
Rayon	1.540 1.520	.02	Positive	No	Yes	Bright yellow
Acrilan (A0118)	1.512 1.520	−.008	Negative	Yes	Yes	Orange, blue
Cantrece (A0025)	1.572 1.520	.052	Positive	Yes	Yes	Rainbow
Cellulose acetate/polyester (A0294)	1.484 1.480	.004	Positive	Yes	Yes	Orange, blue
Triacetate/polyester (A0291)	1.480 1.476	.004	Positive	Yes	Yes	Light rainbow, dark rainbow

Figure 10.11   Micrograph of acrylic (left) and acetate (right) fibers.

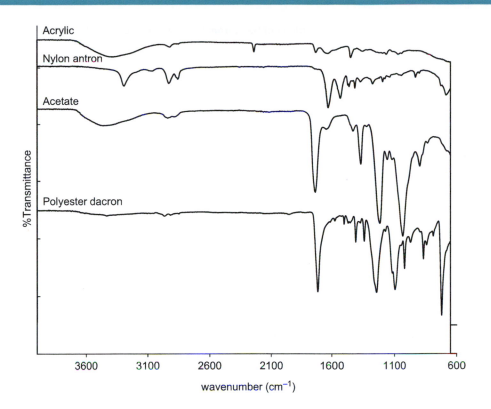

Figure 10.12   ATR FT-IR spectra of synthetic fibers acrylic, Antron nylon, acetate, and Dacron polyester.

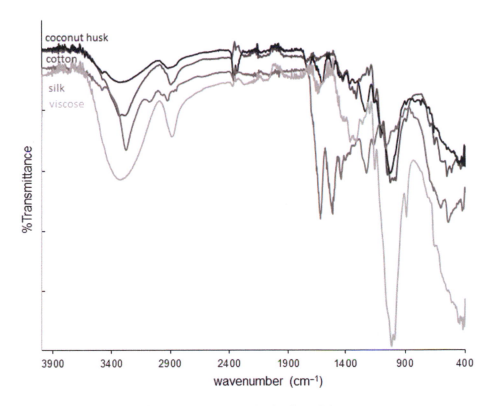

Figure 10.13   ATR FT-IR spectra of natural fibers cotton, coconut husk, silk, and viscose.

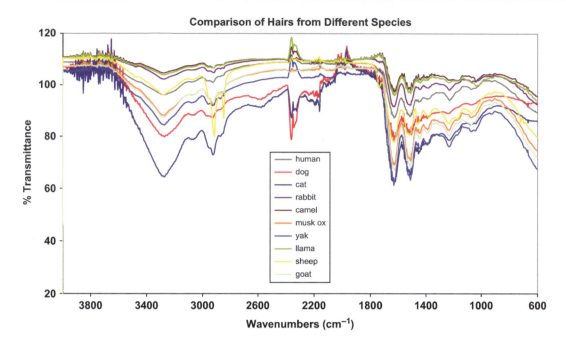

Figure 10.14   ATR FT-IR spectra of hairs from human, dog, cat, rabbit, camel, musk ox, yak, llama, sheep, and goat.

fibers can be identified by their cross section and can be dumbbell, trilobal, round, or flat in shape. *High-performance fibers* are characterized by their exceptional strength and chemical and heat resistance.

Table 10.9 shows fiber characteristics for many fibers. Fibers are uniaxial and have two indices of refraction. Using a variable compensator, the analyst can determine the order. Nylon is a high-birefringent fiber whereas acrylic is a low-birefringent fiber. For acrylic, variable filters cannot be used but one can determine the sign of elongation based on colors. Yellow will be parallel to the slow wave and blue will be perpendicular to the slow wave. Acetate is positive when red is perpendicular to the slow wave. If fibers have the same index of refraction, the analyst can change the mounting medium to see the Becke line. Kevlar fibers exhibit a strong fluorescence. Dyes can be differentiated using a microscope linked to an IR or UV-Vis spectrophotometer. The spectra can also be compared to a database for identification. Figure 10.11 shows acrylic and acetate fibers under the microscope. Figure 10.12 shows ATR FT-IR spectra of four synthetic fibers: acrylic, Antron nylon, acetate, and Dacron polyester. These fibers are easily differentiated by their IR fingerprints. Figure 10.13 shows ATR FT-IR spectra of the natural fibers cotton, coconut husk, silk, and viscose. The natural fibers are also differentiated by their infrared spectra. Figure 10.14 shows the ATR FT-IR spectra of hairs from a human, dog, cat, rabbit, camel, musk ox, yak, llama, sheep, and goat. The hairs are more difficult to differentiate due to their similar composition.

If necessary, solubility is determined. Chemical characteristics can be probed using microscale chemical tests to determine reactivity. From only a 1 cm fiber with a 20 micron diameter, it is possible to determine all of the following: class, polymer composition, cross-sectional shape, melting point, absorption spectrum, fluorescence, refractive index, birefringence, color, finish (e.g., bright/dull), dye class, and dye components. Finally, SEM-EDS may be used as a microanalytical tool for fiber characterization. Surface morphology can be examined with greater resolution using this method. Other methods including thin layer chromatography to compare colorants and dye batch variation and pyrolysis gas chromatography may also be used in fiber analysis.

OTHER TRACE MATERIALS

Other materials that may comprise trace evidence recovered from the crime scene include duct tape, other tape, torn papers, matches and match books, fragments from explosions like tearing components, and materials from manufacturing failure. Physical end matching is termed fracture matching (Figure 10.15). The tape weave is the same in the two pieces and the end fragments can be matched through their edges.

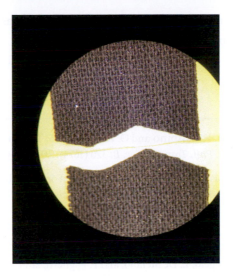

Figure 10.15 Micrograph of tape weave and pattern matching.

Duct tape is composed of three parts: the backing, the scrim, and the adhesive. The *scrim* is the cotton or polyester-blend fabric or yarn used to reinforce the tape structure. The *backing* is most often polyethylene and comes in various colors and thicknesses. Yarn that runs along the length of the tape, or the machine direction, is termed the warp. The yarn that spans across the tape is termed the fill. The number of fibers, diameter, and type of yarn varies between brands and grades of tape. For example, industrial-grade tapes have a higher yarn count and thicker yarn than general retail tape. The *adhesive* components vary by manufacturer but duct tape is composed of elastomers, resins, and fillers. Duct tape fractures are due to faults in the backing and scrim materials, often due to an outside force.

QUESTIONS

1. Which of the following will influence the density of a substance?
   - a. Thermal history
   - b. Chemical composition
   - c. State of matter
   - d. All of the above

2. Which of the following phases of hair growth characterize the last phase in which the hair is naturally shed?
   - a. Telogen
   - b. Anagen
   - c. Catagen
   - d. Haragen
   - e. None of the above

3. Comparing of refractive indices of glass is often done using the ____ method.
   - a. Float
   - b. Submersion
   - c. Immersion

4. In collecting paint evidence from a hit-and-run accident, ____.
   - a. Collect all layers of the evidence sample and the top layer nearby for a reference sample
   - b. Collect all layers of the evidence sample and all layers nearby for a reference sample
   - c. Collect the top layer of the evidence sample and the top layer nearby for a reference sample
   - d. Collect the top layer of the evidence sample and all layers nearby for a reference sample

5. Which of the following parts of a hair contains the hair pigment?
   - a. Cuticle
   - b. Cortex
   - c. Follicular tag
   - d. Medulla

6. Sketch a diagram and label five parts of a human hair.

7. What microscopy method is used to examine natural hairs and fibers?

8. Briefly explain how hairs differ in morphology from synthetic fibers such as acetate and Dacron.

9. What features of a paint chip can be used to individualize the sample?

10. Diagram two bullet holes and label the types of cracks that are typically present and how to determine the sequence of the bullets shot.

References

Roux, C., B. Talbot-Wright, J. Robertson, F. Crispino and O. Ribaux. 2015. The end of the (forensic science) world as we know it? The example of trace evidence. *Phil. Trans. R. Soc. B.* 370 (1674): 20140260.

The Scientific Working Group on Materials Analysis (SWGMAT). 1999. Forensic fiber examination guidelines. *Forensic Sci. Comm.* 1(1) www.fbi.gov/about-us/lab/forensic-science-communications/fsc/april1999/houcktoc.htm (accessed January 31, 2018).

Bibliography

Accardo, G., R. Cioffi, F. Colangelo, R. d'Angelo, L. De Stefano, and F. Paglietti. 2014. Diffuse reflectance infrared Fourier transform spectroscopy for the determination of asbestos species in bulk building materials. *Materials (Basel)* 7:457–470.

American Chemistry Council. Plastic Packaging Resins. https://plastics.americanchemistry.com/Plastic-Resin-Codes-PDF/ (accessed December 2, 2017).

American Fiber Manufacturer's Association, Inc. www.fibersource.com/fiber-products/ (accessed January 31, 2018).

American Society of Trace Evidence Examiners. www.asteetrace.org/ (accessed December 2, 2017).

Armitage, S., S. Saywell, C. Roux, C. Lennard, and P. Greenwood. 2001. The analysis of forensic samples using laser micro-pyrolysis gas chromatography mass spectrometry. *J. Forensic Sci.* 46(5):1043–1052.

Arvidson, S.A., K. Wong, R. Gorga, and S.A. Khan. 2012. Structure, molecular orientation, and resultant mechanical properties in core/sheath poly (lactic acid)/polypropylene composites. *Polym.* 53(3):791–800.

Ball, B. 2016. Chapter 10: Soil and glass analysis. In *Forensic Science for High School*, 3rd edn. Dubuque, IA: Kendall Hunt Publishing Company.

Bianchi, F., N. Riboni, V. Trolla, G. Furlan, G. Avantaggiato, G. Iacobellis, and M. Careri. 2016. Differentiation of aged fibers by Raman spectroscopy and multivariate data analysis. *Talanta* 154:467–473.

Blackledge, R.D. 2007. *Forensic Analysis on the Cutting Edge: New Methods for Trace Evidence Analysis*. Hoboken, NJ: J. Wiley & Sons.

Bottrell, M.C. 2009. Forensic glass comparison: Background information used in data interpretation. *Forensic Sci. Comm.* 11(2). www.fbi.gov/about-us/lab/forensic-science-communications/fsc/april2009/review/2009_04_review01.htm/ (accessed December 2, 2017).

Cantrell, S., C. Roux, P. Maynard, and J. Robertson. 2001. A textile fibre survey as an aid to the interpretation of fibre evidence in the Sydney region. *Forensic Sci. Int.* 123(1):48–53.

Cargille Standards. www.cargille.com/refractivestandards.shtml (accessed January 31, 2018).

Causin, V. 2015. *Polymers on the Crime Scene: Forensic Analysis of Polymeric Trace Evidence*. Vienna: Springer International Publishing.

Chalmers, J.M., G. Howell, M. Edwards, and M. Hargreaves. 2012. Forensic analysis of fibers by vibrational spectroscopy. In *Infrared and Raman Spectroscopy in Forensic Science*, pp. 153–168. Chichester, West Sussex, UK: Wiley.

Cox, R.J., H.L. Peterson, J. Young, C. Cusik, and E.O. Espinoza. 2000. The forensic analysis of soil organic by FTIR. *Forensic Sci. Int.* 108(2):107–116.

Davis, B.J., P.S. Carney, and R. Bhargava. 2011. Theory of infrared microspectroscopy for intact fibers. *Anal. Chem.* 83(2):525–532.

Deedrick, D.W. and S.L. Koch. 2004. Microscopy of hair part 1: A practical guide and manual for human hairs. *Forensic Sci. Comm.* 6(1). https://archives.fbi.gov/archives/about-us/lab/forensic-science-communications/fsc/jan2004/research/2004_01_research01b.htm (accessed January 31, 2018).

Deedrick, D.W. and S.L. Koch. 2004. Microscopy of hair part II: A practical guide and manual for animal hairs. *Forensic Sci. Comm.* 6(3). https://archives.fbi.gov/archives/about-us/lab/forensic-science-communications/fsc/july2004/research/2004_03_research02.htm (accessed January 31, 2018).

Durany, A., N. Anantharamaiah, and B. Pourdeyhimi. 2009. High surface area nonwovens via fibrillating spunbonded nonwovens comprising islands-in-the-sea bicomponent filaments: Structure-process-property relationships. *J. of Mater. Sci.* 44:5926–5934.

Farah, S., T. Tsadok, B. Alfonso, and J.D. Abraham. 2014. Morphological, spectral and chromatography analysis and forensic comparison of PET fibers. *Talanta* 123:54–62.

Flynn, K., R. O'Leary, C. Roux, and B. Reedy. 2006. Forensic analysis of bicomponent fibers using infrared chemical imaging. *J. Forensic Sci.* 51(3):586–596.

Gaudette, B.D. 1999. Evidential value of hair examination. In *Forensic Examination of Hair*, pp. 243–257. London, UK: Taylor and Francis.

Glass Density. https://chem.libretexts.org/Exemplars_and_Case_Studies/Exemplars/Forensics/Glass_Density_Evidence (accessed January 31, 2018).

Govaerta, F. and M. Bernard. 2004. Discriminating red spray paints by optical microscopy, Fourier transform infrared spectroscopy and X-ray fluorescence. *Forensic Sci. Int.* 140:61–70.

Gresham, G.L., G.S. Groenewold, W.F. Bauer, and J.C. Ingram. 2000. Secondary ion mass spectrometric characterization of nail polishes and paint surfaces. *J. Forensic Sci.* 45(2):310–323.

Grieve, M.C. 1995. Another look at the classification of acrylic fibres, using FTIR microscopy. *Sci. Justice* 35(3):179–190.

Grieve, M.C. and R.M.E. Griffin. 1999. Is it a modacrylic fibre? *Sci. Justice* 39(3):151–162.

Hocking, T. Forensic geology: The legal application of earth and soil science. https://slideplayer.com/slide/1551547/

Hopen, T.J., C. Taylor, L. Peterson, and W. Rantanen. 2017. The forensic examination and analysis of paper matches. *NSFTC*, pp. 1–15.

Houck, M.M. and B. Budowle. 2002. Correlation of microscopic and mitochondrial DNA hair comparisons. *J. Forensic Sci.* 47(5):964–967.

Houck, M.M. and J. Siegel. 2006. *Fundamentals of Forensic Science*. Amsterdam: Elsevier/Academic.

Hughes, J.C., T. Catterick, and G. Southeard. 1976. The quantitative analysis of glass by atomic absorption spectroscopy. *Forensic Sci.* 8(3):217–227.

Kammrath, B.W., A. Koutrakos, J. Castillo, C. Langley, and D. Huck-Jones. 2017. Morphologically-directed Raman spectroscopy for forensic soil analysis. *Forensic Sci. Int.* 285:e25–e33. doi:10.1016/j.forsciint.2017.12.034.

Kerney, T. and K.M. Elkins. 2014. Can fibers be differentiated if they share chemical composition, physical and/or optical properties? Using Polarizing Light Microscopy, FTIR and Infrared Chemical Imaging. Unpublished data.

Kirk, P. 1953. *Crime Investigation: Physical Evidence and the Police Laboratory.* New York: Interscience.

Koehler, A. 1935. Who made that ladder? *The Saturday Evening Post.* April 20, 1935. http://www.saturdayeveningpost.com/wp-content/flbk/Ladder_story/#/1/ (accessed July 24, 2018).

Kolowski, J.C., N. Petraco, M.M. Wallace, P.R. DeForest, and M. Prinz. 2004. A comparison study of hair examination methodologies. *J. Forensic Sci.* 49:1253–1255.

Köhler, A. 1893. Ein neues Beleuchtungsverfahren für mikrophotographische Zwecke. *Zeitschrift für wissenschaftliche Mikroskopie und für Mikroskopische Technik* 10(4):433–440.

Köhler, A. 1894. New method of illumination for photomicrographical purposes. *J. of the R. Microscopical Soc.* 14:261–262.

Kubic, T. and N. Petraco. 2009. *Forensic Science Laboratory Manual and Workbook*. Boca Raton: CRC Press.

Lambert, D., C. Muehlethaler, P. Esseiva, and G. Massonnet. 2016. Combining spectroscopic data in the forensic analysis of paint: Application of a multiblock technique as chemometric tool. *Forensic Sci. Int.* 263:39–47.

Lavine, B.K., A. Fasasi, N. Mirjankar, K. Nishikida, and J. Campbell. 2014. Simulation of attenuated total reflection infrared absorbance spectra: Applications to automotive clear coat forensic analysis. *Appl. Spectrosc.* 68(5):608–615.

Lavine, B.K., C.G. White, M.D. Allen, A. Fasasi, and A. Weakley. 2016. Evidential significance of automotive paint trace evidence using a pattern recognition based infrared library search engine for the Paint Data Query Forensic Database. *Talanta* 159:317–329.

Lepot, L., K. De Wael, F. Gason, and B. Gilbert. 2008. Application of Raman spectroscopy to forensic fibre cases. *Sci. Justice* 48(3):109–117.

Locard, E. 1930. The analysis of dust traces, Part I. *Am. J. Police Sci.* 1(3):276–298.

Locard, E. 1930. The analysis of dust traces, Part II. *Am. J. Police Sci.* 1(4):401–418.

Locard, E. 1930. The analysis of dust traces, Part III. *Am. J. Police Sci.* 1(5):496–514.

Long, H., S. Walbridge-Jones, and K. Lundgren. 2014. Synthetic wig fibers: Analysis and differentiation from human hairs. *JASTEE* 5(1):2–21.

Manheim, J., K.C. Doty, G. McLaughlin, and I.K. Lednev. 2016. Forensic hair differentiation using attenuated total reflection Fourier transform infrared (ATR FT-IR) spectroscopy. *Appl. Spectrosc.* 70(7):1109–1117.

McCabe, K.R., F.A. Tulleners, J.V. Braun, G. Currie, and E.N. Gorecho. 2013. A quantitative analysis of torn and cut duct tape physical end matching. *J. Forensic Sci.* 58(S1):S34–S42.

Oien, C.T. 2009. Forensic hair comparison: Background information for interpretation. *Forensic Science Communications* 11(2). www.fbi.gov/about-us/lab/forensic-science-communications/fsc/april2009/review/2009_04_review02.htm/ (accessed January 31, 2018).

Petraco, N. and T. Kubic. 2000. A density gradient technique for use in forensic soil analysis. *J. Forensic Sci.* 45(4):872–873.

Reeve, V., J. Mathiesen, and W. Fong. 1976. Elemental analysis by energy dispersive x-ray: A significant factor in the forensic analysis of glass. *J. Forensic Sci.* 21(2):291–306.

Refractive index. www.gemsociety.org/article/refractive-index-list-of-common-household-liquids/ (accessed January 31, 2018).

Roberts, K., M.J. Almond, and J.W. Bond. 2013. Using paint to investigate fires: An ATR-IR study of the degradation of paint samples upon heating. *J. Forensic Sci.* 58(2):495–499.

Saferstein, R. 2015. *Criminalistics: An Introduction to Forensic Science*, 11th edn. Boston, MA: Pearson.

Sano, T. and S. Suzuki. 2009. Basic forensic identification of artificial leather for hit-and-run cases. *Forensic Sci. Int.* 192(1–3):e27–e32.

Schenk, E.R. and J.R. Almirall. 2012. Elemental analysis of glass by laser ablation inductively coupled plasma optical emission spectrometry (LA-ICP-OES). *Forensic Sci. Int.* 217(1–3):222–228.

Stuart, J.H., J.J. Nordby, and S. Bell. 2014. *Forensic Science: An Introduction to Scientific and Investigative Techniques*, 4th edn. London, UK: Taylor & Francis.

SWGMAT. 2005. Forensic human hair examination guidelines. www.nist.gov/sites/default/files/documents/2016/09/22/forensic_human_hair_examination_guidelines.pdf (accessed January 31, 2018).

SWGMAT. 2013. Guideline for assessing physical characteristics in forensic tape examinations. *JASTEE* 5(1):34–41. www.unitedstatesbd.com/images/unitedstatesbdcom/bizcategories/2961/files/JASTEE_2014_5_1_3.pdf (accessed December 2, 2017).

Taupin, J.M. 2004. Forensic hair morphology comparison—A dying art or junk science? *Sci. Justice* 44(2):95–100.

Trejos, T., R. Koons, S. Becker, T. Berman, J. Buscaglia, M. Duecking, T. Eckert-Lumsdon, T. Ernst, C. Hanlon, A. Heydon, et al. 2013. Cross-validation and evaluation of the performance of methods for the elemental analysis of forensic glass by µ-XRF, ICP-MS, and LA-ICP-MS. *Anal. Bioanal. Chem.* 405(16):5393–5409.

Trzcińska, B., R. Kowalski, and J. Zięba-Palus. 2014. Comparison of pigment content of paint samples using spectrometric methods. *Spectrochim. Acta A Mol. Biomol. Spectrosc.* 130:534–538.

Tulleners, F.A., J. Thornton, and A.C. Baca. 2013. Determination of unique fracture patterns in glass and glassy polymers. NIJ Report. www.ncjrs.gov/pdffiles1/nij/grants/241445.pdf (accessed May 24, 2018).

Tulleners, F.A. and J.V. Braun. 2011. The statistical evaluation of torn and cut duct tape physical end matching. NIJ Report. www.ncjrs.gov/pdffiles1/nij/grants/235287.pdf (accessed May 24, 2018).

USDA, www.nrcs.usda.gov/Internet/FSE_DOCUMENTS/stelprdb1044818.pdf (accessed January 31, 2018).

Vitz, E., J.W. Moore, J. Shorb, X. Prat-Resina, T. Wendorff, and A. Hahn. 2016. Glass Density Evidence. https://chem.libretexts.org/Exemplars_and_Case_Studies/Exemplars/Forensics/Glass_Density_Evidence (accessed January 28, 2018).

Was-Gubala, J. and R. Starczak. 2015. Nondestructive identification of dye mixtures in polyester and cotton fibers using Raman spectroscopy and ultraviolet-visible (UV-Vis) microspectrophotometry. *Appl. Spectrosc.* 69(2):296–303.

Woods, B., C. Lennard, K.P. Kirkbride, and J. Robertson. 2014. Soil examination for a forensic trace evidence laboratory. Part 1: Spectroscopic techniques. *Forensic Sci. Int.* 245:187–194.

Woods, B., C. Lennard, K.P. Kirkbride, and J. Robertson. 2014. Soil examination for a forensic trace evidence laboratory. Part 2: Elemental analysis. *Forensic Sci. Int.* 245:195–201.

Zięba-Palus, J., G. Zadora, J.M. Milczarek, and P. Kościelniak. 2008. Pyrolysis-gas chromatography/mass spectrometry analysis as a useful tool in forensic examination of automotive paint traces. *J. Chromatogr. A* 1179(1):41–46.

Zięba-Palus, J. and B.M. Trzcińska. 2013. Application of infrared and Raman spectroscopy in paint trace examination. *J. Forensic Sci.* 58(5):1359–1363.

CHAPTER 11

Questioned documents and impression evidence

KEY WORDS: questioned documents, impression evidence, physical analysis, chemical analysis, handwriting, physical characteristics, alterations, ink, alternate light source, infrared photography, shoe tread, tire tread, bite mark, tool impression, electrostatic lifting, serial number restoration, forensic odontologist

LEARNING OBJECTIVES

- To define the term questioned documents
- To list examples of questioned documents
- To differentiate between the physical and chemical characteristics of documents
- To explain how to detect alterations in documents including erasures, overwriting, and obliteration
- To explain how to differentiate papers, inks, and pigments by their chemical characteristics
- To identify types of impression evidence and how to preserve and document the evidence
- To explain the use of chemical methods in detecting serial number impressions

OVERWRITING THE US DECLARATION OF INDEPENDENCE

Based on close examination and photography, United States National Archives experts recently reported that the US Declaration of Independence, written on parchment paper measuring 24.25 by 29.75 inches, exhibits evidence of overwriting and the enhancement of famous signatures, including the "J" and "H" in the John Hancock signature. Signatures in the center columns "show evidence of partial enhancement or recreation of missing signatures," according to Mary Lynn Ritzenthaler, retired chief of conservation at the National Archives. A left-hand print, first documented by a conservator from the Fogg Museum at Harvard in 1940, is also suspected to have been left on the document at that time. The damage is not thought to be due to nineteenth-century exhibition or copying.

The US Declaration of Independence was written in 1776. The historic document is now almost 250 years old. Over time, the ink has faded and the text and signatures of the 56 men that signed the document are only faintly readable in some spots. Early documentation of fading of the signatures occurred in 1817, only 40 years after its production. Fading was accelerated in the nineteenth century through press-copying. A damp sheet of thin paper was placed on the document and pressed until enough of the ink was transferred to make a legible copy.

Based on historical records and old photographs, the overwriting alternations would have had to occur between 1903 and 1940. Conservators believe that during that time, drastic steps were taken to improve the readability of the signatures. Document conservation methods and practices have changed considerably since the document was written. Modern historians are referring to the alternations on the Declaration as "defacement" (Ruane 2016).

Twenty-four first printing copies of the Declaration are known to exist. The original Declaration was transferred to the National Archives in 1952, where it is exhibited in its rotunda in Washington, D.C.

OVERWRITING THE US DECLARATION OF INDEPENDENCE (continued)

BIBLIOGRAPHY

Reif, R. 1991. Declaration of independence found in a $4 picture frame. *The New York Times*, www.nytimes.com/1991/04/03/arts/declaration-of-independence-found-in-a-4-picture-frame.html (accessed January 26, 2018).

Ruane, M. 2016. Was the declaration of independence "defaced"? Experts say yes. *The Washington Post*. www.washingtonpost.com/local/was-the-declaration-of-independence-defaced-experts-say-yes/2016/10/21/5bb6efaa-96d9-11e6-bb29-bf2701dbe0a3_story.html?utm_term=.199d7f0c8812 (accessed January 26, 2018).

The Declaration of Independence: A History. www.archives.gov/founding-docs/declaration-history (accessed January 26, 2018).

Questioned documents are documents whose source or authenticity is the subject of an investigation. These include both handwritten and print documents such as a driver's license, passport, lottery ticket, check, contract, will, voter registration form, petition, or suicide note. These also include writing and markings on walls, doors, windows, and other objects. *Impression evidence* is physical evidence characterized by markings or indentations produced by pressure. Impression evidence observed in document evidence includes a bookkeeper's notepad; other impressions include those made by footprints, fingerprints, bite marks, tire tread marks, tool marks, and objects that can impart pressure.

QUESTIONED DOCUMENTS

Questioned documents provide several features for testing by forensic chemistry methods. These include inks that may be differentiated by chromatographic and spectroscopic methods; paper, alterations, erasures, obliterations, overwriting, and watermarks that may be differentiated by microscopic methods; and charred, burnt, and overwriting characteristics that may be investigated by infrared photography. Additionally, handwritten and typed documents may be analyzed for physical features by forensic document examiners, including handwriting, copies, faxes, and printed documents. The paper in questioned documents lends itself to physical and chemical analysis. The *physical analysis* of paper includes its appearance, color, weight, and watermarks. The *chemical analysis* of paper includes fiber composition, pigments, additives, and fillers in the paper.

Many of the other methods described previously in this book can also be applied to questioned documents investigation. These include the use of an alternate light source (ALS), ultraviolet (UV) wand, infrared (IR) photography, white flashlight for reflecting on surfaces, microscopy, Adobe Photoshop, chemical test reagents, and several spectroscopic (i.e., UV-Vis, IR, and Raman spectroscopy) and chromatographic (i.e., thin-layer chromatography [TLC] and high-pressure liquid chromatography [HPLC]) methods. In addition to these methods, electrostatic lifting is used in evaluating impressions.

PHYSICAL ANALYSIS

The *physical characteristics* of handwriting may be attributed to several factors including the style of writing learned (e.g., Palmer, Zaner-Bloser, D-Nealian, etc.), nerve and motor differences that influence small motor skills, and subconscious differences. Variations may also be due to the writing instrument (e.g., pen, pencil, or crayon) used. Additionally, variations may be prevalent in spelling, punctuation, grammar, and phraseology usage as well as personal preferences such as slope, angularity, letter and word spacing, dimensions of letters, connections, and writing style. Further variations are influenced by mental differences, speed, pressure, finger dexterity, margins, crowding, insertions, alignment, writing skill, and the influence of drugs and alcohol. Features like these can be examined by microscopy through the use of stereoscopes and comparison microscopes.

Document comparison and analysis relies on the availability of known writings or reference samples written by a suspect person for comparison to the item of physical evidence. Ideally, the known items for comparison should contain words and phrases that are present in the evidence item. Known documents with few, if any, similar features are less useful. Likewise, questioned documents are difficult to analyze if the item contains only a few words or is in a disguised form. A collection of several known documents, or exemplars, is critical to document analysis as natural variations are documented to occur between sessions copying or recording the same text. While no single handwriting characteristic can be used for a positive handwriting comparison, the final conclusion must be based on a sufficient number of point characteristics in common between the known and questioned samples that is based on the judgment of the forensic document examiner expert. When possible, exemplars should be recorded with the same writing instrument and type of paper or document as the evidence item. They should be recorded within 2–3 years of the questioned document. They may be collected by dictation (harder for suspects to attempt deception techniques) or copying. The words and phrases provided should include those identical to the questioned document.

Variations in typewritten documents are also a subject of forensic document analysis. Typewriters vary in typefaces and wear. Additionally, characters may be aligned up or down or left or right of the correct position, yielding horizontal and vertical misalignment respectively, or aligned at an angle (leaning) yielding perpendicular misalignment. Impressions on the typewriter ribbon may also be consequential to a case. Wear characteristics, differential spacing, and shifted angles of letters may be used to assign individual characteristics to a document/typewriter pair. Typewritten documents are often prepared using a computer and are printed with a printer. Like typewriters, printers may reveal differences in printing due to wear or the damage characteristics of their moving parts. Variations in font type, size and alignment, spacing, margins, and other individual preference characteristics may be documented. Additional differences may be present due to markings produced due to defects, debris, fax machine headers, toner, toner application methods, and brands of printer cartridge inks. Fax machines also have a transmitting terminal identifier (TTI) that can be used for differentiation or association. The TTI styles vary for header and document type and fax machine make and model; the fonts are determined by the sending machine. A side-by-side comparison of the questioned document and at least ten known documents prepared using a typewriter, printer, fax machine, or copier is used to compare if markings produced by the machine are present in the questioned document.

Other differentiating characteristics that may uncover the details and circumstances of a case can include changes introduced to the original document. *Alterations*, or changes, may be made to both handwritten and typed documents. Alterations include erasures, obliterations, and overwriting. Digital alterations meant to improve digital pictures such as lightening, darkening, adding or removing color, and altering contrast can be used to change the meaning of documents and text. Photo editors such as Adobe Photoshop can reveal information including locating writing that has been obscured by the digital alteration of images. Infrared photography and reflecting light at different angles can reveal the contents of charred or darkened documents to enable the writing to be read. Infrared luminescence may detect writings not visible after erasure. Obliteration, or destruction, of the original writing may include overwriting, crossing out, smearing, blotting out, or the heavy destruction of text by placing several lines over it to make the writing unreadable. It is difficult if nearly impossible to detect if the obliteration was performed using the same ink, but a close examination with an ALS, IR light, or UV wand can reveal strokes that are inconsistent with the original writing. Obliterations may be revealed by microscopy or the use of a light source in which infrared radiation is directed at the writing. The light can pass through the upper layer exposing the under layer by absorption in that area. Alterations to the paper can be observed under a microscope and using an ALS with infrared light. If an overwriting alteration to the document is made with a different ink than the original, it can be detected due to differences in the fluorescence properties of the inks, although care must be taken when using an ALS in document analysis since the overwriting fluorescence can be quenched. Raman spectroscopy can be used to differentiate inks and does not suffer from this drawback. Due to the chemical structures of the dyes, some inks, when exposed to blue-green light, absorb the radiation and reradiate infrared light that may be observed by eye or recorded by developing on infrared film. Erasure, or removal of writing or markings, may be produced by physical or chemical methods. Abrasive instruments such as erasers, knives, razor blades, and even sandpaper are known to disturb the upper layer of fibers in the paper as well as remove carbon and ink. Scanning electron microscopy (SEM) can be used to examine the surface of obliterated and altered documents.

CHEMICAL ANALYSIS OF INKS AND PAPER

Inks are primarily composed of organic dyes but may also include inorganic pigments and delusterants. Inorganic pigments include copper phthalocyanine, carmine red, vermillion, azurite, madder lake, and sodium nitroprusside. Organic dyes include crystal violet, fluorescein, Pigment Red 12, and rhodamine B. Inks can fade or oxidize and lose their color over time. Prior to any chemical analysis, especially those that require destructively sampling the evidence, photographs of the evidence must be recorded and absorption of various types of light is attempted. The ALS can help the criminalist to obtain the best contrast for locating overwriting and recording photographs. Owing to the chemical structures of the pigments and dyes, most have fluorescent properties.

For example, US currency has several security features that are only visible under a UV light source. Figure 11.1 shows an image of a US $50 bill under white light. Figure 11.2 shows the same US $50 viewed using an ALS to show the fluorescent fibers in the paper. Figure 11.3 shows the same US $50 viewed under UV light to show the security strip. And Figure 11.4 shows the microwriting that is visible on the US $50 bill on magnification.

The ALS can also be used to visualize identifying characteristics including the presence and locations of moles, scars, and tattoos for the identification of missing persons and the deceased. Environmental factors, putrefaction, and decomposition will make these features less visible. A 2011 paper published in the *Journal of Forensic Identification* by Duncan and Klingle describes the use of infrared lighting techniques for making tattoos visible and easier to photograph.

Another examination of inks using an ALS is shown in Table 11.1. Several red inks from commercial pens fluoresce under various wavelengths. The ALS showed that several different pens from the same brand exhibited similar fluorescence, indicating the same or similar chemical dye and pigment composition. For example, the Pilot Precise V5 and Pilot G-2 07 both fluoresce at all of the wavelengths but the Pilot Precise V5 nearly disappeared into the paper at 415 nm while the Pilot G-2 07 did not. Figure 11.5 shows three original numbers in red ink (top left), the original

Figure 11.1 US $50 viewed under white light. (Courtesy of Tim Phillips.)

Figure 11.2 US $50 viewed using an ALS shows the fluorescent fibers in the paper. (Courtesy of Tim Phillips.)

Figure 11.3 US $50 viewed under UV light shows the security strip. (Courtesy of Tim Phillips.)

Figure 11.4 Microwriting is visible on a US $50 bill on magnification. (Courtesy of Tim Phillips.)

Table 11.1 Fluorescence observed using an ALS on several red inks

Brand	Wavelength (nm)					
	300–400	415	455	CSS	515	555
Office Depot	■	■	■	■	■	■
Pilot Precise V5	■	■	■	■	■	■
Integra Liquid Roller			■	■	■	■
Paper Mate	■	■	■	■	■	■
Paper Mate Write Bros	■	■	■	■	■	■
Pilot G-2 07	■	■	■	■	■	■
Uniball Signo	■	■	■	■	■	■
BIC Grip Roller	■	■	■	■	■	■
Universal Ballpoint Pen				■	■	■

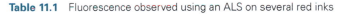

Note: Black signifies when fluorescence was observed.

numbers viewed under the ALS (top right), altered numbers in red ink (bottom left), and altered numbers viewed under the ALS (bottom right). The two inks used in the altered numbers (Figure 11.5) cannot be discriminated by the eye but can be by using the ALS.

Ink and paper color can be matched to Munsell color charts or differentiated using RGB (red-green-blue) charts, CMYB (cyan-magenta-yellow-black) charts, and numerical designations. The color can also be differentiated in

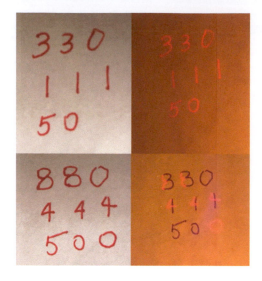

Figure 11.5   Original numbers in red ink (top left); original numbers viewed under the ALS (top right); altered numbers in red ink (bottom left); and altered numbers viewed under the ALS (bottom right).

Adobe Photoshop or using a smartphone app such as ColorAssist or Colorimeter. In a 2010 study by Deitz and Quarino published in the *Journal of Forensic Identification*, a method for discriminating gel inks using Adobe Photoshop was described. Gel inks are insoluble pigmented inks that are not amenable to methods such as TLC and HPLC. The authors report separating 36 blue gel inks into 25 groups using the RGB (red-green-blue), CMYB (cyan-magenta-yellow-black), and Lab modes.

The chemical analysis of paper includes analysis of color, fiber composition, pigments, additives, and fillers. Chemical methods include the application of oxidizing agents to render ink colorless. Chemicals such as lemon juice, acids, bleach, potassium permanganate, and peroxides have been used to erase inks. This can aid the criminalist in the determination of the ink chemical composition. Paper discoloration may be observed using a microscope or ALS with UV light. Charred or burnt documents are those that have been altered through exposure to fire or excessive heat. The resulting documents are darkened or brittle. In order to contrast the writing against the charred background, the criminalist may digitize images to change the contrast and color in order to read the text or drawings. Alternatively, infrared photography and reflecting light at different angles are methods that can be used to reveal the contents of the document. IR spectroscopy can also be used to differentiate colored papers as shown in Figure 11.6. Microscopy can also be used to detect paper discoloration, erasures, overwriting, obliteration, and handwriting characteristics.

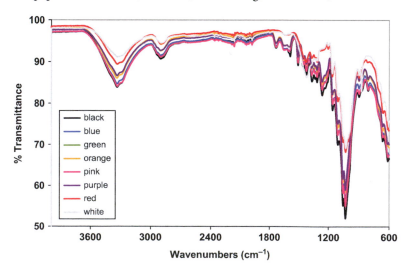

Figure 11.6   IR spectra of color paper.

Inks can be separated and analyzed by TLC and HPLC. TLC is a cheap and easy to perform method that can aid the criminalist in determining if known and questioned documents were prepared using the same pen or same brand/ type of pen and if a document could have been produced before a specific year of interest. Ballpoint inks are mixtures of several organic dyes. The United States Treasury maintains a library of all commercial pen inks produced since 1968. Inks and markers of the same color but produced by different manufacturers consist of different components. The component inks can be separated and resolved by TLC. The source of the ink can be identified if a known pen yields that same ink composition and Rf values as the questioned document. Alternatively, the results can be compared to a TLC Rf database of results for various pen makes/models using a specific mobile phase. In Figure 11.7, four pens (from left to right: Office Depot, Pilot Precise V5, Integra Liquid Roller, Universal Ballpoint Pen Medium) can clearly be differentiated by TLC. The mobile phase consisted of 10 mL of 1-butanol, 1.5 mL of water, and 1.0 mL of ethanol. The color composition, number of dyes, and Rf values for the dyes clearly differ. The same TLC plate viewed under UV light is shown in Figure 11.8. As several of the ink components are fluorescent, this adds an additional method of differentiation. The use of multiple solvent systems and statistics can be used to differentiate most inks.

UV-Vis spectroscopy can be used to differentiate dyes and inks as shown in Figure 11.9. The absorption fingerprint varies for each of the ink combinations in the commercial pens, including those of the same or different color.

Infrared spectroscopy can also be used to differentiate inks. As attenuated total reflectance Fourier transform-infrared spectroscopy (ATR FT-IR) can be performed *in situ* and is nondestructive, it is an ideal method for differentiating inks and pen types. The pen ink mixtures shown in Figures 11.7 and 11.8 have different IR stretch fingerprints. Figure 11.10 shows ATR FT-IR spectra of red inks from Bic, Pentel, and Zebra. Figure 11.11 shows ATR FT-IR spectra of blue inks from Bic, Business Source, and Pilot. Figure 11.12 shows ATR FT-IR spectra for nine different black ink pens from several manufacturers. Although the spectra are very similar, the inks can be differentiated using unique peaks and stretches, although statistical methods such as discriminate function analysis and principal component analysis are often used to aid in the interpretation. The ATR FT-IR spectrum of crystal (Gentian) violet (Figure 11.13), a dye of known structure, shows prominent stretches at 1150 and 1570 cm^{-1}, which are assigned to amine and aromatic stretches respectively (Figure 11.13).

Like IR spectroscopy, Raman spectroscopy can be used to differentiate inks and dyes including crystal violet as shown in Figure 11.14. Different laser options are available including 633, 785, and 1064 nm. Strong peaks were observed at 420, 722, 912, and 1176 cm^{-1}, and medium peaks at 341, 526, 1365, and 1612 cm^{-1}. The peaks can be compared as a fingerprint of the

Figure 11.7   TLC plate containing separations of inks from four red pens.

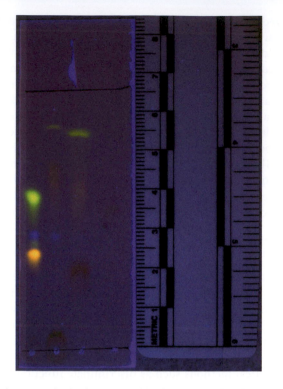

Figure 11.8 TLC of ink from four red pens excited using a UV light. (Courtesy of Tim Phillips.)

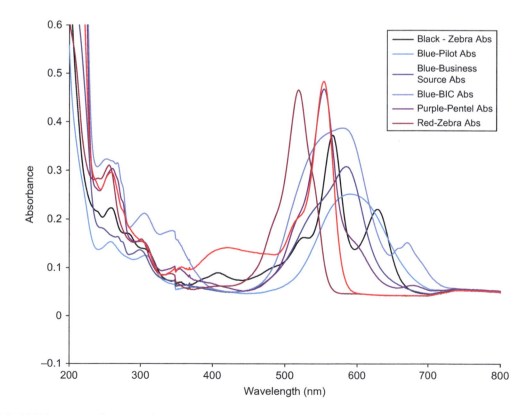

Figure 11.9 UV-Vis spectra of pen inks from various manufacturers.

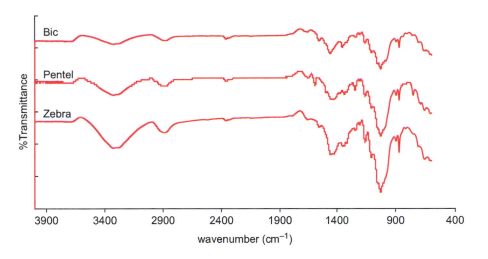

Figure 11.10   ATR FT-IR spectra of red pen inks from various manufacturers.

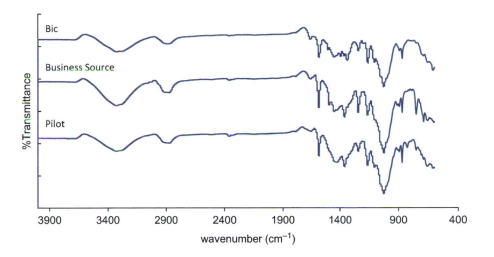

Figure 11.11   ATR FT-IR spectra of blue pen inks from various manufacturers.

dye or assigned to the bonds in the molecule. For example, the $1365\,cm^{-1}$ stretch is assigned to the C=N vibrations and the $1612\,cm^{-1}$ stretch is assigned to the aromatic ring chain vibrations. Surface enhanced Raman spectroscopy using nanoparticles can be used to enhance the Raman signal. Raman is nondestructive and the sample does not even need to touch the sample probe, making the method ideal for the analysis of treasured documents, antiquities, and fine art objects.

In addition to the method discussed previously, several papers have demonstrated that SEM can be used to view inks and obliterations. SEM can differentiate color pigment droplets by particle size and morphology.

IMPRESSION EVIDENCE

Impressions are depressions in a soft surface or material. Impression evidence in document analysis includes indented writings, partially visible depressions on the writing surface or underneath the document. These include unused sheets of a notepad, checkbook, or soft writing surface, such as a desk calendar that remains when the note page or check itself has been removed. The text or drawings that remain can be identified using an electrostatic charge. Impressions can also be made on velour fabrics including those in movie theater seats. Impression evidence is also produced by footwear, tire treads, fingerprints, bite marks, fabrics, serial numbers, and tool mark impressions and may be submitted as evidence in forensic investigations. Figure 11.15 shows the investigation of a footprint in an investigation by the West Midlands Police.

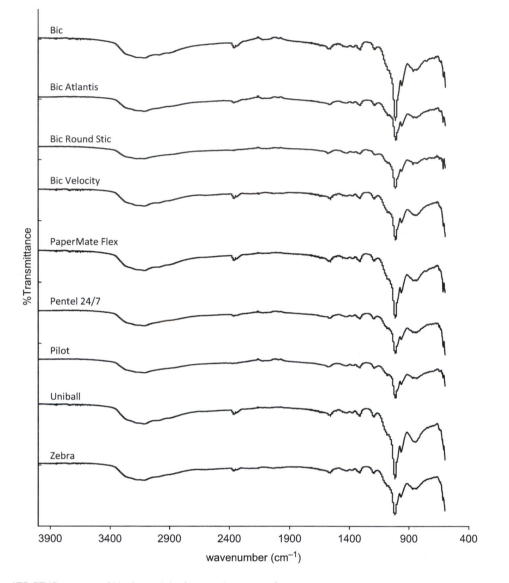

Figure 11.12   ATR FT-IR spectra of black pen inks from various manufacturers.

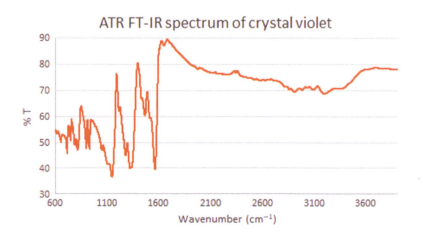

Figure 11.13   ATR FT-IR spectrum of crystal (Gentian) violet dye.

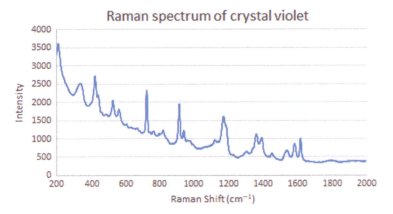

Figure 11.14   Raman spectrum of crystal (Gentian) violet dye.

Figure 11.15   West Midlands Police preserve footprint evidence. (From West Midlands Police, https://www.flickr.com/photos/westmidlandspolice/6352583797/in/photolist-aFmDQR-5eMWpv-55RvMA-7wFGD4-5eMWfX-8MskQy-dZj9BT-7CoWER-4e5GwK-bgnj12-55RvKL-21aUBFn-dJL4yw-8uNpyx-55PrQA-5pnuKB-3KDVfK-55Pvno-4z3pEk-opGBCv-7TUSks-QJ3vFC-TrS-FTV-3Qrdda-5eDQVU-6FoP7r-9kb1jG-oRMk5P-7TRDbk-Tdad4o-3HtVD8-oRMkeX-dGu9CY-eL4quU-3Qvqt1-btdVj6-7CsLbE-67cqGw-TdaiwY-7u3MBX-bS95e-8GTfhw-T9StvZ-fYuMju-8nWeYP-6QWUVh-8ER9An-pBw5y7-6atdQV-6at7oM.)

As with other evidence types, impression evidence should be photographed with a scale to show all visually observable details and the item's location relative to the rest of the crime scene prior to movement for transport to the lab or recovery and preservation by removal, lifting, or casting. Readily recoverable items such as glass or tile can be directly transported and submitted to the laboratory. Electrostatic lifting or physical lifting can be performed using methods employed in fingerprint analysis including the use of J-lifts, tape, or large lifting sheets. The lift material should cover the entire impression. In a physical lift, a fingerprint roller is used to eliminate any air pockets before the impression is lifted. In an electrostatic lift, an electrostatic charge is created using a device such as the Pathfinder causing dust to attach to the lifting film.

Shoe and tire tread marks in soft earth or snow (Figure 11.16) can be casted for transport and analysis by the lab. The casting is performed by applying aerosol hairspray or snow impression wax that is sprayed on the surface, placing a rectangular mold form that encapsulates the evidence, and pouring class I dental stone (mixed with water to form a

Figure 11.16   Suspect footprint impression evidence in the snow (simulated evidence).

Figure 11.17   Cast of a suspect's shoe (simulated evidence) in dental stone.

paste the consistency of batter) into the mold. In the case of the wax, three light coats are sprayed onto the impression at an interval of 1–2 minutes between layers and a final 10 minutes is allowed to let the wax dry prior to pouring the casting material. After the dental stone sets, or hardens, the material can be removed from the impression and submitted with accompanying paperwork to the lab. Figure 11.17 shows a shoe print cast with dental stone. This may take 24–48 hours. Biofoam or ink can be used to cast discarded shoes at the crime scene or shoes from the suspect or victim for comparison as shown in Figures 11.18 and 11.19 respectively.

Footwear, tire tread, fingerprint, and tool mark impressions made in blood can be analyzed using chemical methods including the use of luminol, 1,8-diazafluoren-9-one (DFO) under white light, or a green excitation source to view the fluorescence, and AY7 under blue light. These impressions will be examined further in the following paragraphs with the exception of fingerprint impression evidence that is included in Chapter 12.

Comparison of a known object (i.e., fingerprint, tire tread, tool, sole of a shoe, or bite mark) to an evidence photograph or cast by a criminalist using a point-by-point comparison can lead to the conclusion that the two

Figure 11.18 Biofoam cast of a reference shoe (simulated evidence).

Figure 11.19 Inked suspect shoe prints (simulated evidence).

impressions came from the same source or a different source, or that a conclusion cannot be reached (inconclusive). A sufficient number of points must be compared to support a finding that the questioned and test impressions originated from the same (or different) source. Computer software, databases, and websites can assist in making these comparisons.

Bite mark impressions may be made by the suspect on the victim or in food. In bite mark impression analysis and identification using dental records, comparisons of teeth characteristics, alignment, spacing, angles, cavities, fillings, bridgework, missing teeth, mouth structure, and size are used to compare the evidence to reference impressions. Antemortem data is compared with postmortem data. These comparisons are made by a *forensic odontologist*, a licensed dentist with specialized training including classes and seminars and/or certification from the American Academy of Forensic Sciences. Forensic odontologists examine dental anatomy including primary (deciduous) and secondary (permanent) teeth numbers and characteristics of the maxillary (upper) and mandibular (lower) sets. Teeth are classified into incisors (central and lateral), canines, premolars (bicuspids, first and second), and molars (first, second, and third). Cases may include the identification of individuals found in mass graves or burn victims, missing persons, rape, murder, child abuse, spousal abuse, or dental malpractice cases. A photographic study with image enhancement using computer software such as Adobe Photoshop and dental record comparison can aid in the comparative study of the evidence and a suspect's teeth and mouth characteristics. Molds and casts may also be used. A transparent overlay of the bite perimeter of the teeth taken by scanning the dental casts can be compared to the photographic evidence. Most forensic odontologists will "rule-in" or "include" or "rule-out" or "exclude" evidence but not make a "match" or identification. The Chi Omega murders committed by Ted Bundy in the 1970s in Tallahassee, Florida, are now a well-known example of how bitemark analysis can be used in a criminal case. A wax bite exemplar was made of the suspect's mandibular front teeth. An overlay of the exemplar teeth #22 through #27 on the photographic evidence of bite marks on a victim were found to be consistent with each other.

In shoe tread analysis, it may be determined that a shoe tread is representative of a particular class such as a brand, size, shoe, or sole type. Individualizing characteristics such as wear pattern and damage to the shoe are observable using a stereomicroscope. The results can be compared to a shoe wear database such as Shoeprint Image Capture and Retrieval (SICAR 6), which contains over 22,000 footwear entries (and 7838 tire tread prints) for over 700 brands, or the SoleMate reference database which contains over 37,000 records. Another database that can be used to compare shoe tread evidence is crimeshoe.com, which also has more than 22,000 shoe types in its database. The FBI footwear database has over 14,000 soles. The Everspry EverASM Automated Shoe Impression Matcher software performs matching tasks. The results of the shoe tread analysis including casting, creating inked prints, electrostatic lifting, development using chemical methods, and Biofoam impressions can be used to determine the number of perpetrators, link crime scenes, refute or confirm suspects' alibis, and determine the gait characteristics of the wearer.

In tire tread analysis, class characteristics including the make, tread, width, and diameter of the tire may be determined and noted. As with shoe tread impressions, tire impressions can be cast for further analysis. Wear patterns or damage to the tires can provide individualizing characteristics that should be noted and compared to a suspect vehicle's tires. The suspect tires can be inked and rolled to compare their characteristics (Figure 11.20) to those of the photographed or cast crime scene tire tread markings.

Tool mark impressions may also be encountered in an examination of the crime scene. Screwdrivers and crowbars, among other tools, are often used to force entry to vehicles and residences or turn on the ignition in vehicles. Like all impression evidence, photographs should first be recorded. Soft mold material such as dental rubber or clay may be used to record the impression but the tool should never be directly "fit" or forced into the impression under analysis. This contact could destroy the integrity of the evidence. Test impressions can be made in a separate surface of identical or similar wood, metal, or material under analysis as shown in Figure 11.21. Test impressions, evidence casts, and suspect tools can be examined using the stereomicroscope. Class characteristics including type of tool, manufacturer (if known), size, and spacing between teeth in gripping or cutting tools should be noted. Fine structure including microscopic wear characteristics, dents, and other individualizing characteristics should be recorded for each item. Using a comparison microscope, striation marks from a test impression can be compared to an impression of the suspect tool.

Figure 11.20 Inked tire tread prints.

Figure 11.21 Tool impressions in wood.

Visible prints may be impressed into velvet or the velour fabrics of vehicle interiors or cinema seating. Fabric impressions may be identified by eye, by using a magnifying glass, or by using a stereomicroscope. Prints have also been recovered from nylon fabric. A full handprint including palmar flexion creases was recovered after 21 days using a technique called *vacuum metal deposition*. Vacuum metal deposition uses gold and zinc to recover the prints. It can also be used on smooth surfaces including carrier bags, plastics, and glass. The hand-print may demonstrate the circumstances of the case: "an impression of a palm print on the back of some-one's shirt might indicate that they were pushed off of a balcony, rather than jumping." (www.sciencedaily.com/releases/2011/01/110131073141.htm)

Serial numbers are impressions stamped onto a metal body or frame using hard steel dies (Figure 11.22). This causes a permanent strain on the metal crystals at a certain depth in the metal. Criminalistics may be asked to restore serial numbers that have been damaged by grinding, scraping, rifling, punching, or another method of obliteration. Chemical methods can be used to restore serial numbers. Typically, an etching agent, or strong acid such as hydrochloric acid, is used in the area in which the serial number was applied. The strained area dissolves faster in the acid than the unaltered metal. Different metals will produce different backgrounds based on the chemical used. The original numbers are revealed but must be immediately photographed. This method does not work if the strain zone is removed from the item or the area has received another impression with a different strain pattern. Magnetic particles can also be used to restore serial numbers. In this method, magnets are used to produce a magnetic flow through the metal; particles will display the disruption in the metal where the serial number was originally stamped.

Fingerprint impression detection and recovery will be covered in detail in the next chapter (Chapter 12).

Figure 11.22 Impressed serial number in metal.

QUESTIONS

1. An ink taken from a US Treasury collection for use on a case would be considered a _____ sample.
   - a. Reference
   - b. Standard
   - c. Exemplar
   - d. Evidence
   - e. Negative control

2. Which of the following methods cannot be used to discern abrasions, alterations, and erasures of questioned documents?
   - a. IR spectroscopy
   - b. ALS
   - c. TLC
   - d. Microscopy

3. Which of the following databases should be searched to compare shoe print images?
   - a. IBIS
   - b. NIBIN
   - c. IAFIS
   - d. SICAR

4. Writing on a notepad underneath an important document can be visualized by _____.
   - a. Indentation
   - b. Erasure
   - c. Obliteration
   - d. Electrostatic lifting

5. Which of the following can be run on the same plate to identify the migration of an unknown compound in ink using TLC?
   - a. Second unknown
   - b. Standard
   - c. Reference sample
   - d. Any of the above will work

6. List four factors that result in differences in handwriting between individuals.

7. List two physical analysis methods for paper.

8. List four types of impression evidence.

9. List five questioned documents that may be submitted into evidence to the crime lab. For example, state one feature that you would use to analyze it and one method you would use.

10. Explain how ATR FT-IR microscopy can be used to evaluate questioned documents.

Bibliography

Al-Hadhrami, A.A., M. Allen, C. Moffatt, and A.E. Jones. 2015. National characteristics and variation in Arabic handwriting. *Forensic Sci. Int.* 247:89–96.

Andrasko, J. 2001. HPLC analysis of ballpoint pen inks stored at different light conditions. *J. Forensic. Sci.* 46(1):21–30.

Andrasko, J. 2002. Changes in composition of ballpoint pen inks on aging in darkness. *J. Forensic Sci.* 47(2):324–327.

Barker, J., R. Ramotowski, and J. Nwokoye. 2016. The effect of solvent grade on thin layer chromatographic analysis of writing inks. *Forensic Sci. Int.* 266:139–147.

Bower, N.W., C.J. Blanchet, and M.S. Epstein. 2016. Nondestructive determination of the age of 20th-century oil-binder ink prints using attenuated total reflection Fourier Transform Infrared Spectroscopy (ATR FT-IR): A case study with postage stamps from the Łódź Ghetto. *Appl. Spectrosc.* 70(1):162–173.

Braz, A., M. López-López, and C. García-Ruiz. 2013. Raman spectroscopy for forensic analysis of inks in questioned documents. *Forensic Sci. Int.* 232(1–3):206–212.

Brittain, H.G. 2016. Attenuated total reflection Fourier transform infrared (ATR FT-IR) spectroscopy as a forensic method to determine the composition of inks used to print the United States one-cent blue Benjamin Franklin postage stamps of the 19th century. *Appl. Spectrosc.* 70(1):128–136.

Causin, V., R. Casamassima, C. Marega, P. Maida, S. Schiavone, A. Marigo, and A. Villari. 2008. The discrimination potential of ultraviolet visible spectrophotometry, thin layer chromatography, and Fourier Transform Infrared spectroscopy for the forensic analysis of black and blue ballpoint ink. *J. Forensic Sci.* 53 (6):1468–1473.

Chen, X. 2015. Extraction and analysis of the width, gray scale and radian in Chinese signature handwriting. *Forensic Sci. Int.* 255:123–132.

Chu, P.C., B.Y. Cai, Y.K. Tsoi, R. Yuen, K.S. Leung, and N.H. Cheung. 2013. Forensic analysis of laser printed ink by x-ray fluorescence and laser-excited plume fluorescence. *Anal. Chem.* 85(9):4311–4315.

Deitz, N. and L. Quarino. 2010. Differentiation of blue gel inks using adobe photoshop. *J. Forensic Ident.* 60(3):291–307.

Del Hoyo-Meléndez, J.M., K. Gondko, A. Mendys, M. Król, A. Klisifska-Kopacz, J. Sobczyk, and A. Jaworucka-Drath. 2016. A multi-technique approach for detecting and evaluating material inconsistencies in historical banknotes. *Forensic Sci. Int.* 266:329–337.

de Souza Lins Borba, F., R.S. Honorato, and A. de Juan. 2015. Use of Raman spectroscopy and chemometrics to distinguish blue ballpoint pen inks. *Forensic Sci. Int.* 249:73–82.

Dick, R.M. 1970. A comparative analysis of dichroic filter viewing reflected infrared and infrared luminescence applied to ink differentiation problems. *J. Forensic Sci.* 15(3):357–363.

Djozan, D., T. Baheri, G. Karimian, and M. Shahidi. 2008. Forensic discrimination of blue ballpoint pen inks based on thin layer chromatography and image analysis. *Forensic Sci. Int.* 179(2–3):199–205.

Duncan, C.D. and C. Klingle. 2011. Using reflected infrared photography to enhance the visibility of tattoos. *J. Forensic Ident.* 61(5):495–519.

Gál, L., M. Oravec., P. Gemeiner, and M. Čeppan. 2015. Principal component analysis for the forensic discrimination of black inkjet inks based on the Vis-NIR fibre optics reflection spectra. *Forensic Sci. Int.* 257:285–292.

Galbraith, N.G. 1986. Alcohol: Its effect on handwriting. *J. Forensic Sci.* 31(2):580–588.

Genest S., R. Salzer, and G. Steiner. 2013. Molecular imaging of paper cross sections by FT-IR spectroscopy and principal component analysis. *Anal. Bioanal. Chem.* 405(16):5421–5430.

Gürses, A., M. Açıkyıldız, K. Güneş, and M.S. Gürses. 2016. *Dyes and Pigments*. Basel, Switzerland: Springer International Publishing.

Joshi, B., K. Verma, and J. Singh. 2013. A comparison of red pigments in different lipsticks using thin layer chromatography (TLC). *J. Anal. Bioanal. Tech.* 4:157. doi:10.4172/2155-9872.1000157.

Kopainsky, B. 1989. Document examination: Applications of image processing systems. *Forensic Sci. Rev.* 1(2):85–101.

Miller, L.B. 1969. Restoration of obliterated handwriting. *J. Forensic Sci. Soc.* 9(1):82–83.

Nam, Y.S., J.S. Park, N.K. Kim, Y. Lee, and K.B. Lee. 2014. Attenuated total reflectance Fourier transform infrared spectroscopy analysis of red seal inks on questioned document. *J. Forensic Sci.* 59(4):1153–1156.

Nam, Y.S., J.S. Park, Y. Lee, and K.B. Lee. 2014. Application of micro-attenuated total reflectance Fourier transform infrared spectroscopy to ink examination in signatures written with ballpoint pen on questioned documents. *J. Forensic Sci.* 59(3):800–805.

Poon, N.L., S.S. Ho, and C.K. Li. 2005. Differentiation of coloured inks of inkjet printer cartridges by thin layer chromatography and high performance liquid chromatography. *Sci. Justice* 45(4):187–194.

Rohde, E., C. Vogt, and W.R. Heineman. 1998. The analysis of fountain pen inks by capillary electrophoresis with ultraviolet/visible absorbance and laser-induced fluorescence detection. *Electrophoresis* 19(1):31–41.

Saferstein, R. 2007. *Criminalistics: An Introduction to Forensic Science*, 9th edn. Upper Saddle River, NJ: Pearson Prentice Hall.

Saini, K., H. Jaur, and M. Gupta. 2014. Analyses of blue gel pen inks using thin-layer chromatography and visible spectrophotometry. *J. Forensic Ident.* 64(1):28–42.

Saini, K. and J.S. Saroa. 2011. Differentiation of color photocopy toners using TLC, UV, and FTIR techniques. *J. Forensic Ident.* 61(6):561–580.

Senior, S., E. Hamed, M. Masoud, and E. Shehata. 2012. Characterization and dating of blue ballpoint pen inks using principal component analysis of UV-Vis absorption spectra, IR spectroscopy, and HPTLC. *J. Forensic Sci.* 57(4):1087–1093.

Senior, S., E. Hamed, M. Masoud, and E. Shehata. 2012. Characterization and dating of blue ballpoint pen inks using principal component analysis of UV-Vis absorption spectra, IR spectroscopy, and HPTLC. *J. Forensic Sci.* 57(4):1087–1093.

Sonnex, E., M.J. Almond, J.V. Baum, and J.W. Bond. 2014. Identification of forged Bank of England £20 banknotes using IR spectroscopy. *Spectrochim. Acta A: Mol. Biomol. Spectrosc.* 118:1158–1163.

Srihari, S.N., Meng L., and Hanson L. 2016. Development of individuality in children's handwriting. *J. Forensic Sci.* 61(5):1292–1300.

Stuart, H.J., J.N. Jon, and S. Bell. 2014. *Forensic Science: An Introduction to Scientific and Investigative Techniques*, 4th edn. Boca Raton, FL: CRC Press.

Sun, Q., Y. Luo, Q. Zhang, X. Yang, and C. Xu. 2016. How much can a Forensic laboratory do to discriminate questioned ink entries? *J. Forensic Sci.* 61(4):1116–1121.

Taroni, F., R. Marquis, M. Schmittbuhl, A. Biedermann, A. Thiéry, and S. Bozza. 2012. The use of the likelihood ratio for evaluative and investigative purposes in comparative forensic handwriting examination. *Forensic Sci. Int.* 214 (1–3):189–194.

Taroni, F., R. Marquis, M. Schmittbuhl, A. Biedermann, A. Thiéry, and S. Bozza. 2014. Bayes factor for investigative assessment of selected handwriting features. *Forensic Sci. Int.* 242:266–273.

Tebbett, I.R. 1991. Chromatographic analysis of inks for forensic science applications. *Forensic Sci. Rev.* 3(2):71–82.

Tyrrell, J.F. 1939. Decipherment of charred documents. *J. Crim. L. Criminol.* 30(2):236–242.

Wang, J., G. Luo, S. Sun, Z. Wang, and Y. Wang. 2001. Systematic analysis of bulk blue ballpoint pen ink by FTIR spectrometry. *J. Forensic Sci.* 46(5):1093–1097.

Wang, X.F., J. Yu, A.L. Zhang, D.W. Zhou, and M.X. Xie. 2012. Nondestructive identification for red ink entries of seals by Raman and Fourier transform infrared spectrometry. *Spectrochim. Acta A: Mol. Biomol. Spectrosc.* 97:986–994.

Was-Gubala, J. and R. Starczak. 2015. Nondestructive identification of dye mixtures in polyester and cotton fibers using Raman spectroscopy and ultraviolet-visible (UV-Vis) microspectrophotometry. *Appl. Spectrosc.* 69(2):296–303.

Williamson, R., A. Raeva, and J.R. Almirall. 2016. Characterization of printing inks using DART-Q-TOF-MS and attenuated total reflectance (ATR) FTIR. *J. Forensic Sci.* 61(3):706–714.

CHAPTER 12

Latent print development

KEY WORDS: fingerprint, dermis, dermal papillae, epidermis, visible print, patent print, plastic print, latent print, minutiae, loop, arch, whorl, reflected ultraviolet imaging system, fingerprint powder, nanoparticle powder, ninhydrin, superglue fuming, Sudan Black, Oil Red O, RTX, small particle reagent, iodine fuming, gun blue, silver nitrate, physical developer, lifting

LEARNING OBJECTIVES

- To understand how fingerprints are formed and the basis for their use in individualization
- To be able to differentiate between fingerprint general ridge patterns and locate and identify minutiae in fingerprints
- To explain the difference between visible, plastic, and latent fingerprints
- To list the chemical components of fingerprints
- To list several methods of developing latent prints and be able to determine the best method based on the color and porosity of the surface
- To explain how latent fingerprint development methods work
- To describe how to preserve and document a fingerprint

FINGERPRINTS

Lee Boyd Malvo was charged in a multistate and multi-incident shooting spree from September 5, 2002, to October 22, 2002, in Maryland, Virginia, and Washington, D.C., in which ten people were murdered in 15 shootings. Malvo was 17 years old when the crimes were committed. A Bushmaster XM-15 0.223-caliber rifle was used in most of the shootings. The Bushmaster is a semiautomatic assault rifle similar to the military M-16. It was suspected to have been stolen from Bull's Eye Shooter Supply in Tacoma, Washington, but was not yet reported missing or stolen at the time of the shootings. Malvo had lived in Tacoma, Washington.

Malvo's latent fingerprint was lifted from a gun and an ammunition magazine found near the state liquor store in Montgomery, Alabama, whose employees were his first two victims in 2002. The print matched the prints on the Bushmaster and were used to link the firearm to Malvo and link the cases. Expert witnesses testified that a fingerprint, palm print, and DNA found on the Bushmaster rifle matched Malvo's. The prints also matched Malvo's print on file at the Washington state immigration records and aided in the identification of Malvo through the use of the Integrated Automated Fingerprint Identification System (IAFIS). Malvo had been arrested previously by the Immigration and Naturalization Service. The Federal Bureau of Investigation (FBI), home of IAFIS, and the Department of Homeland Security merged their fingerprint identification systems in September, 2006.

Malvo was convicted and sentenced for the crimes in several trials. On May 26, 2017, a federal district judge in Virginia overturned Malvo's two life sentences without parole and sent his case back to the state courts in Chesapeake and Spotsylvania County in Virginia for resentencing. The judge found the sentences unconstitutional under Miller v. Alabama because Malvo was under 18 years old at the time the crimes were committed.

FINGERPRINTS (continued)

Malvo remains in a Virginia prison based on his previous convictions in Virginia and previous sentences from Maryland that were not overturned.

BIBLIOGRAPHY

Beyerle, D. 2009. D.C. snipers' first two victims were at ABC store in Montgomery. *The Gadsden Times*, November 5, www.gadsdentimes.com/news/20091105/dc-snipers-first-two-victims-were-at-abc-store-in-montgomery (accessed January 27, 2017).

Davis, R. 2016. *Evidence Collection and Presentation*, 2nd edn. Chennai, TN: Lawtech Publishing Co.

Hess, K.M., C.H. Orthmann, and H.L. Cho. 2013. *Criminal Investigation*, 11th edn. Boston, MA: Cengage Learning.

Laptop map shows trail of sniper scenes. *The Baltimore Sun*, November 6, 2003. www.baltimoresun.com/news/maryland/bal-te.md.sniper06nov06-story.html (accessed January 27, 2017).

Fingerprints are produced by the contact of friction ridge skin with a surface; they may form impressions and transfer residue. Fingerprints are used in criminalistics as a tool for the identification of suspects and victims and to link a person to a place. Fingerprint evidence may be useful in corroborating stories or statements, placing someone at the scene of a crime, or giving clues or details as to what occurred at a scene. The position of fingerprints themselves may also be of evidentiary value, for example, fingerprints found on the inside surface of glass at the point of entry. Friction ridge skin is found not only on the last joint of the fingers and the thumb, but also on the length portion of the digits and the palm of the hands as well as on the toe pads and soles of the feet. Fingerprints are formed during fetal development. By the tenth week, the pads are formed on the hands and feet. By the twelfth week, the skin ridges are formed. By the eighteenth week, the ridges are fully formed and will remain unchanged for a person's lifetime. Other physical characteristics, including those once probed using anthropometry, the personal identification system devised by Alphonse Bertillon in 1883, such as height and width of the head are subject to change with age. Additionally, facial characteristics are often strikingly similar between unrelated individuals (Figure 12.1). Names may be identical. Even the genetic makeup may be identical—in the case of identical twins. Although physical characteristics and dental records are used to estimate age, the age of the person who deposits a latent print cannot be determined.

Fingerprints have been used to help solve crimes for over 100 years but have been used in identification (for business transactions, etc) for over 2000 years. The work of Henry Faulds in recording fingerprints using printer's ink in the 1870s, Sir Francis Galton's book *Fingerprints* published in 1892 describing fingerprint minutiae, and a fingerprint classification system presented by Sir Edward Henry in 1897, enabled fingerprinting to be used as an identification tool. The use of fingerprints by the New York City Civil Service Commission started in 1901 to certify civil service applications. A key case in 1903 accelerated fingerprint use in the United States. A man by the name of Will West was sentenced to be imprisoned at the penitentiary at Leavenworth, Kansas. Another man was already in the prison with the name "William West." Strikingly, they shared not only a name but also facial characteristics and Bertillon anthropometric measurements—but their fingerprints were different. Fingerprints began to be used as an identification tool by all major US cities following the training of American police by Scotland Yard representatives in 1904 at the World's Fair. By 1905, fingerprinting was adopted by the US military and the first fingerprint cards for recording prints were produced in 1908. By World War I, most of Europe had also adopted fingerprinting for identification. At the time of its founding in 1924 from the Bureau of Investigation and Leavenworth prison, the FBI merged the fingerprint records from the two entities. The FBI's collection remains the largest in the world. Fingerprints remain accepted as unique and permanent. In over 110 years of use by the FBI and over 50 million fingerprint cards, no two fingers have been found to possess identical skin friction ridge characteristics. Based on theoretical calculations, the probability of two identical fingerprint patterns in the world's population is extremely small (Galton calculated 1 in 64 billion).

The skin includes the *dermis*, *dermal papillae*, and *epidermis*. The dermis is the innermost layer and contains the sweat glands, nerves, dermal proteins, oil glands, and hair follicles. The dermal papillae lies between the dermis and epidermis and is characterized by the shape of the surface skin ridges that lie above it. The surface layer is the epidermis. The epidermis contains the friction ridge skin. Within the friction ridge is a single row of pores, open ducts from sweat glands that discharge perspiration to the skin surface. Specifically, there are three types of sweat glands including eccrine, apocrine, and sebaceous glands. Perspiration from pores in the friction ridge skin on the fingers, palms, toes, and soles of the

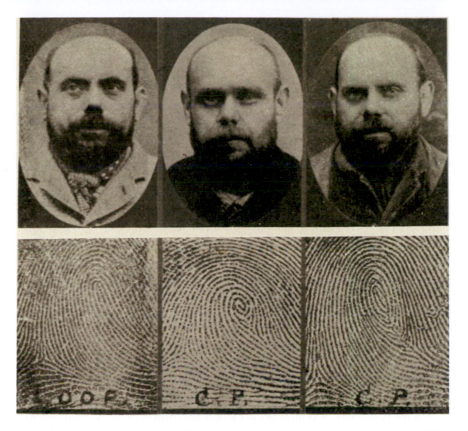

Figure 12.1 Three men share remarkably similar facial characteristics but their fingerprints are different. (From Internet Archive Book Images, Image from page 823 of The American Museum Journal (c. 1900–1918), Taken c. 1825, https://www. flickr.com/photos/internetarchivebookimages/18136021206/in/photolist-bQWj9V-neEJnk-QXGhcZ-5Bt8eE-bqL71n-8n3HwC-2b5sRx-9gmSqG-kZafqV-5Ussby-7jULBe-ornGAC-ouGZie-oduMtv-ot8j8c-otcAfY-ovanY4-ouH2xz-odvA6X-9D3V6q-7j7RGm-68Uoo5-768bS9-7xHtZ2-tEXwHz-tHwh3s-ouZeCt-QCWds8-764hHR-tCBRfN-USDAkz-otAaqo-ovvPQH-oe8nr8-ovC8ND-oe87jL-ovkR3c-wRXeRo-x3tx8Q-odjsCV-5LY7qv.)

feet comes from eccrine glands and contains water, salts, amino acids (especially serine, glycine, ornithine, alanine, and aspartic acid), urea, lactic acid, sugars, creatine, creatinine, glycogen, vitamins, and choline that can be used to detect the fingerprint using chemical methods. Additionally, oils transferred from hairy portions of the body to the fingers can also be deposited in fingerprints and detected using chemical means. Apocrine glands found in the pubic, mammary, and anal areas secrete water, proteins, carbohydrates, and cholesterol. Sebaceous glands found on the forehead, chest, back, and abdomen areas secrete water, glycerides, hydrocarbons, fatty acids, squalene, cholesterol, and wax esters.

Fingerprints are classified in a logical and searchable sequence. The anatomy of fingerprints includes the general ridge patterns for classification including loops, arches, and whorls, as well as minutiae used for identification. Subclassifications also exist and include the plain arch, tented arch, ulnar loop, radial loop, plain whorl, central pocket whorl, double loop, and accidental whorl. In order to properly classify fingerprints, it is important that the fingers are fully rolled. Figure 12.2 shows fingerprints with double loop, whorl, and single loop patterns.

As many fingerprints submitted as evidence are only partial prints, it is often not possible to classify the general ridge pattern. However, the print can still be used for identification. It is estimated that there are as many as 150 minutiae, or friction ridge characteristics, on each finger, on average. *Minutiae* include ridge endings, bifurcations, trifurcations, short ridges, lakes/enclosures, and other ridge details such as short ridges/islands, bridges, opposed bifurcations, ridge crossings, opposed bifurcation with ridge endings, dots, and hook/spurs (Figure 12.3). In fingerprint pattern analysis, the location and type of minutiae must match. Normally, 8–16 minutiae in a point-by-point comparison is considered acceptable for a positive identification, or a match, although there is no set minimum. Figure 12.4 shows fingerprint minutiae points for analysis. The International Association for Identification (IAI) stated in 1973 that the final determination lies with the examiners who will base their judgement on their experience and knowledge.

Figure 12.2 Fingerprints with a (left to right) double loop, whorl, and single loop. (From Internet Archive Book Images, Image from page 194 of The World Book: Organized Knowledge in Story and Picture (1918), Taken circa 1918, https://www.flickr.com/photos/internetarchivebookimages/14784147253/in/photolist-owqBUz-2F9bZV-ejmhnQ-2GbpG7-mfsfX4-2G76Lr-rwFbM3-JXRHcb-2FdFdS-2Fdqtd-2F943v-2FdpFJ-691fJy-4EhfLV-2F9aFa-JA3yPo-2FdHBu-g9x7rd-2Fdyay-p9NLK-2F94fx-2F93sc-2G76QF-2FdFjs-2F9beT-oMYMwH-2F9b4z-2Gbqe1-9fvjkA-5joXXC-2Fdq3E-85ZffM-2FdqC7-pB9cXy-2Fdy61-2F94yi-oMxUXd-oMxoPj-2G7616-2F9jtr-bxZCo7-2Gbqph-9nNHW-2GbqSU-2FdxEA-2F9cfc-2GbpXJ-2Fdqfh-2F9aua-2F9c8i.)

While the *loops* are the most common pattern (60%), approximately a third are whorls (34%) and the remaining are arches (5%) and accidental whorls (1%). All loops have one delta, a ridge point at or directly in front of the point where the ridge lines diverge, and a core, which is located in the approximate center of the pattern. Ridges flow from one side in the space between the delta and the core, form the loop, and exit the side they entered. Ulnar loops (94%) flow toward the little finger; radial loops (6%) flow toward the thumb.

Whorls are characterized by ridges that form spirals or concentric circles and look like a "bull's eye" target. Plain whorls have two deltas and at least one ridge that makes a complete circuit in the path of a spiral, oval, or circle-like form. At least one of the recurving ridges in the circle-type pattern area must touch or cross an imaginary line that could be drawn between the two deltas. A central pocket loop whorl differs from the plain whorl in that the recurving ridges in the circle-type pattern area must not touch or cross an imaginary line that could be drawn between the two deltas and tends to make a complete circle. Double loops are a type of whorl characterized by two separate loop formations in one fingerprint, each with its own separate and distinct shoulders and deltas. Finally, accidental whorls consist of two or more deltas in combination with two or more different types of patterns exclusive of the plain arch; the subclass also includes any unusual patterns that are not placed in any of the other classifications. These may include a loop and a whorl, a loop and a tented arch, a loop and a central pocket loop, or a double loop and a central pocket loop.

Arches lack deltas and cores. Arches can be divided into two main types including the plain arch (60%) and the tented arch (40%). The ridges of the plain arch enter from one side of the pattern in an up-arch and flow down and exit the other side in a wave-like pattern. The difference between the plain arch and the tented arch is that the tented arch ridges in the center form an up thrust with an angle of less than 90°.

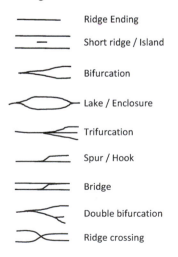

Ridge Ending

Short ridge / Island

Bifurcation

Lake / Enclosure

Trifurcation

Spur / Hook

Bridge

Double bifurcation

Ridge crossing

Figure 12.3 Fingerprint minutiae.

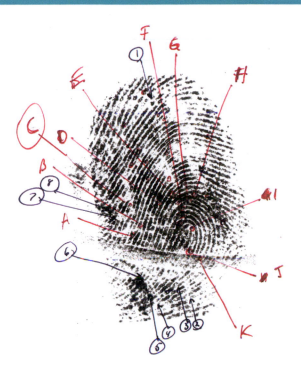

Figure 12.4 Fingerprint points for analysis. (From Vince Alongi, https://www.flickr.com/photos/vincealongi/7075085533/in/pool-csi/.)

Perpetrators, including, notably, the gangster John Dillinger, have several times tried to remove or obscure their fingerprints using acid or cuts. As fingerprints are raised ridges in the epidermis, they are deep features that cannot be erased. Likewise, while cuts and scarring alter fingerprints, they also give rise to new unique points for comparison.

Due to the raised nature of the friction ridge skin, several types of fingerprints may be deposited at the crime scene including visible (patent), plastic, and latent fingerprints. *Visible* or *patent prints* are those that can be observed with the eye without chemical processing. Some examples of visible prints are those deposited in colored materials such as blood, paint, grease, ink, or dust. *Inked fingerprints* recorded on ten-print fingerprint cards are an example of visible prints. *Plastic prints* may also be visible to the eye; these fingerprints are impressions in soft materials including soap, dust, wax, or putty. In contrast, latent prints are not visible to the eye without physical or chemical processing; they can be planted on any surface. It is known that some individuals sweat more and are oilier than others; they will generally leave better latent prints for processing. Cold weather promotes pore closing and less sweating and thus less secreted material to be deposited in a latent print while hot weather promotes sweating and pore opening and a loss of salt electrolytes. As oils deposited in latent prints are not water soluble, they will remain even if items are submerged in water.

Fingerprints are often found at crime scenes and on a wide variety of evidence including paper, checks, tools, firearms, cartridge cases, explosives, glass, tile, wood, doors, and household items such as soda cans and drinking glasses, among others. Fingerprints have even been developed on food items including apples, bananas, guavas, oranges, tomatoes, onions, potatoes, peppers, and eggs. The length of time since a fingerprint was deposited has been noted to be an important variable in the development of latent prints. Over time, the compounds contained in the fingerprint degrade and oxidize.

Criminalists use several physical and chemical processing methods to detect and develop latent prints. However, whenever possible, latent prints will initially be located using noninvasive and nondestructive methods such as the *reflected ultraviolet imaging system* (RUVIS). The RUVIS is a nondestructive, spectroscopic tool with a quartz halogen (argon), xenon arc, or indium arc laser used to locate latent prints on objects and surfaces. Components of the fingerprint including amino acids and proteins will fluoresce when irradiated with a RUVIS system at 525 nm. Observers wear safety goggles that absorb the light from the laser but permit the fluorescence from the latent print residue to pass through and reach the eyes of the observer. The wavelength and color of goggles can be adjusted to detect prints on different surfaces and of different compositions. White light can also be used at an angle to detect prints that are not readily visible.

Table 12.1 lists many types of physical and chemical methods for developing latent prints including black powder, fluorescent powder, nanoparticle powder, Sudan Black, Oil Red O, RTX, small particle reagent (SPR)/molybdenum

Table 12.1 Latent print development methods

Method	Surface	Chemical function	Notes
Black powder	Light-colored, porous or nonporous	Adsorbs to oils, moisture, and dirt in prints	Unabsorbed magnetic powder can be removed without disturbing print
Fluorescent powder	Light or dark color to provide contrast, porous or nonporous	Adsorbs to oils, moisture, and dirt in prints	View under alternate light source or black/UV light
Nanoparticle powder	Light-colored, porous or nonporous	Adsorbs to oils, moisture, and dirt in prints	Small size allows for better definition of minutiae
Sudan Black	Light-colored, porous or nonporous, moist or wet	Stains oils/lipids in prints black	Works on objects that have been submersed in water
Oil Red O	Light-colored, porous or nonporous, moist or wet	Stains oils/lipids in prints red	Works on objects that have been submersed in water and thermal papers
RTX	Light-colored, porous or nonporous, metals including coins, moist or wet	Ruthenium tetroxide reacts with oils/lipids	Use fume hood, avoid breathing fumes
Small particle reagent/ molybdenum disulfide	Wet or moist, nonporous	Reacts with oils to form gray deposit	
Ninhydrin	Porous	Reacts with amino acids in prints to form purple color	Heat with steam to accelerate development
Superglue fuming/cyanoacrylate	Nonporous	Reacts with moisture and alkanes in print to form methyl cyanoacrylate white polymers	Develop in humidity chamber in hood; avoid breathing fumes
Iodine fuming	Porous, light-colored	Iodine sublimes to adsorb to oils, moisture, and dirt in prints	Photograph quickly as iodine will continue to sublime from print when removed from chamber
Gun blue	Metal	Copper selenide blue/black product is formed in reaction with brass	Reacts with metal around fingerprint minutiae
Silver nitrate/physical developer	Porous	Silver reacts with salts in prints to form silver chloride black precipitate	Use last, interferes with DNA typing

disulfide, ninhydrin, superglue fuming/cyanoacrylate, iodine fuming, gun blue, and silver nitrate/physical developer. The techniques discussed are divided into two categories: techniques that involve chemical reactions and physical techniques based on intermolecular forces in order to create adhesion or adsorption or a stain that changes color as it changes environment. The physical application of fingerprint powder is preferred for processing latent prints deposited on hard, nonporous surfaces including glass, mirror, tile, metal, or painted wood. Traditional physical latent print detection methods include magnetic and nonmagnetic powders of various colors such as black, gray, silver, and orange, among other colors, in addition to newer fluorescent, magnetic, and nanoparticle fingerprint powders. The fat-soluble dyes Sudan Black and Oil Red O are used to stain lipids in the prints and work on nonporous surfaces. Chemical processing is usually optimal for fingerprints deposited on porous surfaces including paper, cardboard, and cloth, although if the chemical reacts with the substrate it may be nonporous. Chemical methods include ninhydrin spray, cyanoacrylate fuming, iodine fuming, silver nitrate, physical developer, RTX, SPR/molybdenum disulfide, and gun blue. Of the chemical reactions, the methods are divided into two categories: those that react with the secretions and material transferred in the fingerprint deposit and those that react with the surface or substrate the fingerprint is impressed on.

One of the oldest techniques used by latent print examiners is simply dusting fingerprints with fine powder using a brush or applying and removing the powder with a magnetic wand. In fingerprint studies, this method is reported to yield the best results on most materials. *Fingerprint powders* include bichromatic, black magnetic, gray, colored, fluorescent, and dual-purpose powder. In this physical method, the brush serves to distribute the particles evenly over

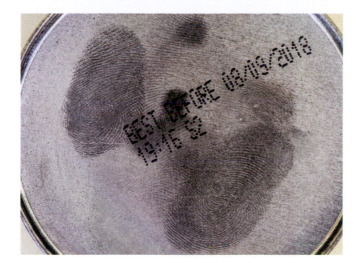

Figure 12.5 Fingerprint dusted with black magnetic fingerprint powder.

the print. Unlike brushes, the magnetic wand used with magnetic powder never comes into contact with the print, thereby reducing the potential for damage and contamination from the brush. When black powder is lightly passed over the latent print, it adheres to the residues left in the latent print on a surface. Black powder is comprised of potassium nitrate, sulfur, and activated black carbon or charcoal and is available from several commercial suppliers. To provide the best contrast, dark colored powders should be used when enhancing prints on a light surface; conversely, light-colored powders work well when enhancing prints on a dark surface. Gray powder is finely ground aluminum. Gray powder works well on polished metal surfaces. Figure 12.5 shows a fingerprint on a can developed with black magnetic powder. Figure 12.6 shows a fingerprint on currency using fluorescent fingerprint powder viewed under UV light. In a study in my lab, black powder and a nanoparticle powder performed best for developing latent prints on fruits and vegetables.

Nanoparticles are emerging for use in latent print development. Gold, cadmium sulfide, zinc oxide, CdSe/ZnS, iron, and functionalized CdTe nanoparticles have been tested for this use. Due to their small size, nanoparticles have been shown to better resolve the materials deposited from the friction skin ridges in fingerprints than traditional fingerprint powders. They can be dusted on the print or the print can be dipped in a nanoparticle solution. Gold nanoparticles have also been tested in conjunction with silver developer; this preparation further enhanced the latent print as compared to the nanoparticles alone in my lab. Physical developer with nanoparticles shows even smaller silver nucleation sites. Powders infused with nanoparticles have been shown to yield a 30% improvement in latent print definition. Bovine serum albumin can also be complexed with gold nanoparticles and conjugated to an antibody

Figure 12.6 Fingerprint developed on currency by dusting with fluorescent fingerprint powder and viewing under UV light. (From Jack Spades, https://www.flickr.com/photos/jackofspades/1376867166/in/photolist-2qFZPX-36END5-tALECJ-7NMqCV/.)

Figure 12.7   Superglue fuming humidity chamber.

to form bioconjugate nanoparticles; this can be used in the lifestyle analysis of fingerprints, for example, to test for nicotine or other drug use.

In some cases, fingerprints on nonporous objects are developed prior to dusting with fingerprint powder. *Super Glue®* *(cyanomethacrylate) fuming* is performed by placing the evidence items and a tray of Super Glue® on a hot plate in a humidity chamber in a fume hood. The cyanoacrylate ester in the glue evaporates in the chamber and preferentially condenses on the water portion of the sweat and alkanes in the fingerprint deposited, forming a hard, whitish poly-cyanoacrylate deposit. In an investigation of the cyanoacrylate fuming mechanism using MALDI-MS, a nucleophile was found to initiate the reaction and cyanoacrylate was found to form three dimers and a trimer. Qualifying masses included 177, 224, 235, and 302 amu. Figure 12.7 shows a superglue fuming chamber. Figure 12.8 shows a fingerprint on a can developed using superglue fuming.

Sudan Black is a dye that stains the lipids in sebaceous perspiration black. It can be used on light-colored, porous, and nonporous substrates that are dry or moist or wet. The reagent solution is made using Sudan Black powder, an organic solvent and water. In one study, fingerprints made on glass slides and plastic cards were submerged in tap water for a range of 1 to 15 days. The fingerprints dried and were subsequently developed using black powder, silver metallic powder, fluorescent powder, Sudan Black (powder and solution), and SPR. Black Powder, Sudan Black (both powder and solution), and SPR worked the best for developing the latent prints on the glass slides, while the black powder worked best for the plastic cards (Castelló et al., 2013).

Figure 12.8   Fingerprint developed using superglue fuming.

Oil Red O is a dye that stains lipids and lipoproteins red. This reagent has two components: the stain solution and a phosphate buffer. The stain solution is prepared using Oil Red O powder, sodium hydroxide, and water. In 2006, Rawji et al. evaluated the Oil Red O for developing latent prints on porous surfaces including white and thermal papers that were submerged in water for 2 hours and found that it worked well.

RTX, ruthenium tetroxide, reacts with unsaturated fatty acids to form black or brownish-black ridges in the prints. It can be used on light-colored, porous, or nonporous substrates including metals such as coins or cartridge casings. The reagent should be applied in the fume hood to avoid breathing its vapors.

Molybdenum disulfide (MoS_2) particles form one type of SPR. The SPR binds to the fatty substance found in fingerprint residue forming a gray precipitate. It is used on nonporous surfaces such as glass and can be used on wet or moist surfaces. In a study, it performed better than superglue fuming (Cucè et al., 2004).

The use of *gun bluing* results in a color change on the surface that is not in contact with the latent fingerprint; this is the inverse of black powder and the other methods that outline the contact of the friction skin ridges with the surface. Commercially, gun bluing is a process used to give an even finish to guns. Gun blue applied to a brass cartridge casing with a latent print darkens the exposed brass, leaving the brass covered by the fingerprint residues bright and shiny. Gun blue is composed of copper sulfate and selenous acid as an oxidizing solution to yield a blue-black color on the formation of copper selenide (CuSe). In a study in my lab, the black powder and gun blue worked best for developing latent prints on brass cartridge casings submerged in fresh water. Gun blue also gave the best results for latent print development on brass cartridge casings submerged in salt water. For nickel cartridge casings, black powder worked best overall for samples submerged in salt and fresh water.

Other dyes can be used to stain components of fingerprints in certain conditions. For example, Coomassie Blue has been shown to develop bloody fingerprints.

Iodine fuming is performed by placing solid iodine in a covered chamber with a porous, light-colored evidence item in a fume hood. Alternatively an iodine fume gun can be used. The iodine sublimes in the chamber and adsorbs to the oils and grease that may be present in latent prints. The developed print takes on a yellow-brown color that fades rapidly, requiring a photograph to be taken immediately.

Ninhydrin, also called Ruhemann's purple, is a chemical agent that can also develop latent prints on porous surfaces. Ninhydrin reacts with amino acids and proteins in the prints. It works well even on old or weak prints. Ninhydrin can be either sprayed on or the evidence item can be dipped in the liquid. The chemical reaction is accelerated by heat and steam; an iron is often used above the print. The developed print takes on a bluish-purple color and with the addition of a ninhydrin fixative, it becomes reddish purple. 1,2-Indanedione is another dye that also detects amino acids. Figure 12.9 shows a fingerprint developed using ninhydrin.

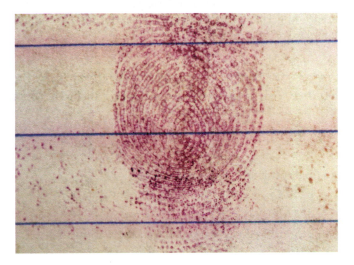

Figure 12.9   Fingerprint developed using ninhydrin.

Figure 12.10 Comparison of fingerprints using a comparison microscope. (From Tamasflex, https://commons.wikimedia.org/w/index.php?curid=10614628.)

Silver nitrate is a chemical that reacts with the salts, such as sodium chloride, in the latent print. A resulting salt, silver chloride, turns brown when exposed to light. Small articles can be dipped in the liquid while a paintbrush can be used to brush the silver nitrate on larger items. As the silver-containing solution is destructive to the print, for further analysis including DNA typing, silver nitrate and physical developer are used last.

Physical developer is a silver-based liquid reagent that also contains rhodamine 6G, which reacts with lipids, fats, oils, and waxes present in fingerprint residue and is used as a last resort to develop prints when other chemical methods have been ineffective. It is used on porous surfaces. In this method, Ag^+ is reduced to Ag^0. Exposing prints to gold nanoparticles, and washing the excess residue prior to using physical developer, can help preferentially reduce the physical developer. Any unwashed nanoparticles facilitate large silver deposits. This method is used last as it has been shown to interfere with DNA typing.

Once the latent print has been developed and detected, it must be preserved for future comparison, as needed. On nonporous surfaces, this is achieved through the use of *lifting* (e.g., J-lift, adhesive tape) and photography. Submitted fingerprint evidence and reference fingerprint cards can be compared using a comparison microscope as shown in Figure 12.10 to facilitate the comparison of minutiae.

QUESTIONS

1. Superglue fuming allows a crime scene investigator to visualize _____ that were not previously apparent.
    a. Visible prints
    b. Latent prints
    c. Plastic prints

2. Which of the following is a fingerprint identification database?
    a. CODIS   b. PDQ
    c. NIBIN   d. IAFIS

3. Which of the following is a chemical in fingerprints that reacts with ninhydrin?
    a. Amylase   b. Amino acids
    c. Fatty acids   d. Ethanol

4. Which of the following is a chemical that reacts with lipids in fingerprints?

   a. RTX   b. Sudan Black

   c. SPR   d. Oil Red O   e. All of the above

5. Which of the following fingerprint development methods should be performed last, if necessary?

   a. Physical developer   b. Iodine fuming

   c. Fluorescent powder   d. Ninhydrin

6. Name a method to preserve a fingerprint and remove it with the evidence.

7. Name a method to develop latent prints on light-colored surfaces on a small area using a physical technique.

8. Name a chemical method of developing prints on colorful aluminum surfaces.

9. What is a latent fingerprint?

10. List and briefly describe five methods that can be used to detect and develop latent fingerprint evidence and which biological component of the fingerprint the chemical reacts with or binds to.

References

Castelló, A., F. Francesc, and V. Fernando. 2013. Solving underwater crimes: Development of latent prints made on Submerged Objects. *Sci. Justice* 53(3):328–331.

Cucè, P., G. Polimeni, A.P. Lazzaro, and G. De Fulvio. 2004. Small particle reagents technique can help to point out wet latent fingerprints. *Forensic Sci. Int.* 146(Suppl.):S7–S8.

Bibliography

Adair, T.W. and R. Shaw. 2007. The dry-casting method: A reintroduction to a simple method for casting snow impressions. *J. Forensic Ident.* 57(6):823–831.

Badiye, A. and N. Kapoor. 2015. Efficacy of Robin® powder blue for latent fingerprint development on various surfaces. *Egypt. J. Forensic Sci.* 5:166–173.

Bailey, M. and K.M. Elkins. 2014. The effect of time, temperature and topography on the detection of latent prints on fruit and vegetables. Unpublished data.

Barros, R., B. Faria, and S. Kuchelhaus. 2013. Morphometry of latent palm prints as a function of time. *Sci. Justice* 53–54:402–408.

Beaudoin, A. 2004. New technique for revealing latent fingerprints on wet, porous surfaces: Oil Red O. *J. Forensic Ident.* 54:413–421.

Bentsen, R.K., J.K. Brown, A. Dinsmore, K.K. Harvey, and T.G. Kee. 1996. Post firing visualization of fingerprints on spent cartridge cases. *Sci. Justice* 36:3–8.

Book, M.K. and J. Tullbane. 2008. Detection of latent prints on handguns after submersion in water. *Evidence Technol. Mag.* www.evidencemagazine.com/index.php?option=com_content&task=view&id=658 (accessed December 3, 2017).

Budowle, B., J. Buscaglia, and R.S. Perlman. 2006. Review of the scientific basis for friction ridge comparisons as a means of identification: Committee findings and recommendations. *Forensic Sci. Commun.* 8. www.fbi.gov/about-us/lab/forensic-science-communications/fsc/jan2006/research/2006_01_research02.htm (accessed December 3, 2017).

Burgess, A. and K.M. Elkins. 2017. Latent Print Identification on Tightly Woven Fabrics including Under Armour. Unpublished data.

Cantú, A. 2014. The physical principles of the reflected ultraviolet imaging systems. *J. Forensic Ident.* 64(2):123–141.

Centoricka, M. and B. Rone. 2003. Development of latent fingerprints on various surfaces by using the RTX method. *Probl. Forensic Sci.* 51:155–157.

Cheng, K.H., J. Ajimo, and W. Chen. 2008. Exploration of functionalized CdTe nanoparticles for latent fingerprint detection. *J. Nanosci. Nanotechnol.* 8(3):1170–1173.

Choi, M.J., K.E. McBean, P.H.R. Ng, A.M. McDonagh, P.J. Maynard, C. Lennard, and C. Roux. 2008. An evaluation of nanostructured zinc oxide as a fluorescent powder for fingerprint detection. *J. Mater. Sci.* 43(2):732–737.

Crown, D.A. 1969. The development of latent fingerprints with ninhydrin. *J. Crim. Law Criminol. Police Sci.* 60(2):258–264.

Czekanski, P., M. Fasola, and J. Allison. 2006. A mechanistic model for the superglue fuming of latent fingerprints. *J. Forensic Sci.* 51(6):1323–1328.

Davis, R. 2016. *Evidence Collection and Presentation*, 2nd edn. Chennai, TN: Lawtech Publishing Co.

Dominick, A.J., N. NicDaeid, S. Bleay, and V.G. Sears. 2009. The recoverability of fingerprints on paper exposed to elevated temperatures—Part 2: Natural fluorescence. *J. Forensic Ident.* 59(3):340–355.

Enustun, B.V. and J. Turkevich. 1963. Coagulation of colloidal gold. *J. Am. Chem. Soc.* 85(21):3317–3328. doi:10.1021/ja00904a001

Faulds, H. 1880. On the Skin-furrows of the hand. *Nature* 22:605.

Federal Bureau of Investigation US Department of Justice. History of the "west brothers" identification. https://82141360.weebly.com/uploads/1/6/9/5/16958524/9115685_orig.jpg?418 (accessed December 3, 2017).

Ferguson, S., L. Nicholson, K. Farrugia, D. Bremner, and D. Gentles. 2013. A preliminary investigation into the acquisition of fingerprints on food. *Sci. Justice* 53(1):67–72.

Fisher, B.A.J. and D.R. Fisher. 2012. *Techniques of Crime Scene Investigation*, 8th edn. Boca Raton, FL: CRC Press.

Fraser, J., P. Deacon, S. Bleay, and D.H. Bremner. 2013. A comparison of the use of vacuum metal deposition versus cyanoacrylate fuming for visualization of fingermarks and grab impressions on fabrics. *Sci. Justice* 54:133–140.

Friesen, J.B. 2015. Forensic chemistry: The revelation of latent fingerprints. *J. Chem. Educ.* 92(3):497–504.

Galton, F. 1892. *Finger Prints*. London: MacMillan and Co.

Gaskell, C., S.M. Bleay, and J. Ramadani. 2013. Natural yellow 3: A novel fluorescent reagent for use on grease-contaminated fingermarks on nonporous dark surfaces. *J. Forensic Ident.* 63(3):274–285.

Girod, A., R. Ramotowski, and C. Weyermann. 2012. Composition of fingermark residue: A qualitative and quantitative review. *Forensic Sci. Int.* 223(1–3):10–24.

Guigui, K. and A. Beaudoin. 2007. The use of Oil Red O in sequence with other methods of fingerprint development. *J. Forensic Ident.* 57:550–581.

Hays, M. 2013. An identification based on palmar flexion creases. *J. Forensic Ident.* 63(6):633–641.

Hess, K.M., C.H. Orthmann, and H.L. Cho. 2013. *Criminal Investigation*, 11th edn. Boston, MA: Cengage Learning.

Holder, E.H., L.O. Robinson, and J.H. Laub. 2011. The fingerprint handbook. *NIJ.* www.ncjrs.gov/pdffiles1/nij/225320.pdf (accessed December 3, 2017).

International Association for Identification. 2017. Journal of forensic identification abstracts 1998–2017. www.theiai.org/jfi/jfi_titles.php/ (accessed January 29, 2018).

Jaber, N., A. Lesniewski, H. Gabizon, S. Shenawi, D. Mandler, and J. Almog. 2012. Visualization of latent fingermarks by nanotechnology: Reversed development on paper—A remedy to the variation in sweat composition. *Angew. Chem. Int. Ed.* 51:1–5.

Jasuja, O.P., M.A. Toofany, G. Singh, and G.S. Sodhi. 2009. Dynamics of latent fingerprints: The effect of physical factors on quality of ninhydrin developed prints—A preliminary study. *Sci. Justice* 49:8–11.

Jones, B.J., R. Downham, and V.G. Sears. 2010. Effect of substrate surface topography on forensic development of latent fingerprints with iron oxide powder suspension. *Surf. Interface Anal.* 42(5):438–474.

Kabklang, P., S. Riengrojpitak, and W. Suwansamrith. 2009. Latent fingerprint detection by various formulae of SPR on wet nonporous surfaces. *J. Sci. Res. Chula Univ.* 34(2): 59–64. http://forensic.sc.mahidol.ac.th/proceeding/50_Phatwalan.pdf.

Keeping track of the criminal by his finger prints: The wonderful art, long used in China, rapidly being adopted by the police of this country, with the New York force leading. *The New York Times*. 1911.

Knighting, S., J. Fraser, K. Sturrock, P. Deacon, S. Bleay, and D.H. Bremner. 2012. Visualization of fingermarks and grab impressions on dark fabrics using silver vacuum metal deposition. *Sci. Justice* 53:309–314.

Kücken, M. and A. Newell. 2005. Fingerprint formation. *J. Theor. Biol.* 235:71–83.

Leban. D.A. and R.S. Ramotowski. 1996. Evaluation of gun blueing solutions and their ability to develop latent prints on cartridge cases. *CDBIAI* www.cbdiai.org/Articles/leben_ramotowski_10-96.pdf (accessed January 29, 2017).

McRoberts, A., ed. 2011. *The Fingerprint Sourcebook*. Washington, DC: US Dept. of Justice, Office of Justice Programs, National Institute of Justice.

Meloan, C.E., R.E. James, and R. Saferstein. 2004. Experiments 13 and 14. In *Lab Manual—Criminalistics: An Introduction to Forensic Science*, 8th edn. Upper Saddle River, NJ: Pearson Prentice Hall.

Miller, A. 2008. Choosing the best fingerprint powder for your scene. *Evidence Technol. Mag.* www.evidencemagazine.com/index.php?option=com_content&task=view&id=1387 (accessed December 3, 2017).

Nause, L.A. and M.P. Souliere. 2008. Recording a known tyre impression from a suspect vehicle. *J. Forensic Ident.* 58(3):305–314.

Nixon, C., M.J. Almond, J.V. Baum, and J.W. Bond. 2013. Enhancement of aged and denatured fingerprints using the cyanoacrylate fuming technique following dusting with amino acid-containing powders. *J. Forensic Sci.* 58(2):508–512.

Norlin, S., M. Nilsson, P. Heden, and M. Allen. 2013. Evaluation of the impact of different visualization techniques on DNA in fingerprints. *J. Forensic Ident.* 63(2):189–204.

O'Neill, K. C., P. Hinners, and Y. J. Lee. Chemical imaging of cyanoacrylate-fumed fingerprints by matrix-assisted laser desorption/ionization mass spectrometry imaging. *J Forensic Sci.* 2018 Mar 23. doi:10.1111/1556-4029.13773.

Peeler, G., S.J. Gutowski, H. Wrobel, and G. Dower. 2008. The restoration of impressed characters on aluminum alloy motorcycle frames. *J. Forensic Ident.* 58(1):27–32.

Rawji, A. and A. Beaudoin. 2006. Oil Red O versus physical developer on wet papers: A comparative study. *J. Forensic Ident.* 56:33–54.

Robbins, L. and K.M. Elkins. 2015. The use of various techniques to detect latent prints on spent cartridge casings submerged in water over time. Unpublished data.

Ross, E. and M. Gorn. 2010. A study of pyridyldiphenyl-triazine as a chemical enhancement technique for soil and dust impressions. *J. Forensic Ident.* 60(5):532–546.

Saferstein, R. 2007. Exercise 5. In *Basic Laboratory Exercises for Forensic Science*. Upper Saddle River, NJ: Pearson Prentice Hall.

Sametband, M., I. Shweky, U. Banin, D. Mandler, and J. Almog. 2007. Application of nanoparticles for the enhancement of latent fingerprints. *Chem. Commun. (Camb)*. 12(11):1142–1144.

Science Daily. February 2, 2011. Forensic breakthrough: Recovering fingerprints on fabrics could turn clothes into silent witnesses. www.sciencedaily.com/releases/2011/01/110131073141.htm (accessed December 2, 2017).

Singh, G., G.S. Sodhi, and O.P. Jasuja. 2006. Detection of latent fingerprints on fruits and vegetables. *J. Forensic Ident.* 56(3):374–381.

Sodhi, G.S. and J. Kaur. 2001. Powder method for detecting latent fingerprints: A review. *Forensic Sci. Int.* 120:172–176.

Sokolov, K., K. Follen, J. Aaron, I. Pavlova, A. Malpica, R. Lotan, and R. Richards-Kortum. 2003. Real-time vital optical imaging of precancer using anti-epidermal growth factor receptor antibodies conjugated to gold nanoparticles. *Cancer Res.* 63:1999–2004.

The laying on of hands for fingerprints: Woman expert thinks system will not be confined to criminals, but will become universal—Chinese used it for identification sixteen centuries ago. *New York Times*. Jun 29, 1919. p. 80.

Trapecar, M. and M.K. Vinkovic. 2008. Techniques for fingerprint recovery on vegetable and fruit surfaces used in Slovenia—A preliminary study. *Sci. Justice* 48(4):192–195.

Turkevich, J. and B.V. Enustun. 1963. Coagultion of colloidal gold. *J. Am. Chem. Soc.* 85(21):3317–3328.

Ulery, B.T., R.A. Hicklin, G.I. Kiebuzinski, M.A. Roberts, and J. Buscaglia. 2013. Understanding the sufficiency of information for latent fingerprint value determination. *Forensic Sci. Int.* 230(1–3):99–106.

Ulery, B.T., R.A. Hicklin, M.A. Roberts, and J. Buscaglia. 2014. Measuring what latent fingerprint examiners consider sufficient information for individualization determinations. *PLoS ONE* 9(11): e110179. doi:10.1371/journal.pone.011017.

Velthuis, S. and M. de Puit. 2011. Studies toward the development of a positive control test for the cyanoacrylate fuming technique using artificial sweat. *J. Forensic Ident.* 61(1):16–29.

Wolfe, J. 2008. Sulfur cement: A new material for casting snow impression evidence. *J. Forensic Ident.* 58(4):485–498.

CHAPTER 13

Firearms

KEY WORDS: firearms, ammunition, bullet, cartridge casing, primer, gunpowder, magazine, firing chamber, firing pin, barrel, breechblock, blowback, recoil, gas piston, handgun, revolver, rifle, shotgun, gauge, semiautomatic, lands, grooves, broach, mandrel, cutter, caliber, bullet wipe, ballistics, test fire, Integrated Ballistic Identification System, gunshot residue, Griess test, diphenylamine test, dermal nitrate test

LEARNING OBJECTIVES

- To list types of firearms and briefly explain how the different types function
- To explain how rifling and microgrooving features are imparted to barrels
- To list and identify features that can be used to differentiate spent bullets and cartridge casings shot with different weapons
- To explain how test fires are performed
- To explain how to determine the distance a shot was fired from the target
- To explain the use and value of databases in firearms analysis
- To list chemical tests to detect and identify propellants and explosives used in firearms

SOLDIER ATTACKS SOLDIERS READYING TO DEPLOY AT FORT HOOD US ARMY BASE

In the afternoon of November 5, 2009, a US Army soldier shouted "Allahu Akbar" and opened fire at the crowded Fort Hood's Soldier Readiness Processing Center, leading to the worst mass murder at a US military installation to that time. Soldiers were at the Soldier Readiness Processing Center to receive medical screening before a deployment. In about 10 minutes, 13 people were killed, including 12 soldiers and 1 Department of Defense employee, and more than 30 others were wounded before civilian police arrived, shot the gunman, and took him into custody. Criminal Investigation Division special agents recovered 214 5.7 × 28 mm caliber casings from the Soldier Readiness Processing Center. A witness reported seeing blue-tipped ammunition.

Nidal Malik Hasan, a 39-year-old US Army Major, purchased a FN Five-seveN® semiautomatic pistol on August 1, 2009, from Guns Galore in Killeen, Texas. According to an employee, Fredrick Brannon, Hasan entered the store on July 31, 2009, and asked for the most high-tech handgun that they sold that could hold the most magazines. In addition to Brannon, the store manager and a regular customer at the store, Specialist William Gilbert, were also present. Based on Hasan's requested specifications, the two employees and Gilbert all recommended the FN Five-seveN®. Gilbert was familiar with firearms and owned an FN Five-seveN®. He explained that the firearm has a low recoil, is light, and is "very, very easy to fire with one hand." When Hasan returned to purchase the weapon on August 1, 2009, Brannon processed Hasan for a Federal Bureau of Investigation (FBI) background check. To prove their Texas residency, soldiers must show a Texas identification card or a military ID

SOLDIER ATTACKS SOLDIERS READYING TO DEPLOY AT FORT HOOD US ARMY BASE (continued)

and permanent change of station orders for a base in Texas. Hasan passed the background check and purchased the weapon, which had a manufacturer's suggested retail price of $1349. He returned to Guns Galore weekly to purchase more ammunition and clips, starting off with four or five boxes each visit but later purchasing six to eight boxes of ammunition each visit. Each box contained 50 rounds. Hasan proceeded to take a concealed handgun license certification class on October 10, 2009, at Stan's Outdoor Shooting Range in Florence, Texas. Hasan also bought a membership to the range at Stan's and practiced shooting his firearm there.

The FN Five-seveN® pistol fires 5.7×28mm ammunition. It was developed in 1989 by FN and fires projectiles very fast for a pistol. It was designed to penetrate soft body armor and is used by military and law enforcement personnel around the world. It is made of polymer on the outside and has steel interior components, giving it a light weight of only 23 ounces unloaded. It is easy to load—the cartridge only needs to be pushed down and not slid back into place—and can be taken apart with one hand. It holds magazines that hold 20 rounds of ammunition but Hasan extended them to hold 30 rounds. The SS197R Sporting Round of ammunition is a blue-tipped bullet weighing 40 grains.

Major Hasan was a US Army psychiatrist who treated soldiers returning from war with post-traumatic stress disorder. Although concerns were raised about his behavior in performance reviews, he was continually promoted. After the attack, he was charged with 13 counts of premeditated murder and 32 counts of premeditated attempted murder and was tried in an Article 32 proceeding. Hasan was found guilty by a jury for 45 counts of premeditated murder and attempted premeditated murder on August 23, 2013. He was sentenced to death.

BIBLIOGRAPHY

Army major kills 13 people in Fort Hood shooting spree. www.history.com/this-day-in-history/army-major-kills-13-people-in-fort-hood-shooting-spree (accessed June 21, 2017).

Huddleston, S. 2010. Hasan sought gun with "high magazine capacity." http://blog.mysanantonio.com/military/2010/10/hasan-sought-gun-with-high-magazine-capacity/ (accessed June 21, 2017).

McHale, T. 2016. FN Five-SeveN Pistol: Gun Review. www.range365.com/fn-five-seven-pistol-gun-review (accessed June 21, 2017).

Prosecutors end case in Hasan Article 32 hearing. http://kdhnews.com/news/prosecutors-end-case-in-hasan-article-hearing/article_c2cea5e3-063c-5bb4-a978-3cb69bae0501.html (accessed June 21, 2017).

Forensic scientists assigned to the firearms unit of the crime laboratory must have a working knowledge of all types of weapons and how they work as they and/or their ammunition (new or fired) may be encountered at the crime scene or seized as evidence. As the ammunition for different firearms and different manufacturers varies greatly, forensic scientists may be asked to determine what type and brand of weapon it came from and if a fired bullet or spent cartridge casing submitted into evidence was used with a specific, suspected weapon. Unique impression markings on the bullets and cartridge casings can be individualizing. Examiners may also be asked to conduct a ballistics investigation to estimate the distance from the muzzle of the firearm to the target or characterize gunpowder residue recovered from the suspect's hands or garments, or the area around the gunshot wound.

FIREARMS

Firearms may be single-action, automatic, or semiautomatic weapons. In all of these types, *ammunition* consists of a *bullet* packed in a *cartridge casing* with *primer* and *gunpowder propellant* (Figure 13.1). A *magazine* holds several cartridges of ammunition. The cartridge is loaded into the *firing chamber*. The *firing pin* in the chamber strikes and ignites the primer which, in turn, ignites the gunpowder. Ammunition types include centerfire and rimfire types. The gases produced in the combustion reaction propel the bullet forward through the *barrel* of the firearm toward the target and push the fired cartridge case back against the *breechblock*.

For analysis purposes, there are three main types of operating systems for firearms including blowback, recoil, and gas piston. In *blowback* weapons, the fired cartridge exerts pressure to push the breechblock back against the spring. The cartridge is extracted and ejected from the weapon by the moving breechblock. The compressed spring propels the breechblock forward and a new cartridge is removed from the magazine and inserted into the firing chamber. The barrel of the weapon and the breechblock recoil together in *recoil* weapons. Then the breechblock unlocks from

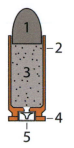

Figure 13.1   Components of a bullet (1-bullet; 2-casing; 3-propellant; 4-rim; 5-primer). (From Quadrell [modifications made by Indy muaddib], https://commons.wikimedia.org/w/index.php?curid=10792980.)

the barrel and recoils further backward and compresses a spring. The compressed spring returns the breechblock to its starting position, loading another cartridge into the chamber. In *gas-piston* weapons, there is a small hole in the barrel leading to a gas piston. On operation, a small amount of the propellant gases passes through the hole and exerts pressure on the piston. As the piston is connected to the breechblock, it is also forced to the rear.

HANDGUNS

Handguns are firearms that are designed to be operated with one hand and include revolvers and pistols; examples are shown in Figures 13.2 and 13.3 respectively. Ammunition is described in caliber; a 0.50 caliber round is approximately 0.50 inches in diameter. Revolvers have a revolving cylinder while pistols do not. There are two types of revolver handguns including single-action and double-action revolvers. *Single-action revolvers* are fired by manually cocking the hammer, which rotates the cylinder to place one of the chambers directly in line with the hammer and cocks the firing mechanism; pulling the trigger leads it to make contact with the cartridge portion to ignite the primer and fire the bullet. *Double-action revolvers* are fired by a long trigger pull. The trigger pull cocks the hammer, rotates the cylinder in line with the hammer, drops the hammer, and fires the bullet. Most modern handguns are pistol types. Pistol subtypes include single-shot pistols, semiautomatic pistols, and machine pistols. In semiautomatic pistols, the firing of a round will cycle the pistol, eject the fired cartridge casing, and insert a new cartridge into the chamber. Machine pistols are automatic pistols.

Figure 13.2   Revolver handgun (0.357 Magnum). (From Mitch Barrie, https://www.flickr.com/photos/simonov/351717261/in/photolist-x5DhH-oCB9o5-cRdSxf-bvt7Z1-ojtQEf-d12ZeU-dipqwy-bvt7Nd-oQB3Wm-qjduL4-VNmUQe-g1F917-9pvmNp-3par6A-ojtnZd-4tYb4v-Tag1XT-4JshQr-5rWQUE-diDxYq-dk4p8B-j7rCTo-piSj9S-dZPLwJ-nFMtt-nNiArK-dk4qnd-dJZ8uY-oFA7rz-oDymRd-dZJrDP-VXGeFD-WeyK7h-oFysHs-o25VfG-op6pW3-9URYNo-oFyqUC-op77vg-af6BEp-9qPEZW-9qPEXf-ovKiTX-9qPEXN-62hqiQ-9qPEY7-9qLFhn-9qPEWY-4fyVR1-22jwW.)

Figure 13.3   Pistol. (From Mitch Barrie, https://www.flickr.com/photos/simonov/351717147/in/photolist-x5DfK-YhAD3Q-Upx-ABm-fiSUWC-Vj9cDU-CKTLjC-fiCAF8-f3UUkC-nB3ha-cHPK-c9Pw8S-93PS7-9QcLJn-ftzHNL-fiSN4S-co7FhY-cHQa-iTW4fw-f3UKpu-9QcLeX-asoRWD-giNB7p-39oWnb-9uT3xF-fiQXTq-PsJan-PseVs-iXdeR-PsbtW-PsMUH-rkGkdy-PsFMc-PsAun-PsDvt-ftkrji-w8Qmnk-PskHz-Psxqn-PrL6Q-PsubM-asrKgo-PsjFV-iEciut-PsrZD-fiSBw3-iEdXpd-Psgr8-fiCpnp-iEgpvu-61YiaK.)

RIFLES

Rifles are also known as long guns. A rifle is shown in Figure 13.4. There are five types of single-barrel rifles: single-action, lever-action, bolt-action, semiautomatic, and automatic. In *lever-action rifles*, a lever is located below the weapon's receiver and the magazine is a tube positioned under the barrel. Raising the lever results in a new round entering the chamber of the rifle. Dropping the lever moves the breechblock back and cocks the firing mechanism. *Slide or pump-action rifles* are a type of lever-action rifle that uses a slide located under the barrel to extract and eject the fired cartridge case and cock the firing mechanism. In a *straight-pull bolt-action rifle*, drawing a straight-pull bolt to the rear causes the fired cartridge case to be extracted and ejected while moving it forward removes a cartridge from the magazine, inserts it into the chamber, and cocks the rifle. Other bolts are turned down to lock the bolt closed; *turn-bolt rifles* can fire more powerful ammunition. Semiautomatic rifles are the rifle equivalent to semiautomatic pistols. Most *semiautomatic rifles* are gas-piston weapons, but they may also be blowback or recoil in function. Ammunition is most commonly stored in box magazines within the weapon or detached but may also be stored in drum or helical magazines. In semiautomatic rifles, squeezing the trigger results in many actions. Once a round is fired, the fired cartridge case is extracted and ejected, a fresh cartridge is loaded, and the firing mechanism is cocked for another shot.

Another type of rifle is the double or combination rifle, a type of sporting rifle that has two rifled barrels or one rifled barrel and one shotgun barrel. The double barrels may be positioned side to side or over and under.

SHOTGUNS

A *shotgun* is a type of long gun with a smooth barrel as shown in Figure 13.5. Shotguns use pellet or small lead ball ammunition. A wad of paper packaged with explosive is used to propel the lead balls or pellets. The *gauge* is used to describe the diameter of the shotgun barrel. For example, a 12-gauge shotgun has a 0.730-inch barrel. As the gauge increases, the diameter decreases. For example, a 16-gauge shotgun's barrel is only 0.670 inch in diameter.

ASSAULT RIFLES

Assault rifles have a detachable magazine and use an intermediate cartridge that is less powerful than a full battle cartridge but more powerful than a rifle cartridge. Assault rifles are characterized by having at least two of the three following features: a sight mechanism, grenade launcher, a bayonet mount, a pistol grip beneath the chamber, and a flash guard that cools or disperses exiting gases from the muzzle. The most commonly encountered assault rifles include the US military M-16 and the Russian Kalashnikov AK-47 weapons. They fire an intermediate reduced-charge rifle cartridge. The AR-15 (Figure 13.6) is a civilian version of the M-16 assault rifle; when retrofitted with a bump stock kit, it fires essentially like an automatic assault rifle.

FIREARMS MANUFACTURING METHODS

The inner surface of the weapon's barrel has intended and unintended markings on it from the manufacturer. The barrel may be formed from a solid bar of steel that is hollowed out by drilling, cutting, or high pressure. The drills used vary in size and can leave random and non-systematic marks on the inside of the barrel. Additionally, a broach cutter can be used to impress the inner surface of the barrel with spiraling grooves. *Grooves*, or spiral impressions, are cut or impressed into the inner surface of the barrel metal by the manufacturer (Figure 13.7). *Lands* are the remaining raised surfaces produced by the original drill bore or cutter when making the grooves. The lands and grooves (rifling) serve to rapidly spin the fired bullet through the barrel for better accuracy at the target. Buttons or plugs, cutters, and mandrels are three types of *broaches*. *Buttons* contain the groove pattern (desired width and number) that is impressed onto the

Figure 13.4   Rifle. (From Antique Military Rifles, https://www.flickr.com/photos/36224933@N07/4876767851/in/photolist-XZaUn-7FcbQS-8qWHbk-frn6i3-9USYkZ-emtYkr-dUGzrs-frncSd-bmDz5W-4RC7qj-dpXxeU-emqU2z-bXnxm-emHUTL-dpXx8C-6dV4Yd-emHonj-76aeZh-7Fcc4Q-ExP2x8-8pqUzA-fso6Rs-bXnxn-eNNvo5-emHnQN-emtYSP-2JRvaJ-gpH3q-4RYrtm-7FcceY-7uhfnv-nVu2na-dUGzof-cxciZG-9M7Cky-iATjwd-6RDEA8-6VNFLH-9dcUhQ-nBdrAh-nREcDE-oopYWR-emvZ5A-fsnPyC-8KtXe4-emty1H-7um8VA-emu1wF-8KtXdV-87Whst.)

Figure 13.5 Shotgun. (From Mesa Tactical, https://www.flickr.com/photos/mesatacticalphotos/8726559989/in/photolist-6ZE8Pq-eDbFHJ-9jQyG4-eie4Y5-ei8UfR-ei7ek8-eieu7Q-ppbcae-ei8i9z-ehYSAy-dyaW9o-ei7gPD-ei7yhk-ei75ft-8h9a2h-fVecjH-ei8am6-ei7QcH-ei7yVD-eidiYG-eieons-ei8deK-eiegGG-ei6UYx-eidxpC-fFaK8f-eieDEs-5pMsKS-ei8Ndx-6ZA6EV-p84KDi-ei8Kxz-dooxt9-eicEtq-eicX7h-ei6VMe-ei8Gqa-eidtsj-ei8TTK-eidT9A-eie8tU-ppvVMm-ei7mdg-cEKFZ-eicTHE-T5s9T-eid7go-p7Yat5-ei7prg-ei8bbn.)

barrel or passed through the barrel using high pressure. *Cutters* are concentric steel rings that were used in the production of pre-1940 weapons. Smaller cutter rings are rotated and passed through the barrel to cut the grooves into the barrel at the desired direction and twist; then larger rings are used for the same purpose until the barrel features are properly formed. *Mandrels* are hardened steel rods with the rifling pattern that are inserted into the bore and impressed into the barrel by hammering or compression with heavy rollers. Like buttons, mandrels do not cut the metal of the barrel. Once completed, the manufactured barrel will retain its rifling characteristics including the direction of twist as well as the number and dimensions of its lands and grooves. The diameter of the barrel measured between opposite lands is termed the *caliber*. The number of lands and grooves, caliber, and direction of twist are all class characteristics of modern guns (except shotguns where gauge replaces caliber). Metal fragments may also be impressed at random into the metal in the barrel manufacturing process. These can impart striations on a bullet that passes through the barrel. Wear and tear from use and improper cleaning can impart other unique characteristics. Together, many of these irregular markings and microgroovings intentionally imparted by the manufacturer to the barrel form unique individual characteristics that are present in each gun. Even two firearms manufactured successively will have unique individualizing characteristics. The impressions of the lands and grooves with their direction of twist, random impressions in the metal, and microgrooves will all be impressed on bullets traversing the barrel when fired.

EXPLOSIVES AND PROPELLANTS

Gunshot primer residue (GSR) contains lead styphnate (lead 2,4,6-trinitrobenzene-1,3-diolate), barium nitrate ($Ba(NO_3)_2$), and antimony sulfide (Sb_2S_3) (Figure 13.8). Gunpowder residue is nitrocellulose-containing smokeless powder that produces gases on ignition that propel the bullet forward. Single-base powder contains only nitrocellulose while double base powder contains nitrocellulose and nitroglycerin. Triple-base powder used in artillery ammunition contains nitrocellulose, nitroglycerin, and nitroguanidine. Some gunpowder propellant structures are shown in Figure 13.9. After the weapon is fired, traces of GSR and unburned smokeless powder are often recovered from the shooter, the victim, and/or the target surface if the shot was fired at close range.

BALLISTICS

Firearms project ammunition at high velocity and over long distances if their path is uninterrupted. *Ballistics* is a science concerned with the study of a projectile in motion. A target hit at close range will often result in *stippling*,

Figure 13.6 AR-15 rifle. (From Mitch Barrie, https://www.flickr.com/photos/simonov/30656308722/in/photolist-NGZAwW-bkrYfa-8X163Z-bkrXaz-p3QjVL-jj9FYV-AaUAJR-4QoASk-dvTeFT-We9B9H-ipMZaG-qMTfPV-7LHpDL-r34KQJ-7T8ZRL-7LH929-r31JVd-7T5K1R-au53H3-eLn7UU-bks5Ev-r5fY43-7T8ZAw-eiWKrq-5vpXCj-e7Ahn-dvEpzf-qMTeqv-iuJwSs-r5a4AX-95Lwrd-dqWjTY-r5kNR8-pp2Bea-qMM1hj-iuK2Kj-f4ynzZ-dvTgGv-HW2bT-qMV48x-USmUkX-bks4ha-qMHZyb-cPfbT3-q8iocb-USmU6D-p3RqXr-q8ioxw-7LCHUB-q8kb7q.)

Figure 13.7   Lands and grooves in a barrel. (From baku13, https://commons.wikimedia.org/w/index.php?curid=853147.)

whereby traces of impacting gunshot residue result in bruises and traces of gunshot powder result in skin burns in a circular pattern. If the barrel was pressed against the skin of the victim's skull on discharge, a contact wound with a star pattern would be exhibited. On other skin, a burn imprint of the gun muzzle may be recorded. The shape and density distribution of the GSR and gunpowder residue particles allow ballistics experts to determine the distance from which a weapon was fired. For example, when the weapon is fired while in contact with the victim or a target surface, a starburst pattern results with a smoke halo of lead or other GSR particles. When fabric is present, the fibers around the hole are burned. If the weapon is fired at a distance of approximately 12–18 feet, a smoke halo stippling of GSR will be present on the target. At a distance of 25–36 feet, unburned gunpowder residues can be found on the target in the form of scattered specks but no pattern. At distances greater than 36 feet, no powder residue will be recovered but a *bullet wipe*, or dark ring around the hole, will be present.

Precise determinations of distances can only be made if the powder residue patterns made on the victim's skin or clothing and test patterns made on similar fabric by the suspect weapon and same ammunition at a known distance are similar. Test fires are required in order to determine distances involving shotguns. For example, the spread for a 12-gauge shotgun increases by one inch for each yard in distance the shooter is from the target.

FIREARMS EVIDENCE HANDLING AND LABELING

As firearms can be deadly weapons, firearms examiners must take extreme caution to protect themselves from accidental discharge or misfire while examining weapons submitted as evidence. In most cases, it will be necessary to

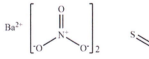
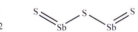

barium nitrate   antimony sulfide

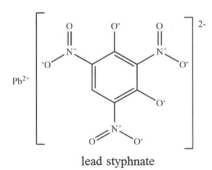

lead styphnate

Figure 13.8   Structures of primer explosives. (Structure prepared by Ashley Cowan.)

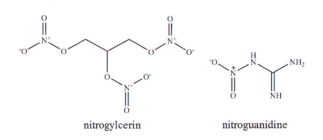

Figure 13.9 Structures of gunpowder propellants. (Structure prepared by Ashley Cowan.)

unload the weapon. Prior to unloading revolvers, the cylinders containing cartridges and their positions must be recorded. In other cases, supporting documentation should include an indication of which cylinder was loaded at the time of seizure and the seized ammunition should be packaged separately but within the case box and be made available for examination. Firearm evidence is handled by holding the weapon at the edge of the trigger guard or the checkered portions of the grip (as these regions are the least likely to provide useable fingerprints). Firearms are tagged on the trigger guard and then packaged in cardboard boxes and secured with zip ties.

Bullets recovered from the crime scene must be handled so as not to introduce additional impressions or markings. If they were recovered underwater, they should be transported in a container in the same water to reduce the potential for rusting. Bullets should be marked on the nose or the base, but not the sides, so as not to obliterate striations and other impressions present. Dry bullets should be carefully protected by packaging in cotton or tissue paper and placing the bundle in a pillbox. Fired cartridge casings should be labeled inside the shell. Discharged shotgun shells can be labeled on the paper or plastic tube remaining on the shell or on the metal on the mouth end of the shell.

FIREARMS COMPARISONS

For firearms comparisons, there are several databases and comparison tools used around the world and by different agencies. The FBI hosts a General Rifling Characteristic File that catalogs class characteristics of known weapons. Recorded class characteristics include the manufacturer, caliber, direction of twist, number, and size of lands and grooves. The shape, diameter, length, composition, and density vary significantly among bullets (Figure 13.10). Firearms examiners can mass the bullets on a balance and measure the diameter using calipers to aid in the determination of the caliber.

A *test fire* is a method that firearms examiners employ to compare an evidence bullet with a reference bullet from their ammunition case that has been fired with a weapon in question that may have been used in a crime. In order to enable the recovery of the bullet and not introduce other markings, the test fire is directed into a long water tank or cotton-filled box. After using the weapon to perform a test fire with new ammunition, the ejected cartridge case and fired bullet will be carefully examined and compared to evidence submitted from the victim or crime scene.

Firearms examiners employ stereomicroscopes and ballistics microscopes (comparison microscopes) to examine weapons and fired pellets, lead balls, bullets and spent cartridge cases, and items recovered from test fires. Shotgun pellets are difficult to individualize as no markings are impressed that can lead directly back to the weapon(s) from

Figure 13.10 Selected cartridges with bullets (left to right): (1) .17 HM2; (2) .17 HMR; (3) .22LR; (4) .22 WMR; (5) .17 SMc; (6) 5mm/35 SM4; (7) .22 Hornet; (8) .223 Remington; (9) .223 WSSM; (10) .243 Winchester; (11) .243 Winchester Improved (Ackley); (12) .25–06; (13) .270 Winchester; (14) .308; (15) .30–06; (16) .45–70 Govt; (17) .50–90 Sharps. (From Arthurrh, https://commons.wikimedia.org/w/index.php?curid=2411382.)

which they were fired. Nonetheless, the crime scene investigator can recover the wad of plastic or paper used to package the explosives (that propelled the lead balls or pellets) and the pellets and balls themselves.

As a bullet passes through the barrel, its surface is cast with impressions of the rifle markings, grooves, twist, striations, and wear characteristics of the weapon. Evidence and test-fire bullets and cartridge casings can be examined simultaneously in the same field of view using a comparison microscope. These cylindrical items under evaluation are mounted in the same direction on rotatable wax holders instead of a stage below the microscope's lenses. Lands, grooves, twist, microgrooves, and striation markings will be observed on the bullets as shown in Figure 13.11. The items can be rotated to locate a well-defined section where the markings can be directly compared and aligned side by side in the field of view. Individual characteristics such as longitudinal striations, scratches, nicks, abrasion, or wear can be used to determine if bullets were fired with a specific suspect weapon used in a test fire. If these differ between the evidence item and the test-fire item, the suspect weapon can be eliminated from further consideration. For a match, the evidence and test-fire bullet lands and grooves must have identical widths, the twist must be identical, and the striations must coincide with the same intensity in the same lighting. Grit and rust can alter the markings impressed on bullets, even those fired from the same firearm.

While the bullet is marked by the interior features of the barrel, the soft metal of the cartridge case is marked with features of the loading and firing mechanisms through its contact with these metal parts. As described at the beginning of the chapter, pulling and releasing the trigger causes the firing pin to release and strike the primer in the primer cup of the cartridge casing. The ignition of the primer causes the gunpowder in the primer cup to ignite. The burning gases propel the bullet forward while the cartridge casing is pushed backward against the breechblock. In this process, the firing pin marking may be centered, off-center, or distorted, and striations and impressions associated with its contact with the cartridge casing can be individualizing. The cartridge casing may also be impressed with markings from the surface of the breechblock including machining imperfections. Extractor markings are impressed by removing the cartridge from the firing chamber. Ejector markings on the edge of the cartridge casing face may be present due to the mechanism used to eject, or throw, the fired cartridge case from the firearm. Other markings on the cartridge casing include those impressed from the magazine or clip used to hold the ammunition or from imperfections in the wall of the chamber. There are no minimum number of points needed for a match; this judgement is left up to the expert examiner.

Automated firearms search systems such as the Integrated Ballistic Identification System (IBIS) hosted by the Bureau of Alcohol, Tobacco, Firearms and Explosives (ATF) incorporate a firearms database and comparison and image retrieval tools. Bulletproof contains bullet striation and markings data while Brasscatcher is a database of cartridge

Figure 13.11 Micrograph of lands, grooves, twists, and striations on 7.62×51 mm NATO bullets and an unfired cartridge, with counterclockwise rifle marks. (From btr, https://commons.wikimedia.org/w/index.php?curid=474096.)

Figure 13.12 Dermal nitrate test. (From Jack Spades, https://www.flickr.com/photos/jackofspades/4233679992/in/photolist-7s7HyC.)

casing markings. Networked labs can share firearms examination casework and comparison data by attaching a video camera or camera to their microscope or through the use of the Vision X comparison microscope. A new evidence bullet striation or cartridge case firing pin photo can be searched against and compared to other submitted photos in the database and compared side by side with the hits. NIST hosts a 3D ballistics database. Drugfire is a database of cartridge casing markings maintained by the FBI. Another system is the National Integrated Ballistics Information Network, which is jointly organized by the ATF and FBI. It includes a database of bullet and cartridge case images for evidence and test-fire items. It has been used to link specific weapons to multiple crimes. The final comparison and report is conducted by a firearms examiner using microscopic methods.

GUNSHOT RESIDUE ANALYSIS

GSR and gunpowder residue is projected forward toward the target from the barrel and blown back onto the shooter's firing hand when a weapon is fired. Thus, partially burned and unburned gunpowder particles may be recovered both from the target and the shooter by swabbing. GSR should be recovered from the hands of suspects ideally less than 2 hours after they fire a weapon. The sampled portion should include the thumb web, palm, and back of the suspect's hands. GSR can also be recovered from the surfaces surrounding the entry point of the bullet on victims and other targets by swabbing or through the use of adhesive disks that capture the gunpowder and gunshot residue. The sampling and subsequent analysis can help to determine if a suspect has recently fired, handled, or was near a recently fired weapon.

Fabrics and surfaces of items receiving the bullet are examined using an infrared light microscope to locate any traces of gunpowder residue. Colorimetric chemical tests such as the Griess test for nitrites can be used to tentatively identify gunpowder residue. The test reagent can be impregnated into photographic paper that can be pressed by ironing it onto the suspect fabric or surface to transfer the particles to the paper. A pink color is indicative of gunpowder residue. The dermal nitrate test can be used to test for the presence of unburned nitrates using diphenylamine (Figure 13.12). Hot paraffin or wax is applied to the suspect's hand using a paintbrush. On solidifying, it is removed and tested with the reagent. A blue color indicates a positive reaction. However, false positives may occur with fertilizers, urine, cosmetics, and tobacco.

Raman, infrared, and mass spectrometry have been used to detect primer particles and gunshot residue. A scanning electron microscope can be used to identify GSR by the characteristic size and shape of the particles and differentiate it from brake residue. The EDX attachment is used to perform an elemental analysis of the GSR residue and can identify barium, lead, and antimony present in the sample. Atomic absorption spectroscopy and neutron activation analysis can also be used to analyze the elemental composition and concentration of primer residue on evidence.

QUESTIONS

1. Which microscope is best suited to the initial evaluation of evidence such as cartridge cases?
   a. Scanning electron microscope
   b. Compound light microscope
   c. Comparison microscope
   d. Stereoscope
2. The lands and grooves in a gun's barrels are formed by a process termed _____.
   a. Calibering
   b. Rifling
   c. Breechblocking
   d. Microgrooving

3. Which of the following materials will have only class characteristics?
   a. Unfired bullets
   b. Toolmarks
   c. Fingerprints
   d. Fired bullets

4. To determine if a bullet's breech face marks originated from a specific firearm, what is the testing procedure? Perform a ____ for comparison to the evidence markings.
   a. Test-fire with the evidence bullet
   b. Test-fire with a bullet of the same caliber from the same manufacturer as the evidence
   c. Test-fire with a bullet of the same caliber of any manufacturer
   d. Compare the breech marks from the weapon and evidence directly

5. In firearms analysis, firing pin markings would be found on which of the following?
   a. Fired bullets
   b. Cartridge casing
   c. Paper wad
   d. Gunpowder
   e. All of the above

6. How does shotgun ammunition differ from that fired with a revolver? How do the resulting markings on the spent ammunition differ?

7. List three features of bullets used in individualization.

8. List three features of cartridge casings used in individualization.

9. Explain the use and value of databases in firearms analysis.

10. List chemical tests to detect and identify propellants and explosives used in firearms.

Bibliography

Arndt, J., S. Bell, L. Crookshanks, M. Lovejoy, C. Oleska, T. Tulley, and D. Wolfe. 2012. Preliminary evaluation of the persistence of organic gunshot residue. *Forensic Sci. Int.* 222(1–3):137–145.

Bachrach, B. 2002. Development of a 3D-based automated firearms evidence comparison system. *J. Forensic Sci.* 47(6):1253–1264.

Bandodkar, A.J., A.M. O'Mahony, J. Ramírez, I.A. Samek, S.M. Anderson, J.R. Windmiller, and J. Wang. 2013. Solid-state forensic finger sensor for integrated sampling and detection of gunshot residue and explosives: Towards "lab-on-a-finger." *Analyst* 138(18):5288–5295.

Banno, A., T. Masuda, and K. Ikeuchi. 2004. Three dimensional visualization and comparison of impressions on fired bullets. *Forensic Sci. Int.* 140(2–3):233–240.

Bell, S. 2012. *A Dictionary of Forensic Science*. Oxford, UK: Oxford University Press.

Berendes, A., D. Neimke, R. Schumacher, and M. Barth. 2006. A versatile technique for the investigation of gunshot residue patterns on fabrics and other surfaces: m-XRF. *J. Forensic Sci.* 51(5):1085–1090.

Berryman, H.E., A.K. Kutyla, and J.R. Davis, II. 2010. Detection of gunshot primer residue on bone in an experimental setting—An unexpected finding. *J. Forensic Sci.* 55(2):488–491. doi:10.1111/j.1556-4029.2009.01264.x.

Bolton-King, R.S. 2016. Preventing miscarriages of justice: A review of forensic firearm identification. *Sci. Justice* 56(2):129–142.

Bueno, J., V. Sikirzhytski, and I.K. Lednev. 2013. Attenuated total reflectance-FT-IR spectroscopy for gunshot residue analysis: Potential for ammunition determination. *Anal. Chem.* 85(15):7287–7294.

Cooper, R., J.M. Guileyardo, I.C. Stone, V. Hall, and L. Fletcher. 1994. Primer residues deposited by handguns. *Am. J. Forensic Med. Pathol.* 15(4):325–327.

Fowler, W., A. North, C. Stronge, and P. Sweeney. 2016. *The Illustrated World Encyclopedia of Guns: Pistols, Rifles, Revolvers, Machine and Submachine Guns Through History in 1100 Clear Photographs*. Dayton, OH: Lorenz Books.

French, J. and R. Morgan. 2015. An experimental investigation of the indirect transfer and deposition of gunshot residue: Further studies carried out with SEM-EDX analysis. *Forensic Sci. Int.* 247:14–17.

Goddard, C.H. 1980. A history of firearms identification to 1930. *Am. J. Forensic Med. Pathol.* 1(2):155–168.

Greely, D. and E. Weber. 2017. Transfer and distribution of gunshot residue through glass windows. *J. Forensic Sci.* 62(4):869–873.

Haag, L.C. 2013. The forensic aspects of contemporary disintegrating rifle bullets. *Am. J. Forensic Med. Pathol.* 34(1):50–55.

Heard, B. 2008. *Handbook of Firearms and Ballistics: Examining and Interpreting Forensic Evidence*, 2nd edn. Hoboken, NJ: Wiley.

Hinrichs, R., P.R.O. Frank, and M.A.Z. Vasconcellos. 2017. Short range shooting distance estimation using variable pressure SEM images of the surroundings of bullet holes in textiles. *Forensic Sci. Int.* 272:28–36.

James, S.H., J.J. Nordby, and S. Bell. 2014. *Forensic Science: An Introduction to Scientific and Investigative Techniques*, 4th edn. Boca Raton: CRC Press.

Kersh, K.L., J.M. Childers, D. Justice, and G. Karim. 2014. Detection of gunshot residue on dark-colored clothing prior to chemical analysis. *J. Forensic Sci.* 59(3):754–762. doi:10.1111/1556-4029.12409.

Kilty, J.W. 1975. Activity after shooting and its effect on the retention of primer residue. *J. Forensic Sci.* 20(2):219–230.

Leifer, A., Y. Avissar, S. Berger, H. Wax, Y. Donchin, and J. Almog. 2001. Detection of firearm imprints on the hands of suspects: Effectiveness of PDT reaction. *J. Forensic Sci.* 46(6):1442–1446.

Li, D. 2006. Ballistics projectile image analysis for firearm identification. *IEEE Trans Image Process* 15(10):2857–2865.

Lindström, A.C., J. Hoogewerff, J. Athens, Z. Obertova, W. Duncan, N. Waddell, and J. Kieser. 2015. Gunshot residue preservation in seawater. *Forensic Sci. Int.* 253:103–111.

Ma, L., J. Song, E. Whitenton, A. Zheng, T. Vorburger, and J. Zhou. 2004. NIST bullet signature measurement system for RM (Reference Material) 8240 standard bullets. *J. Forensic Sci.* 49(4):649–659.

Md Ghani, N.A., C.Y. Liong, and A.A. Jemain. 2010. Analysis of geometric moments as features for firearm identification. *Forensic Sci. Int.* 198(1–3):143–149.

Morris, K.B., E.F. Law, R.L. Jefferys, E.C. Dearth, and E.B. Fabyanic. 2017. An evaluation of the discriminating power of an Integrated Ballistics Identification System® Heritage™ system with the NIST standard cartridge case (Standard Reference Material 2461). *Forensic Sci. Int.* 280:188–193.

Nichols, R.G. 2003. Firearm and toolmark identification criteria: A review of the literature, part II. *J. Forensic Sci.* 48(2):318–327.

Nichols, R.G. 2007. Defending the scientific foundations of the firearms and tool mark identification discipline: Responding to recent challenges. *J. Forensic Sci.* 52(3):586–594.

Saferstein, R. 2015. *Criminalistics: An Introduction to Forensic Science*, 11th edn. Boston: Pearson.

Schehl, S.A. 2000. Firearms and toolmarks in the FBI laboratory. *Forensic Science Communications* 2. www.fbi.gov/about-us/lab/forensic-science-communications/fsc/april2000/schehl1.htm (accessed December 12, 2017).

Shen, Y., Q. Zhang, X. Qian, and Y. Yang. 2015. Practical assay for nitrite and nitrosothiol as an alternative to the griess assay or the 2,3-diaminonaphthalene assay. *Anal. Chem.* 87(2):1274–1280.

Song, J., E. Whitenton, D. Kelley, R. Clary, L. Ma, S. Ballou, and M. Ols. 2004. SRM 2460/2461 standard bullets and casings project. *J. Res. Natl. Inst. Stand. Technol.* 109(6):533–542.

Song, J., T.V. Vorburger, S. Ballou, R.M. Thompson, J. Yen, T.B. Renegar, A. Zheng, R.M. Silver, and M. Ols. 2012. The National Ballistics Imaging Comparison (NBIC) project. *Forensic Sci. Int.* 216(1–3):168–182.

Tagliaro, F., F. Bortolotti, G. Manetto, V.L. Pascali, and M. Marigo. 2002. Dermal nitrate: An old marker of firearm discharge revisited with capillary electrophoresis. *Electrophoresis* 23(2):278–282.

Taudte, R.V., A. Beavis, L. Blanes, N. Cole, P. Doble, and C. Roux. 2014. Detection of gunshot residues using mass spectrometry. *Biomed. Res. Int.* 2014:1–16.

Thomas, F. 1967. Comments on the discovery of striation matching and on early contributions to forensic firearms identification. *J. Forensic Sci.* 12(1):1–7.

Tong, M., J. Song, and W. Chu. 2015. An improved algorithm of congruent matching cells (CMC) method for firearm evidence identifications. *J. Res. Natl. Inst. Stand. Technol.* 120:102–112.

Vorburger, T.V., J. Yen, J.F. Song, R.M. Thompson, T.B. Renegar, A. Zheng, M. Tong, and M. Ols. 2015. The second National Ballistics Imaging Comparison (NBIC-2). *J. Res. Natl. Inst. Stand. Technol.* 119:644–673.

Warlow, T. 2011. *Firearms, the Law, and Forensic Ballistics*, 3rd edn. Boca Raton: CRC Press.

Weber, I.T., A.J. Melo, M.A. Lucena, E.F. Consoli, M.O. Rodrigues, G.F. de Sá, A.O. Maldaner, M. Talhavini, and S. Alves, Jr. 2014. Use of luminescent gunshot residues markers in forensic context. *Forensic Sci. Int.* 244:276–284.

Zeichner, A., B. Eldar, B. Glattstein, A. Koffman, T. Tamiri, and D. Muller. 2003. Vacuum collection of gunpowder residues from clothing worn by shooting suspects, and their analysis by GC/TEA, IMS, and GC/MS. *J. Forensic Sci.* 48(5):961–972.

CHAPTER 14

Fire, arson, and explosives

KEY WORDS: fire, ignition temperature, activation energy, oxidized, reduced, heat, heat of combustion, conduction, convection, radiation, direct flame contact, flash point, pyrolysis, ignitable liquid, fuel, temperature, flame point, flammable, combustible, spontaneous combustion, glowing combustion, accident, incipient stage, growth stage, rollover, flashover, smoldering, backdraft, accelerant, origin, SPME, arson, explosives, low explosive, high explosive, incendiary weapon, IED, flammable material, dynamite, Molotov cocktail, ion mobility spectrometer

LEARNING OBJECTIVES

- To list the requirements to initiate and sustain a fire
- To define the terms ignition temperature, flash point, flame point, flammable, combustible, pyrolysis, and spontaneous combustion
- To list the three stages of a fire
- To define arson and list some potential evidence of arson in a case
- To explain how solid phase microextraction and gas chromatography–mass spectrometry can be used to detect and identify ignitable liquids
- To explain the difference between a low and high explosive and blast fragments and the effects each will produce
- To give examples of low and high explosives and how they can be combined in weapons
- To explain how the ion mobility spectrometer works and how it can be used in investigations of explosions

1995 BOMBING OF THE ALFRED P. MURRAH FEDERAL BUILDING IN OKLAHOMA CITY

At 9:02 a.m. on April 19, 1995, a Ryder rental truck was packed with explosives and parked in a drop-off zone on the north side of the Alfred P. Murrah Federal Building located in downtown Oklahoma City, Oklahoma, where it exploded. The blast could be felt up to 55 miles away; seismometers registered a 3.0 on the Richter scale. A 30-foot wide and 8-foot deep crater formed at the blast site. One-third of the Murrah building was destroyed. The blast damaged or destroyed 324 other buildings and shattered glass in 258 buildings in a 16-block radius. In addition, 86 cars burned or were destroyed in the vicinity of the site. At the time of the explosion, 646 people were estimated to have been in the building including 21 children at the America's Kids day care center located on the second floor of the building. Fifteen of the children in attendance at the day care center that morning were killed, as were three of their teachers; the others suffered severe burns and broken bones and skulls. In total, 168 people were killed and 680 others were injured in the attack.

The nine-story glass-front building was built in 1977 and was named for attorney and US Federal Judge Alfred Paul Murrah. The building housed US government agencies including the Bureau of Alcohol, Tobacco, Firearms and Explosives (ATF), the Drug Enforcement Agency (DEA), the Social Security Administration, and recruiting offices for the US Army and Marine Corps. The truck was situated directly beneath the government building's day care center.

Later that day, Timothy McVeigh, a former US Army soldier, was stopped for driving without a license plate in Noble County, Oklahoma, only 90 minutes from the Murrah building and was arrested for having a concealed weapon. Numerous leads implicated him in the bombing case. Six days later, Terry Nichols learned he was being sought by authorities and turned himself in. Both men were found to have been associated with the extreme right-wing Patriot Movement.

The bomb was found to have contained over 4800 pounds of ammonium nitrate (fertilizer) (AN), nitromethane (NM), metal cylinders of acetylene, and diesel fuel. It was ignited remotely using a two-fuse ignition system that ignited blasting caps which would detonate stolen Tovex Blastrite Gel "sausages"—primary explosives stolen from a rock quarry—which would, in turn, detonate the ANNM mixture. The explosives were nailed to the floor boards of the truck and were balanced so that it could be driven.

In the investigation of the case, it was determined that three of McVeigh's friends and family members served as accomplices. Jennifer McVeigh illegally mailed bullets to her brother. Michael and Lori Fortier knew of the attack—Michael, also a former US Army soldier, had helped McVeigh to scout the target building and Lori helped McVeigh make and laminate a fake license he used in the attack to rent the Ryder truck.

The Oklahoma City Bombing remains as the United States' worst case of domestic terrorism. It was the largest criminal case in US history. The FBI led an investigative team of over 900 local, state, and federal law enforcement and military personnel. Investigators amassed 3.5 tons of evidence, conducted 28,000 interviews, and collected almost 1 billion pieces of information.

Timothy McVeigh and Terry Nichols were indicted on murder and conspiracy charges on August 11, 1995. McVeigh was tried in Denver, Colorado, and was convicted of the bombing using a weapon of mass destruction, conspiracy, and 11 counts of murder on June 2, 1997; Nichols was tried in McAlester, Oklahoma, and was convicted of conspiracy and eight counts of involuntary manslaughter on December 23, 1997. McVeigh was sentenced to death; he was executed by lethal injection on June 11, 2001. In 2004, in Oklahoma State Court, Nichols was tried and found guilty of 161 counts of murder and sentenced to 161 consecutive life terms in prison without parole. Michael and Lori Fortier and Jennifer McVeigh testified against the perpetrators. Both women received immunity for their testimony and Michael received a reduced sentence. Michael Fortier was sentenced to 12 years in prison and fined $75,000 for failing to warn the authorities of the attack.

BIBLIOGRAPHY

Bragg, R. 1995. Terror in Oklahoma: The children: Tender memories of day-care center are all that remain after the bomb. *The New York Times.* www.nytimes.com/1995/05/03/us/terror-oklahoma-children-tender-memories-day-care-center-are-all-that-remain.html (accessed February 1, 2018).

Oklahoma City Bombing. www.history.com/topics/oklahoma-city-bombing (accessed February 1, 2018).

Oklahoma City Bombing. https://en.wikipedia.org/wiki/Oklahoma_City_bombing (accessed February 1, 2018).

FIRE

Fire is the rapid oxidation of matter in an exothermic combustion decomposition reaction resulting in gases, heat, light (flame), and other reaction products. Fires also leave unreacted fragments including soot, carbon black, and particulate matter. For a fire to occur, there must be an ignition source or sufficient heat, fuel, and oxygen. These are depicted in the fire triangle shown in Figure 14.1. The ignition source ignites the gaseous fuel, which reacts with the oxygen causing the fuel to oxidize and decompose and heat and light to be generated. Piloted ignition sources include open flames, sparks, matches, and lighters. The matter, or fuel, must be in the gas phase and at its ignition temperature. The *ignition temperature* is the minimum temperature that must be attained to ignite the fuel vapors. This is the temperature needed to overcome the *activation energy* of the reaction. The heat and sparks produced once the fire is active continue to ignite other fuel molecules and sustain the chain reaction. Finally, molecular oxygen, or another oxidizer such as a perchlorate, nitrate, or peroxide, is a required reactant. The concentration of atmospheric oxygen is 20.8% and 15%–16% is required to sustain a flaming combustion reaction.

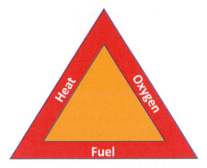

Figure 14.1 Fire triangle.

As an example, butane fuel is initially held together by covalent bonds and is *oxidized*, or loses electrons, in a combustion reaction. The charge on the O atoms in molecular oxygen is zero. The oxygen is *reduced*, meaning the atoms gain electrons, to 2-charge in carbon dioxide and water. Hydrogen, oxidized to H^+, and carbon, oxidized to C^{4+}, form new ionic bonds with the oxygen in new charge-balanced molecules. The balanced combustion reaction for butane with molecular oxygen to form carbon dioxide, water, light, and heat is shown in Equation 14.1.

$$2C_4H_{10} + 13O_2 \rightarrow 8CO_2 + 10H_2O + \text{light} + \text{heat} \tag{14.1}$$

The heat generated when the fuel burns is called the *heat of combustion*. *Heat* is defined as the energy needed to maintain or change the temperature of the fuel or other matter. In a fire, when heat is transferred to a solid, liquid, or gas, the temperature of the object or fuel increases or is maintained. The chemical reaction is sustained by heating more material to the gas phase to burn. Heat can be transferred by conduction, convection, radiation, or direct flame contact. *Conduction* is the transfer of heat by direct contact with the material such as a metal fire poke in the heat from a fire. Heat is transferred to the fuel by conduction such as a pot in contact with a gas flame on a stove. *Convection* occurs when the density of liquids and gases is reduced causing them to expand and rise on heat transfer. Infrared *radiation* causes the radiative transfer of heat until the vaporized or pyrolyzed fuel is ignited by the flames. Finally, in *direct flame contact*, the hot gases and flames from the reaction come into contact with the fuel.

For a liquid, the temperature needed for vaporization is the *flash point*, while for a solid, the phase change from a solid to a gas by heat alone is *pyrolysis*. Table 14.1 shows the flash point and autoignition temperature for several ignitable liquids. For pyrolysis, as little as 8% oxygen is needed and smoldering and charring is observed. In contrast, forced pyrolysis results when solid fuel comes in contact with an external heat source; no smoldering or flames are

Table 14.1 Flash point and autoignition temperature for selected ignitable liquids

Fuel	Flash point (°C)	Autoignition temperature (°C)
Acetone	−17.78	465
Acetylene	−18.15	300
Butane	−60	405
Carbon disulfide	−30	102
Ethyl alcohol (70%)	16.6	365
Isopropyl alcohol	12–13	399
Methyl alcohol	11–12	470
Methyl ethyl ketone	−9	505
Propane	−104	470
Toluene	4	535
Xylene	25	530

Source: PubChem.

produced. In forced pyrolysis, wood or other solid material degrades, gasifies, and releases vapors in a non-oxidative, endothermic process. In an endothermic process, heat is transferred from the environment to an object.

The heat of combustion is the energy released as heat when an ignitable liquid, fuel, or material undergoes complete combustion with oxygen. Materials vary widely in their heats of combustion as shown in Table 14.2. The rate of the reaction, or flame, will be faster as the surface area (e.g., wood shavings as compared to a wood log), temperature, and pressure is increased. For most reactions, a 10°C increase in temperature increases the reaction rate by two to three times. Reducing the volume will also increase the rate of the reaction. The surface area will also affect the ignition temperature or amount of energy required for ignition. Low-density materials, such as pine, allow the energy to remain at the surface while high-density materials, such as oak, draw energy away from the surface by acting as insulators. This also makes high-density materials more difficult to ignite as the energy is drawn away from the ignition source.

Fires are grouped into classes by fuel types. *Class A* fires are fueled by wood, paper, cloth, plastics, and rubber. *Class B* fires are fueled by flammable and combustible liquids, greases, and gases while energized electrical equipment fuels *Class C* fires. Combustible metals such as magnesium, sodium, potassium, titanium, and zirconium fuel *Class D* fires.

Referring back to the example with butane, the heat of combustion, or enthalpy of combustion (ΔH), for the combustion of butane is −2877.5 kJ/mol. So, for 1 mole of butane (58.1 g), −2877.5 kJ of energy would be released. To evaluate the effect of heat produced, we can calculate the final temperature of water after it is heated by the heat of the reaction using calorimetry ($q = mC_p\Delta T$). *Temperature* is defined as the molecular motion of a material as compared to a reference point such as 0°C, the freezing point of water. If 1000 g (1 L) of water (4.184 J/g°C) was originally at 0°C, its final temperature would be 687.7°C. This kind of heat drives the reaction by propagating the phase change of the fuel, such as wood, from the solid to gas phase.

The *flame point*, a temperature a few degrees above the flash point, is the temperature necessary to produce sufficient vapors and thus sustain the flame. The fire tetrahedron includes the features of the fire triangle, heat, oxygen, and fuel, and adds this fourth variable—an uninhibited chain reaction. Fire suppression is based on controlling or removing one of these components.

Materials are said to be *flammable* if they are capable of burning with a flame and have a flashpoint of less than 37.8°C (100°F). Gasoline is flammable: the flashpoint is only −42.8°C (−45°F). (The ignition temperature is always higher than the flashpoint; for low octane gasoline it is 280°C [536°F].) Its flammable range is 1.4%–7.6% in concentration at standard temperature and pressure. Propane is flammable from 2.15% to 9.6% concentration. Acetylene and hydrogen's flammable range is 2.5%–100% and 4%–75% respectively, with the highest percentage capable of burning at a high temperature. Conversely, materials are termed *combustible* if they are liquids with a flash point above 37.8°C (100°F) and burn under ambient temperature and pressure. *Spontaneous combustion*, a rare event, occurs in poorly ventilated containers or areas as a result of a natural heat-producing process when appropriate proportions of fuel and oxygen are present. Due to bacterial activity in a hot storage area, hay may spontaneously combust in barns. Linseed oil or other highly unsaturated oils (but not animal fats and household oils or lubricants) stored with air in

Table 14.2   Heats of combustion for various fuels

Fuel	Heat of combustion (kJ/mol)
Acetone	−1789
Acetylene	−1310
Butane	−2878
Carbon disulfide	−1687
Ethyl alcohol (70%)	−1337
Isopropyl alcohol	−2005
Methyl alcohol	−726.1
Methyl ethyl ketone	−2444
Propane	−2220
Toluene	−3910
Xylene	−4335

Source: PubChem.

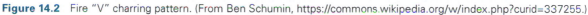

Figure 14.2   Fire "V" charring pattern. (From Ben Schumin, https://commons.wikipedia.org/w/index.php?curid=337255.)

tightly sealed containers are also susceptible to spontaneous combustion. In contrast, *glowing combustion*, or smoldering, is burning at the fuel–air interface or surface (e.g., cigarette, coals) without producing a flame. In glowing combustions, there is not sufficient heat to pyrolize the fuel.

Fires may burn and consume cars, businesses, barns, and homes. Fires may be caused by *accidents* including faulty wiring, cigarette smoking, improperly cleaned heating systems, dryer lint, and overheated electrical motors. The evidence associated with these events will not contain traces of purposefully placed chemicals or ignitable liquids.

There are three stages of fire as shown in Figure 14.3. The first stage is the *incipient*, or growth, stage. It begins at ignition in which the fuel reaches the flash point. Initially, the fire is localized to the ignition source, fuel configuration (mass and geometry), and immediate surroundings. On ignition, hot gases including carbon monoxide and sulfur dioxide rise and fill the room, convection draws the oxygen to the bottom of the flames, and, if there is solid fuel above the flame, both convection and direct contact methods will transfer heat. The second stage is the *free-burning*, growth or development, stage. In this stage, more fuel and oxygen are consumed, the flames continue to spread up and outward from the ignition source. In some fires a characteristic "V"-shaped pattern may be formed by convection, conduction, and direct flame methods. The "V" charring pattern is observed in many fires, as most fires move upward, and the point of the "V" indicates the origin and ignition point (Figure 14.2, arson). If present, the point in the "V" should be searched for the ignition source and any indication that arson may be involved. The oxygen concentration in the air is reduced to approximately 16%. The hot gases including carbon monoxide, hydrochloric acid, and hydrogen cyanide and smoke that collect at the ceiling or high points of the room or structure begin to radiate downward and spread throughout the room and beyond, unless the room is tightly sealed. The high area at the ceiling will be hot and the floor or low area will be cool. Secondary fires may break out. *Rollover* occurs when particulate

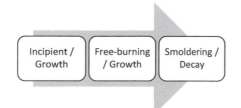

Figure 14.3   Developmental stages of a fire.

matter and fuel at the upper layer or level of the room ignite into flames and spread the fire across the ceiling level. When the temperature of the upper level reaches 1100°F, sufficient heat is generated to simultaneously ignite all of the remaining fuels in the room in an event called a *flashover*, leading to temperatures of 2000°F at the ceiling level and over 1000°F at the floor. Such a rapid temperature rise and increase in pressure results in the breaking of window glass, if present, the intense burning of all remaining fuel, and even the burning of the undersides of objects if oxygen is not limited. The third, and final, stage of fire is the *smoldering*, or decay, stage in which the final oxygen concentration is approximately 13%. This stage occurs when the oxygen concentration drops below the 15%–16% necessary to sustain the fire and results in significant smoldering and charring. More toxic compounds are released in this stage than in the previous stages with a flaming combustion. New hazards can present themselves if a new source of oxygen is introduced or enters the area. The oxygen can cause the accumulated soot, particulate matter, and gases to ignite and can cause an explosion in a *backdraft*. A backdraft can lead to significant structural damage. The stages of fire development are shown in Figure 14.3.

ARSON AND ACCELERANTS

For a fire or burn act to be considered *arson*, there must be criminal intent and burning with damage to property. Law enforcement must prove that the fire occurred as a result of arson and not by accident. Some indications of arson include an isolated fire or smoke in one part of a structure, "streamers" such as gasoline or paper to spread the fire from one room or area to another and connect fires, an abnormally bright fire, a colored fire, white smoke, and the presence of out-of-place ignitable liquids such as *accelerants* and/or their containers and/or severe burning on the floor where such a liquid was poured (Figure 14.4). Accelerants include inexpensive and easily obtained ignitable liquids including hydrocarbon fuels, oils, and cleaning agents. Some examples of accelerants are gasoline, kerosene, diesel fuel, methyl ethyl ketone, acetone, ethanol, motor oil, mineral spirits, turpentine, propane, and butane. Other indicators of arson include the use of a timer or sensor time delay device and evidence of breaking and entering or theft. These indicators of arson should be documented and collected as evidence for processing a potential arson case.

A perpetrator may have started or conspired to start the fire, aided or abetted another perpetrator, or been involved in another way. It may be difficult to determine the motive, *modus operandi*, and who started the fire and locate the suspect. The destruction may be extensive and the act may have been planned. The perpetrator may or may not be present to watch the burning, although the arsonist may stay to watch the fire burn.

Figure 14.4   Ignitable liquids: methyl ethyl ketone, charcoal lighter fluid, and turpentine.

The investigation of suspected arson must begin immediately after the fire has been extinguished and it is safe to enter the area. There is no requirement to obtain a warrant as the evidence will be lost without immediate investigation (as ruled by the US Supreme Court). The investigation should include photographs, sketches, and notes. Identification of the *origin* of the fire, ignition device(s), accelerants, and other evidence must be collected prior to clean-up as accelerants will evaporate quickly. Ignition devices may include a match, cigarette, firearms, electrical sparking devices, or even a "Molotov cocktail" consisting of a flammable liquid in a bottle with a wick that can be lit.

Fortunately for forensic scientists and law enforcement, accelerants are rarely entirely consumed in a fire. Traces of accelerant chemicals on burned items such as ash, soot, burned upholstery, wood, carpet, or other materials collected into evidence can be identified using chemical methods. A portable vapor "sniffer" instrument can be used to detect the presence of accelerants and determine what materials to collect. The instrument contains a heating filament that oxidizes the accelerant. This increases the temperature of the filament and leads to detection. A digital output reports the rise in temperature. It is not selective and cannot differentiate between which accelerant was used.

In addition to the burned materials listed previously, soil and vegetation from outdoor burned areas should also be sampled. The clothing from the suspect should also be collected, if possible. Burned or charred material should be collected in new, quart-sized or larger clean paint cans. Samples can be scooped up or cut to fit the container. The containers should be filled half to two-thirds full and covered with a tightly fitted cap. Alternatively, another air-tight but non-petrochemical (i.e., plastic) container can be used. Upholstery, rags, plaster, wallboards, carpet, and other porous surface reference samples should be collected as well as the burned samples to test for the presence of unburned accelerants. Fluids found in open bottles should be collected and sealed. Evidence samples should be refrigerated or stored under cool conditions to reduce evaporation.

The paint can lids can be punctured with a small hole to sample the headspace for the presence of accelerant. Solid phase microextraction (SPME) fibers can be used to concentrate and sample accelerants. The filter is a charcoal-coated strip that can be inserted through the hole in the lid, sealed, and left to adsorb the accelerant; alternatively, the paint can can be actively heated gently on a hot plate or heating mantle to promote more accelerant transitioning to the gas phase. The accelerant adsorbed to the fiber can be removed by washing with a solvent.

Gas chromatography–mass spectrometry, gas chromatography-flame ionization detector, and gas chromatography-infrared spectroscopy can be used to detect and identify partially burned and unburned accelerants such as those shown in Table 14.3. The American Society for Testing and Materials (ASTM) method E1618-01 is focused on the analysis of ignitable liquids. Hydrocarbon fuels are classified into three classes—light, medium, and heavy—based on their carbon chain length; examples are given in Table 14.4 and three ignitable liquids

Table 14.3 Boiling point and molecular mass of common ignitable liquid accelerants or major components

Accelerant	Molecular mass (g/mol)	Boiling point (°C)
Acetone	58.08	56
Acetylene	26.038	−84
Butane	58.124	−1
Carbon disulfide	76.131	46
Diesel (1-dodecene, $C_{12}H_{24}$)	168.324	214
Ethyl alcohol	46.069	79
Gasoline (octane, C_8H_{18})	114.232	126
Isopropyl alcohol	60.096	83
Kerosene (dodecane, $C_{12}H_{26}$)	170.34	216
Methyl alcohol	32.042	65
Methyl ethyl ketone	72.107	80
Propane	44.097	−42
Toluene	92.141	111
Turpentine (dipentene, $C_{10}H_{16}$)	136.238	177
Xylene	106.168	144

Source: PubChem, https://pubchem.ncbi.nlm.nih.gov/compound/.

Table 14.4   Ignitable liquids in the light, medium, and heavy classes

Light (C_4–C_9)	Medium (C_8–C_{13})	Heavy (C_8–$C_{36}+$)
Methane	Gasoline	Kerosene
Propane	Turpentine	Diesel fuel
Butanes	Lighter fluid	Motor oil
Pentanes		Many insecticides
Hexanes		Many lamp oils
Heptanes		
Toluene		
Alcohols including methanol and ethanol		
Ketones including acetone and methyl ethyl ketone		

Source: ASTM E1618-01, www.astm.org/cgi-bin/resolver.cgi?E1618-14.

including methyl ethyl ketone, charcoal lighter fluid, and turpentine purchased from a paint and home goods supply store are shown in Figure 14.4. In general, hydrocarbons with a lower molecular mass will elute earlier than those with a higher molecular mass (i.e., butane < octane < dodecane), although accelerants containing functional groups such as alcohols e.g., (ethanol) and ketones (e.g., acetone) will have a higher boiling point due to their polarity and capacity for forming dipole–dipole interactions. Figure 14.5 shows the gas chromatogram for 1-butanol, a pure substance alcohol, and Figure 14.6 shows the gas chromatogram for unburned lighter fluid, a complex mixture. Figure 14.7 shows the gas chromatogram for unburned gasoline, a medium ignitable liquid and complex mixture of hydrocarbons. For comparison, the chromatogram for turpentine, also a medium-class flammable liquid, is shown in Figure 14.8. The chromatogram for kerosine, a heavy-class flammable liquid, is shown in Figure 14.9. The chromatogram for lamp oil, a heavy-class flammable liquid, is shown in Figure 14.10. Note that the heavy ignitable liquids take longer to elute than those that contain liquids in the light and medium classes.

The mass spectra for ignitable liquids will have peaks for the major ions present for different compound types composing the accelerant. For example, major ions produced from alkanes include 43, 57, 71, and 85 atomic mass units (amu).

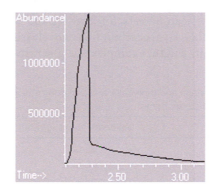

Figure 14.5   Gas chromatogram of 1-butanol. The sample was run on an Agilent 7890A and 0.2 μL of a 1 μL/mL solution prepared in methanol was injected at 280°C and cooled to 50°C for 2 minutes prior to separation with a 50°C–280°C temperature ramp at a rate of 60°C/min. The final temperature was held for 4 minutes; the run time was 9.83 min.

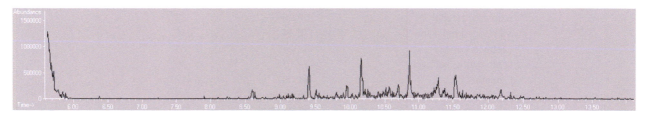

Figure 14.6   Gas chromatogram of unburned lighter fluid.

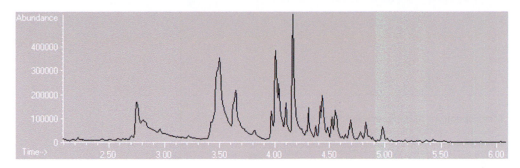
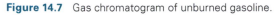

Figure 14.7   Gas chromatogram of unburned gasoline.

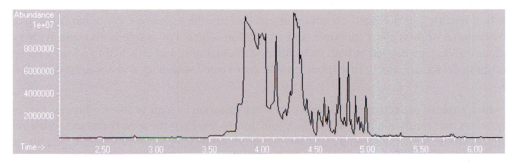

Figure 14.8   Gas chromatogram of unburned turpentine.

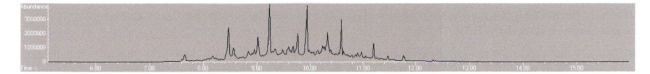

Figure 14.9   Gas chromatogram of unburned kerosene.

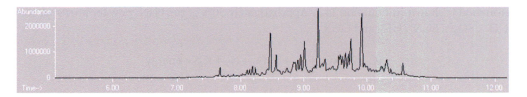

Figure 14.10   Gas chromatogram of unburned lamp oil.

The mass spectrum for 1-butanol ($C_4H_{10}O$) shows the qualifying peaks of 31 and 45 amu characteristic of alcohols are present as well as the molecular ion peak of 74 amu. Ketones such as methyl ethyl ketone contain the major ions 43, 58, 72, 86 amu; the molecular ion peak is 72 amu. The mass spectrum for dodecane from kerosene is shown in Figure 14.11. It has the characteristic peaks of alkanes including 43, 57, 71, and 85 amu (Table 14.5).

In a burned sample or if the accelerants have been weathered, the shorter chain hydrocarbons will burn off or evaporate more easily than the longer chain hydrocarbons. In a gas chromatogram, the earliest eluting components may not be present or their percentage may be very small. A test burn may be conducted to analyze if a particular accelerant found at a scene of suspected arson was used to start the fire using unburned materials recovered from the scene. The test burn sample can be collected and analyzed using the same parameters as the evidence sample for direct comparison. This is an ongoing area of research.

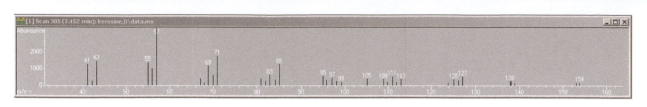

Figure 14.11 Mass spectrum of dodecane found in kerosene.

EXPLOSIVES

Explosives are substances that undergo a rapid oxidation reaction with the production of large quantities of gases. It is this sudden buildup of gas pressure that results in an explosion. An explosion is defined as a chemical reaction or mechanical action resulting in the rapid expansion of gases within the explosives' container. An explosive consists of an explosive material or mixture, a container, and an ignition source. For example, an explosive may consist of a container such as a PVC pipe capped at both ends, filled with dynamite, and attached to a lead wire ignited with a spark from a timing device or a safety fuse (black powder core coated with fiber and a waterproofing outer layer) lit with a match.

The speed at which explosives decompose underlies their classification as high or low explosives. *Low explosives* are defined as explosives that result in a blast of less than 1000 m/s or less than the speed of sound. With low explosives, there is deflagration (burning), a very rapid oxidation that produces heat, light, and a subsonic pressure wave, but not detonation. *High explosives* result in a blast detonation or chemical explosion of more than 1000 m/s velocity (to 8500 m/s) with a smashing or shattering effect.

Low explosives include black powder and smokeless powder (e.g., nitrated cotton or nitrocellulose single-base powder or nitroglycerin mixed with nitrocellulose double-base powder), chlorate mixtures (potassium chlorate and sugar or carbon, sulfur, starch, phosphorous, magnesium filings, and initiated by sulfuric acid), and gas–air mixtures using natural gas (such as methane 5.3%–13.9%).

While fuels that power a fire may be hydrocarbons without their own oxygen, explosives characteristically contain oxygen atoms in their structures, especially in the form of reactive nitrates, chlorates, perchlorates, and peroxides. The oxygen atoms serve as the final electron receptor. (Other atoms may also be used as the electron receptor, although oxygen is very common.) For example, black powder contains 75% potassium nitrate, 15% charcoal, and 10% sulfur. Gunpowder contains 15 parts potassium nitrate, 3 parts charcoal, and 2 parts sulfur. The chemical reaction on its detonation is shown in Equation 14.2. Approximately 50% of the black powder burns, producing residue traces of K^+, NO_3^-, SO_4^{2-}, SCN^-, HS^-, and OCN^-.

$$3C + 2KNO_3 + S \rightarrow 3CO_2 + N_2 + K_2S + energy \tag{14.2}$$

Table 14.5 Major mass spectrometry ion peaks for different hydrocarbon compound cation types. More masses are included in Table 5.2

Possible hydrocarbon	Ion peak
CH_3^+	15
$C_2H_5^+$	29
CH_2OH+	31
$C_3H_7^+$, CH_3CO^+	43
$CH_3C(H)OH^+$	45
$C_4H_9^+$	57
$CH_3C(O)CH_3^+$	58
$C_5H_{11}^+$	71
$C_6H_{13}^+$, $C_4H_9CO^+$	85

Equation 14.3 shows the chemical reaction that occurs when nitroglycerin explosive is used.

$$4C_3H_5(NO_3)_3 \rightarrow 12CO_2 + 10H_2O + 6N_2 + O_2 + energy \tag{14.3}$$

The majority of commercial and military explosives are high explosives. Commercial explosives are manufactured products used for blasting rock for construction applications including preparations for roads, basements, and parking garages. High explosives consist of primary and secondary types. Primary explosives include mercury fulminate ($Hg(CNO)_2$), silver fulminate (AgCNO), lead azide ($Pb(N_3)_2$), lead styphnate ($PbC_6HN_3O_8$), diazonitrophenol ($C_6H_2N_4O_5$), and silver azide (AgN_3). Primary high explosives are ultrasensitive to shock, friction, and heat. These are used as a detonating charge. Secondary high explosives include nitroglycerin, ammonium nitrate/fuel oil (ANFO), ethylene glycol dinitrate (EGDN), trinitrotoluene (TNT), tetraethyl pyrophosphate (TEPP), trinitrophenylmethylnitramine (Tetryl), cyclotrimethylenetrinitramine (RDX), high melting-point explosive (HMX), and triacetone triperoxide (TATP). They are more stable to shock, heat, and friction than primary explosives.

The most famous commercial high explosive is dynamite, created by Alfred Nobel (namesake of the Nobel Prizes) in 1867 when he adsorbed nitroglycerin onto kieselguhr or diatomaceous earth to desensitize it. The resulting material, "pulp dynamite," was easier and less dangerous to handle and ship. Several chemical structures of explosives are shown in Figure 14.12.

INCENDIARY WEAPONS

Incendiary weapons are *improvised explosive devices* (IEDs) made using flammable material, a container, and an ignition source. *Flammable materials* include common household chemicals and those discussed previously including gasoline, kerosene, and diesel fuel, among others. They are produced using instructions from the internet or "underground publications," typically with inaccurate instructions and incomplete or missing safety information. A pipe bomb is a type of illegal improvised explosive device. Normally, when a pipe bomb consists of black powder, the pipe's seam will split and its face plates on the capped ends will be pushed out. Another IED is a Molotov cocktail consisting of a glass bottle containing a flammable liquid and a rag as a wick (Figure 14.13). Another type of simple IED is black powder rolled in paper with a string lead. Containers may include light bulbs, propane and butane cylinders, plastic pipes, bottles, and cans. Ignition sources can include matches, gas lighters, fireworks, roadway flares, and electrical

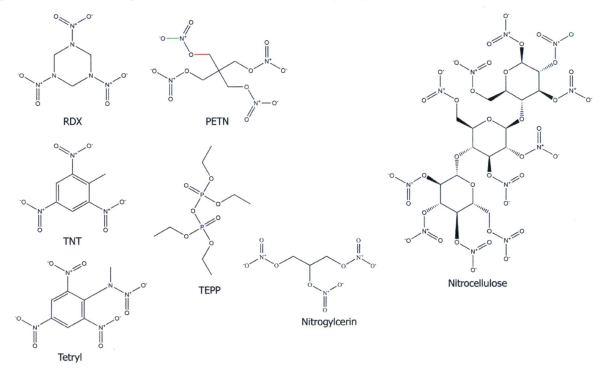

Figure 14.12 Chemical structures of explosives. (Structure prepared by Ashley Cowan.)

Figure 14.13 Molotov cocktail. (From 7th Army Training Command, https://www.flickr.com/photos/7armyjmtc/24716093859/in/photolist-vxBJa-4JaktA-88mSs4-NqroJg-9fiTKY-m1evdB-RVuVoT-vxBt5-6GEhHY-vxCsH-4CnLPR-9fiU6G-4A8SAG-c3zZo1-64jnYD-Jhgs6m-6GAfTn-MoB98g-6GEUrS-H1uufN-Dgb2Qa-DgAAmp-DE5y1v-Dgb4XM-Dgbc7X-E5KnPi-DfQz3b-E5jCPi-DE5rBi-DQ2JCu-DjqyrL-E5K276-Djqz3W-DHEtMV-E6H3ZJ-DjLc9V-DjLdwz-DjqDtu-DjLdTX-E5Jgnc-DQ2F71-E8UTnR-EcqHb5-zusyx5-Ac2Q4E-LTYvw6-PM8Rjx-PXmapk-9ffM9k-9fiSC5.)

components and devices. Ignition sources may be paired with timing devices, sensors (e.g., sound, magnetic, or pressure), a radio-operated remote control, or a cell phone. Figure 14.14 shows an IED found by Australian soldiers in the Mirabad Valley Region in Afghanistan.

IDENTIFICATION OF EXPLOSIVES

An ion mobility spectrometer may be used to search the crime scene for explosives residue. In an explosion, investigators must search the bomb seat or crater and extend outward in circles of larger diameter for fragments of wire, explosive, wrappers, duct tape, pipe, or other pieces. In a high-explosive detonation, fragments may need to be dug up from deep within the bomb seat. Color tests such as the Griess test may also be used in their presumptive identification. Chemical instrumentation methods including thin-layer chromatography, attenuated total reflectance Fourier transform–infrared spectroscopy (ATR FT-IR), and Raman spectroscopy are used to identify explosives. Figure 14.15 shows how ATR FT-IR spectroscopy is used to identify the explosives NH_4NO_3, KNO_3, $NaNO_3$, $KClO_4$, and $NaClO_4$.

Figure 14.14 Improvised explosive device. (From ResoluteSupportMedia, https://www.flickr.com/photos/isafmedia/4258450647/in/photolist-7uiF1T-8RQ2sD-bBnQxm-VZg87e-VMQu7x-cCVPhu-ebbQ61-n3aw6s-9811Jr-bmXZKy-48wAno-W6emuq-pfR8Ez-8d4d4y-8XHpFS-2xEV4U-8FDs53-W1Xuq9-bEBEhc-dyhvYu-jEqkLY-hjb7ek-fmu9tB-rvt575-bBnQtd-ifo3TK-8ZUzLm-6zdu3s-bk2CjW-VW1aj9-nyntiZ-6hehxq-VsXaNu-4Q7mGG-etdYVN-ifniAz-9jnjAp-8AUiPV-aikuZT-kwPBRr-goPYfJ-bJLvpX-ULrWKp-br7KrE-YDHJgc-aMMHqn-Zn8pDH-S4zPgy-huQBTY-VZdiLr.)

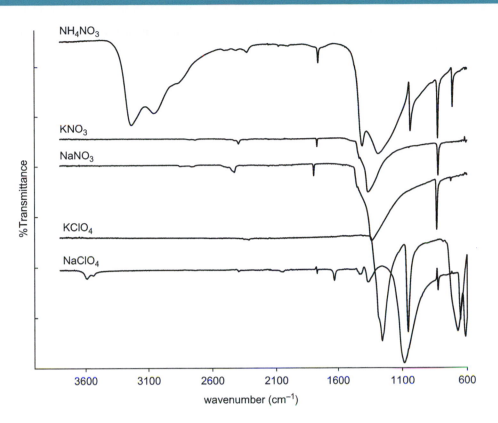

Figure 14.15   ATR FT-IR spectra of explosives NH_4NO_3, KNO_3, $NaNO_3$, $KClO_4$, and $NaClO_4$.

QUESTIONS

1. In order to start a fire, heat, fuel, and _____ are required.

   a. Wood   b. Lighter fluid

   c. Gasoline   d. Oxygen

2. To start a fire, the minimum temperature at which a liquid fuel produces enough vapor to burn, known as the _____, must be reached.

   a. Ignition temperature   b. Heat of combustion

   c. Pyrolysis temperature   d. Flame point   e. Flash point

3. What is the first phase in a fire's progression?

   a. Incipient   b. Free burning

   c. Emergent smoldering   d. Oxygen-regulated smoldering

4. Where is the fire's greatest damage?

   a. The lowest point   b. The point of origin

   c. The ceiling   d. The point at which the fire was extinguished

5. An analyst compared a mixture of methyl ethyl ketone, lighter fluid, and gasoline to an unknown accelerant from an arson scene. Which of the following techniques would be best for the comparison of the two samples?

   a. Ultraviolet-visible spectroscopy   b. Infrared spectroscopy

   c. Fluorescence spectroscopy   d. Gas chromatography

6. Explain how you would process a crime scene of suspected arson in terms of the evidence to collect and what to look for in processing. What can help you to determine if the crime was arson or not?

7. Briefly explain how you can use the GC spectrum to determine that a sample is an accelerant and the class/type of accelerant present.

8. What evidence would you seek to collect in an investigation of an intentional explosion?

9. What chemical technique(s) would be appropriate to use to determine the identity of an explosive?

10. What would you collect at a crime scene in which an improvised exploded device was detonated? What point of interest would you be sure to examine?

Bibliography

Akhavan, J. 2011. *The Chemistry of Explosives*, 3rd edn. Cambridge, UK: RSC Publishing.

ASTM International. 2014. Standard test method for ignitable liquid residues in extracts from fire debris samples by gas chromatography–mass spectrometry. *ASTM E1618–14*. www.astm.org/cgi-bin/resolver.cgi?E1618-14 (accessed December 3, 2017).

Bertsch, W. 1997. Analysis of accelerants in fire debris—Data interpretation. *Forensic Sci. Rev.* 9(1):1–22.

Bertsch, W. and Q. Ren. 1999. Contemporary sample preparation methods for the detection of ignitable liquids in suspect arson cases. *Forensic Sci. Rev.* 11(2):141–156.

Bors, D. and J. Goodpaster. 2017. Mapping smokeless powder residue on PVC pipe bomb fragments using total vaporization solid phase microextraction. *Forensic Sci. Int.* 276:71–76.

Borusiewicz, R., J. Zieba-Palus, and G. Zadora. 2006. The influence of the type of accelerant, type of burned material, time of burning and availability of air on the possibility of detection of accelerants traces. *Forensic Sci. Int.* 160(2–3):115–126.

Bruno, T.J. and S. Allen. 2013. Weathering patterns of ignitable liquids with the advanced distillation curve method. *J. Res. Natl. Inst. Stand. Technol.* 118:29–51.

Cacho, J.I., N. Campillo, M. Aliste, P. Viñas, and M. Hernández-Córdoba. 2014. Headspace sorptive extraction for the detection of combustion accelerants in fire debris. *Forensic Sci. Int.* 238:26–32.

Caddy, B., F.P. Smith, and J. Macy. 1991. Methods of fire debris preparation for detection of accelerants. *Forensic Sci. Rev.* 3(1):57–69.

Chemical Explosives. 1998. ES310 introduction to naval weapons engineering. http://fas.org/man/dod-101/navy/docs/es310/chemstry/chemstry.htm (accessed January 28, 2018).

Dhole, V.R., M.P. Kurhekar, and K.A. Ambade. 1995. Detection of petroleum accelerant residues on partially burnt objects in burning/arson offences. *Sci. Justice* 35(3): 217–221.

Fernández de la Ossa, M.Á., J.M. Amigo, and C. García-Ruiz. 2014. Detection of residues from explosive manipulation by near infrared hyperspectral imaging: A promising forensic tool. *Forensic Sci. Int.* 242:228–235.

Fisher, B.A.J. and D.R. Fisher. 2012. *Techniques of Crime Scene Investigation*, 8th edn. Boca Raton, FL: CRC Press.

Houck, M.M., ed. 2015. *Forensic Chemistry*. Cambridge, MA: Elsevier Academic Press.

Icove, D. and G.A. Haynes. 2017. *Kirk's Fire Investigation*, 8th edn. London, UK: Pearson.

International Association of Arson Investigators. 2015. *Fire Investigator: Principles and Practice to NFPA 921 and NFPA 1033*, 4th edn. Burlington, MA: Jones & Bartlett Learning.

Kaneko, T., H. Yoshida, and S. Suzuki. 2008. The determination by gas chromatography with atomic emission detection of total sulfur in fuels used as forensic evidence. *Forensic Sci. Int.* 177(2–3):112–119.

Laska, P.R. 2015. *Bombs, IEDs, and Explosives*. Boca Raton, FL: CRC Press.

Lennard, C.J., V. Tristan Rochaix, P. Margot, and K. Huber. 1995. GC–MS database of target compound chromatograms for the identification of arson accelerants. *Sci. Justice* 35(1): 19–30.

Li, F., J. Tice, B.D. Musselman, and A.B. Hall. 2016. A method for rapid sampling and characterization of smokeless powder using sorbent-coated wire mesh and direct analysis in real time–mass spectrometry (DART-MS). *Sci. Justice* 56(5):321–328.

Matyáš, R. and J. Pachman. 2013. *Primary Explosives.* New York, NY: Springer.

McCurdy, R.J., T. Atwell, and M.D. Cole. 2001. The use of vapour phase ultra-violet spectroscopy for the analysis of arson accelerants in fire scene debris. *Forensic Sci. Int.* 123(2–3):191–201.

PubChem. https://pubchem.ncbi.nlm.nih.gov/ (accessed January 31, 2018).

Roberts, K., M.J. Almond, and J.W. Bond. 2013. Using paint to investigate fires: An ATR-IR study of the degradation of paint samples upon heating. *J. Forensic Sci.* 58(2): 495–499.

Romolo, F.S., E. Ferri, M. Mirasoli, M. D'Elia, L. Ripani, G. Peluso, R. Risoluti, E. Maiolini, and S. Girotti. 2015. Field detection capability of immunochemical assays during criminal investigations involving the use of TNT. *Forensic Sci. Int.* 246:25–30.

Romolo, F.S., L. Cassioli, S. Grossi, G. Cinelli, and M.V. Russo. 2013. Surface-sampling and analysis of TATP by swabbing and gas chromatography/mass spectrometry. *Forensic Sci. Int.* 224(1–3):96–100.

Saferstein, R. 2015. *Criminalistics: An Introduction to Forensic Science*, 11th edn. Boston, MA: Pearson.

Sodeman, D.A. and S.J. Lillard. 2001. Who set the fire? Determination of arson accelerants by GC-MS in an instrumental methods course. *J. Chem. Educ.* 78(9):1228.

Steinfeld, J.I. and J. Wormhoudt. 1998. Explosives detection: A challenge for physical chemistry. *Annu. Rev. Phys. Chem.* 49:203–232.

Stuart, J.H., J.J. Nordby, and S. Bell. 2014. *Forensic Science: An Introduction to Scientific and Investigative Techniques*, 4th edn. London, UK: Taylor & Francis.

Touron, P., P. Malaquin, D. Gardebas, and J. Nicolai. 2000. Semi-automatic analysis of fire debris. *Forensic Sci. Int.* 110(1):7–18.

Wright, J.D. and J. Singer. 2008. *Fire and Explosives (Forensic Evidence).* Abingdon, UK: Routledge.

Yoshida, H., T. Kaneko, and S. Suzuki. 2008. A solid-phase microextraction method for the detection of ignitable liquids in fire debris. *J. Forensic Sci.* 53(3):668–676.

Zapata, F., M.Á.F. de la Ossa, E. Gilchrist, L. Barron, and C. García-Ruiz. 2016. Progressing the analysis of improvised explosive devices: Comparative study for trace detection of explosive residues in handprints by Raman spectroscopy and liquid chromatography. *Talanta* 161:219–227.

Zapata, F. and C. García-Ruiz. 2017. Analysis of different materials subjected to open-air explosions in search of explosive traces by Raman microscopy. *Forensic Sci. Int.* 275:57–64.

CHAPTER 15

Chemical, biological, radiological, nuclear, and explosives (CBRNE)

KEY WORDS: weapons of mass destruction, CBRNE, natural disaster, technical disaster, terrorism, bioterrorism, overt attack, covert attack, chemical weapons, blood agent, pulmonary agent, blister agent, nerve agent, nettle agent, incapacitating agent, vomiting agent, riot agent, biological weapons, bacteria, anthrax, Glanders, fungi, viruses, Ebola, smallpox, influenza, protein toxins, ricin, abrin, botulinum toxin, epsilon toxin, diphtheria toxin, *Staphylococcal* enterotoxin B, mycotoxins, nuclear weapons, nuclear chemistry, radioactive decay, ionizing radiation, radioactive particles, alpha particle, beta particle, deuteron, positron, gamma rays, radioisotope, radionuclide, half-life, isotope, belt of stability, nuclear fusion, nuclear fission, chain reaction, critical mass, enriched, depleted uranium, Geiger–Mueller counter, dirty bombs, radiation poisoning, nuclear power plants, atomic bomb, dual use research, designer weapons

LEARNING OBJECTIVES

- To define the term CBRNE
- To differentiate between a covert and overt attack
- To list the types of chemical, biological, radiological, and nuclear weapons
- To explain the biochemical effects of chemical, biological, radiological, and nuclear weapons
- To explain the use of chemical and instrumental methods in detecting and identifying chemical, biological, radiological, and nuclear weapons
- To describe the function and use of a Geiger–Mueller counter
- To differentiate between nuclear fusion and fission and explain the concept of an uninhibited chain reaction in nuclear chemistry
- To list the types of particles produced in nuclear chemistry reactions and describe the features and risks of each
- To describe the value of and threats from dual use research

WEAPONS OF MASS DESTRUCTION

Weapons of mass destruction (WMD) are those that, when used, have a relatively large-scale impact on people, property, food sources, and/or infrastructure.

According to 18 US Code § 2332a (Use of WMD), WMD are defined as

"(A) any destructive device as defined in section 921 of this title;

(B) any weapon that is designed or intended to cause death or serious bodily injury through the release, dissemination, or impact of toxic or poisonous chemicals, or their precursors;

(C)   any weapon involving a biological agent, toxin, or vector (as those terms are defined in section 178 of this title); or

(D)   any weapon that is designed to release radiation or radioactivity at a level dangerous to human life."

Destructive devices (18 US Code § 921) include explosives, incendiary devices, poison gas, bombs, grenades, mines, rockets powered by a propellant charge of more than four ounces, missiles powered by an explosive or incendiary charge of more than one-quarter ounce, or similar devices. *CBRNE* is an acronym that describes the types of WMD agents: chemical, biological, radiological, nuclear, and explosives.

Disasters caused by WMD agents must be differentiated from other types of hazards and disasters including natural disasters and technical disasters. *Natural disasters* include tornados, earthquakes, tsunamis, floods, storms, influenza, SARS, AIDS, and so on. *Technical disasters* are typically caused by human error and include nuclear power plant meltdowns, bridge collapses, accidental transfer of biological agents from secure stocks, and laboratory chemical explosions. Natural and technical disasters may cause limited or large numbers of casualties depending on the event, time of day, location, and emergency response. *Terrorism* is defined as activities that threaten the civilian population, critical infrastructure, property, and key resources in the pursuit of political aims or government coercion. *Bioterrorism* is a form of terrorism in which microorganisms, viruses, and naturally produced proteins and small molecule toxins are used to cause disease, debilitation, and/or death to humans as well as plants and animals that we rely on for food. Like terrorist attacks, natural and technical disasters (e.g., Chernobyl in 1986, Hurricane Katrina in 2005, Hurricane Maria in 2017) can lead to severe medical consequences, a lasting memory, and a loss of confidence.

The United States' national policy restricts its use of CBRNE agents. The United States will not strike first against (attacking) nations using chemical agents. Counter-use of chemical weapons on attacking nations or groups requires authorization from the President of the United States. The United States has pledged not to employ biological agents, including toxins, in warfare. In 1969, Richard Nixon announced at Fort Detrick in his Remarks Announcing Decisions on Chemical and Biological Defense Policies and Programs that "First, in the field of chemical warfare, I hereby reaffirm that the United States will never be the first country to use chemical weapons to kill. And I have also extended this renunciation to chemical weapons which incapacitate." And "Second, biological warfare, which is commonly called germ warfare—this has massive, unpredictable, and potentially uncontrollable consequences. It may produce global epidemics and profoundly affect the health of future generations. Therefore, I have decided that the United States of America will renounce the use of any form of deadly biological weapons that either kill or incapacitate." "Our bacteriological programs in the future will be confined to research in biological defense, on techniques of immunization, and on measures of controlling and preventing the spread of disease." The Department of Homeland Security Science and Technology Directorate Chemical and Biological Defense Division is focused on such research to reduce the potential consequences of chemical and biological weapons attacks on the civilian population, infrastructure, and agriculture. Research priorities include urban monitoring systems, new detection and identification technologies, bioassays, bioforensics capabilities, emergency response tools, and restoration technologies.

WMD attacks may be covert or overt. *Overt attacks* occur with warning and/or notification by the perpetrator. *Covert attacks* are characterized by the silent release of a WMD agent into a population. These attacks often employ biological agents but radiological and chemical agents could also be used. Surveillance of similar symptoms and malaise and early reporting is paramount to an appropriate response by emergency responders, hospitals, medical clinics, and testing facilities.

Michael Aiscough is credited with the following statement: "The 20th century was dominated by physics, but recent breakthroughs indicate that the next 100 years will likely be 'the Biological Century.'" Sir William Stewart, President of the British Association for the Advancement of Science and former Chief Advisor to the British government, spoke these words to fellow scientists: "There are those who say the First World War was chemical, the Second World War was nuclear, and that the third world war—God forbid—will be biological. Information on the potential use of biological agents is widely available in the published literature. The offensive use of biological weapons is forbidden by international convention. Yet, the published literature lists around 30 conventional microbes as potential BW [biological warfare] agents." Although biological agents were among the earliest WMD agents used by man and nations, this chapter will first focus on chemical agents and their uses beginning in World War I, then biological agents and uses of biological weapons, and will finish with nuclear weapons and radiological agents and continued threats. Chemical, biological, and nuclear agents' structures, chemical properties, biochemical properties and function, historical uses, antidotes and decontamination, detection techniques, and dual use research will be covered.

CHEMICAL WEAPONS IN TERRORISM

In March and April 2013, sarin was used in several (small-scale) chemical attacks in Syria. On August 21, 2013, its use in the Eastern Ghouta suburb of Damascus led to the deaths of 1429 people (as calculated by United States Central Intelligence Agency analysts)—including more than 400 children. In April 2017, the use of sarin as a chemical weapon in Syria caused the deaths of 89 civilians, primarily women and children. Reports have suggested that sarin was used in 2016 in Aleppo. Blood and urine samples taken from ten of the victims, including three during autopsy and seven at the hospital, indicated that they had been exposed to sarin or sarin-like substances. A lab in Turkey identified isopropyl methylphosphonic acid or IMPA, a degradation product of sarin.

Sarin is a highly volatile, clear, colorless liquid at room temperature. It is referred to as an organophosphorus compound because of its chemical structure. It is a nerve agent that targets the acetylcholinesterase receptor at neuromuscular junctions. Its effects can be lessened through the use of atropine, a chemical competitor of sarin at the acetylcholine receptors.

Initial symptoms of sarin use among victims include runny nose, eye irritation, shrunken pupils and blurred vision, drooling, tightness in the chest, nausea, and vomiting. Breathing becomes slow and shallow. Victims may vomit, defecate, and urinate. Twitching, jerking, and then convulsions start due to the constant firing of nerves and not being able to switch them off. The lungs secrete fluids. If a lethal dose is inhaled (5 mg is the LD_{50} for a 70 kg person), death can occur in less than 10 minutes due to the paralysis of lung muscle leading to suffocation.

Sarin is easily synthesized from methylphosphonyl difluoride and isopropyl alcohol that form a racemic mixture of sarin and hydrogen fluoride as a byproduct. Islamist rebels found and arrested in Turkey in May 2013 were found to have been carrying chemicals that could be used to make sarin. Sarin has a short shelf life for use; it begins degrading immediately. Once sarin is loaded to a chemical warhead, it must be used within a few days or less. Syria has indicated that terrorists such as al-Nusra rebel forces and its Islamist allies are responsible for the attacks. MIT Professor Theodore Postol, an expert in technology and national security and Scientific Advisor to the Chief of Naval Operations at the Pentagon, reviewed United Nations photos of the warhead used in the 2013 attack with a group of colleagues. He concluded that it was an improvised munition that could be produced locally in a machine shop. It did not match the specifications of rockets and warheads in the Syrian arsenal.

BIBLIOGRAPHY

Dewan, A. and H. Alkashali. 2017. Syria chemical attack: Authority finds 'incontrovertible' evidence of Sarin. www.cnn.com/2017/04/20/middleeast/syria-chemical-attack-sarin-opcw/index.html (accessed February 1, 2018).

Hersh, S.M. 2013. Whose sarin? *London Review of Books* 35(24): 9–12. www.lrb.co.uk/v35/n24/seymour-m-hersh/whose-sarin (accessed February 1, 2018).

Sample, I. 2013. Sarin: The deadly history of the nerve agent used in Syria. www.theguardian.com/world/2013/sep/17/sarin-deadly-history-nerve-agent-syria-un (accessed February 1, 2018).

CHEMICAL WEAPONS

Chemical weapons are chemicals used for the purpose of inflicting physical or psychological harm or death to an enemy. The first use of chemical weapons was by the French in 1912; they used the lacrimator ethyl bromoacetate to catch a bank robber. Another early chemical weapon was xylyl bromide, also called white cross, "weiss Kreuze," by Germany against Russia. Grun Kreutz and gelb Kreutz, chlorine-based and mustard gas agents respectively, were other chemical agents evaluated. Chemical weapons are classified into eight categories: blood agents, blister agents, pulmonary/choking agents, nerve agents, nettle/urticant agents, incapacitating agents, vomiting agents, and riot/tear agents. Chemical weapons may be disseminated using improvised explosive devices, spray tanks, bombs, rockets, artillery shells, or missile warheads.

BLOOD AGENTS

Blood agents (Figure 15.1) cause respiratory damage and asphyxiation. These include cyanogen chloride (NCCl, CK) and hydrogen cyanide (HCN, AC). Cyanogen chloride is a colorless gas with an acrid odor. It is slowly hydrolyzed by water at neutral pH to form chloride ions and cyanate and forms explosive polymers in storage. Hydrogen cyanide is a colorless liquid that boils at warm room temperatures (25.6°C or 78.1°F). As it was first isolated from Prussian blue, it became known as prussic acid. It hydrolyzes in water and is weakly acidic ($pK_a = 9.2$). Both are scheduled by the

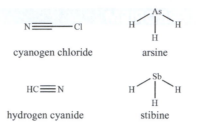

Figure 15.1   Chemical structures of blood agents. (Structure prepared by Ashley Cowan.)

Chemical Weapons Convention as Schedule III agents that have uses in industry but are toxic chemicals. Hydrogen cyanide is lethal in one minute at 3500 parts per million (ppm). Cyanide binds in the oxygen binding site in hemoglobin and thus reduces oxygen uptake and transfer to tissues leading to suffocation. The cyanide ion also binds to iron in cytochromes, including cytochrome c oxidase, in the mitochondria, disrupting electron transfer. These are especially dangerous as they penetrate gas-mask filters. Hydrogen cyanide was used in World War I by the United States and Italy. A second-generation agent, Zyklon B (after Zyklon A, an insecticide invented by Fritz Haber), was used in German concentration camps for extermination. Hydrogen cyanide has also been used by Iraq against the Kurds in the 1980s, by the Aum Shinrikyo cult in an attempt to release it into a Tokyo subway station in 1995, and was considered for release in a New York subway station by Al Qaeda.

Another set of blood agents are arsine (AsH_3) and stibine (SbH_3, SA). These agents are colorless, flammable, pyrophoric, toxic gases that are more dense than air (arsine is 2.5 times denser than air). Arsine has a slight garlic odor from oxidation with air, although it is toxic before it is detectable by smell. Stibine is less stable and decomposes more quickly than arsine. Arsine and stibine bind to hemoglobin in red blood cells. The cells are destroyed by the body, which leads to anemia and nephropathy. The LC_{50} for stibine is 100 ppm (mice). Arsine was investigated for battlefield use in World War II but not used. Stibine and arsine are detectable using the Marsh test but modern methods including atomic absorption spectroscopy, ICP-MS, neutron activation analysis, or x-ray fluorescence analysis are now employed.

PULMONARY/CHOKING AGENTS

Pulmonary agents cause severe irritation to the lungs, eyes, and skin, as well as choking and vomiting. Examples of chemicals used as pulmonary agents include chlorine gas, phosgene, diphosgene, disulfur decafluoride, perfluoroisobutene, mercury, nitric oxide, red and white phosphorous, titanium tetrachloride, zinc oxide, and sulfur trioxide (Figure 15.2). Pure chlorine gas is a greenish yellow in color with a pungent irritating odor. Chlorine gas was used in 1915 during World War I by the Germans against the French at the Second Battle of Ypres, Belgium. Soldiers reported that it stung the back of the throat and chest, smelled of pepper and pineapple, and tasted metallic. Chlorine gas acts by its reaction with water in the lungs to form hydrochloric acid, which destroys living tissue. Gas masks to protect soldiers from chlorine gas were equipped with activated charcoal filters. Chlorine was used as a weapon by the Liberation Tigers of Tamil Eelam in 1990 in its assault on a Sri Lankan Armed Forces camp at East Kiran, by insurgents in 2007 in Anbar province in Iraq against the local civilian population and coalition forces, and again in 2014 in Iraq.

Phosgene and mustard gas soon replaced the use of chlorine in World War I. Phosgene is listed on the Controlled Weapons Convention Schedule III. Phosgene (carbon dichloride, CG) ($COCl_2$) is a colorless toxic industrial gas with a musty hay-like, suffocating odor. It forms a white to pale-green cloud on release. It was used extensively in World War I including the Battle of Verdun in 1917; it is credited with causing 85% of the 100,000 deaths due to chemical

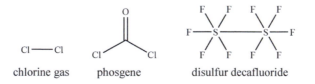

Figure 15.2   Chemical structures of pulmonary/choking agents. (Structure prepared by Ashley Cowan.)

weapons in World War I. It was used in 1987 in the Iran–Iraq War. Diphosgene is a colorless liquid at room temperature, which makes it more convenient to handle than phosgene, a gas at room temperature. It was used by the Germans in 1916 during World War I because it was found to destroy the filters in the gas masks used at that time.

Disulfur decafluoride (S_2F_{10}) is a highly toxic, colorless gas with a sulfur dioxide-like odor considered for use as a chemical weapon in World War II. It is four times as toxic as phosgene. As it does not cause tearing or skin irritation, it provides little warning of exposure. It decomposes to SF_6 and SF_4 above 150°C and by a disproportion reaction in the lungs. SF_4 reacts with water to form sulfurous acid and hydrofluoric acid.

Other choking agents include nitric oxide (NO), a toxic, colorless gas that is a strong oxidizer and acts as a vasodilator. It forms nitrogen dioxide on air contact. Perfluoroisobutene (1,1,3,3,3-pentafluoro-2-(trifluoromethyl)prop-1-ene) is a strong electrophile and undergoes rapid hydrolysis in water. It is ten times as toxic as phosgene and is a Chemical Weapons Convention Schedule II agent. Mercury (Hg_2) is a silvery-colored liquid at room temperature that was used as a chemical weapon in 2012 at Albany Medical Center where it was used to contaminate food.

Red and white phosphorous (P4) are also choking agents. White phosphorus is a choking agent that is characterized by white to yellow acrid fumes. It is used in smoke, tracer, illumination, and incendiary weapons and can be used to mask troops. It is a water-soluble waxy solid that is volatile, hydroscopic, and flammable. It was used by British, US and Japanese forces in World War II. Later uses were reported in Korea, Vietnam, Chechnya, Iraq, and Israel. More recently, use is suspected by Yemen (2009) and Libya (2011). It can be detected by its chemiluminescence under UV light. Red phosphorous is used by some countries in place of white phosphorous as it burns cooler. When exposed to sunlight, red phosphorous becomes amorphous and crystallizes on further heating.

Similar to the phosphorus compounds, titanium tetrachloride is used to make smoke screens to hide troops and objects. Due to its absorbance of infrared radiation, it can also hide objects from radar detection. Titanium tetrachloride is corrosive and reacts strongly with water to produce hydrochloric acid, forming a colorless to light-yellow liquid. As a result, it has a penetrating acid odor and irritates and burns the eyes and skin. Zinc oxide is a nontoxic white powder that is insoluble in water. Aerosolized forms that are inhaled are dangerous. Sulfur trioxide is a hydroscopic, corrosive, crystalline white solid that causes severe burns. It was used in the first known application of chemical weapons. Around 256 AD, the Roman fort at Dura was seized by the Persian Empire; sulfur trioxide is thought to have filled a mine and killed everyone inside.

BLISTER AGENTS

Blister agents cause severe and painful burns that require immediate medical attention to reduce the likelihood of infection and to rehabilitate exposed tissue. Blister agents include ethyldichloroarsine, methyldichloroarsine, phenyldichloroarsine, lewisite, sulfur mustards, and nitrogen mustards (Figure 15.3). When used at close range, white phosphorous is also a blister agent. Ethyldichloroarsine (Ethylarsonous dichloride, ED) is a colorless, volatile liquid that was used as a blister agent during World War I. Methyldichloroarsine (dichloromethlarsane, MD) is also a colorless, volatile liquid weaponized by Germany in 1917–1918 during World War I. It causes eye and nose irritation; the blisters may not arise for hours and convulsions, abdominal pain, coughing, and shortness of breath may occur after 3 to 5 days. Activated charcoal masks offer protection if they do not employ rubber seals as the chemical agent will penetrate rubber. It can be decontaminated using bleach and a base such as sodium hydroxide. Phenyldichloroarsine (phenylarsonous dichloride, PD) is a colorless, odorless, oily liquid that can form hydrochloric acid and phenylarsenious acid, which irritate and burn skin and the mucous membranes of the eyes, lungs, and throat. The arsenic can displace calcium in bone and tissue, causing bone marrow damage.

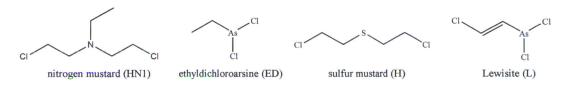

nitrogen mustard (HN1)   ethyldichloroarsine (ED)   sulfur mustard (H)   Lewisite (L)

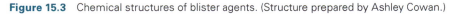

Figure 15.3   Chemical structures of blister agents. (Structure prepared by Ashley Cowan.)

Lewisite, sulfur mustard, and nitrogen mustard are Chemical Weapons Convention Schedule I agents. Lewisite (2-chloroethenylarsonous dichloride, L) is a colorless, odorless-geranium-smelling, oily liquid that is ten times more volatile than mustard gas. It penetrates latex gloves and ordinary clothing and lab coats. It causes stinging, burning, and itching pain and a rash of fluid-filled blisters. It damages capillaries and causes hypoxia and hypovolemia, nosebleeds, respiratory problems and failure, and may cause blindness with acute exposure. Lewisite acts by inhibiting the thiol group of metabolism enzymes and is a suicide inhibitor of the metabolic enzyme pyruvate dehydrogenase. An antidote is dimercaprol that binds the arsenic in lewisite. It can be neutralized with bleach. It was stockpiled by the United States for use in World War II but was not used as it decomposes by hydrolysis and soldiers rapidly responded to the smell in tests by donning gas masks. It was reportedly used in 1980 during the Iran–Iraq War. Accidental exposure to lewisite in stockpiles has caused deaths in recent years in China.

Sulfur mustard (Bis(2-chloroethyl) sulfide, H) is an oily, yellow-brown, highly persistent liquid with a garlic smell that forms blisters on the skin and in the lungs of exposed persons. Pure, distilled (HD) sulfur mustards are colorless to yellow viscous oily liquids that are highly fat soluble, leading to rapid absorption through the skin. Sulfur mustard used in conjunction with lewisite is referred to as HL; the properties of both would be felt. Large-scale production of sulfur mustards was developed by Germany in 1916. They were used by Germany against Allied soldiers in 1917 near Ypres, Belgium, and later that year by the Allies at Cambrai, France. Since World War I, it has seen use in several conflicts. Its presence is detectable by soldiers and civilians by its garlic-like smell. Gas masks were ineffective as the skin was not protected to prevent blister formation. It is an alkylating agent as well as a blister agent. The compound causes cellular toxicity when a chloride ion is eliminated and the cyclic sulfonium ion produced causes permanent alkylation of guanine nucleotides in DNA, preventing DNA replication and cell division, thus leading to cell death or cancer. It can be decontaminated using bleach. Exposure can be determined by the presence of 1,1′-sulfonylbis-methylthioethane (SBMTE), a conjugation product with glutathione, excreted in the urine of victims.

Nitrogen mustards include bis(2-chloroethyl)ethylamine (HN1), bis(2-chloroethyl)methylamine (HN2), and tris (2-chloroethyl)amine (HN3). Like sulfur mustard, nitrogen mustards are also cytotoxic and lead to dangerous blisters. A cyclic aminium ion is formed that also alkylates guanine in DNA, causing interstrand crosslinks to form. This leads to apoptosis via the p53 pathway. It was also stockpiled for use in World War II but not used in combat.

NERVE GASES

Nerve gases target the nervous system and cause loss of muscle control, respiratory failure, and even death. These include the G-series agents tabun (GA), sarin (GB), soman (GD), cyclosarin (GF), GV, methyl fluorophosphoryl homocholine iodide (MFPhCh), and the V-series agents EA-3148, VE, VG, VM, VP, VR, and VX (Figure 15.4). The G-series nerve agents are organophosphates that were synthesized in German insecticide research and inhibit acetylcholinesterase.

Tabun ((RS)-Ethyl N,N-Dimethylphosphoramidocyanidate, GA) is a colorless, tasteless, and highly toxic liquid with a faint fruity odor. Although it can be decontaminated with bleaching powder, this results in the production of the poisonous gas cyanogen chloride. It was used in 1984 in the Iran–Iraq war.

Sarin ((RS)-Propan-2-yl methylphosphonofluoridate, GB) is a colorless, odorless liquid. The LD_{50} is 5 mg for a 70 kg person. It is often deployed as a binary weapon in missiles or even spray cans for ease of use. Due to the high acidity and reactivity of the products, it has a poor shelf life and corrodes containers unless neutralized with compounds such as tributylamine or trimethylamine. At a high pH, it rapidly decomposes to nontoxic phosphonic acids including

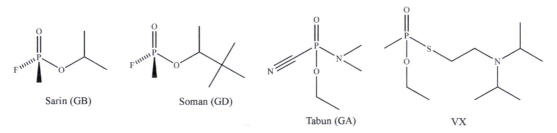

Sarin (GB)   Soman (GD)   Tabun (GA)   VX

Figure 15.4 Chemical structures of nerve agents. (Structure prepared by Ashley Cowan.)

IMPA and subsequently methylphosphonic acid (MPA) which are used in its detection. Tabun and sarin were synthesized in 1936 and 1938 respectively by Gerhard Schrader in Germany. Sarin has been used several times in chemical warfare including by Iraq in the Iran–Iraq war (1980–1988), in an attack on Nagano (1994), Matsumoto (1994), a Tokyo subway (1995) by the Aum Shinrikyo cult, and in Syria (2013).

Soman (3,3-Dimethylbutan-2-yl methylphosphonofluoridate, GD) is a volatile, corrosive, colorless liquid that appears yellow-brown in color and smells like camphor when impurities are present. It was synthesized in 1944 in Germany by Nobel Laureate Richard Kuhn and Konrad Henkel. It inhibits both acetylcholinesterase and butylrylcholinesterase. It can be deployed as a binary munition with methylphosphonyl difluoride in one capsule and pinacolyl alcohol and an amine in another.

Cyclosarin (Cyclohexyl methylphoshonofluoridate, GF) was discovered in 1936 in Germany in work on organophosphorous insecticides. It is a colorless, persistent, flammable, toxic liquid at room temperature that evaporates 69 times slower than sarin. It has a sweet and musty odor of peaches or shellac. The LD_{50} is 1.2 mg. It can be deployed as a binary munition with methylphosphonyl difluoride in one capsule and cyclohexanol (or a mixture of cyclohexylamine and cyclohexanol) in another. Iraq has manufactured and used cyclosarin; it was used with sarin in the Iran–Iraq war (1980–1988). Another G-series agent is 2-(Dimethylamino)ethyl N,N-dimethylphosphoramidofluoridate (GV). Sarin, soman, and tabun are listed on the Chemical Weapons Convention Schedule I.

V-series nerve agents are also organophosphate compounds. They were studied by the United States and USSR during the Cold War and are listed on the Chemical Weapons Convention Schedule I. VX is odorless, tasteless, and amber-colored like motor oil. It is highly viscous and highly persistent. It is toxic with an LD_{50} of 10 mg for humans. Binary VX is prepared by mixing O-(2-diisopropylaminoethyl) O′-ethyl methylphosphonite with elemental sulfur. EA-3148 is reportedly 50% more potent than VX but weaponization was not pursued. V-series agents can be mostly decontaminated using concentrated sodium hydroxide or strong nucleophiles as it cleaves the P-S bond and P-O bond (some of the products are also toxic). Affected persons can be decontaminated by removing contaminated clothing and paraphernalia, washing with household bleach, and flushing with clean water. VM and VR (LD_{50} of 10–50 mg) are similar in toxicity to VX but more effective in that antidotes such as pralidoxime must be given immediately to work. VG (reported in a 1955 paper) has a toxicity about the same as sarin (approximately one tenth of VX) and is classified under Schedule II of the Chemical Weapons Convention, unlike other nerve agents. It was sold under the trade name Amiton (Russian name is Tetram) in 1954 but was found to be too toxic for safe use.

Organophosphates are acetylcholinesterase inhibitors. On stimulation of a motor neuron, acetylcholine is released and taken up by the nearby muscle cell, stimulating muscle contraction. While acetylcholinesterase normally breaks down the acetylcholine neurotransmitter into acetic acid and choline after muscle contraction, organophosphates covalently inhibit acetylcholinesterase. This results in an accumulation of acetylcholine in the intercellular space between the neuron and muscle cells and results in uncontrolled muscle contraction followed by flaccid paralysis. Death is caused by asphyxiation. Antidote treatment (or pre-treatment) includes atropine, pyridostigmine, benactyzine, obidoxime, pralidoxime, and asoxime chloride (HI-6). Atropine works by blocking the activity of muscarinic acetylcholine receptors. Oxime drugs act by reversing the binding of organophosphate compounds to acetylcholinesterase; they have a greater affinity for the organophosphate than the enzyme. By displacing the phosphate from the enzyme, its activity is restored.

NETTLE OR URTICANT AGENTS

Nettle or urticant agents are corrosive and injurious to skin but without blister formation. An example of a nettle agent is phosgene oxime (dichloroformaldoxime, CX). It was used by the USSR in Afghanistan from 1979 to 1989. It is a colorless solid that may be yellow due to impurities. It has a strong odor and violently irritating vapor. Hydrazine degrades phosgene oxime into HCN and nitrogen gas. Contact with metals also degrades the agent. It is soluble in water and hydrolyzed by an alkaline solution (Figure 15.5).

INCAPACITATING AGENTS

Incapacitating agents are used to slow the movement of, debilitate, and confuse their victims. These include QNB (BZ, 3-Quinuclidinyl benzilate), LSD (lysergic acid diethylamide), and fentanyl and its derivatives including levofentanyl,

phosgene oxime (CX)

Figure 15.5 Chemical structure of a nettle/urticant agent. (Structure prepared by Ashley Cowan.)

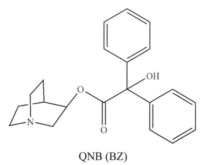

QNB (BZ)

Figure 15.6 Chemical structure of an incapacitating agent. (Structure prepared by Ashley Cowan.)

carfentanil, and remifentanil. QNB (Figure 15.6) is odorless and a Chemical Weapons Convention Schedule II agent that is dispersed as an aerosolized solid. It leads to stupor and confusion, and illusions and hallucinations. Its military code BZ is derived from the "buzz" it gave recipients. QNB is an anticholinergic that acts as a competitive inhibitor of acetylcholine at muscarinic receptors. As physostigmine increases the concentration of acetylcholine in synapses, neuromuscular and neuroglandular junctions, it is used as a specific antidote. LSD is a psychedelic drug that causes psychological effects and alters awareness of surroundings, perceptions, and feelings and is used to confuse victims when used as a chemical warfare agent. It binds to all dopamine (D_{1-5}) and most serotonin receptors (5-HT1A,1B,2A,2B,2C,5A5B,6). Levofentanyl was used in an attempt by Mossad agents to kill the Hamas leader Khalid Mishal in 1997. Kolokol-1 is a synthetic opioid used by Russia against Chechen attackers in the 2002 Moscow Theatre hostage crisis (15% mortality). The specific antidote for fentanyl is naloxone and for carfentanil is naltrexone.

Other incapacitating agents that were investigated by the US military include dimethylheptylpyran, a synthetic analogue of THC, pelargonic acid vanillylamide, a synthetic capsaicinoid pepper spray that causes severe pain to the eyes, and sleeping gases.

VOMITING AGENTS

Vomiting agents are used to induce soldiers to remove gas masks, making them more susceptible to other chemical agents. An example is chloropicrin (trichloro(nitro)methane, PS). Chloropicrin is a colorless liquid that was first synthesized in 1848 through the reaction of sodium hypoclorite, NaOCl, and picric acid, $HOC_6H_2(NO_2)_3$. In World War I, Germans employed chloropicrin against Allied forces, causing them to vomit, remove their gas masks, and expose themselves to even more toxic chemical weapons. Another vomiting agent is Adamsite (10-Chloro-5,10-dihydrophenarsazinine, DM). It was synthesized in 1918 and was later used in the Vietnam War. Adamsite is odorless and forms bright green/yellow crystals. The structures of these agents are shown in Figure 15.7.

RIOT/TEAR AGENTS

Riot/tear agents are used to control crowds and cause temporary debilitation. Chloroacetophenone (CN, 2-chloro-1-phenylethanone), 2-chlorobenzylidene malononitrile (CS), capsicum spray (OC, "pepper spray"), and dibenzoxazepine (CR) gases are members of this category (Figure 15.8). Chloroacetophenone is the active ingredient in Mace™. Chloroacetophenone has a sharp, irritating odor and is a colorless to gray or white crystalline solid that may appear as a blue-white cloud at the point of release. 2-chlorobenzylidene malononitrile was first discovered by American scientists Ben Corson and Roger Stoughton in 1928. Dibenzoxazepine and 2-chlorobenzylidene malononitrile were developed by the British Ministry of Defense at Porton Down in the late 1950s to early 1960s. At room temperature,

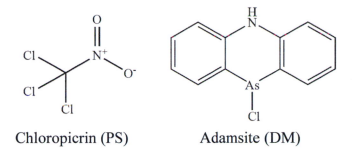

Chloropicrin (PS)   Adamsite (DM)

Figure 15.7 Chemical structures of vomiting agents. (Structure prepared by Ashley Cowan.)

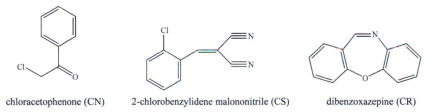

chloracetophenone (CN)   2-chlorobenzylidene malononitrile (CS)   dibenzoxazepine (CR)

Figure 15.8 Chemical structures of riot/tear agents. (Structure prepared by Ashley Cowan.)

2-chlorobenzylidene malononitrile is a white crystalline solid and dibenzoxazepine is a pale-yellow crystalline solid. It has a pepper-like odor. These agents act by irritating the mucous membranes in the eyes, lungs, nose, and mouth. CR is the most potent tear gas and has the least systemic toxicity. CS is ten times more potent than CN but less systematically toxic. CN may even cause temporary blindness. Capsicum spray is derived from *Capsicum sp.* chili peppers' capsaicinoids (six types). A probable target of dibenzoxazepine and chloroacetophenone is the transient receptor potential cation channel (TRPA1) ion channel's protein sulfhydryl groups. Riot/tear agents can be used to contaminate water, food, and air. CS was used by the police in Bahrain, Israel, and the United States, in Canada for riot control, and to control crowds in Hong Kong in 2014.

TRENDS IN CHEMICAL CHARACTERISTICS OF CHEMICAL WARFARE AGENTS

The chemical weapons agents are highly toxic due to their halogen and heavy metal constituent atoms. Several of the chemical warfare agents are gaseous at room temperature, although some are liquids. Arsenic compounds are reactive with water. Nitrogen mustards and other chlorine-containing compounds such as cyanogen chloride are soluble in water. With the exception of hydrogen cyanide, the chemical agents are more dense than air and will settle in trenches or low-lying areas. As with other compounds, melting point and boiling point tend to increase with molecular weight. The lower molecular weight compounds with heavy atoms (e.g., halogens, etc.) tend to have higher densities. More hydrocarbon-rich incapacitating and riot agents are insoluble. Table 15.1 lists the chemical properties of several chemical weapons agents.

TOXIC INDUSTRIAL CHEMICALS

A continued, easily accessible threat that could be purposed as chemical weapons are toxic industrial chemicals including chlorine gas, phosgene, cyanogen chloride, hydrogen cyanide, and chloropicrin (trichloronitromethane). Recent accidents remind us how injurious and deadly exposure to these chemicals can be.

DETECTION AND IDENTIFICATION METHODS

The chemical warfare agents are detectable on-site using M8 and M9 paper (a type of pH paper) and Dräger colorimetric tubes. Chemical identification is performed using gas chromatography–mass spectrometry (GC-MS), high-performance liquid chromatography (HPLC), Fourier transform-infrared spectroscopy (FT-IR), and Raman spectroscopy equipped with databases. Handheld Raman spectrometers can detect agents through their containers.

Table 15.1 Chemical properties of chemical weapons

Type	Chemical agent	Chemical formula	Appearance	MW (g/mol)	Density @25°C (g/mL)	MP (°C)	BP (°C)	Solubility in water
Blood	Cyanogen chloride (CK)	NCCl	Colorless	61.47	2.7683 (0°C)	−6.55	13	Soluble
	Hydrogen cyanide (AC)	HCN	Pale blue transparent to colorless	27.03	0.687	−14	25.6–26.6	Miscible
Blister	Ethyldichloroarsine (ED)	$C_2H_5AsCl_2$	Colorless	174.89	1.1420 (14.5°C)	−65	155.3	Reactive
	Methyldichloroarsine (MD)	CH_3AsCl_2	Colorless	160.86	1.836	−55	133	Reactive
	Phenyldichloroarsine (PD)	$C_6H_5AsCl_2$	Colorless	222.93	1.65 (20°C)	−20	252–255	Reactive
	Lewisite (L)	$C_2H_2AsCl_3$	Colorless	207.32	1.89	−18	190	Reactive
	Sulfur mustard (HD, H, HT, HL, HQ)	$C_4H_8Cl_2S$	Colorless	159.07	1.27	14.4	217	Practically insoluble
	Nitrogen mustard (HN1)	$C_6H_{13}Cl_2N$	Colorless	170.08	1.09	−34	88.5	Soluble
	Nitrogen mustard (HN2)	$C_5H_{11}Cl_2N$	Colorless	156.05	1.118	−76	75	Soluble
	Nitrogen mustard (HN3)	$C_6H_{12}Cl_3N$	Colorless	204.53	1.2347	−4	143	Soluble
Nerve	Tabun (GA)	$C_5H_{11}N_2O_2P$	Colorless to brown	162.13	1.0887	−50	247.5	Miscible
	Sarin (GB)	$C_4H_{10}FO_2P$	Colorless	140.09	1.0887	−56	158	Miscible
	Soman (GD)	$C_7H_{16}FO_2P$	Colorless	182.18	1.022	−42	198	Miscible
	Cyclosarin (GF)	$C_7H_{14}FO_2P$	Colorless	180.16	1.1278	−30	239	Sparingly soluble
	GV	$C_6H_{16}FN_2O_2P$	Colorless	198.18	1.1		207.1	Sparingly soluble (?)
	Methyl fluorophosphoryl homocholine iodide (MFPhCh)	$C_7H_{18}FINO_2P$	Colorless	325.1				Sparingly soluble (?)
	EA-3148	$C_{12}H_{26}NO_2PS$	Colorless	279.38	1.1		353.2	Sparingly soluble (?)
	VE	$C_{10}H_{24}NO_2PS$		253.34				
	VG	$C_{10}H_{24}NO_3PS$		269.34	1.1		315.1	
	VM	$C_9H_{22}NO_2PS$		239.32				
	VR	$C_{11}H_{26}NO_2PS$		267.37	1		323.5	
	VX	$C_{11}H_{26}NO_2PS$	Amber-like color	267.37	1.00083	−3.9	300	
Nettle	Phosgene oxime (CX)	$CHCl_2NO$	Colorless to yellow-brown	113.93		35–40	128	Freely soluble
Pulmonary	Chlorine	Cl_2	Yellow-green	35.45	3.2 (0°C)	−101.5	−34.04	Reactive
	Phosgene oxime (CG)	$CHCl_2NO$	Colorless to yellow-brown	113.93		30–40	128	Freely soluble
	Diphosgene (DP)	$C_2Cl_4O_2$	Colorless	197.82	1.65	−57	128	Insoluble

(Continued)

Table 15.1 (Continued) Chemical properties of chemical weapons

Type	Chemical agent	Chemical formula	Appearance	MW (g/mol)	Density @25°C (g/mL)	MP (°C)	BP (°C)	Solubility in water
	Disulfur decafluoride	S_2F_{10}	Colorless	254.1	2.08	−55	29	Insoluble
Incapacitating	QNB/Agent 15 (BZ)	$C_{21}H_{23}NO_3$	Colorless solid	337.41		164–165	322	Insoluble
	Dimethylheptylpyran (DMHP)	$C_{25}H_{38}O_2$	Pale yellow	370.57	1		481.9	Insoluble
	EA-3167	$C_{20}H_{29}NO_3$		329.43	1.2		472.9	Insoluble
	PAVA	$C_{17}H_{27}NO_3$	White	293.41	1.1	54		Insoluble
Vomiting	Chloropicrin (PS)	CCl_3NO_2	Colorless	164.375	1.692	−69	112	Very slightly soluble
Riot	OC (capsaicin pepper spray)	$C_{18}H_{27}NO_3$	Colorless	305.42		62–65	210–222	Practically insoluble
	CS	$C_{10}H_5ClN_2$	White crystalline	188.6	1.04	93	310	Insoluble
	CN (mace)	C_8H_7ClO	Colorless	154.59	1.324	54–56	244.5	Insoluble
	CR (dibenzoxazepine)	$C_{13}H_9NO$	Pale-yellow crystalline solid	195.22	1.16	73		Insoluble

FOODBORNE PATHOGENS: ACCESSIBLE BIOTERRORISM AGENTS

In 1984, a community outbreak of salmonellosis was reported at a restaurant in The Dalles, Oregon. In 1996, a shigellosis outbreak in a large medical center was reported in Georgia. In 2015, an outbreak of listerosis caused deaths and hospitalizations across the United States and Canada. In each of the three cases, illness and deaths were caused by bacterial agents. The first two cases are documented cases of bioterrorism—perpetrators knowingly planted bacteria onto uncooked food items including salad, muffins, and donuts. The third case is representative of many outbreaks of foodborne pathogens from contaminated food processing plants. The salmonellosis outbreak was caused by members of the Rajneeshpuram religious commune who sought to alter election results by incapacitating members of the adult voting population who ate at a restaurant salad bar in The Dalles with *Salmonella typhimurium*. Restaurant staff and patrons fell ill. This case was the largest outbreak of foodborne illness reported to the United States Centers for Disease Control in that year. The *Shigella dysenteriae* type 2 strain that was used to contaminate muffins and donuts in a hospital break room originated from the hospital lab and the perpetrator was found to be a hospital employee. While the third case was not caused by a nefarious perpetrator, it serves as a reminder of the trust placed in companies producing food for a nation. The plant had tested positive for listeria nine times due to contamination for more than a year before the US Food and Drug Administration (FDA) showed up to test a plant; the results had not been reported. In the listerosis case, four people had died and another thirty-three people were hospitalized across the United States and Canada. Companies are expected to notify the FDA when they test and determine their products have a "reasonable probability" of causing serious adverse health consequences and must immediately perform a corrective action. In these cases, the source of the illnesses was caught relatively quickly and was not too widespread; however, foodborne pathogens are highly accessible and can be used as bioterror agents to quickly incapacitate and invoke fear in a large population. Although salmonella, shigella, and listeria do not cause infections that can be transmitted between sick individuals, other infectious agents would cause more widespread cases. For example, another recent outbreak of a foodborne pathogen involved norovirus, a virus that is easily transmitted between sick individuals. More than 200 people were recently sickened at a restaurant where food was contaminated with the virus; the public and public health officials were only informed after the restaurant had closed and cleaned. Not informing public health officials led to criminal investigations by the Justice Department.

BIBLIOGRAPHY

Kolavic, S.A., A. Kimura, S.L. Simons, L. Slutsker, S. Barth, and C.E. Haley. 1997. An outbreak of Shigella dysenteriae type 2 among laboratory workers due to intentional food contamination. *JAMA* 278(5):396–398.

Storm, S. 2016. Dole knew about listeria problem at salad plant, F.D.A. report says. www.nytimes.com/2016/04/30/business/dole-knew-about-listeria-problem-fda-report-says.html (accessed February 1, 2018).

Török, T.J., R.V. Tauxe, R.P. Wise, J.R. Livengood, R. Sokolow, S. Mauvais, K.A. Birkness, M.R. Skeels, J.M. Horan, and L.R. Foster. 1997. A large community outbreak of salmonellosis caused by intentional contamination of restaurant salad bars. *JAMA* 278(5):389–395.

BIOLOGICAL WEAPONS

Biological weapons include bacteria, fungi, viruses, and proteins that can be used to incapacitate, sicken, or kill humans and agricultural products including plants and animals. In addition to physical harm, biological weapons inflict immeasurable psychological harm to individuals and communities. Biological weapons differ from nuclear weapons in that they are often readily available. In fact, due to the ease of access, biological weapons were the first employed WMDs.

Biological weapons may be disseminated to be taken in by inhalation or ingestion using traditional weapons including rockets and warheads but they are relatively fragile and often destroyed by heat and explosion. Perpetrators have used biological weapons to contaminate food at restaurants to sicken civilians by ingestion or by coating them on sharp objects for injection. The skin provides a good barrier for biological agents and they are not absorbed directly through it.

HISTORY OF BIOLOGICAL WEAPONS USE

It has been documented that in the sixth century BC, the Assyrians poisoned their enemy's drinking water by contaminating their wells with the fungus, rye ergot. In the fourth century BC, Scythians dipped their bows' arrows in the blood, manure, and tissues of decomposing bodies. In 1340 AD, attackers of the castle of Thun L'Eveque in Hainault (northern France) hurled dead horses and other animals inside by catapult and, similarly, in 1422 AD, decaying cadavers were launched over castle walls at Karlstein in Bohemia by attackers with the hope of spreading illness. In 1495 AD, the Spanish contaminated French wine with the blood of lepers and in the mid-1600s, a Polish military general loaded saliva from rabid dogs into hollow artillery shells for use against enemies. In 1763, British soldiers gave blankets contaminated with smallpox to Native Americans during the French and Indian War.

In 1915, Anton Dilger produced anthrax and Glanders, which were used to infect 3000 horses used to transport supplies to the frontlines in World War I. As a result of these events and to limit the use of biologicals in warfare internationally, on June 17, 1925, the "Protocol for the Prohibition of the Use in War of Asphyxiating, Poisonous or Other Gases, and of Bacteriological Methods of Warfare," or the Geneva Protocol treaty, was signed by 38 nations, banning the use of chemical and biological methods in warfare. This was the first international effort to limit the use of biological WMDs in warfare.

Even with an international treaty to limit the use of biological weapons, nations began formal programs to conduct research on biological weapons including weaponization, deployment, human effects, and vaccine and therapeutic response research. Great Britain began its biological weapons research in 1934 and they tested anthrax on sheep on Gruinard Island off the Scottish coast on July 15, 1942. Japan began its biological weapons program with Unit 731 and performed testing on prisoners of war in Manchuria that led at least 1000 prisoners to be sickened by aerosolized anthrax. Japan poisoned a Soviet water supply in 1939 using typhoid bacteria and dropped rice and wheat mixed with plague-carrying fleas over China and Manchuria from the air in 1940. In 1942, German and Soviet soldiers were sickened with inhalational tularemia before the battle of Stalingrad; it is suspected that the rare agent was intentionally released as the USSR had developed a tularemia weapon in 1941.

The United States' bioweapons program was approved under President Franklin Delano Roosevelt in November 1942 and began at Camp Detrick, Maryland, in Spring 1943. The US program focused on the bioweapon development of anthrax, plague, tularemia, Q fever, Venezuelan equine encephalitis, brucellosis, and botulinum toxin. After World War II, in May 1949, a US Army Chemical Corps Special Operations Division conducted field tests with bioweapons formulations. In 1950, biological weapons tests from Naval warships sprayed Norfolk, Hampton, and Newport News, Virginia, and San Francisco, California, to demonstrate the feasibility of large-scale deployment from the water. In 1953, simulated anthrax was distributed in the air by aerosol generators on cars in St. Louis, Missouri, Minneapolis, Minnesota, and Winnipeg to demonstrate the feasibility of large-scale land deployment. In 1957, biological aerosols distribution from airplanes was simulated from South Dakota to Minnesota, Ohio to Texas, and Michigan to Kansas to demonstrate the feasibility of large-scale deployment from the air.

US biological weapons research and production ended on November 25, 1969, when President Richard Nixon announced that the United States would renounce the use of biological weapons. After being opened for signature on April 10, 1972, finally on March 26, 1975, the multilateral disarmament Biological Weapons Convention (BWC), or Convention on the Prohibition of the Development, Production and Stockpiling of Bacteriological and Toxin Weapons and on their Destruction, went into effect, banning the development, production, and stockpiling of biological weapons. Seventy-nine nations were signatories to the treaty.

MODERN THREAT CLASSIFICATION

The United States Centers for Disease Control (CDC) has classified potential biological terrorism weapons agents and diseases into three categories (A, B, and C) by potential to cause harm. Category A agents have the greatest potential to cause harm, are the easiest to disseminate in a designated area, and have the potential for major public health impact. Category A agents are characterized by high mortality rates and the potential to cause panic and significant disruption to society. Included in Category A are anthrax, Ebola, and smallpox. Category B agents, while not as threatening as Category A agents, still pose serious threats to animal and human health. They could be easily disseminated and their effects would be incapacitating to individuals and groups; they include cholera, *E. coli*, and ricin. Category C agents are emerging or reemerging disease threats. Category C agents include diphtheria, SARS, and West Nile virus. Category A, B, and C agents are listed in Table 15.2.

Table 15.2 US CDC Category A, B, and C agents

Category A	Category B	Category C
Anthrax	Abrin	Diphtheria
Botulism	Cholera	Exotic Newcastle
Ebola	*Cryptosporidium parvum*	Foot and mouth
Marburg	*E. coli*	Hantavirus
Plague	Epsilon toxin	MRSA
Smallpox	Glanders	Nipah
Tularemia	Malta fever	SARS
Viral hemorrhagic fevers	Meliodosis	West Nile virus
	Psittacosis	
	Q fever	
	Ricin	
	Salmonella	
	Shigella	
	Staphylococcal enterotoxin B	
	Typhoid fever	
	Typhus fever	
	Viral encephalitis	

Source: CDC, https://emergency.cdc.gov/agent/agentlist-category.asp.

RECENT BIOLOGICAL WEAPONS CASES AND ONGOING THREATS

The BWC did not halt bioweapons research and production by governments of nation states. Additionally, since 1972, several rogue militant and cult groups and individuals have carried out numerous bioterrorism attacks using a variety of bioweapon agents. Interest in the use of anthrax has intensified. In 1979, after the BWC was signed, anthrax spores were accidentally released from a Soviet military bioweapons facility in Sverdlovsk, USSR, as a result of a faulty exhaust filter. This led to an anthrax outbreak that sickened 94 people and caused 64 civilian deaths. From 1978 to 1980, Rhodesian and South African apartheid forces used anthrax on Black tribal lands; thousands of cattle and 182 people are known to have died as a result. In 1993, Aum Shinrikyo, a Japanese religious cult group, released anthrax from an office building that led to pet deaths but no human deaths. In October 2001, four letters containing anthrax spores were sent from Trenton, New Jersey, from a US military research facility stock at Fort Detrick, to the media including the New York Post and Tom Brokaw at NBC and government-elected employees including Senators Tom Daschle and Patrick Leahy; twenty-two people became infected (eleven each with cutaneous and inhalational anthrax) and five died (all from inhalational anthrax).

There have also been several reported cases of ricin being used as a biological weapon. In 1978, ricin was used to poison and assassinate Georgi Markov, a Bulgarian defector, by the use of a metal injection weapon inserted at the end of an umbrella which injected the weapon into his calf muscle; he died 3 days after the attack. In the 1980s, Iraq attempted to weaponized ricin. In 1983, the FBI arrested two people in the United States for possessing an ounce of nearly pure ricin. In 1991, four members of the Patriots Council, an antigovernment extremist group, were arrested for plotting to kill a US marshal by planning to mix ricin with DMSO and smear it on his car door handle; the four perpetrators were arrested for their crime. In 2002, six people were arrested in Manchester, England, for using an apartment as a ricin laboratory and plans were found near Kabul prepared by al-Qaida to weaponize ricin. In 2003, ricin was found in a South Carolina post office letter with a threat to poison water supplies. Ricin was also found in the White House mailroom in a package to a Mississippi judge and a vial sent to the White House. In 2003, Chechen separatists possessed castor seeds and recipes for making ricin in a London apartment that they planned to use on the Russian Embassy. In 2004, ricin was found in a letter addressed to Senate Majority Leader Bill Frist. In 2008, a man in Las Vegas manufactured ricin in a hotel room and was convicted of possessing the toxin. In 2011, in Georgia, four men were arrested for plotting to use ricin in Atlanta and beyond. In 2013, ricin was found in letters addressed to US Senator Wicker and US President Obama.

In addition to those discussed previously, other agents were also used as bioweapons in the Cold War. In 1972, the right-wing militant group Order of the Rising Sun was found to possess 30–40 kg of typhoid bacteria cultures that they planned to use to contaminate city water supplies in several cities across the Midwestern United States. Technical manuals on how to produce biological weapons were found in the possession of the Symbionese Liberation Army in 1975. Botulinum toxin was found in a makeshift laboratory in a Red Army Faction Paris apartment in 1980. In 1984, *Salmonella* bacteria cultures were purchased from biological research supplier American Type Culture Collection by the Bhagwan Shree Rajneesh cult and were used to contaminate salad bars in Oregon; this led to 751 reported cases of foodborne poisoning and salmonellosis. In January 2014, a woman attempted to murder her mother with abrin purchased from the internet by poisoning her Diet Coke soda.

As a result of these events and the increasing frequency of bioterrorism cases, forensic chemists need to be able to sample, test, and identify biological warfare agents. They are classified into five categories: bacteria, viruses, fungi, protein toxins, and small molecule toxins. In the following sections, the five types of biological weapons are described in greater detail and methods for their detection are listed.

BACTERIA

Bacteria are single-celled organisms that may infect humans, animals, and plants. They range in shape from spherical (cocci), spiral-shaped (spirilla), to rod-shaped (bacilli). Bacteria have DNA genomes in the form of a circular chromosome. Pathogenic bacteria include *Bacillus anthracis* (anthrax), *Brucellosis melitensis* (Malta Fever), *Burkholderia mallei* (Glanders), *Burkholderia pseudomallei* (Meliodosis), *Chlamydophila psittaci* (Psittacosis), *Escherichia coli* O157:H7 (*E. coli*), *Coxiella burnetii* (Glanders), *Rickettsia rickettsii* (rickets), *Salmonella enterica* serotype typhimurium (Salmonella), *Salmonella enterica* serotype Typhi, (Typhoid fever), *Francisella tularensis* (tularemia), *Shigella flexneri* (Shigella), (Methicillin-resistant) *Staphylococcus aureus* (MRSA), *Ricksettia prowazakii* (Typhus fever), *Vibrio cholerae* (Cholera), and *Yersinia pestis* (plague). Bacterial infections can be highly contagious. Although anthrax is not contagious, the plague, Glanders, Typhoid fever, and Q fever are contagious. Malta fever can be transmitted through direct contact with infected body fluids. Psittacosis can be transmitted from birds but not between humans. Antibiotics can be used to treat bacterial infections. There are vaccines for anthrax (since 1881) and cholera to prevent infections and cholera outbreaks.

Bacteria grow on a variety of nutrient rich sources including dead bodies, food, and laboratory growth media. By culturing them in the lab, their color, shape, and other morphological characteristics can be examined under a microscope and be used to differentiate the bacterial species. Some bacterial species, including *B. anthracis*, may form spores, a dormant form of the bacterium that germinate when conditions are favorable. Spores are resistant to cold, heat, drying, chemicals, and radiation. Bacteria are routinely classified by the presence of one (Gram-positive) or two (Gram-negative) cell membranes and can be sorted into two groups based on the ability of the cultured cells to take up the Gram stain or not as shown in Table 15.3. Gram-negative bacteria have the cytoplasmic membrane, a thicker peptidoglycan layer and an outer membrane containing lipopolysaccharides. Unlike gram-negative bacteria, Gram-positive bacteria have teichoic acids in the cell wall. While most of the pathogenic bacteria listed earlier are Gram-negative (pink), four stain with the Gram stain (purple color). Figure 15.9 shows *B. anthracis* stained with the Gram stain and Figure 15.10 shows a culture of *E. coli*.

FUNGI

Fungal pathogens include molds and yeasts. Fungi are single-celled microbes that reproduce by budding. They cause human diseases such as ringworm and athlete's foot as well as plant diseases including rusts, smuts, and leaf, stem and root rots. Some fungi, including *Histoplasma* and *Coccidoides*, are highly pathogenic. Mycotoxins include aflatoxins produced by *Aspergillus flavus* and *Aspergillus parasiticus*; fumonisins are produced by Fusarium molds such as *Fusarium verticillioides*; ochratoxins are produced by *Aspergillus ochraceus*, *Aspergillus carbonarius*, and *Penicillium verrucosum*; zearalenone is produced by some *Fusarium* and *Gibberella* species including *Fusarium graminearum*, *Fusarium culmorum*, *Fusarium cerealis*, *Fusarium equiseti*, *Fusarium verticillioides*, and *Fusarium incarnatum*; and deoxynivalenol and nivalenol are produced by *Fusarium graminearum*-infected wheat. Fungal infections are extremely difficult to treat, although there are antifungal agents available for some uses.

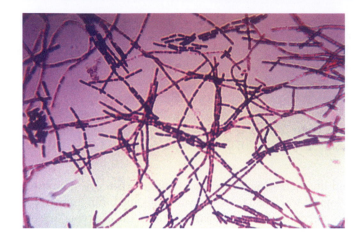

Figure 15.9 *Bacillus anthracis* stained with the Gram stain. (From CDC Public Health Image Library, Photo ID# 2226, https://phil. cdc.gov/details_linked.aspx?pid=2226.)

VIRUSES

Viruses are infectious agents that are smaller in size and genome than bacteria. As they require a host cell for replication, they are not live agents. Viruses may consist of single-stranded RNA or double-stranded DNA genomes and an external protein coat (or nucleocapsid) and an outer lipid layer. Several viruses that may be used as biological weapons are listed in Table 15.4. Smallpox is a double-stranded DNA virus. The Ebola, Exotic Newcastle, Foot and Mouth, Hantavirus, Hepatitis A, human immunodeficiency virus (HIV), Influenza (e.g., H1N1, H3N2, H2N2, pH1N1, H7N9, H5N1, H7N7, H7N2, H7N3 and H9N2), Marburg, Nipah, SARS, viral encephalitis, viral hemorrhagic fever, and West Nile virus are all single-stranded RNA viruses.

Viral infections tend to be highly contagious. Foot and mouth virus is contagious to animals through direct contact and the viral equine encephalitis viruses are contagious between horses. Exotic Newcastle is contagious to birds or humans from birds but is not transmitted between humans, and West Nile virus is also transmitted from birds. Viral infections can be prevented by vaccination and by the use of antiviral agents, where available. Vaccines are available for smallpox and strains of the flu virus. The smallpox vaccine was the first vaccine developed; it was developed in 1796 by Edward Jenner, although it is not routinely given anymore since the World Health Organization declared the disease eradicated on May 8, 1980 (the United States ended routine smallpox vaccination in 1972). Two smallpox samples remain: one at the US CDC in Atlanta, GA, and the second at Vector in Moscow, Russia. A vaccine is currently being tested to prevent Ebola virus infections.

Table 15.3   Categories of bacteria by Gram stain

Gram-positive bacteria	Gram-negative bacteria
Bacillus anthracis	*Burkholderia pseudomallei*
Clostridium perfringens	*Brucellosis melitensis*
Cornebacterium diphtheriae	*Chlamydophila psittaci*
Staphylococcus aureus	*Clostridium botulinum*
	Coxiella burnetii
	Escherichia coli
	Francisella tularensis
	Ricksettsia prowazakii
	Salmonella enterica
	Shigella flexneri
	Vibrio cholera
	Yersinia pestis

Figure 15.10 *E. coli* culture. (From VeeDunn, https://www.flickr.com/photos/micronerdbox/6684097669/in/photolist-bbDKng-723LtT-fjuzei-4qoPye-9tif9A-fjuzdB-FaqfE-RntrPh-65WzQC-dXWMBK-hkPZvk-boXZ6G-6pvw2c-AisGBn-83MXrk-gx42LD-dWxzEU-9kWhLX-9Q2Ns1-fuUV2g-pXSPvJ-obJAvn-5U6ywF-DMBCN-ag7Moy-7kmo98-nPuCpV-rbDFrG-5p3fkU-9pwwRk-e8SXsd-6gHsjB-49oVTJ-TtuPjq-cAdF9b-b1rekg-ftXvC7-4HH38B-6MhG8z-9esPBk-3p5REb-9mMpEM-oNLqb-5DAg9s-7cqR4k-hUngC-9qzGfP-6UeJQ9-hxzF44-br4Qw2.)

Table 15.4 Viruses by type and lethal dose (where reported)

Virus	Type	Lethal dose
Ebola	ssRNA	1–10 viral particles
Exotic Newcastle	ssRNA	
Foot and Mouth	ssRNA	10^9 PFU[a]
Hantavirus	ssRNA	
Marburg	ssRNA	1–10 viral particles
Nipah	ssRNA	
SARS	ssRNA	
Smallpox	dsDNA	~670 PFU[a]
Viral encephalitis	ssRNA	10–100 viral particles
Viral hemorrhagic fever	ssRNA	1–10 viral particles
West Nile virus	ssRNA	

[a] Particle forming units.

PROTEIN TOXINS

Protein toxins are biomacromolecules produced by living organisms such as bacteria, viruses, plants, and animals. Although produced by biological organisms, they share many similarities to chemical weapons. Toxins categorized by the CDC (Table 15.5) include the botulinum toxin produced by *Clostridium botulinum*, the diphtheria toxin produced by *Cornebacterium diphtheria*, the epsilon toxin produced by *Clostridium perfringens*, Staphylococcal enterotoxin B produced by *Staphylococcus aureus*, the ricin toxin produced by the castor bean plant *Ricinus communis*, and the abrin toxin produced by the *Abrus precatorius* plant. Related toxins include volkensin (*Adenia volksenii*) and viscumin (*Viscum album*), a toxin from mistletoe. Several of the species produce multiple types, or isoforms, of the toxins. The bacteria have to be cultured to obtain significant amounts of the bacterial toxins, which are purified by solubility and size at low temperature. Toxins can be digested (or cut up) by proteases and inactivated by heat, which unfolds the proteins. Unlike the bacteria themselves, the toxins, when purified out of the bacteria, are not contagious. However, the toxins are extremely toxic—they are more toxic per mol than chemical agents—so detection methods have to be extremely sensitive. The sizes and toxicity of these protein toxins are also shown in Table 15.5. Their effects may be reversed using antitoxins. Antitoxins are available for diphtheria and botulism and an antitoxin is under development for ricin. A vaccine was first developed by Behring in 1913 for diphtheria. Where applicable, the bacterial infections can be treated using antibiotics.

In addition to its classification as a CDC Category B agent, ricin is listed as a Schedule I controlled substance under the Chemical Weapons Convention. Additionally, ricin is a select agent, and its use, possession, or transfer is regulated under US and international law. The major reason ricin is a documented public health threat and has been used

Table 15.5 Sizes and toxicity of protein toxins (in kilo Daltons, kDa)

Protein toxin	Size (kDa)	Toxicity (LD50)
Abrin	60 kDa (A&B chains linked by a disulfide bond)	0.7 µg/kg
Botulinum	150 kDa (single chain)	1.3–2.1 ng/kg (intravenous), 0.7–0.9 µg or 10–13 ng/kg (inhalation), 70 µg (ingestion)
Diptheria	63 kDa (535 amino acids, A&B chains linked by a disulfide bond)	0.1 µg/kg
Epsilon	32.9 kDa prototoxin	70–100 ng/kg (intravenous)
Enterotoxin B	28.366 kDa (239 amino acid, single chain)	0.02 µg/kg
Ricin	66 kDa (A&B chains linked by a disulfide bond)	350–700 µg/70 kg (inhalation)

in so many cases is that it is easy to obtain. The seeds (Figure 15.11) are legal for purchase and available on the internet and from nurseries. The castor bean plant that produces ricin is legal and a common ornamental. The plant can be grown at home without any special precautions. Abrin, a related toxin, is produced by the rosary pea, another plant that is also legal and easy to obtain, although the worldwide quantity is small as compared to the castor bean plant.

Ricin and abrin share the same mode of action biochemically: they irreversibly block protein synthesis at the cellular level. The ricin and abrin protein toxins are heterodimers consisting of two chains, the A and B chains, which are linked by a disulfide bond and fold into a globular structure; their structures are shown in Figures 15.12 and 15.13 respectively. Both toxins are type 2 ribosome inactivating proteins with essentially the same mechanism. Ricin is part of a family of proteins coded from a 28-gene region; the ricin-coding region is annotated in the National Center for Biotechnology Information's Genbank. It is a lectin-type glycoprotein that contains a large number of mannose sugars that enable it to bind to cells harboring mannose receptors, especially endothelial cells. The B chain (262 amino acid residues, 34 kDa) can dock the protein to the cell surface using its carbohydrates that contain terminal N-acetyl galactosamine or β-1,4-linked galactose residues. Ricin toxin A and B chains were observed to translocate across endosome membranes. After the ricin toxin B chain facilitates the A chain's crossing into the cytosol of the cell, the A chain cleaves from the B chain and is taken up by the Golgi apparatus and disables the ribosomes of the rough endoplasmic reticulum. The ricin A chain (267 amino acid residues, 32 kDa) was observed to inhibit protein synthesis by interfering with the function of the ribosome while the B chain bound to carbohydrates. After the protein toxin is brought into the cell, the A chain is an N-glycoside that as a type 2 ribosomal inactivating protein (or type 1 without the B chain) by depurinating adenine cleaves a specific glycosidic bond in the 28S subunit of the ribosomal RNA, thereby inhibiting protein synthesis, and initiating apoptosis, or programmed cell death. Apoptosis can lead to fluid and protein leakage from the cell membrane or membrane edema.

Figure 15.11 Castor beans containing ricin. (From Muséum de Toulouse, https://www.flickr.com/photos/museumdetou-louse/4092392632/in/photolist-7eCzKE-fWubB4-9dPBoA-Z5y9XW-9dLxZT-3n7bh7-9dLy3k-8wA2UP-yes7Mh-psdwNU-jdYBL-fEzXeV-59mxNB-daGk6x-QUVraH-8wA2GP-9dLyoX-7PwR2k-5qPu6C-5tYs96-cuiGaU-Ss1zQp-XAGSjB-a7BrAY-qBXSs9-7jc5K1-4SWMTi-6dPFhv-q7MN1X-7yNFxg-9JmNK9-UMHRxe-5YZxmA-92SNKJ-7yBMbA-bAMxMs-2gjUJv-9geFAB-5FtxVj-Wpmj86-XBd6t6-bwC9vZ-kFLP4-Xp22Gi-mZowXP-8UDxRs-YVcXT1-U6gTUN-4fmdN-5BBT2i.)

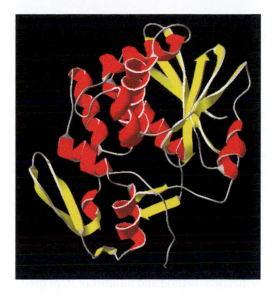

Figure 15.12 Ricin A chain structure (1ift.pdb).

In 1888, a German scientist, Peter Hermann Stillmark, extracted and described the ricin lectin toxin and disseminated his findings in his thesis. The extraction procedures of ricin and abrin are characterized by common features of protein extraction methods. Crude extraction methods have been published in rogue publications and over the internet and analytical procedures are available in libraries in peer-reviewed biochemical research journals to isolate functional ricin. The methods range from simple home-kitchen methods to more sophisticated laboratory-based methods and range from the "poor man's James Bond," "silent death" method, and the US Army patented method, to name a few. Even the crude methods could yield a deadly mixture. The simplest methods for purifying ricin require widely available components and chemicals. While more sophisticated methods require various forms of chromatography including gel filtration, affinity chromatography, ion exchange chromatography, and fine protein liquid chromatography and yield purer forms of the chemical target, several methods, especially those available open-source on the internet, rely on solubility coupled with paper towels or coffee filters for the separation.

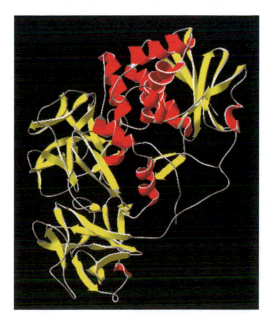

Figure 15.13 Abrin A chain structure (1abr.pdb).

Figure 15.14   Ricin purification scheme (10 step).

A general purification scheme that could be used by a perpetrator is shown in Figure 15.14. The globular, water-soluble protein is extracted from the mash that remains from isolation of the castor bean oil.

Although the goal of the methods is the same—to extract ricin from castor beans—the variations in chemical reagents and filter papers can aid the forensic scientist in determining which method was used and in demonstrating links to a perpetrator. Possession of these papers or pamphlets and the chemicals and castor beans needed to extract ricin as shown in Table 15.6 may be incriminating. While their individual presence may not be remarkable, collectively, they could demonstrate a method or intent. Characteristic paraphernalia of a perpetrator using this method would include sodium hydroxide or lye, acetone, glass jars, coffee filters, a glass or metal blender, a glass, an opaque plastic or metal funnel, Epsom or table salt, paper towels, and a scraping tool. Typically, two ricin toxins are recovered from the bean (termed by various authors as ricin d and ricin e, RCA II, ricin peak II, ricin D and OR, ricin R and ricin Q, RCL III & IV). Depending on the purification method and number of steps used, the ricin may contain considerable salt. For use as a WMD agent, however, ricin (or abrin) would most likely be introduced to water sources, cold foods, or beverages or be aerosolized and dispersed.

Botulinum toxin (Figure 15.15) is also a Category B agent. There are seven types of botulinum toxin, botulinum A, B, C, D, E, F, and G. Types A and B are used medically under the brand name Botox to treat muscle spasms and neuromuscular diseases because they have the longest lasting effects. Botulinum toxin acts biochemically by being transported in a vesicle to nerve endings that use the neurotransmitter acetylcholine. Under the acidic conditions in the nerve cell, the toxin is converted to the active form by its self-proteolytic function and exits the vesicle membrane. In the cytoplasm, botulinum toxin is an enzyme that cleaves the SNAP-25 proteins of the SNARE protein family. The cleaved SNAP-25 cannot mediate vesicle fusion with the host cell membrane, preventing acetylcholine release from axon cell endings and resulting in paralysis.

The function of *Staphylococcal* enterotoxin B is to make the infected individual prone to infection by acting as a superantigen that activates T-cells and renders them unable to fight infections that may be caused by other antigens. Enterotoxin B stimulates an inflammatory response by cross-linking major histocompatibility complex class II molecules with T-cell receptors to activate a large number of T-cells (up to 20%).

Diphtheria toxin functions by inhibiting protein synthesis. Diphtheria is another protein toxin with two domains, R (C-terminal domain) and T (Figure 15.16). The R domain binds to a receptor on the cell's surface and a change in pH induces a conformational change in the T domain to trigger its insertion into the endosomal membrane. Then the R domain is transferred into the cytoplasm. In the cytoplasm, the toxin then transfers ADP-ribose from NAD to a diphthamide residue of the eukaryotic elongation factor 2 (eEF-2).

Table 15.6   Methods and chemicals used in ricin extraction and crude and analytical purification schemes

Step	Methods employed
1	Obtain castor beans: Internet, local sources
2	Soaking in base/lye, removal using tweezers/fingernail/razor blade/pliers/hammer
3	Grinding, blending, mashing with stick/spoon
4	Acetone, hexane, ether, diethyl ether, ethyl acetate
5	Filtration using coffee filters, Buchner funnel, or fluted filter paper, decanting
6	Size exclusion chromatography, FPLC, HPLC, affinity chromatography, coffee filter
7	Epsom salt (magnesium sulfate), ammonium sulfate, sodium sulfate
8	Dialysis, wash steps
9	Room temperature, heat (inactivates)
10	Scraping, pressure grinding

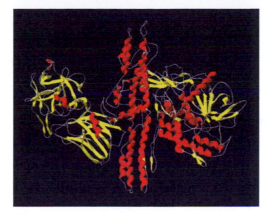

Figure 15.15   Botulinum toxin (1bta.pdb).

Viscotoxins are protein toxins produced by the European mistletoe (*Viscum album*) plant. The most toxic is visco-toxin A3 and the least toxic is viscotoxin B. Viscotoxins are toxic to their target cells as a result of their ability to destabilize and disrupt the cell membrane, especially those containing negatively charged phospholipids.

Prions are a subset of protein toxins. Prions were first discovered in 1982 by Stanley Pruisiner. They consist of mutated or misfolded proteins that have been documented to cause fatal neuro-degenerative disorders with symptoms of dementia. Kuru, Creutzfeldt–Jacob disease (CJD), Gerstmann–Straussler syndrome (GSS), fatal familial insomnia, scrapie in sheep and goats, and bovine spongiform encephalopathy (BSE, also known as Mad Cow disease) are all diseases with a prion pathogen. Prions are proteins that are highly insoluble and highly resistant to proteases. They are also resistant to disinfectants, heat and sterilization processes (as they are already misfolded proteins), and sunlight. They evoke no immune or inflammatory response in the victim. There is no treatment yet for prion diseases.

SMALL MOLECULE TOXINS

Small molecule toxins include mycotoxins, secondary metabolites produced by several fungal molds and spores and some animals. Mycotoxins are produced by over 50 fungal species including the following: *Fusarium* species *Fusarium moniliforme, F. equiseti, F. oxysporum, F. culmorum, F. solani, F. avenaceum, F. graminearum, F. roseum,* and *F. nivale; Aspergillus* species *A. flavus, A. parasiticus, A. niger, A. alternata, A. ochraceus, A.clavatus, A. fumigatus, A. versicolor,* and *A. ustus; Monographella nivalis; Acremonium crotocinigenum; Chaetomium* species *C. globosum,* and *C. cochliodes; Penicillium* species *P. citreoviride, P. citrinum, P. islandicum, P. expansum, P. roquefortii, P. claviforme, P. aurantiogriseum, P. crustosum, P. griseofulvum, P. rubrum, P. brunneum, P. kloeckeri, P. rugulosum, P. hirsutum, P. viridictum, P. verrucosum; Myrothecium* species *M. roridum* and *M. verrucaria; Dendrodochium* spp.; *Cylindrocarpon* spp.; *Stachybotrys chartarum; Eurotium chevalieri; Rhizoctonia leguminicola; Alternaria alternata;*

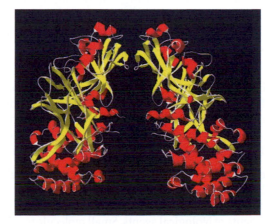

Figure 15.16   Diphtheria toxin (1f0l.pdb).

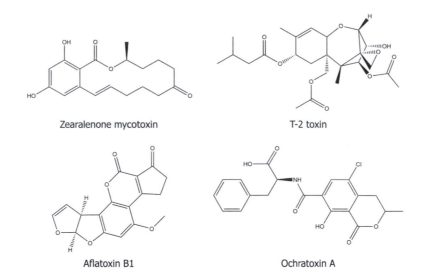

Zearalenone mycotoxin   T-2 toxin

Aflatoxin B1   Ochratoxin A

Figure 15.17 Structures of biological toxins.

Trichothecium roseum; and *Trichoderma viride*. Mycotoxins vary in size from a few hundred to approximately 1000 Da and vary in their toxicities. Unlike protein toxins, small molecule toxins may be absorbed by the skin. Toxin weapons may also be ingested, injected, or inhaled. Mycotoxins are cytotoxic to bacterial, animal, and plant cells; they disrupt cellular reproduction processes (e.g., protein, DNA, and RNA synthesis) and membrane function.

An example of a small molecule toxin is the T-2 mycotoxin ((2α,3α,4β,8α)-4,15-bis(acetyloxy)-3-hydroxy-12,13-epoxytrichothec-9-en-8-yl 3-methylbutanoate) produced by the *Fusarium* fungus species that may be ingested from moldy grains; its toxicity is relatively low at 1 mg/kg compared to other toxins. When isolated, it is a yellow oily liquid. The United States accused the USSR of using T-2 as a biological weapon delivered by low-flying aircraft from the 1970s through the 1990s in Laos, Kampuchea, and Afghanistan. It may also have been in an Iraqi missile that detonated in a US military camp in Saudi Arabia in 1991 during the Gulf War.

Aflatoxin is a small molecule toxin produced by *Aspergillus* species *A. flavus* and *A. parasiticus* fungi. Ochratoxin is a small molecule toxin produced by *Aspergillus ochraceus* and *Penicillium viridictum*. Selected biological toxins are shown in Figure 15.17.

METHODS OF DETECTION AND IDENTIFICATION

There are several methods used to detect and identify biological weapon agents. These include biochemical, molecular biological, and microbiological methods such as culturing, colony staining, negative staining electron microscope visualization, plaque assays, immunochromatographic tests, enzymatic activity assays, enzyme-linked immunosorbent assays (ELISA), polymerase chain reaction (PCR) assays, restriction fragment length polymorphism (RFLP) assays, immunofluorescence assays, and genome sequencing. These methods are beyond the scope of this book but are discussed in detail elsewhere.

Plaque assays are used to detect viruses including influenza. Culturing and colony staining are used to differentiate bacteria species. PCR and RFLP assays can be used with any biological weapon source DNA using primers that bind to highly specific targets in the genomic DNA; species with an RNA genome can be probed using reverse-transcriptase PCR and then RFLP, if desired. For example, three foodborne pathogens were shown to be identified and differentiated using PCR and post-PCR high-resolution melt curves of the copied DNA amplicons.

As they are large molecules, protein toxins are not volatile and thus are not amenable to investigation using GC-MS, although chemical methods including liquid chromatography–mass spectrometry (LC-MS), and especially matrix-assisted laser desorption/ionization–mass spectrometry (MALDI-MS) are frequently used to determine the size and identity of toxins and protein components of biological agents. Small molecule toxins are also separated and identified using LC-MS techniques. In a study, HPLC was used with electron impact ionization and tandem mass spectrometry to detect and identify botulinum neurotoxins A, B, E, and F by their different masses which eluted at different retention times. The drawback to MS methods is that they are complicated and expensive, require highly trained scientists to operate, and are nonspecific. ELISA and immunochromatographic assays are also used to detect protein toxins.

DEATH BY RADIONUCLIDE

Alexander Litvinenko was a former lieutenant colonel for the Russian Federal Security Service (FSB) and KBG officer. Facing prosecution by Russian courts for accusing Russian President Vladimir Putin and the FSB of staging and orchestrating several events including human rights violations, he found political asylum from Russia in the United Kingdom.

On November 1, 2006, he met with two former KGB officers in London and had lunch with Mario Scaramella, an Italian officer and "nuclear expert." After lunch, he began having severe diarrhea and vomiting. Later, he could not walk without assistance and even became unconscious and experienced extreme pain and was transported to the hospital.

Litvinenko died only 3 weeks later on November 22, 2006. On his death, he was confirmed to have died of acute radiation syndrome caused by Polonium-210 poisoning. The Polonium-210 had destroyed his organs and he died of heart failure. Polonium-210, like other fissionable elements, spontaneously disintegrates and releases harmful gamma radiation. Gamma radiation is highly energetic and causes the ionization of atoms in the body which causes chemical changes to the body's organ systems and metabolic processes. As the radiation is tasteless, colorless, and odorless, it was not detected until the symptoms presented themselves, causing doctors to look for a root cause, however Polonium-210 releases such a small amount of gamma radiation that Geiger counters failed to detect. In addition to gamma radiation, Polonium-210 disintegrates to Lead-206 with the release of primarily alpha particles. Its half-life is 138 days. The cause of death was determined using blood and urine samples sent to the United Kingdom's Atomic Weapons Establishment that were tested using gamma spectroscopy. Although the scientists did not detect gamma rays, they did detect a small spike of 803 keV attributed to the gamma ray signal from Polonium-210. Spectroscopic methods also detected alpha radiation in a larger urine sample. Although alpha particles do not penetrate human skin or even a sheet of paper and are not considered harmful when handling radioactive elements, alpha particles can be fatal if ingested or inhaled. Also, alpha emitters usually do not trigger radiation detectors and can more easily be smuggled. Polonium-210 is a highly controlled and extremely rare element with limited industrial uses—only about 100 g are produced annually worldwide. Most is produced through neutron bombardment in nuclear reactors. The National Research Council licenses it for use in antistatic devices—but it would require dismantling 30,000 such devices to obtain a lethal dose of the element.

Polonium is considered to be 250,000 times more poisonous, by mass, than hydrogen cyanide. The radioactivity detected in Litvinenko's body corresponded to a dose of approximately 10 μg of Polonium-210—more than 200 times the lethal dose of 50 ng. As he became ill gradually, it was difficult to determine the cause of death. Investigations found that he was poisoned through his cup of tea by the Russian KGB officers he had met with, Andrey Lugovoy and Dmitry Kovtun, and the chemical signature of the Polonium-210 matched that produced in a Russian nuclear reactor.

BIBLIOGRAPHY

Blum, D.A. 2012. Poison for assassins. www.wired.com/2012/07/polonium-210-and-assassinations/ (accessed February 1, 2018).

Litvinenko: A deadly trail of polonium. 2015. www.bbc.com/news/magazine-33678717 (accessed February 1, 2018).

Poisoned Russian ex-spy "sadistic, slow murder." 2006. www.spiegel.de/international/poisoned-russian-ex-spy-sadistic-slow-murder-a-450538.html (accessed February 1, 2018).

NUCLEAR WEAPONS

Nuclear chemistry is defined as the study of reactions that result from changes in the nuclei of atoms. *Radioactive decay* is defined as the spontaneous disintegration of unstable isotopes resulting in the release of radiation or energy. Radiation is given off from natural sources including soil, rocks, water, and the atmosphere as well as man-made sources including cell phone batteries, microwave ovens, and lasers. Other sources of radiation include x-rays, radioactive tracers used in medicine, and smoke detectors. Radiation is odorless, colorless, and tasteless. *Ionizing radiation* is that which can pull electrons from atoms and compounds such as water to form hydroxyl radicals (OH•) which can impart chemical changes to DNA, proteins, and other molecules that make up body tissue when stealing electrons to complete its octet. An example of ionizing radiation is x-ray radiation.

NUCLEAR CHEMISTRY REACTIONS AND TYPES OF RADIATION

Nuclear fusion and fission reactions result in the formation of new elements and may result in the emission of particles, including alpha and beta particles, and gamma radiation. These types of particles and radiation are shown in Table 15.7. Radioactive decay can be written using nuclear equations. In balancing nuclear equations, balance the mass numbers (superscripts) so that the masses on the left of the arrow and the masses on the right of the arrow are equal. The (electrical) charges (subscripts) should be balanced using stoichiometric coefficients so that the charges on the left of the arrow and the right of the arrow add up to zero.

Alpha (α) particles are the heaviest of the particles and are equivalent to a helium nucleus. They consist of two protons and two neutrons. Due to their size and mass, they can be stopped relatively easily with only a single sheet of paper, thin clothing, or even one's own skin to provide protection from these particles in the air. However, alpha particles can be very dangerous if they are ingested as they, like other particles, can act as ionizing radiation and lead to cancer formation.

Beta (β) particles are formed when neutrons spontaneously decompose to form a proton by ejecting a high-energy electron. Iodine-131 decay leads to beta particle formation. The other product is xenon-131. Although the penetrating power of beta particles is limited, it is greater than that of alpha particles. Beta particles are stopped by a layer of clothing, firefighter's gear, less than an inch of plastic, and 5 mm of aluminum. Like alpha particles, beta particles are dangerous if taken internally.

Other nuclear particles include protons, neutrons, deuterons, and positrons. As the name suggests, protons consist of one proton. Neutrons consist of one neutron but can spontaneously decompose to form a proton and a beta particle (high-energy electron) in β-decay.

$$\ce{^1_0 n} \rightarrow \ce{^1_1 p} + {}^{0}_{-1}B$$

A deuteron is a deuterium atom that consists of one proton, one neutron, and one electron. They can be formed by the collision of a proton and a neutron.

A positron is a positively charged particle similar to an electron. At high energy, protons can fuse together to form a deuteron and emit a positron. Positrons can also be emitted when elements, such as carbon-11, decompose.

$$^1_1 p + {}^1_1 p \rightarrow {}^2_1 d + {}^0_1 B$$

$$^{11}_{6} C \rightarrow {}^{11}_{5} B + {}^0_1 B$$

Table 15.7 Radioactive particles

Particle	Common name	Protection
$^4_2 He$ or $^4_2 a$	Alpha particle or helium nucleus	Single sheet of paper, thin clothing, skin
$^2_1 H$ or $^2_1 d$	Deuteron	—
$^0_{-1} B$ or $^0_{-1} e$	Beta particle or electron	Layer of clothing, firefighter's gear, less than an inch of plastic, 5 mm aluminum
$^0_1 B$ or $^0_1 e$	Positron	—
$^1_0 n$	Neutron	—
$^0_0 y$	Gamma rays	Several feet of concrete, inches of steel, or an inch of lead

Deuterons can fuse to form alpha particles or the nuclei of helium-4 atoms.

$$^2_1d + {}^2_1d \rightarrow {}^4_2He$$

Positrons can collide with electrons to give off energy in the form of gamma radiation. Gamma radiation is extremely penetrating and several feet of concrete, inches of steel, or an inch of lead is needed to stop it. Exposure to gamma radiation is extremely dangerous if it is injected, inhaled, ingested, or absorbed. Gamma (γ) radiation does not have an associated particle but is high-energy electromagnetic radiation similar to x-rays released when radionuclides decay. Positrons that collide with electrons cause both to be annihilated but give off energy in the form of gamma rays.

$$^0_1B + {}^0_{-1}B \rightarrow 2{}^0_0\gamma$$

Table 15.7 shows the many types of radioactive particles.

Metastable cobalt-60 can give off gamma radiation without decomposing to form a new element or isotope.

$$^{60}_{27}Co^m \rightarrow {}^{60}_{27}Co + {}^0_0\gamma$$

Table 15.8 shows the radiation emitted by commonly encountered radionuclides.

RADIOACTIVE DECAY AND HALF-LIFE

Radioactive decay, or a decrease in radioactivity, follows first-order kinetics. The *half-life* ($t_{1/2}$) is the time it takes an isotope's activity to be reduced by half. The disintegration of an isotope may occur frequently or rarely. Iodine-131 has an 8-day half-life while uranium-238 has a 4.47-billion-year half-life. The half-life can be determined by

$$\frac{N_t}{N_0} = 0.5^n \quad \frac{N_t}{N_0} = 0.5^{t/t_{1/2}} \quad t_{1/2} = 0.693/k \tag{15.1}$$

where:
- N_0 represents the number of neutrons present at $t = 0$
- N_t represents the number of neutrons present at time t
- n is equivalent to the number of half-lives as shown in Equation 15.1

Table 15.8 Radiation emitted by commonly encountered radionuclides; all nuclides with more than 83 protons are radioactive

Isotope	Atomic number	Alpha	Beta	Gamma
Tritium (H-3)	1		•	
Phosphorus-32	15		•	
Cobalt-60	27		•	•
Strontium-90	38		•	
Technetium-99	43		•	•
Iodine-131	53		•	•
Cesium-131	55			•
Cesium-137	55		•	•
Iridium-192	77		•	•
Polonium-210	84	•		
Radon-222	86	•		
Radium-226	88	•		•
Thorium-232	90	•		•
Uranium-235	92	•		•
Plutonium-239	94	•	•	•
Americium-241	95	•		•

Radioactive elements are termed *radioisotopes*, although the term *radionuclide* is used when discussing specific radioisotopes, such as uranium-235.

HISTORY OF NUCLEAR CHEMISTRY AND RADIOACTIVITY

Uranium was discovered by Martin Heinrich Klaproth from the mineral pitchblende in 1789 but its radioactive properties were elucidated by Henri Becquerel over 100 years later in 1896. X-rays were discovered in 1895 by Wilhelm Conrad Roentgen. During this time, Maria "Marie" Skłodowska-Curie and Pierre Curie showed that the minerals pitchblende and chalcolite were complex mixtures and contained multiple radioactive elements including thorium. These were exciting times in chemistry and physics. Roentgen was awarded the Nobel Prize in Physics in 1901 for the discovery of x-rays. Becquerel and Pierre and Marie Curie shared the 1903 Nobel Prize in Physics for their discoveries of spontaneous radioactivity and the radiation phenomena of other elements respectively.

In 1898, Marie Curie discovered the radioactive isotopes radium and polonium while studying radiation from pitchblende and chalcolite during her research for her Ph.D. thesis; for this work, she was awarded the 1911 Nobel Prize in Chemistry. (In the 1920s, radium was sold and used as a "rejuvenating" tonic in bottled water. In 1933, Drs. Gettler and Norris from the Office of the Chief Medical Examiner in New York City reported on a case involving a 52-year-old man who was poisoned after consuming about 1400 bottles of "Radithor," which contained $2\,\mu g/60\,mL$ of radium per bottle.)

In the years leading up to World War II, several new elements were discovered. Others were prepared by bombarding naturally occurring elements with neutrons, helium atoms, and deuterons. Several of these discoveries and innovations were celebrated with the awarding of Nobel Prizes as shown in Table 15.9. Scientists began synthesizing nuclides in 1919. Ernest Rutherford reported his synthesis of oxygen-17 from nitrogen-14 on bombardment with alpha particles in that year. (Rutherford was previously awarded the Nobel Prize in Chemistry in 1908 for his studies on the decay of elements and chemistry of radioactive elements). Irène Joliot-Curie (daughter of Marie and Pierre Curie) and her husband Frédéric Joliot-Curie synthesized phosphorus-30, the first nonnatural radionuclide, by bombarding aluminum-27 with alpha particles using a linear accelerator. They shared the 1935 Nobel Prize in Chemistry for their work synthesizing new radioactive elements.

Otto Hahn, Lise Meitner, and Fritz Strassmann discovered nuclear fission in 1938 by bombarding uranium with neutrons (discovered in 1932 by James Chadwick which led to his 1935 Nobel Prize in Physics); Otto Hahn won the 1944 Nobel Prize in Chemistry for this contribution. Starting in the 1930s and 1940s, over 100 new elements and isotopes

Table 15.9 Nobel prizes awarded in nuclear chemistry

Award recipient	Nobel prize, year	Noted contribution
Wilhelm Conrad Röntgen	Nobel Prize in Physics, 1901	"in recognition of the extraordinary services he has rendered by the discovery of the remarkable rays subsequently named after him"
Antoine Henri Becquerel, Pierre Curie, Marie Curie	Nobel Prize in Physics, 1903	"in recognition of the extraordinary services he has rendered by his discovery of spontaneous radioactivity" (Becquerel), "in recognition of the extraordinary services they have rendered by their joint researches on the radiation phenomena discovered by Professor Henri Becquerel" (Curies)
Ernest Rutherford	Nobel Prize in Chemistry, 1908	"for his investigations into the disintegration of the elements, and the chemistry of radioactive substances"
Marie Curie	Nobel Prize in Chemistry, 1911	"in recognition of her services to the advancement of chemistry by the discovery of the elements radium and polonium, by the isolation of radium and the study of the nature and compounds of this remarkable element"
Frédéric Joliot and Irène Joliot-Curie	Nobel Prize in Chemistry, 1935	"in recognition of their synthesis of new radioactive elements"
James Chadwick	Nobel Prize in Physics, 1935	"for the discovery of the neutron"
Otto Hahn	Nobel Prize in Chemistry, 1944	"for his discovery of the fission of heavy nuclei"
Edwin Mattison McMillan and Glenn Theodore Seaborg	Nobel Prize in Chemistry, 1951	"for their discoveries in the chemistry of the transuranium elements"

Source: Nobel Media, https://www.nobelprize.org/.

Figure 15.18   Uranium cake, a form of uranium oxide. (From Nuclear Regulatory Commission, https://www.flickr.com/photos/nrcgov/14492248719/in/photolist-6QvFY-o5CyDe-qpkBr9-UpyxVY-9r6YPw-8LxCPz-7mkJt-7mkHz-RMAiwt-GrKFz9-7mkMA-6QvFZ-7mkKG-bnfvYg-7CTYvE-7mkM9-6QvG1-bnfUdn-dco2vD-7mkLx-HoncPz-6hSft3-7mkNm-6RtTw-pUPHbV-YkWBt5-RuCfvN-Hq8Q1t-BNXQpL-GrKAnq-Mw1Dfs-GqXsJF-GXgXJA-GzeJ3U-BhKABH-GNYTA9-GRaPEt-He3bhq-Tnn4kW-UAc6tE-TqdxZi-7V8YVE-GMXidi-wmLaHk-Mc7Prq-vprFG8-FYtZFH-vprzfF-wmjiLa-w4PuEc.)

were prepared by Glenn Seaborg and his colleagues at the University of California at Berkeley. There, cobalt-60, plutonium-239, all transuranium elements through 102, and several isotopes were prepared using a linear accelerator or a cyclotron. Plutonium-239 and neptunium were synthesized in 1940. Seaborg and Edwin McMillan were awarded the 1951 Nobel Prize in Chemistry for their work in the discovery, synthesis, and investigation of ten transuranium elements. A photo of uranium cake, a form of uranium oxide, is shown in Figure 15.18.

MODES OF RADIOACTIVE DECAY

Both uranium-235 and uranium-238 can decay to thorium by alpha decay. Thorium and the subsequent decay products also decay by alpha decay until lead-214 is reached; finally, two cycles of beta decay are followed by alpha decay until the stable isotope lead-206 is reached.

Plutonium-239 decays to uranium-235 with the emission of an alpha particle (He nucleus).

$$^{239}_{94}Pu \rightarrow\ ^{235}_{92}U + ^{4}_{2}He$$

Uranium-235 decays to thorium-231, also with the emission of an alpha particle.

$$^{235}_{92}U \rightarrow\ ^{231}_{90}Th + ^{4}_{2}He$$

The natural decay of uranium-238 to lead-206 proceeds via alpha decay and beta decay.

$$^{238}_{92}U \rightarrow\ ^{234}_{90}Th + ^{4}_{2}He$$

$$^{234}_{90}Th \rightarrow\ ^{234}_{91}Pa + ^{0}_{-1}B$$

$$^{234}_{91}Pa \rightarrow\ ^{234}_{92}U + ^{0}_{-1}B$$

$$^{234}_{92}U \rightarrow\ ^{230}_{90}Th + ^{4}_{2}He$$

$$^{230}_{90}Th \rightarrow\ ^{226}_{88}Ra + ^{4}_{2}He$$

$$^{226}_{88}Ra \rightarrow\ ^{222}_{86}Rn + ^{4}_{2}He$$

$$^{222}_{86}Rn \rightarrow\ ^{218}_{84}Po + ^{4}_{2}He$$

$$^{218}_{82}Po \rightarrow\ ^{214}_{82}Pb + ^{4}_{2}He$$

$$^{214}_{82}\text{Pb} \rightarrow {}^{214}_{83}\text{Bi} + {}^{0}_{-1}B$$

$$^{214}_{83}\text{Bi} \rightarrow {}^{210}_{82}\text{Po} + {}^{0}_{-1}B$$

$$^{214}_{84}\text{Po} \rightarrow {}^{210}_{82}\text{Pb} + {}^{4}_{2}\text{He}$$

$$^{210}_{82}\text{Pb} \rightarrow {}^{210}_{83}\text{Bi} + {}^{0}_{-1}B$$

$$^{210}_{83}\text{Bi} \rightarrow {}^{210}_{84}\text{Po} + {}^{0}_{-1}B$$

$$^{210}_{84}\text{Po} \rightarrow {}^{206}_{82}\text{Pb} + {}^{4}_{2}\text{He}$$

ISOTOPES AND NUCLEAR REACTIONS

Stable isotopes have a ratio of neutrons to protons of one; these isotopes lie on the *belt of stability* on the Periodic Table. Lighter elements and those below the belt of stability undergo *nuclear fusion* reactions in which heavier nuclei are produced from the joining of lighter nuclei. Isotopes that lie below the belt of stability are neutron poor and are characterized by proton-decay or electron capture.

Nuclear fission is a type of nuclear reaction characterized by the splitting of a heavy nucleus. These heavy nuclei isotopes lie above the belt of stability and are not stable. They split into two lighter nuclei with the release of one or more neutrons through nuclear fission reactions. The neutrons released can initiate a *chain reaction* creating self-sustaining nuclear fission reactions that initiate additional fission events until the fissionable material is consumed and reaches the critical mass. The *critical mass* is the minimum amount of fissionable material that can sustain a chain reaction. Heavier elements undergo fission with the dividing point at iron-56.

Other neutron-rich nuclides above the belt of stability undergo beta decay. Uranium-235 and plutonium-239, among others, undergo nuclear fission. These two nuclei were used in the first two atomic bombs produced by the United States. They are also used in nuclear power plants.

In the nuclear fission reaction of uranium-235, the neutrons produced in the reaction propagate a chain reaction in which uranium-235 captures a neutron and undergoes nuclear fission. The reaction continues as long as there is enough uranium-235 to absorb the neutrons produced.

$$^{1}_{0}\text{n} + {}^{235}_{92}\text{U} \rightarrow {}^{92}_{36}\text{Kr} + {}^{141}_{56}\text{Ba} + 3{}^{1}_{0}\text{n}$$

In the nuclear fission reaction of plutonium-239, the neutrons produced propagate a chain reaction and significant amounts of gamma radiation are produced in the reaction.

$$^{1}_{0}\text{n} + {}^{239}_{94}\text{Pu} \rightarrow \left[{}^{240}_{94}\text{Pu} \right] \rightarrow {}^{134}_{54}\text{Xe} + {}^{103}_{40}\text{Zr} + 3{}^{1}_{0}\text{n} \qquad 73\%$$

$$^{1}_{0}\text{n} + {}^{239}_{94}\text{Pu} \rightarrow \left[{}^{240}_{94}\text{Pu} \right] \rightarrow {}^{240}_{94}\text{Pu} + {}^{0}_{0}\text{y} \qquad 27\%$$

USES OF RADIONUCLIDES IN BOMBS

The atomic bomb, "Little Boy," that was detonated over Hiroshima, Japan, on August 6, 1945, was a 12-kiloton bomb composed primarily of uranium-235. Uranium-235 has a natural abundance of only 0.7200% while uranium-238 has a natural abundance of 99.2745%. Uranium-235 can be purified from uranium ore by gas centrifugation, which uses differences in its mass as compared to uranium-238 to separate it. Owing to the low natural abundance, tons of uranium ore are used in processing to obtain kilograms of uranium-235. Material that has undergone this process to obtain uranium with higher than natural quantities of uranium-235 is termed *enriched*. Conversely, the uranium remaining after this separation process is known as *depleted uranium*. The uranium for the "Little Boy" bomb was produced at Oak Ridge National Lab in Tennessee as part of the Manhattan Project. In contrast to the uranium used in power plants for energy production that is enriched to 3%–5%, uranium enriched for use in nuclear weapons is enriched to 85% or more. "Little Boy" consisted of 64 kg of 80% enriched uranium-235. The effects of this atomic bomb were devastating to the human, plant, pet, and livestock population in the area where it was detonated. Over 66,000 lives were lost in less than a minute and over 69,000 people were injured from the blast. Several thousand suffered from short- and long-term effects of radiation poisoning.

Weapons-grade plutonium-239 was produced by bombarding neptunium-237 with neutrons. The second atomic bomb, "Fat Man," that the United States detonated over Japan's city of Nagasaki on August 9, 1945, was a 20-kiloton bomb powered by plutonium-239 and killed in the range of 39,000–80,000 people. It weighed 10,300 pounds and was 128 inches in length and 60 inches in diameter. The 14 pounds of fissile material was packed in the center and a layer of explosives surrounded it. Detonation of the explosives exerted pressure on the core and reduced the volume of the plutonium, thereby increasing its density so that it reached a critical mass. Over 100,000 more people were injured by the blast, radiation burns, and shrapnel damage.

Nuclear weapons are currently held by (in order of acquisition) the United States of America (1945), Russia (1949), the United Kingdom (1952), France (1960), and the People's Republic of China (1964). These five nations are signatories to the 1968 Nuclear Non-Proliferation Treaty along with 184 other nations. Nuclear, non-signatory states that have conducted nuclear tests include India (1974), Pakistan (1998), and North Korea (2006-present). (North Korea signed and withdrew in 2003.) Israel also did not sign and is believed to have nuclear weapons but this has not yet been declared. Several other nations have a goal of acquiring nuclear weapons; this capability needs to be considered by forensic scientists. The 1953 detonation of a 23-kiloton nuclear bomb XX-34 BADGER as part of the Operation Upshot-Knothole test at the Nevada Test Site is shown in Figure 15.19.

RADIOACTIVITY UNITS

Several radioactivity units are encountered in nuclear chemistry literature and publications to describe the quantity of energy produced from the radioactive material and the amount of radiation consumed on exposure. The SI unit of radioactivity, becquerel (Bq), was introduced in 1975. It is defined as energy from radioactive material produced on the decay of one nucleus per second.

$$1\,Bq = 1\,s^{-1}$$

Figure 15.19 23 kiloton XX-34 BADGER 1953 US nuclear bomb test at the Nevada Test Site. (From The Official CTBTO Photostream, https://www.flickr.com/photos/ctbto/4926598654/in/photolist-9o3GNk-ctwL97-5tr7yA-5m5Pz3-5TQrA4-5Kycny-56SZw6-5EVNgC-7NJPbm-65FRKi-5BUKP8-5yuxru-6LJc2a-6DR1Tw-6oshc5-69ncLP-eeZcZ6-66bShT-5BZ2ti-qmT7s4-65xwn1-au2E3G-5syMqM-63mmAe-6STkCC-aeMwEh-6AafCv-8d9pon-cjVxdo-5S2oZ7-5vSRXB-5YAHRq-6TL9R7-67TV3j-7nuSJp-69gtED-5m5AcM-5yopMG-66vxqU-4QT56H-5uLk9i-5YwggJ-62JiQz-8vm797-WDWscd-k5WjR-9pNRCa-9qzJLe-GqC8W-62JiN8.)

A curie (Ci, decay activity) was originally defined as "the quantity or mass of radium emanation in equilibrium with one gram of radium" and was used prior to 1975. It corresponds to 3.66×10^{10} Bq/g. Thus,

$$1\,Ci = 3.7 \times 10^{10}\,Bq$$

Also, prior to 1975, the term *rad* (radiation absorbed dose) was used to define human and animal exposure to radiation. The Grey (Gy) is another unit; it has a value of 100 rad.

$$1\,rad = 0.01\,J/kg$$

$$1\,Gy = 100\,rad$$

The *rem* was introduced in 1976 as a unit for quantifying the biological effects of radiation and the SI unit of an ionizing radiation dose is the Sievert (Sv), defined as the biological effect from the deposit of a joule of radiation energy in a kilogram of human tissue. The ionizing radiation of interest in biology is that which can remove an electron from water. The cation H_2O^+ reacts with water to form hydronium ion, H_3O^+, and a highly reactive hydroxyl free radical ($OH^\bullet$).

$$1\,rem = \text{Biological damage from } 0.01\,J/kg$$

$$1\,Sv = 100\,rem = 1\,J/kg$$

It is not only the emitted radiation that can lead to human health issues from working with or being exposed to radioactive isotopes—plutonium and the other metals are also toxic because they are heavy metals.

DETECTION AND IDENTIFICATION OF RADIOACTIVE MATERIAL

The Geiger–Mueller counter (Figure 15.20) is a portable and inexpensive tool that can be used to detect the presence of radioactive material such as that found naturally occurring in rocks, in discarded medical equipment, and released by the burning of coal in coal-fired power plants. The counter determines radiation levels by measuring alpha and beta particles and gamma rays that enter the test window in the gaseous phase by their abilities to ionize argon (Ar) gas to Ar^+ and free electrons sealed in the detector. The ionized electrons migrate in a negatively charged shell to the positively charged anode (and the argon ions migrate toward the negatively charged shell), creating an electric current that is amplified and counted in the detector. The output is reported in counts per minute or Becquerels and is proportional to the number of radioactive decay events resulting from the radioisotope. The drawback of this method is that it does not identify the type of radiation being emitted. However, the risk associated with suspected radioactive sites can be ascertained quickly. For example, Geiger counters are used to detect radon seeping in from veins in basements in the intermountain west—especially in and around Denver, in western parts of Washington State, Colorado, and on the east coast of the United States along the Appalachian Mountains. Coal-fired power plants used around the world release polonium, radon, and thorium radionuclides contained in coal. Radionuclide contamination on responders, victims, and equipment can also quickly be determined using the Geiger counter. The radiation level quantity can be determined if the instrument is calibrated with standards.

A scintillation counter is another tool that can be used to detect and quantitate the radiation emitted by a substance. It can be used too in the identification of gamma rays using a luminescent material that absorbs energy of incoming material and re-emits the energy in the form of light. In a scintillation counter, the level of radioactivity is proportional to the intensity of the light emitted by the "phosphors" in contact with the samples. Gaseous radioactive particles enter the instrument in the sodium-iodide crystal scintillator and make optical contact, producing a luminous photon "glow" or "scintillation" that activates the photo cathode. A photoelectron is produced that, in turn, produces secondary photons. The signal is amplified with a photomultiplier and is detected using a voltmeter or digital counter. Cobalt-60 can be detected and identified by its characteristic peaks in its gamma spectrum, one at 1173.2 KeV and another at 1332.5 KeV. A good scintillation detector can easily resolve the two peaks. Scintillators

Figure 15.20 Geiger counter.

are used by the Department of Homeland Security, first responders, and nuclear power plant monitors for on-site uranium detection.

Dosimeter badges employ photographic film and are worn by scientists and workers that are routinely exposed to radiation in the workplace. These badges are used to detect if a person has been exposed to radioactivity and if the dosage exceeds allowed quantities in a given period as the radioactivity "clouds" or partially develops the film.

Portable radioisotope identification can also be performed using a smartphone attached to a device. One device detects 115 isotopes to 393 isotopes from 20 KeV to 3.0 MeV. These can simultaneously detect and identify cesium-134 at 605 and 796 KeV and cesium-137 at 662 KeV.

Other instruments, such as the AN/PDR 77 and ADM 300, can capture, analyze, and report alpha, beta, and gamma radiation and aid in the determination of the isotope present by analyzing the decay rate. For example, plutonium-239 has a decay rate of 1.923 microcurie/m^2. Even with the availability of these instruments and tools, new detection tools are needed to detect and identify the source of gamma radiation through shipping containers and at a distance.

CASES OF ACCIDENTAL POISONING WITH RADIOLOGICAL MATERIAL

Radiological material is used in nuclear power plants and medical equipment, among other uses. Radioactive material is used by over 10,000 hospitals worldwide in nuclear medicine equipment (e.g., magnetic resonance imaging machines), isotopic generators, therapy units, radiopharmaceuticals, and computed tomography imaging equipment. Ionizing radiation can be used in chemotherapy to kill cancerous tissue using radiation sources administered externally or internally. For example, iodine-131 is used in thyroid therapy, phosphorus-32 is used in leukemia therapy, cobalt-60 is used in cancer therapy, cesium-131 is used in therapy for prostate cancer, and inidium-192 is used to treat coronary disease. Radioisotopes are also used in diagnostic radiology as "tracers" to generate an "image" of an organ or tissue. Technitium-99 is the most commonly used radioisotope in diagnosis and is used to image the bones, circulatory system, and various organs. Gallium-67 is used in brain tumor and other imaging, iodine-123 is used to image thyroxine production in the thyroid gland, and tellurium-201 is used in heart muscle and coronary artery imaging. Radiological material is also used in public health applications including food irradiation for sterilization (cobalt-60), well logging, detectors for radioactive material, x-ray fluorescence instruments for lead paint analysis, and smoke detectors. Radiological material is also used in business applications including thickness gauges, rifle sights, static eliminators, moisture and density gauges, and luminous dials. Radioisotopes have also been used for centuries in colored glass and ceramic glazes and even makeup in the Victorian era.

There are two potential origins of radiological material that may cause accidental exposure. The material may be intentionally stolen or be found to have accidentally caused exposure. Medical equipment in labs is not guarded and may be acquired by thieves and disgruntled employees. In 2013, Mexican thieves stole a truck containing 40 g of cobalt-60 that was being transported to a storage facility; police were able to recover the vehicle but six people were taken to the hospital to be tested for radiation poisoning. There have been several reported cases of accidental exposure from exposure to radioisotopes including cobalt-60 and cesium-137 from improperly disposed of medical and other equipment. Cobalt-60 is used in equipment for medical and food sterilization. In 1998, eight people developed acute radiation exposure after handling two cobalt-60 radiotherapy machines that were sold as scrap metal. In 2000, a cobalt-60 radiotherapy instrument was sold as scrap metal causing ten people to become ill with radiation syndrome including three people who later died as a result of the exposure. It was dismantled and the radiation source was left unprotected and exposed for several days at the junk yard.

Another accidental exposure, and the most likely exposure, could be caused by a transportation (truck, train, ship, etc.) accident involving radioactive material. Nearly two million kilograms of fissile material has been produced and exists worldwide. Only 3%–4% of shipping containers passing through US ports each day are inspected. However, state departments of transportation employ hazardous material (HazMat) specialists and the transport of hazardous waste material in the United States is highly regulated. Consequences from attacks on nuclear reactors resulting in the release of radioisotopes would vary depending on the location of the reactor, wind direction and velocity, and the extent of the disaster.

"DIRTY BOMBS" AND ACCESSIBILITY OF RADIOACTIVE MATERIAL

Radioactive material may be used to produce "dirty bombs," small incendiary devices containing radiological material. Terrorists may use the "dirty bombs" to disperse the material and expose more people to radioactivity. A suitcase bomb or dirty improved explosive device would require only a small amount of radioactive material. Radionuclides are widely available in household and commercial products. For example, radionuclides are used in smoke detectors (americium-241), lantern mantles (thorium-232), and clocks (radium). In a case in 1991, 14-year-old boy scout David Hahn obtained all of these items in his quest to construct a nuclear reactor in a shed in his backyard. In that case, bomb squads were eventually brought in to dismantle and decontaminate the area.

Other sources of radionuclides include medical and scientific equipment and improperly disposed of medical and scientific equipment. While they may require a larger budget to obtain than the items mentioned previously, these items are all available for purchase on the secondary market.

"Dirty bombs" may also be conventional bombs surrounded by, or filled with, radiological (non-fissile) materials to form a radiological dispersion device. The physical devastation results of these weapons will vary based on the quantity of explosives and type of detonation used, but their use would create fear and chaos and the population would be expected to experience significant long-term psychological effects. Therefore, some experts refer to such weapons as weapons of mass "disruption" rather than "destruction."

TREATMENT OF RADIATION POISONING

Radiation exposure can cause symptoms including headache, fatigue, weakness, vomiting, thermal burn-like effects, bleeding and hair loss, severe diarrhea and electrolyte loss, and secondary infections through broken skin. Internal or external exposure may occur. Internal exposure results from inhalation of radiological particles, exposure through wounds, or eating contaminated food. External exposure includes deposition of radiological particles and/or high-energy radiation on skin and/or clothing. First responders need to evacuate civilians from sites of bomb detonation, improper disposal, or accidental releases of radiological material. While external exposure can be mitigated by removing contaminated clothing and washing away the particles with soap and water, chelation therapy is used for radiation that has entered the body. External decontamination should be performed immediately, even before patients are transported to the hospital. Radiological material can be spread to people, objects, and locations by direct transfer or secondary transfer. Exposure can be reduced by increasing distance from the site, the use of appropriate shielding, and limiting time in an affected area (while monitoring cumulative dose).

In the case of a nuclear bomb, fallout will take 15–20 minutes to arrive. Civilians should be instructed to evacuate quickly, seek cover underground, or shelter indoors with doors and windows closed and HVAC systems off.

Accidental ingestion of iodine-131 can be treated by potassium iodide tablets to displace the radioactive iodine and replace it with the nonradioactive isotope. Exposure to nuclear fallout of cesium-134 or cesium-137 can be treated by the administration of Prussian blue ($[Fe(CN)_6]^{4-}$), which chelates the radioactive metal and prevents its absorption. Strontium-89 and strontium-90 internal exposure can be treated using antacids to decrease absorption, strontium lactate to block absorption, oral phosphates to displace the metal ions, and ammonium chloride to mobilize excretion. Plutonium-239 is chelated by zinc or calcium diethylenetriaminepentaacetic acid (DTPA). The effects of tritium are mitigated by dilution so copious quantities of fluids should be ingested.

USE OF RADIOACTIVE ISOTOPES IN NUCLEAR POWER PLANTS

Nuclear power plants are primarily powered by the fission of uranium-235 but may also be powered by plutonium-239 produced in "breeder" reactors by uranium-238 absorbing neutrons. Nuclear power plants employ plutonium-239 and uranium-235 to produce heat energy that powers steam turbines which, in turn, power electrical generators. Nuclear power plants require much less fissionable material than nuclear bombs. The first nuclear power plant, Obninsk Nuclear Power Plant, opened in Russia on June 26, 1954. Its design was a breeder reactor; while producing energy to produce electricity, it also produces ("breeds") its own fuel. To achieve the critical mass in fuel rods, uranium-235 is enriched to only 3%–4%. Uranium-238 can be mixed with plutonium-239 in "breeder" reactors to produce more fissionable plutonium-239 fuel. In only 10 years, a breeder reactor can produce enough fuel to refuel itself and fuel another reactor. However, this design is not used in the United States due to safety concerns in the handling, transfer, and storage of the breeder fuel as plutonium-239 is highly toxic and carcinogenic. It is, however, a design used in at least seven other countries.

$$\begin{smallmatrix}1\\0\end{smallmatrix}n + \begin{smallmatrix}238\\92\end{smallmatrix}U \rightarrow \begin{smallmatrix}239\\92\end{smallmatrix}U \rightarrow \begin{smallmatrix}0\\0\end{smallmatrix}\gamma + \begin{smallmatrix}239\\94\end{smallmatrix}Pu + 2\begin{smallmatrix}0\\-1\end{smallmatrix}B$$

A control rod (typically boron or cadmium) is positioned between each fuel rod to control the nuclear fission reactions in the reactor core. Water is used not only in the production of energy but also as a coolant that circulates in the reaction core to cool the fuel rods. As a result, nuclear power plants are positioned on or near water sources including oceans and rivers. For safety, the reactor core and steam generator are contained in "containment shells" of thick concrete and there are backup generators that control the system. Nuclear power plant structures are strong enough to protect the public from the impact of commercial aircraft. Nuclear power plants do not explode but these containment shells have proved essential in reducing the release of radionuclides in cases of nuclear power plant meltdown. The detection of characteristic radionuclides that may be released into the air is used to differentiate a nuclear power plant meltdown as opposed to the detonation of a nuclear bomb. Investigators have been asked to differentiate these radionuclides in some past, highly unfortunate, but accidental, cases.

A nuclear power plant meltdown occurred on March 28, 1979, at Three-Mile Island in the United States. This accident was rated a 5 (out of 7) on the International Nuclear Event Scale. The accident was caused by a failure that caused the main pumps to stop and prevented the steam generators from removing heat and subsequently a valve functioned improperly, which caused the plant to lose coolant. One half of the core melted as a result of the overheating. However, the containment building walls remained intact. Still, iodine-131, xenon-135, and krypton were detected after the incident and high levels of cesium-137 were found in deer. The civilian exposure was estimated at 1.4–8 mrem and the water in the reactor building was found to have a radioactivity 300 times the normal level.

A nuclear power plant disaster rating 7 on the International Nuclear Event Scale occurred on April 26, 1986, at Chernobyl, Ukraine (former USSR), reactor Unit 4 (Figure 15.21). A series of events including a power surge, attempted plant shutdown, and a large power strike led to the meltdown of the plant and a fire. The reactor vessel ruptured and caused steam explosions that exposed the graphite moderator of the reactor to the air. This caused a fire to ignite and radioactive fallout to be carried by the air. As a result of this accident, an 18-mile radius area around the plant was closed to the public and workers and their families (135,000 people) were evacuated and relocated; more were relocated later from surrounding areas. Several radioisotopes including iodine-131, cesium-137, krypton-85, xenon-133, tellurium-132, and strontium-90 were detected in the air. Thirty-one deaths were recorded among reactor and emergency workers and thousands of deaths were later attributed to cancer and leukemia that resulted from radiological poisoning from the event.

Figure 15.21   Chernobyl Unit 4 reactor damaged by nuclear meltdown. (From IAEA Imagebank, https://www.flickr.com/photos/
iaea_imagebank/5394791059/in/photolist-9dHHqX-9dGEHv-a1fgSP-8AEz79-a1fiEv-a1fspK-4fBG9p-9dKJpb-9dKKRQ-a19d12-
a1ihrG-9dGMHg-6qxqWJ-8oXU4y-8oXTJ1-k9841Q-5m9ufo-5m9u4b-5m5d8a-ikXDDc-5m5d6p-4mFLr5-f6iiXD-8oXNtN-
k9grwX-5m9u2w-a1c8r3-a19kQX-a1frp8-GqN5c-9dHn44-a19hDX-9dHK6a-9dHFUZ-8AEAqw-a1idD3-qBVTho-8oUJAZ-dMhewP-
7d6vuY-2cMJbu-6F7af1-GqJ4F-7d2nqX-a1fkLT-7d2jKx-6qxr9d-9y1Fzm-7d6eVb-6qxsW1.)

Another accident rating a 7 on the International Nuclear Event Scale occurred on March 12, 2011, at the nuclear
power plant in Fukushima Dai-ichi, Japan. A 9.0 Richter scale earthquake and a 15-meter tsunami flooded the generators and triggered a loss of coolant, which resulted in the meltdown of reactors 1, 2, and 3 (Figure 15.22). The
nuclear fallout was similar in composition to Chernobyl but a cesium-137 to cesium-134 ratio of 0.5–0.6 was recorded
at Chernobyl and a ratio of 1 was recorded at Fukushima. Iodine-131 was approximately 12 times lower at Fukushima
than Chernobyl. In total, 154,000 people were relocated.

IDENTIFICATION OF A NUCLEAR BOMB DETONATION VERSUS A NUCLEAR POWER PLANT MELTDOWN

The ratios of cesium-134 to cesium-137 and iodine 133 to iodine 131 can be used to determine if radioactive fallout is
due to the detonation of a nuclear bomb or a nuclear power plant meltdown. The lack of detectable neutrons indicates
a nuclear power plant meltdown as opposed to the detonation of a nuclear bomb. However, if the concrete sarcopha-
gus of the nuclear power plant is breached, numerous radionuclides can be detected in the environment including
iodine-131, iodine-132, iodine-133, cesium-136, cesium-137, barium-140, cerium-141, cerium-144, technium-132, lan-
thanum-140, krypton-85, xenon-133, zirconium-95, niobium-95, ruthenium-103, ruthenium-106, and neptunium-239.

DUAL USE RESEARCH: IMPACTS AND PUBLICATION

Several chemical warfare agents were developed in the process of conducting other chemical research. Chlorine is
used as a cleaning agent and disinfectant (e.g., bleach) in coolants and as an oxidizing reagent and a reagent in substi-
tution reactions. It is used to disinfect swimming pools and drinking water. It is used in making paper and textiles as
well as insecticides. About 20% of chlorine is used in the production of PVC polymer. It is also used in the production
of chloroform and carbon tetrachloride. Arsine and stibine are used in the semiconductor industry. Chloropicrin is
an antimicrobial agent and is used as a fungicide, herbicide, insecticide, and nematicide. It is injected into the soil
to fumigate it. Nitroglycerin is unstable and explosive, even by shaking, but Swedish engineer Alfred Nobel (who
established the Nobel Prizes in peace, chemistry, physiology or medicine, literature, physics, and later, economics)
stabilized it in 1867 by adding silica. The safer preparation was used in road and basement construction as a blasting

Figure 15.22   Fukushima Dai-ichi Nuclear Power Plant near reactor Unit 3 damaged by a 2011 earthquake and 15-meter tsunami. (From IAEA Imagebank, https://www.flickr.com/photos/iaea_imagebank/6234779912/in/photolist-auWSRo-f5mgxE-9MsLjj-9MpYcg-b2v28n-nr4RSN-f571Nr-f5mgTN-ebYEBg-amVogx-f572bg-brSfjT-f5mgZS-f5mgNy-e2in52-e3GtAp-amWiDP-eba4oh-ec5k69-eioUze-hUMLTc-i6eTHU-befRme-f573e4-9r2xA8-9qApAg-9w7Ek7-dTYfow-hTqcmb-e2igir-dTSBuk-amZudj-bny8Gd-a5SAxw-bvMvK2-g2PZr8-qUsR7d-9qDqqY-kbv84T-auWUqC-e2iiSv-9qDqBA-b2BMHg-e2ighk-ojEv55-f7Swo7-b2uTvg-omsA6a-bvMtU2-9w4AXe.)

agent. Nitroglycerin was also found to treat angina and heart disease as workers in his factories reported feeling better during the week than on the weekends. In the body, nitroglycerin slowly decomposes to nitric oxide (NO) that relaxes smooth muscles in blood vessels causing them to widen, which allows more blood to pass through, brings oxygen to peripheral cells, and circulates to the heart where the heart muscle also benefits from increased oxygen. Together, Robert F. Furchott, Louis J. Ignarro, and Ferid Murad won the 1998 Nobel Prize in Physiology or Medicine "for their discoveries concerning nitric oxide as a signaling molecule in the cardiovascular system."

In other cases, research into chemical warfare agents led to important discoveries that aided the development of medical treatments. Research into vesicant mustard gases led to the discovery that nitrogen mustard is an anti-cancer agent. HN2 is used as the anti-cancer drug mustine. It was discovered that exposure to nitrogen mustard caused people to develop significantly reduced white blood cell counts due to damaged bone marrow and that the agent can be used to treat lymphoma. This work was held for publication from 1943 when completed until after World War II in 1946 when the work on nitrogen mustard was declassified.

After the discovery of x-rays in 1895, Marie Curie immediately recognized that the radioactive elements could be useful in medicine and used radium to sterilize infected tissue and x-rays to detect broken bones in mobile units in World War I. Lise Meitner served in World War I as an x-ray nurse for the Australian Army. Radium was used to treat tuberculosis using its bactericidal properties.

Seaborg's work, although submitted for publication in 1941, was held in secret until its eventual publication in 1946 in *Physical Review Letters*. He and his team had determined that plutonium-239 was fissionable and held potential as a second (after uranium-235), and even more powerful, nuclear energy source for an atomic bomb. He supported the United States' effort in World War II to bring a plutonium bomb to fruition, leading a team of researchers at University of Chicago's Metallurgical Laboratory (known as the Met Lab) in developing the optimal separation process to produce pure plutonium-239 for the war effort. After the war, radioisotopes were used to power nuclear power plants. US production of plutonium ceased in 1988; the United States purchased it from Russia in 1993 to power equipment for space missions. However, in 2013, the Department of Energy announced that small amounts of plutonium-238 would be produced annually to support the space exploration program. It has been used multiple times since 1961 to power US spacecraft. Plutonium-238 is currently used to power the Mars rover Curiosity (which uses 4.4 kg of the radioactive material) and New Horizons (which uses 11 kg).

Table 15.10 CDC's dual use research of concern (DURC) experiments

Disrupt immunity or the effectiveness of an immunization against the agent or toxin without clinical and/or agricultural justification
Confer to the agent or toxin resistance to clinically and/or agricultural useful preventative or treatment interventions against that agent or toxin, or facilitate their ability to evade methods of detection
Increase the stability, transmissibility, or ability to disseminate the agent or toxin
Alter the host range or tropism of the agent or toxin
Enhance the susceptibility of a host population to the agent or toxin
Generate or reconstitute an eradicated or extinct agent or one of the 15 DURC toxins or agents

Source: CDC, www.cdc.gov/flu/avianflu/avian-durc-qa.htm.

Potential bioweapons include both emerging threats and designer superbug weapons. For example, the H5N1 flu virus has killed 300 million birds and 200 humans since 1997. Between November 2002 and July 2003, SARS sickened 8098 people worldwide and killed 774 people. The CDC's dual use research of concern (DURC) experiments are listed in Table 15.10. The CDC will approve these experiments on a case-by-case basis on board review and if the work is performed in a secure facility. Several studies have been conducted to identify determinants of pathogenicity and transmissibility for several flu viruses including H1N1, H5N1, and H3N2. However, this research has dual use. On the positive side, this research can be used to develop vaccines (e.g., universal flu vaccine), detection methods, and treatments.

EMERGING THREATS AND DESIGNER WEAPONS

Information including blueprints and black-market supplies needed to design and produce an atomic bomb is available on the internet and in the open-source literature. Terrorists, cult groups, extremist organizations, apocalyptic groups, national/separatist groups, politico-religious groups and even a lone individual with extreme objectives and technical and financial resources may be motivated to pursue nuclear terrorism. Although there are strategic considerations, there is certainly an aura of fear and vulnerability with nuclear weapons that is deeply rooted in the general public and these attacks certainly may be quite lethal. Additionally, land would be contaminated, markets would be disrupted, and long-term cancer incidences would increase.

The 1997 JASON Defense Advisory Panel report documents four potential bioterrorism weapons threats including: (1) binary weapons; (2) designer diseases; (3) host-swapping diseases; and (4) stealth viruses. Binary weapons could combine two agents that are safe for handling alone but deadly when they are mixed. Genetic engineering could be used to produce protein toxins with increased toxicity by using site-directed mutagenesis to change the gene. Molecular biology techniques including cloning methods could be used to extract genes of interest from one (or more) organism(s) and insert them into another organism to create a designer disease. A perpetrator could genetically engineer the flu virus to have enhanced virulence, pathogenicity, and/or transmissibility using data collected and information gathered from these studies. The superbug could be designed to be resistant to antivirals or be immunosuppressant. Bacterial superbugs could be created to be antibiotic resistant. Alternatively, a flu virus, such as avian flu, could be engineered to both infect humans and use humans as hosts. Stealth viruses could be created to infect a host but not make the host ill until activated by a physiological or environmental trigger.

One example of a superbug is the *B. anthracis* superbug that was engineered by the USSR using genes from the smallpox and encephalomyelitis viruses that is resistant to approximately ten antibiotics. This work is relatively inexpensive and easily performed by undergraduate biology students. Anthrax is accessible as it is found naturally in soil. An aerosolized anthrax superbug could easily be sprayed over huge swaths of land, as has been demonstrated, and would infect huge numbers of people (and potentially animals too). Biological weapons can be released covertly without a detectable taste or smell; the incident would be unknown until people start to fall ill. New superbugs would be difficult to diagnose, detect, and treat and could overwhelm the medical and emergency management system including hospitals and emergency medicine teams. Drug-resistant tuberculosis and methicillin-resistant *Staphylococcus aureus* (MRSA) agents are also widely accessible from infected individuals and would be deadly if used as a biological weapon. Use of these agents would cause fear, panic, social disruption, and mass casualties to civilian populations.

QUESTIONS

1. Which of the following is not a biodefense research priority?
   a. New therapeutics
   b. Genomics research
   c. Handgun safety
   d. Vaccines

2. Which of the following is a source of radioisotopes for WMD?
   a. Medical equipment
   b. Bacteria
   c. Chlorine
   d. Castor beans

3. Which of the following is an emerging agent that might be used for biological warfare?
   a. *B. anthracis*
   b. *Y. pestis*
   c. H5N1
   d. *Salmonella sp.*

4. Which of the following is not a societal effect of WMD?
   a. Fear
   b. Calm
   c. Case management
   d. Mass casualties

5. Which of the following affect the ability of a microorganism to infect a host?
   a. Age of host
   b. Health of potential host
   c. Mode of entry
   d. Strain of microorganism
   e. All of the above

6. Which of the following is not a category of chemical military weapons?
   a. Nerve agents
   b. Diuretic agents
   c. Blister agents
   d. Blood agents
   e. Choking agents

7. For which of the following WMDs is the onset time the longest?
   a. Hydrogen cyanide
   b. Phosphine
   c. Anthrax
   d. Sarin

8. Which of the following is susceptible to a CBRNE threat?
   a. Humans
   b. Animals
   c. Crops
   d. All of the above

9. Which of the following is not a route of entry for a bioweapon?
   a. Desorption
   b. Injection
   c. Absorption
   d. Ingestion
   e. Inhalation

10. Which of the following is not a property of biological weapons?
    a. Pathogen has no taste or smell
    b. Highly contagious
    c. Easy to disseminate
    d. Easy to detect
    e. Difficult to treat

11. Differentiate the uranium used in a nuclear power plant from that used in an atomic bomb.

12. What are the most accessible CBRNE agents? Why?

13. How can an investigator determine if a nuclear bomb was dropped on a city versus a nuclear power plant exploding?

14. How does decontamination differ for biological agents as compared to chemical or nuclear agents?

15. How long will it take for 1 gm of cesium-137 to decay to 1 mg? ($t_{1/2} = 30.1$ years)

16. Give three examples of industrial chemicals or commercial off-the-shelf products that have been used as chemical weapons and list the cases.

17. What type of evidence is collected in a terror attack? What evidence would lead you to conclude that the attack was chemical, biological, or nuclear? Give at least one specific example for each.

18. Define false negative and false positive. Which is worse for WMD detection?

19. What is a "dirty bomb"?

20. List the four types of particles or radiation given off in nuclear reactions and what material first responders and crime scene investigators can use to protect themselves from each of them.

Bibliography

Amon, S.S., R. Schecter, T.V. Inglesby, D.A. Henderson, J.G. Bartlett, M.S. Ascher, E. Eitzen, A.D. Fine, J. Hauer, M. Layton, et al. 2001. Botulinum toxin as a biological weapon: Medical and public health management. *JAMA* 285(8):1059–1070.

Anderson, G.P., J.L. Liu, M.L. Hale, R.D. Bernstein, M. Moore, M.D. Swain, and E.R. Goldman. 2008. Development of antiricin single domain antibodies toward detection and therapeutic reagents. *Anal. Chem.* 80:9604–9611.

Anderson, G.P., R.D. Bernstein, M.D. Swain, D. Zabetakis, and E.R. Goldman. 2010. Binding kinetics of antiricin single domain antibodies and improved detection using a B chain specific binder. *Anal. Chem.* 82:7202–7207.

Arthur, C. 2001. Top scientist warns Britain to be prepared for biological warfare. www.independent.co.uk/news/science/top-scientist-warns-britain-to-be-prepared-for-biological-warfare-9132615.html (accessed January 30, 2018).

Baldoni, A.B., A.C.G. Araújo, M.H. de Carvalho, A.C.M.M. Gomes, and F.J.L. Aragão. 2010. Immunolocalization of ricin accumulation during castor bean (Ricinus communis L.) seed development. *Int. J. Plant Biol.* 1:61–65.

Ball, P. 2004. Out in the open. *Nature*, 29 Oct 2004. doi:10.1038/news041025-22.

Barr, J.R., H. Moura, A.E. Boyer, A.R. Woolfitt, S.R. Kalb, A. Pavlopoulos, L.G. McWilliams, J.G. Schmidt, R.A. Martinez, and D.C. Ashley. 2005. Botulinum neurotoxin detection and differentiation by mass spectrometry. *Emerging Infectious Diseases* 11(10):1578–1583. doi:10.3201/eid1110.041279.

Beaumelle, B., M. Alami, and C.R. Hopkins. 1993. ATP-dependent translocation of ricin across the membrane of purified endosomes. *J. Biol. Chem.* 268:23661–23669.

Becher, F., E. Duriez, V. Holland, J.C. Tabet, and E. Ezan. 2007. Detection of functional ricin by immunoaffinity and liquid chromatography—Tandem mass spectrometry. *Anal. Chem.* 79:659–665.

Beviacqua, V.L.H., J.M. Nilles, J.S. Rice, T.R. Connell, A.M. Schenning, L.M. Reilly, and H.D. Durst. 2010. Ricin activity assay by direct analysis in real time mass spectrometry detection of adenine release. *Anal. Chem.* 82:798–800.

Blum, D. 2011. Castor beans. In *The Poisoner's Handbook: Murder and the Birth of Forensic Medicine in Jazz Age New York*, 7–9. London, UK: Penguin Books.

Bradberry, S. 2007. Ricin and abrin. *Medicine* 35:576–577.

Brandon, D.L. 2011. Detection of ricin contamination in ground beef by electrochemiluminescence immunosorbent assay. *Toxins* 3:398–408.

Brinkworth, C.S., E.L. Pigott, and D.J. Bourne. 2009. Detection of intact ricin in crude and purified extracts from castor beans using matrix-assisted laser desorption ionization mass spectrometry. *Anal. Chem.* 81:1529–1535.

Brinkworth, C.S. 2010. Identification of ricin in crude and purified extracts from castor beans using on-target tryptic digestion and MALDI mass spectrometry. *Anal. Chem.* 82(12):5246–5252.

Brodwin, E. 2013. Is it too late to determine which chemical weapons were used in Syria? *Nature*, 29 Aug. doi:10.1038/nature.2013.13639.

Budowle, B., S. Schutzer, R. Breeze, P. Keim, and S. Morse, eds. 2010. *Microbial Forensics*. Burlington, MA: Elsevier Academic Press.

Bullock, J., G. Haddow, D.P. Coppola. 2016. *Introduction to Homeland Security*, 5th edn. Waltham, MA: Butterworth-Heinemann.

California Institute of Technology. December 22, 2015. U.S. demonstrates production of fuel for missions to the solar system and beyond. www.jpl.nasa.gov/news/news.php?feature=4806 (accessed December 3, 2017).

CDC. Dual use research of concern and bird flu: Questions & answers. www.cdc.gov/flu/avianflu/avian-durc-qa.htm (accessed January 30, 2018).

CDC. 2017. Bioterrorism Agents/Diseases. https://emergency.cdc.gov/agent/agentlist-category.asp (accessed January 31, 2018).

Chan, A.P., J. Crabtree, Q. Zhao, H. Lorenzi, J. Orvis, D. Puiu, A. Melake-Berhan, K.M. Jones, J. Redman, G. Chen, et al. 2010. Draft genome sequence of the oilseed species Ricinus communis. *Nat. Biotechnol.* 28:951–959.

Choppin, G.R., J.-O. Liljenzin, and J. Rydberg. 2002. Chapter 8: Detection and measurement techniques. In *Radiochemistry and Nuclear Chemistry*, 3rd edn, 192–237. Waltham, MA: Butterworth-Heinemann.

Clark, D.L. and D.E. Hobart. 2000. Reflections on the legacy of a legend, Glenn T. Seaborg, 1912–1999. *Los Alamos Sci.* 26:56–61.

Colburn, H.A., D.S. Wunschel, H.W. Kreuzer, J.J. Moran, K.C. Antolick, and A.M. Melville. 2010. Analysis of carbohydrate and fatty acid marker abundance in ricin toxin preparations for forensic information. *Anal. Chem.* 82:6040–6047.

Cook, D.L., J. David, and G.D. Griffiths. 2006. Retrospective identification of ricin in animal tissues following administration by pulmonary and oral routes. *Toxicology* 223:61–70.

Coopman, V., M. De Leeuw, J. Cordonnier, and W. Jacobs. 2009. Suicidal death after injection of a castor bean extract (ricinus communis L.). *Forensic Sci. Int.* 189:e13–e20.

Craig, H.L. 1962. Patent US3060165: Preparation of toxic ricin. www.google.com/patents/US3060165 (accessed January 30, 2018).

Crouch, J.H., H.K. Crouch, A. Tenkouano, and R. Ortiz. 1999. VNTR-based diversity analysis of 2x and 4x full-sib *Musa* hybrids. *Electron. J. Biotechnol.* 2(3):99–108.

Daffonchio, D., N. Raddadi, M. Merabishvili, A. Cherif, L. Carmagnola, L. Brusetti, A. Rizzi, N. Chanishvili, P. Visca, R. Sharp, et al. 2006. Strategy for identification of bacillus cereus and bacillus thuringiensis strains closely related to bacillus anthracis. *Appl. Environ. Microbiol.* 72:1295–1301.

Darby, S.M., M.L. Miller, and R.O. Allen. 2001. Forensic determination of ricin and the alkaloid marker ricinine from castor bean extracts. *J. Forensic Sci.* 46:1033–1042.

Despott, E. and M.J. Cachia. 2004. A case of accidental ricin poisoning. *Malta Med. J.* 16(4):39–41.

Devell, L., H. Tovedal, U. Bergstrom, A. Appelgren, J. Chyssler, and L. Andersson. 1986. Initial observations from the reactor accident at Chernobyl. *Nature* 321:192–193.

Drosten, C., S. Göttig, S. Schilling, M. Asper, M. Panning, H. Schmitz, and S. Günther. 2002. Rapid detection and quantification of RNA of Ebola and Marburg viruses, Lassa virus, Crimean-Congo hemorrhagic fever virus, reverse Rift Valley fever virus, dengue virus, and yellow fever virus by real-time reverse transcription-PCR. *J. Clin. Microbiol.* 40(7):2323–2330.

Elkins, K.M., A.C.U. Perez, and K.C. Sweetin. 2016. Rapid and inexpensive species differentiation using a multiplex real-time polymerase chain reaction high-resolution melt assay. *Anal. Biochem.* 500:15–17.

Epstein, J. 2009. Weapons of mass destruction: It is all about chemistry. *J. Chem. Educ.* 86(12):1377–1381.

Felder, E., I. Mossbrugger, M. Lange, and R. Wölfel. 2012. Simultaneous detection of ricin and abrin DNA by real-time PCR (qPCR). *Toxins* 4:633–642.

Fester, U. 1997. Ricin: Kitchen improvised devastation. In *Silent Death*, 107–118. Green Bay, WI: Festering Publications.

Fitzgerald, G.J. 2008. Chemical warfare and medical response during World War I. *Am. J. Public Health* 98(4):611–625.

Forest, J. 2012. *The Terrorism Lectures: A Comprehensive Collection for Students of Terrorism, Counterterrorism, and National Security*. Orange County, CA: Nortia Press.

Frankel, M.S. 2012. Regulating the boundaries of dual-use research. *Science* 336:1523–1525.

Frawley, D.A., M.N. Samaan, R.L. Bull, J.M. Robertson, A.J. Mateczun, and P.C.B. Turnbull. 2008. Recovery efficiencies of anthrax spores and ricin from nonporous or nonabsorbent and porous or absorbent surfaces by a variety of sampling methods. *J. Forensic Sci.* 53:1102–1107.

Fredriksson, S.-Å., A.G. Hulst, E. Artursson, A.L. de Jong, C. Nilsson, and B.L.M. van Baar. 2005. Forensic identification of neat ricin and of ricin from crude bean extracts by mass spectrometry. *Anal. Chem.* 77:1545–1555.

Ganesan, K., S.K. Raza, and R. Vijayaraghavan. 2010. Chemical warfare agents. *J. Pharm. Bioallied Sci.* 2(3):166–178.

Gilman, A. 1946. Symposium on advances in pharmacology resulting from war research: Therapeutic applications of chemical warfare agents. *Fed. Proc.* 5:285–292.

Gilman, A. and F.S. Philips. 1946. The biological actions and therapeutic applications of the h-chloroethylamines and sulfides. *Science* 103:409–415.

Girard, J.E. 2007. *Criminalistics: Forensic Science and Crime*. Burlington, MA: Jones and Bartlett Learning.

Goldman, E.R., A.R. Clapp, G.P. Anderson, H.T. Uyeda, J.M. Mauro, I.L. Medintz, and H. Mattoussi. 2004. Multiplexed toxin analysis using four colors of quantum dot fluororeagents. *Anal. Chem.* 76:684–688.

Goldman, E.R., G.P. Anderson, J.L. Liu, J.B. Delehanty, L.J. Sherwood, L.E. Osborn, L.B. Cummins, and A. Hayhurst. 2006. Facile generation of heat-stabile antiviral and antitoxin single domain antibodies from a semisynthetic llama library. *Anal. Chem.* 78:8245–8255.

Goodman, L.S., M.M. Wintrobe, W. Dameshek, M.J. Goodman, A. Gilman, and M.T. McLennan. 1946. Nitrogen mustard therapy: Use of methyl-bis (h-chloroethyl) amine hydrochloride and tris (h-chloroethyl)amine hydrochloride for Hodgkin's disease, lymphosarcoma, leukemia, and certain allied and miscellaneous disorders. *JAMA* 132:126–132.

Gopalakrishnakone, P., M. Balali-Mood, L. Llewellyn, and B.R. Singh, eds. 2015. *Biological Toxins and Bioterrorism*. New York, NY: Springer.

Halford, B. 2012. Tracing a threat. *C&ENews* 90:10–15.

Harber, D. 1993. Ricin. In *Assorted Nasties*. El Dorado, AR: Desert Publications.

Hellin, H. Der giftige eiweisskörper abrin und seine wirkung auf das blut. PhD dissertation, Dorpat Universität.

Herfst, S., E.J.A. Schrauwen, M. Linster, S. Chutinimitkul, E. de Wit, V.J. Munster, E.M. Sorrell, T.M. Bestebroer, D.F. Burke, D.J. Smith, et al. 2012. Airborne transmission of influenza A/H5N1 virus between ferrets. *Science* 336:1534–1541.

Imai, M., T. Watanabe, M. Hatta, S.C. Das, M. Ozawa1, K. Shinya, G. Zhong, A. Hanson, H. Katsura, S. Watanabe, et al. 2012. Experimental adaptation of an influenza H5 HA confers respiratory droplet transmission to a reassortant H5 HA/H1N1 virus in ferrets. *Nature* 486:420–430.

Ishiguro, M., T. Takahashi, G. Funatsu, K. Hayashi, and M. Funatsu. 1964. Biochemical studies on ricin: I. Purification of ricin. *J. Biochem.* 55:587–592.

Kalb, S.R. and J.R. Barr. 2009. Mass spectrometric detection of ricin and its activity in food and clinical samples. *Anal. Chem.* 81:2037–2042.

Kean, S. 2010. *The Disappearing Spoon*. New York, NY: Back Bay Books/Little, Brown and Company.

Kim, J.S., G.P. Anderson, J.S. Erickson, J.P. Golden, M. Nasir, and F.S. Ligler. 2009. Multiplexed detection of bacteria and toxins using a microflow cytometer. *Anal. Chem.* 81:5426–5432.

Kolavic, S.A., A. Kimura, S.L. Simons, L. Slutsker, S. Barth, and C.E. Haley. 1997. An outbreak of Shigella dysenteriae type 2 among laboratory workers due to intentional food contamination. *JAMA* 278(5):396–398.

Kreuzer, H.W., J.B. West, and J.R. Ehleringer. 2013. Forensic applications of light-element stable isotope ratios of ricinus communis seeds and ricin preparations. *J. Forensic Sci.* 58:S43–S50.

Kreuzer, H.W., J.H. Wahl, C.N. Metoyer, H.A. Colburn, and K.L. Wahl. 2010. Detection of acetone processing of castor bean mash for forensic investigation of ricin preparation methods. *J. Forensic Sci.* 55:908–914.

Kull, S., D. Pauly, B. Störmann, S. Kirchner, M. Stämmler, M. Dorner, P. Lasch, D. Naumann, and B.G. Dorner. 2010. Multiplex detection of microbial and plant toxins by immunoaffinity enrichment and matrix-assisted laser desorption/ionization mass spectrometry. *Anal. Chem.* 82:2916–2924.

Kumar, O., A.B. Nashikkar, R. Jayaraj, R. Vijayaraghavan, and A.O. Prakash. 2004. Purification and biochemical characterisation of ricin from castor seeds. *Defence Sci. J.* 54:345–351.

Kumar, O., S. Pradhan, P. Sehgal, Y. Singh, and R. Vijayaraghavan. 2010. Denatured ricin can be detected as native ricin by immunological methods but nontoxic in vivo. *J. Forensic Sci.* 55:801–807.

Kurt, S. 1991. *The Poor Man's James Bond*, Vol. 1. El Dorado, AR: Desert Publications.

Kwon, Y., C.A. Hara, M.G. Knize, M.H. Hwang, K.S. Venkateswaren, E.K. Wheeler, P.M. Bell, R.F. Renzi, J.A. Fruetel, and C.G. Bailey. 2008. Magnetic bead based immunoassay for autonomous detection of toxins. *Anal. Chem.* 80:8416–8423.

Ladant, D., J.E. Alouf, and M.R. Popoff. 2006. *The Comprehensive Sourcebook of Bacterial Protein Toxins*, 3rd edn. New York, NY: Elsevier Science.

Leffel, E.K. and D.S. Reed. 2004. Marburg and Ebola viruses as aerosol threats. Biosecurity and bioterrorism: Biodefense strategy. *Practice Sci.* 2(3):186–191.

Lin, T.T. and S.L. Li. 1980. Purification and physicochemical properties of ricins and agglutinins from ricinus communis. *Eur. J. Biochem.* 105:453–459.

Lipsitch, M., J.B. Plotkin, L. Simonsen, and B. Bloom. 2012. Evolution, safety, and highly pathogenic influenza viruses. *Science* 336:1529–1531.

Madden, L.V. and M. Wheelis. 2003. The threat of plant pathogens as weapons against U.S. crops. *Annu. Rev. Phytopathol.* 41:155–176.

Masson, O., A. Baeza, J. Bieringer, K. Brudecki, S. Bucci, M. Cappai, F.P. Carvalho, O. Connan, C. Cosma, A. Dalheimer, et al. 2011. Tracking of airborne radionuclides from the damaged Fukushima Dai-Ichi nuclear reactors by European networks. *Environ. Sci. Technol.* 45:7670–7677.

McGrath, S.C., D.M. Schieltz, L.G. McWilliams, J.L. Pirkle, and J.R. Barr. 2011. Detection and quantification of ricin in beverages using isotope dilution tandem mass spectrometry. *Anal. Chem.* 83:2897–2905.

McLaughlin P.D., B. Jones, and M.M. Maher. 2012. An update on radioactive release and exposures after the Fukushima Dai-ichi nuclear disaster. *Br. J. Radiol.* 85(1017): 1222–1225.

Mei, Q., C.K. Fredrickson, W. Lian, S. Jin, and Z.H. Fan. 2006. Ricin detection by biological signal amplification in a well-in-a-well device. *Anal. Chem.* 78:7659–7664.

Meselson, M., J. Guillemin, M. Hugh-Jones, A. Langmuir, I. Popova, A. Shelokov, and O. Yampolskaya. 1994. The Sverdlovsk anthrax outbreak of 1979. *Science* 266:1202–1208.

Millard, J.T. 2006. *Adventures in Chemistry*. Boston, MA: Cengage Learning.

Monath, T.P. and L.K. Gordon. 1998. Strengthening the biological weapons convention. *Science* 282:1423.

Moran, J.J., C.J. Ehrhardt, J.H. Wahl, H.W. Kreuzer, and K.L. Wahl. 2013. Integration of stable isotope and trace contaminant concentration for enhanced forensic acetone discrimination. *Talanta* 116:866–869.

Mosquera, B., R. Veiga, L. Mangia, C. Carvalho, L. Estellita, D. Uzeda, A. Facure, B. Violini, and R.M. Anjos. 2004. 137Cs distribution in Guava trees. *Brazilian J. Phys.* 34:841–844.

Mulchandani, P., A. Mulchandani, I. Kaneva, and W. Chen. 1999. Biosensor for direct determination of organophosphate nerve agents 1. *Potentiometric Enzyme Electrode Biosens. Bioelectron.* 14:77–85.

Murch, R.S. 2011. "Amerithrax": The investigation of bioterrorism using bacillus anthracis spores in mailed letters. In *Encyclopedia of Bioterrorism Defense* (eds R. F. Pilch and R. A. Zilinskas). Wiley: doi:10.1002/0471686786.ebd0138.

Nature Editorial. 2003. Statement on the consideration of biodefense and biosecurity. *Nature* 421:771.

Nenot, J.-C. 2009. Radiation accidents over the last 60 years. *J. Radiol. Prot.* 29:301–320.

Nixon, R. Remarks announcing decisions on chemical and biological defense policies and programs. www.presidency.ucsb.edu/ws/?pid=2344 (accessed January 30, 2018).

Nobel Media AB. 2014a. The Nobel Prize in Chemistry 1908. www.nobelprize.org/nobel_prizes/chemistry/laureates/1908/ (accessed January 28, 2018).

Nobel Media AB. 2014b. The Nobel Prize in Chemistry 1911. www.nobelprize.org/nobel_prizes/chemistry/laureates/1911/ (accessed January 28, 2018).

Nobel Media AB. 2014c. The Nobel Prize in Chemistry 1935. www.nobelprize.org/nobel_prizes/chemistry/laureates/1935/ (accessed January 28, 2018).

Nobel Media AB. 2014d. The Nobel Prize in Chemistry 1944. www.nobelprize.org/nobel_prizes/chemistry/laureates/1944/ (accessed January 28, 2018).

Nobel Media AB. 2014e. The Nobel Prize in Chemistry 1951. www.nobelprize.org/nobel_prizes/chemistry/laureates/1951/ (accessed January 28, 2018).

Nobel Media AB. 2014f. The Nobel Prize in Physics 1901. https://www.nobelprize.org/nobel_prizes/physics/laureates/1901/ (accessed January 28, 2018).

Nobel Media AB. 2014g. The Nobel Prize in Physics 1903. www.nobelprize.org/nobel_prizes/physics/laureates/1903/ (accessed January 28, 2018).

Nobel Media AB. 2014h. The Nobel Prize in Physics 1935. www.nobelprize.org/nobel_prizes/physics/laureates/1935/ (accessed January 28, 2018).

Organisation for the Prohibition of Chemical Weapons. Schedule 1. Chemical weapons convention. www.opcw.org/chemical-weapons-convention/annexes/annex-on-chemicals/schedule-1/ (accessed December 3, 2017).

Ovendun, S.P.B., S.-Å. Fredriksson, C.K. Bagas, T. Bergström, S.A. Thomson, C. Nilsson, and D.J. Bourne. 2009. De novo sequencing of RCB-1 to -3: Peptide biomarkers from the castor bean plant ricinus communis. *Anal. Chem.* 81:3986–3996.

Paatero, J., K. Hämeri, M. Jantunen, P. Hari, C. Persson, M. Kulmala, R. Mattsson, H.-C. Hansson, and T. Raunemaa. 2011. Chernobyl: Observations in Finland and Sweden. In *Aerosol Science and Technology: History and Reviews*, D.S. Ensor, ed., 339–366. Research Triangle Park, NC: RTI Press. doi:10.3768/rtipress.2011.bk.0003.1109.

Pavlovic, M., A. Luze, R. Konrad, A. Berger, A. Sing, U. Busch, and I. Huber. 2011. Development of a duplex real-time PCR for differentiation between E. coli and Shigella spp. *J. Appl. Microbiol.* 110:1245–1251.

Pigott, E.J., W. Roberts, S.P.B. Ovendon, S. Rochfort, and D.J. Bourne. 2012. Metabolomic investigations of Ricinus communis for cultivar and provenance determination. *Metabolomics* 8:634–642.

Pradhan, S., M. Boopathi, O. Kumar, A. Patel, P. Pandey, T.H. Mahato, B. Singh, and R. Vijayaraghavan. 2009. Molecularly imprinted nanopatterns for the recognition of biological warfare agent ricin. *Biosens. Bioelectron.* 25:592–598.

PubChem. https://pubchem.ncbi.nlm.nih.gov/ (accessed January 31, 2018).

Rappuoli, R. and P.R. Dormitzer. 2012. Influenza: Options to improve pandemic preparation. *Science* 336:1531–1533.

Rasko, D.A., P.L. Worsham, T.G. Abshire, S.T. Stanley, J.D. Bannan, M.R. Wilson, R.J. Langham, R.S. Decker, L. Jiang, T.D. Read, et al. 2011. Bacillus anthracis comparative genome analysis in support of the Amerithrax investigation. *PNAS* 108(12):5027–5032.

Roberts, D.J. 2011. Technology is playing an expanding role in policing. *Police Chief* 78:72–73. www.nxtbook.com/nxtbooks/naylor/CPIM0111/#/72 (accessed January 30, 2018).

Ryan, J. and J. Glarum. 2008. *Biosecurity and Bioterrorism: Containing and Preventing Biological Threats*, 1st edn. Oxford, UK: Butterworth-Heinemann.

Schieltz, D.M., S.C. McGrath, L.G. McWilliams, J. Rees, M.D. Bowen, J.J. Kools, L.A. Dauphin, E. Gomez-Saladin, B.N. Newton, H.L. Stang, et al. 2011. Analysis of active ricin and castor bean proteins in a ricin preparation, castor bean extract, and surface swabs from a public health investigation. *Forensic Sci. Int.* 209:70–79.

Seaborg, G.T., E.M. McMillan, J.W. Kennedy, and A.C. Wahl. 1946. Radioactive element 94 from deuterons on uranium. *Phys. Rev.* 69(7–8):366–367.

Sehgal, P., O. Kumar, M. Kamewararao, J. Ravindran, M. Khan, S. Sharma, and R. Vijayaraghavan. 2011. Differential toxicity profile of ricin isoforms correlates with their glycosylation levels. *Toxicology* 282:56–67.

Seto, Y. 2001. The sarin gas attack in Japan and the related forensic investigation. *OPCW Synth.*:14–17. www.opcw.org/news/article/the-sarin-gas-attack-in-japan-and-the-related-forensic-investigation/ (accessed May 25, 2018).

Sferopoulos, R. 2009. A review of chemical warfare agent (CWA) detector technologies and commercial-off-the-shelf items: A review of chemical warfare agent (CWA) detector technologies and commercial-off-the-shelf items. *Aust. Government Department Defence*. Victoria, AU: Human Protection and Performance Divison. http://digext6.defence.gov.au/dspace/bitstream/1947/9902/1/DSTO-GD-0570%20PR.pdf (accessed December 3, 2017).

Shea, D.A. 2013. Chemical weapons: A summary report of characteristics and effects. *Congressional Res. Service* R42862:1–12.

Shia, W.W. and R.C. Bailey. 2013. Single domain antibodies for the detection of ricin using silicon photonic microring resonator arrays. *Anal. Chem.* 85:805–810.

Sigma-Aldrich. Solubility information. www.sigmaaldrich.com/united-kingdom/technical-services/solubility.html (accessed December 3, 2017).

Silva, B.A., M.P. Stephan, M.G.B. Koblitz, and J.L.R. Ascheri. 2012. Influence of NaCl and pH concentration on the extraction of ricin in castor bean (Ricinus communis L.) and its characterization by electrophoresis. *Ciênc. Rural* 42(7):1320–1326.

Silverstein, K. 2004. *The Radioactive Boy Scout: The True Story of a Boy and His Backyard Nuclear Reactor*. New York, NY: Random House.

Simonova, M.V., T.L. Valyakina, E.E. Petrova, R.L. Komaleva, N.S. Shoshina, L.V. Samokvalova, O.E. Lakhtina, I.V. Osipov, G.N. Philipenko, E.K. Singov, et al. 2012. Development of xMAP assay for detection of six protein toxins. *Anal. Chem.* 84:6326–6330.

Skogen, V., P.A. Jeiium, E.N. Kovoleva, E. Danilova, D.S. Halvorsen, N. Maksinzova, and H. Sjursett. 1999. Detection of diphtheria antitoxin by four different methods. *Clin. Microbiol. Infect.* 5:628–633.

Spiers, E.M. 2010. *A History of Chemical and Biological Weapons*. Islington, UK: Reaktion Books.

Steiner, W.E., C.S. Harden, F. Hong, S.J. Klopsch, H.H. Hill, Jr., and V.M. McHugh. 2006. Detection of aqueous phase chemical warfare agent degradation products by negative mode ion mobility time-of-flight mass spectrometry [IM(tof)MS]. *Am. Society Mass Spectrom.* 17:241–245.

Stephanov, A.V., L.I. Marinin, A.P. Pomerantsev, and N.A. Staritsin. 1996. Development of novel vaccines against anthrax in man. *J. Bacterio.* 44:155–160.

Stillmark, P.H. 1889. *Ueber Ricin. Arch. Pharmakol. Dorpat* 3:59–70.

Stine, R., M.V. Pishko, and C.-L. Schengrund. 2005. Comparison of glycosphingolipids and antibodies as receptor molecules for ricin detection. *Anal. Chem.* 77:2882–2888.

Sturm, M.B. and V.L. Schramm. 2009. Detecting ricin: Sensitive luminescent assay for ricin a-Chain ribosome depurination kinetics. *Anal. Chem.* 81:2847–2853.

Surveillance for Bioterrorism. https://emergency.cdc.gov/bioterrorism/surveillance.asp (accessed January 30, 2018).

Taitt, C.R., G.P. Anderson, B.M. Lingerfelt, M.J. Feldstein, and F.S. Ligler. 2002. Nine-analyte detection using an array-based biosensor. *Anal. Chem.* 74:6114–6120.

Thullier, P. and G. Griffiths. 2009. Broad recognition of ricin toxins prepared from a range of ricinus cultivars using immuno-chromatographic tests. *Clin. Toxicol.* 4:643–650.

Torok, T.J., R.V. Tauxe, R.P. Wise, J.R. Livengood, R. Sokolow, S. Mauvais, K.A. Birkness, M.R. Skeels, J.M. Horan, and L.R. Foster. 1997. A large community outbreak of salmonellosis caused by intentional contamination of restaurant salad bars. *JAMA* 278(5):389–395.

Wade, M.M, T.D. Biggs, J.M. Insalaco, L.K. Neuendorff, V.L.H. Bevilacqua, A.M. Schenning, L.M. Reilly, S.S. Shah, E.K. Conley, P.A. Emanuel, and A.W. Zulich. 2011. Evaluation of handheld assays for the detection of ricin and staphylococcal enterotoxin B in disinfected waters. *Int. J. Microbiol.*: 1–5. doi:10.1155/2011/132627.

Wald, C. 2004. Focus: Landmarks: The physical review's explosive secret. *Phys. Rev. Focus* 14(17). https://physics.aps.org/story/v14/st17 (accessed December 3, 2017).

Wall, M.M. 2013. US Makes First Plutonium in 25 Years, for Spacecraft. www.space.com/20290-plutonium-spacecraft-nasa-fuel.html (accessed December 3, 2017).

Warden, C.J.H. and L.A. Waddell. 1884. Non-bacillar nature of abrus or jequirity poison. *Ind. Med. Gaz.* 19(6):151–162.

Webb-Robertson, B.-J., H. Kreutzer, G. Hart, J. Ehleringer, J. West, G. Gill, and D. Duckworth. 2012. Bayesian integration of isotope ratio for geographic sourcing of castor beans. *J. Biomed. Biotechnol.* 2012:1–8.

Weerth, C. 2009. The cross-border detection of radiological, biological, and chemical active and harmful terrorist devices. *World Customs J.* 3(2):93–106.

Wettasinghe, R.C., M. Zabet-Moghaddam, G. Ritchie, and D.L. Auld. 2013. Relative quantitation of ricin in ricinus communis seeds by image processing. *Ind. Crops Prod.* 50:654–660.

Wolinetz, C.D. 2012. Implementing the new U.S. dual-use policy. *Science* 336:1525–1527.

Worbs, S., K. Köhler, D. Pauly, M.-A. Avondet, M. Schaer, M.B. Dorner, and B.G. Dorner. 2011. Ricinus communis intoxications in human and veterinary medicine—A summary of real cases. *Toxins (Basel)* 3:1332–1372.

Wu, J., P. Jia, C. Wang, Y. Zhao, H. Peng, W. Wei, and H. Li. 2011. Immunochromatography detection of ricin in environmental and biological samples. *Nano Biomed. Eng.* 3:169–173.

Yang, M., M. Goolia, W. Xu, H. Bittner, and A. Clavijo. 2013. Development of a quick and simple detection methodology for foot-and-mouth disease virus serotypes O, A and Asia 1 using a generic RapidAssay Device. *Viro J.* 210:125. doi:10.1186/1743-422X-10-125.

Zhao, S., W-S. Liu, M. Wang, J. Li, Y. Sun, N. Li, F. Hou, J.-Y. Wan, Z. Li, J. Qian, and L. Liu. 2012. Detection of ricin intoxication in mice using serum peptide profiling by MALDI-MS. *Int. J. Mol. Sci.* 13:13704–13712.

Östin, A., T. Bergström, S.-Å. Fredriksson, and C. Nilsson. 2007. Solvent-assisted trypsin digestion of ricin for forensic identification by LC-ESI MS/MS. *Anal. Chem.* 79:6271–6278.

CHAPTER 16

Environmental forensics

KEY WORDS: environmental forensics, pesticides, herbicides, organochlorines, DDT, organophosphates, carbamates, pyrethroids, Agent Orange, fungicides, antimicrobials

LEARNING OBJECTIVES

- To describe the field of environmental forensics and what it encompasses
- To describe the use of pesticides and herbicides in warfare
- To list some pesticides and herbicides and their biochemical effects
- To explain how chemical instrumentation methods can be used to detect and identify herbicides and pesticides

ENVIRONMENTAL AND HUMAN EFFECTS OF DDT USE

In her book, *Silent Spring*, which was published in 1962, Rachel Carson, a former marine biologist with the US Fish and Wildlife Service and renowned book and magazine article author, exposed the broad effects of DDT beyond its broad insecticide applications. She described how DDT entered the food chain from its crop application and accumulated in the fatty tissues of mammals, including humans, and caused tumors and cancer among other effects.

DDT, as the organochlorine chemical dichlorodiphenyltrichloroethane is known, was first synthesized by Othmar Zeidler in 1874. In 1948, Paul Müller was awarded the Nobel Prize in Physiology or Medicine for his 1939 discoveries that DDT had a high efficiency against arthropods as a contact poison or insecticide. He was awarded a Swiss patent for the chemical in 1940. Trade names for DDT include Anofex, Cesarex, Chlorophenothane, Dedelo, Dinocide, Didimac, Digmar, ENT 1506, Genitox, Guesapon, Guesarol, Gexarex, Gyron, Hildit, Ixodex, Kopsol, Neocid, OMS 16, Micro DDT 75, Pentachlorin, Rukseam, R50, and Zerdane. DDT is listed as Environmental Protection Agency (EPA) toxicity Class II (toxic). It is a white crystalline solid with no odor or taste. It has been used by the military and in agriculture for eradicating pests, especially lice and mosquitos, and is applied as a dust or by spraying an aqueous suspension. DDT was banned from further use in Sweden in 1970. In the United States, DDT was banned by the EPA from further use in 1972. As a persistent organic pollutant, its use is restricted by the United Nations Environment Programme.

DDT is absorbed through the small intestine and accumulates in fat tissues. It is not absorbed readily through the skin unless in a solution. Acute DDT toxicity leads to tingling sensations in the mouth, nausea, vomiting, diarrhea, increased liver enzyme activity, decreased thyroid function, irritation of membranes, disturbed gait, excitability, malaise, fatigue, tremors, and convulsions. Oral LD_{50}s range from approximately 100 mg/kg in rodents to 1000 mg/kg in larger mammals. Exposure to DDT has been linked to breast and pancreatic cancer, among others, male infertility, miscarriages and low birth weight, developmental delays, and nervous system and liver damage. DDT ingestion causes sterilization in rats, irregular reproductive cycles in mice, and thinning of eggshells and deaths of embryos in birds. It is highly toxic to fish and moderately toxic to amphibians and humans.

DDT is very persistent in mammalian tissues due to its fat solubility and in the environment with a half-life of 2–15 years. Owing to its chemical structure, it is immobile in most soils but may accumulate in the top layer, which is

ENVIRONMENTAL AND HUMAN EFFECTS OF DDT USE (continued)

rich in organic matter, and has a low solubility in water (<1 mg/mL at room temperature). The half-life of DDT in an aquatic environment is 150 years. DDT breakdown products include 1,1-dichloro-2,2-bis(p-dichlorodiphenyl) ethylene (DDE), 1,1-dichloro-2,2-bis(p-chlorophenyl)ethane (DDD), and bis(dichlorodiphenyl)acetic acid (DDA). DDA is readily excreted through the urine. DDT is broken down by photodegradation and lost by volatilization. DDT adsorbs to natural organic matter (NOM). DDD and DDE breakdown products are also found in soil; both have similar chemical characteristics and persistence as DDT in the environment. The United States Department of Agriculture detected DDT breakdown products in many foods including 60% of heavy cream samples tested, 42% of kale greens tested, and 28% of carrots tested. The Centers for Disease Control and Prevention (CDC) found DDT breakdown products in the blood of 99% of the people tested.

DDT continues to be used elsewhere around the world—especially Africa and Asia; the levels sprayed continue to increase annually to combat malaria. In 2007, the United Nations Environment Programme reported that at least 3950 tons of DDT were sprayed in Africa and Asia. In South Africa and North Korea, DDT is sprayed inside homes.

BIBLIOGRAPHY

Carson, R. 1962. *Silent Spring.* Boston, MA: Houghton Mifflin Company.

Cone, M. 2009. Should DDT be used to combat malaria? *Sci. Am.* www.scientificamerican.com/article/ddt-use-to-combat-malaria/ (accessed January 24, 2018).

Fiedler, H. 2003. *Persistent Organic Pollutants.* Heidelberg: Springer-Verlag.

Dichlorodiphenyltrichloroethane (DDT) factsheet. www.cdc.gov/biomonitoring/DDT_FactSheet.html (accessed January 24, 2018).

DDT, A brief history and status, EPA. www.epa.gov/ingredients-used-pesticide-products/ddt-brief-history-and-status (accessed January 24, 2018).

DDT, National Pesticide Information Center. http://npic.orst.edu/factsheets/archive/ddttech.pdf (accessed January 24, 2018).

Public health statement for DDT, DDE, and DDD. www.atsdr.cdc.gov/phs/phs.asp?id=79&tid=20 (accessed January 24, 2018).

The DDT story. www.panna.org/resources/ddt-story (accessed January 24, 2018).

The Nobel Prize in Physiology or Medicine 1948, Paul Müller. www.nobelprize.org/nobel_prizes/medicine/laureates/1948/ (accessed January 24, 2018).

Environmental forensics is focused on detecting and identifying pollutants and contaminants released into the environment in order to assess the extent of damage, facilitate cleanup, and levy fines on responsible parties. The materials can range from chemicals in industrial waste, detergents used in natural gas fracking, oil from spills, radioactive species released in burning coal by power plants, and herbicides and pesticides used in agriculture and warfare. Samples may include air, water, and soil from the affected area and reference samples from an adjacent area that was not affected by the event. This chapter will focus on herbicides and pesticides used in agriculture and warfare and their chemical structures, activity, human risks, and chemical identification. These materials are applied in large quantities directly to the environment. Nuclear and radiological forensics is covered in Chapter 15 and other molecules including drugs, plastics, and pathogens and toxins that may be detected in environmental forensics are covered in Chapters 8, 10, and 15, respectively.

As the environmental field is so broad, many regulations have been put into place and laws have been passed to delineate the legal acceptable limits and chemicals allowed for use in the United States and in other countries. For example, pesticide and herbicide use has been controlled for over a century. The first US Federal pesticide act, the Federal Insecticide Act, became law in 1910 and was followed by the Federal Insecticide, Fungicide, and Rodenticide Act in 1947. A revised law, the Federal Environmental Pesticide Control Act, was passed in 1972 and the regulation was transferred to the US EPA. In addition, the US Toxic Substances Control Act was passed in 1976 and places restrictions on some chemicals. Some amendments were included in the Food Quality Protection Act of 1996. Much of the detection and identification testing performed to ensure compliance with the law is conducted by the EPA but some is also contracted to private contractors.

Pesticides and herbicides have seen wide use. The earliest-applied herbicides were salts used to kill agricultural crops as written in the Bible and historical documents. For example, copper and arsenic salts and carbon bisulfide were used by the Romans to conquer Carthage in 164 BC. They have been used as chemical warfare agents to defoliate areas to improve access for troops, kill jungle crops grown by insurgents, reduce time and effort needed to grow crops for farming, and kill insects known to spread disease. Minimal to no long-term effects are expected for the troops and civilians with limited exposure, but long-term effects including allergies, nerve and liver damage, cancer, and genetic changes have been observed under certain conditions, especially including high or long-term exposure. These chemicals also have been shown to cause birth defects. They have relatively low toxicity for fish and invertebrates, birds, and mammals but impacts are observed when the pesticide or herbicide is applied directly to water bodies. As herbicides and pesticides are used on huge swaths of land around the world, it is common to detect one or more of the following substances or their relatives in the soil, water, and even air.

PESTICIDES

Pesticides are substances intentionally released into the environment to kill, repel, or control weed (plant), insect, fungi, and rodent pests. Pesticides include insecticides, herbicides, antimicrobials, and fungicides. Representative organochlorine, organophosphate, carbamate, and pyrethroid compounds used as pesticides are shown in Figure 16.1. The first pesticide developed was tetraethyl pyrophosphate (TEPP), an organophosphate. TEPP is composed of two central phosphate groups with two ethyl group arms hanging from each. Other organophosphates, such as parathion and malathion, have a phosphorous atom bound to an oxygen that is attached to a leaving group, usually an alkyl or aryl group, that is the active site. The oxygen can attack groups and become covalently, and permanently, attached to the target: the enzyme acetylcholinesterase in the body. Organophosphate pesticide development continued during World War II with nerve agent research as the compounds are chemically similar (refer to Chapter 15 for further information on chemical warfare agents).

The most famous pesticide is an organochlorine compound, dichlorodiphenyltrichloroethane (DDT), which was synthesized in 1874 by Othmar Zeidler under Adolf von Baeyer. DDT is a colorless, crystalline, almost odorless organochlorine compound (Figure 16.2). Organochlorines such as DDT have a negative temperature coefficient. DDT and other organochlorine compounds can be dissolved in solvents such as xylene and petroleum distillates; formulated in lotions or liquid suspensions; emulsified; released as an aerosol; formulated into dry formulation granules, pellets, or dusts; and incorporated into smoke candles. Organochlorines are not easily absorbed through the skin and are not acutely toxic when applied in the manner but can cause skin irritation. DDT was used as an insecticide to control malaria and typhus starting in 1939 and was widely used in World War II to control disease. Paul Hermann Mueller was awarded the 1948 Nobel Prize in Physiology or Medicine for his 1939 discoveries of insecticide properties.

DDT was banned in the United States in 1972. Related pesticides include 4,4′-DDT, and 2,4′-DDT. If ingested, they can cause dizziness, headaches, and disorientation. They stimulate the central nervous system and are toxic due to their effect on sodium and chloride ion channels in nerve cells. Since the 1970s, pyrethroids including permethrin and tetramethrin have been widely adopted to control insects.

Other organochlorines (Figure 16.3) include aldrin (1,2,3,4,10,10-Hexachloro-1,4,4a,5,8,8a-hexahydro-1,4:5,8-dimethanonaphthalene), chlordane (1,2,4,5,6,7,8,8-Octachloro-3a,4,7,7a-tetrahydro-4,7-methanoindane), dieldrin

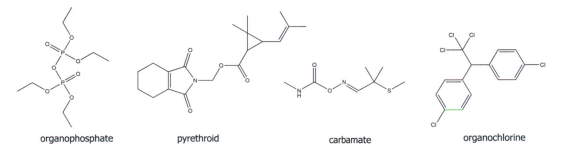

organophosphate   pyrethroid   carbamate   organochlorine

Figure 16.1 Structures of pesticides by class: organophosphates, pyrethroids, carbamates, and organochlorines.

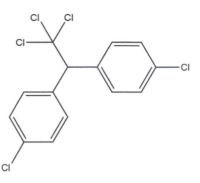

Figure 16.2   Structure of DDT.

((1aR,2R,2aS,3S,6R,6aR,7S,7aS)-3,4,5,6,9,9-hexachloro-1a,2,2a,3,6,6a,7,7a-octahydro-2,7:3,6-dimethanonaphtho[2,3-b]oxirene), endosulfan (6,7,8,9,10,10-Hexachloro-1,5,5a,6,9,9a-hexahydro-6,9-methano-2,4,3-benzodioxathiepine-3-oxide),endrin((1R,2S,3R,6S,7R,8S,9S,11R)-3,4,5,6,13,13-Hexachloro-10-oxapentacyclo[6.3.1.13,6.02,7.09,11]tridec-4-ene), heptachlor (1,4,5,6,7,8,8-Heptachloro-3a,4,7,7a-tetrahydro-4,7-methano-1H-indene), kepone (decachloropentacyclo[5.3.0.02.6.03.9.04.8]decan-5-one),   metachlor   ((RS)-2-Chloro-N-(2-ethyl-6-methyl-phenyl)-N-(1-methoxypropan-2-yl)acetamide), methoxychlor (1,1,1-Trichloro-2,2-bis(4-methoxyphenyl)ethane), mirex (1,1a,2,2,3,3a,4,5,5,5a,5b,6-dodecachlorooctahydro-1H-1,3,4-(methanetriyl)cyclobuta[cd]pentalene), lindane (1,2,3,4,5,6-Hexachlorocyclohexane), and toxaphene (Polychloro-2,2-dimethyl-3-methylidenebicyclo[2.2.1]heptane). Organochlorines are synthesized using Diels–Alder reactions and are characterized by covalently bounded chlorines in organic molecules that result in compounds that are more dense than water. Organochlorine insecticides function by binding at the gamma-aminobutyric acid chloride channel $GABA_A$ site, inhibiting chloride flow into nerve cells. As insecticides, lindane is used to kill lice and scabies; aldrin and dieldrin have been used to kill termites; and endrin is used to kill rodents and birds.

Organophosphate pesticides include chlorpyrifos (O,O-Diethyl O-3,5,6-trichloropyridin-2-yl phosphorothioate), diazinon (O,O-Diethyl O-[4-methyl-6-(propan-2-yl)pyrimidin-2-yl] phosphorothioate), dichlorvos (2,2-Dichlorovinyl dimethyl phosphate), ethion (O,O,O',O'-Tetraethyl S,S'-methylene bis(phosphorodithioate), fenthion (O,O-Dimethyl O-[3-methyl-4-(methylsulfanyl)phenyl]phosphorothioate), malathion (Diethyl 2-[(dimethoxyphosphorothioyl)sulfanyl]butanedioate), and parathion (O,O-Diethyl O-(4-nitrophenyl)phosphorothioate). Their structures, shown in Figure 16.4, and functions are similar to the organophosphate chemical weapons covered in Chapter 15. Both organophosphate pesticides and chemical weapons function as acetylcholinesterase inhibitors.

Carbamates were introduced in the 1950s. They contain a central amide bond with several R-groups and include aldicarb (Temik, 2-Methyl-2-(methylthio)propanal O-(N-methylcarbamoyl)oxime), carbaryl (Sevin, 1-naphthyl meth-

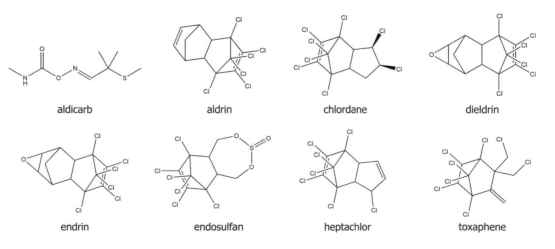

aldicarb   aldrin   chlordane   dieldrin

endrin   endosulfan   heptachlor   toxaphene

Figure 16.3   Structures of organochlorine pesticides.

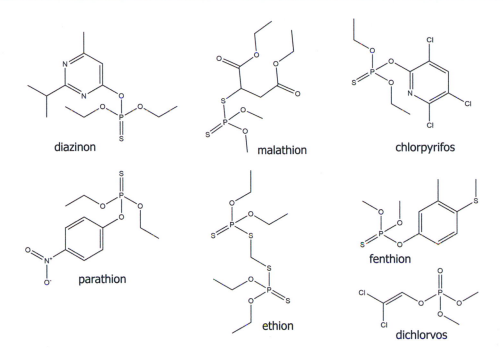

Figure 16.4 Structures of organophosphate pesticides.

ylcarbamate), carbofuran (Furadan, 2,2-Dimethyl-2,3-dihydro-1-benzofuran-7-yl methylcarbamate), ethienocarb, fenobucarb ((2-Butan-2-ylphenyl) N-methylcarbamate), methomyl ((E,Z)-methyl N-{[(methylamino)carbonyl]oxy} ethanimidothioate), and oxamyl (methyl 2-(dimethylamino)-N-[(methylcarbamoyl)oxy]-2-oxoethanimidothioate). Structures of these compounds are shown in Figure 16.5. Carbamate pesticides are used in dusts or sprays. Carbamates have a low vapor pressure, are water soluble, and rapidly decompose in aqueous conditions. Biologically relevant carbamates include the valine residues in the α- and β-chains of deoxyhemoglobin that stabilize the deoxy form and lysine residues in several proteins including urease, phosphotriesterase, and ribulose 1,5-bisphosphate carboxylase by reaction with carbon dioxide from air. Carbamate insecticides act by inhibiting acetylcholinesterase reversibly by binding at the active site. The carbamate binds to the catalytic serine residue but is reversible as the bond is broken by spontaneous hydrolysis in 30–40 minutes. In contrast, organophosphates act by phosphorylating the catalytic site serine residues in acetylcholinesterase non-reversibly. Carbamate therapeutic drugs are used to treat glaucoma and Parkinson's disease, among others. Carbamates and organophosphates reversibly inhibit esterase that causes neuropathy unlike organophosphates that can dealkylate the inhibited enzyme.

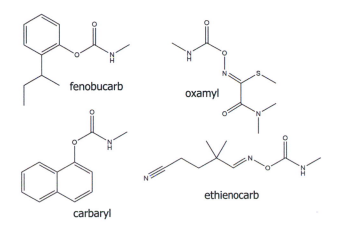

Figure 16.5 Structures of carbamates.

Pyrethroid insecticides are synthetic insecticides similar in structure and function to the natural product pyrethrins extracted from the flowers of pyrethrum plants (*Chrysanthemum cinerariaefolium* and *C. coccineum*). Pyrethroids form the basis of common pesticides marketed and sold for household use to combat chiggers, bedbugs, lice, flies, wasps, and crop pests. They include permethrin, the most commonly used pyrethroid, and related compounds allethrin I and II, bifenthrin (2-Methyl-3-phenylphenyl)methyl (1S,3S)-3-[(Z)-2-chloro-3,3,3-trifluoroprop-1-enyl]-2,2-dimethylcyclopropane-1-carboxylate), cyfluthrin (([R]-cyano-[4-fluoro-3-(phenoxy)phenyl]methyl] (1R,3R)-3-(2,2-dichloroethenyl)-2,2-dimethylcyclopropane-1-carboxylate), cypermethrin([Cyano-(3-phenoxyphenyl)methyl]3-(2,2-dichloroethenyl)-2,2-dimethylcyclopropane-1-carboxylate), cyphenothrin (Cyano(3-phenoxyphenyl)methyl 2,2-dimethyl-3-(2-methylprop-1-en-1-yl)cyclopropanecarboxylate), deltamethrin ([(S)-Cyano-(3-phenoxyphenyl)-methyl] (1R,3R)-3-(2,2-dibromoethenyl)-2,2-dimethyl-cyclopropane-1-carboxylate), fenpropathrin, flumethrin (Cyano(4-fluoro-3-phenoxyphenyl)methyl 3-[2-chloro-2-(4-chloro phenyl)vinyl]-2,2-dimethylcyclopropanecarboxylate),imiprothrin((2,5-Dioxo-3-prop-2-ynylimidazolidin-1-yl)methyl 2,2-dimethyl-3-(2-methylprop-1-enyl)cyclopropane-1-carboxylate),lambda-cyhalothrin(3-(2-chloro-3,3,3-trifluoro-1-propenyl)-2,2-dimethyl-cyano(3-phenoxyphenyl)methyl cyclopropanecarboxylate), metofluthrin (2,3,5,6-Tetra fluoro-4-(methoxymethyl)benzyl 2,2-dimethyl-3-(prop-1-en-1-yl)cyclopropanecarboxylate), resmethrin ([5-(pheny lmethyl)-3-furanyl]methyl 2,2-dimethyl-3-(2-methyl-1-propen-1-yl)cyclopropanecarboxylate), phenothrin ((3-Phen oxyphenyl)methyl 2,2-dimethyl-3-(2-methylprop-1-enyl)cyclopropane-1-carboxylate), prallethrin (2-methyl-4-oxo-3-prop-2-yn-1-ylcyclopent-2-en-1-yl 2,2-dimethyl-3-(2-methylprop-1-en-1-yl)cyclopropanecarboxylate), tetramethrin ((1,3-dioxo-4,5,6,7-tetrahydroisoindol-2-yl)methyl-2,2-dimethyl-3-(2-methylprop-1-enyl)cyclopropane-1-carboxyl-ate), and tralomethrin ((1R,3S)-2,2-Dimethyl-3-(1,2,2,2-tetrabromoethyl)-1-cyclopropanecarboxylic acid [(S)-cyano-[3-(phenoxy)phenyl]methyl] ester). Several structures are shown in Figure 16.6. Pyrethroids function by binding to voltage-gated sodium channels in nerve axonal membranes and keeping them open to allow a flow of sodium ions. The membrane cannot build up a charge potential so the nerve becomes paralyzed which leads to spasms, paralysis, and death.

Immediate treatment of people exposed to pesticides includes washing contaminated skin with soap and water, relieving respiratory exposure by moving to fresh air or the administration of oxygen, rinsing eyes for 15 minutes with clean water, and inducing vomiting or administering activated charcoal to reduce ingestion effects. Antidotes include atropine sulfate, 2-pyridine aldoxime methyl chloride (2-PAM) chloride, and diazepam. Atropine sulfate competes for acetylcholine receptor sites and binds the sites to reduce the effects of the pesticide. 2-PAM chloride reverses the effect of acetylcholinesterase inhibition caused by several pesticides. Diazepam prevents seizures.

HERBICIDES

Herbicides are substances also referred to as "weedkillers." Herbicides (Figure 16.7) are toxic to plants and are administered to kill weeds and other unwanted vegetation. Herbicides are widely used to enable row-crop farming and control aquatic weeds, and are used in forest, lawn, park, and golf course management. Three of the most commonly used herbicides are the chlorinated compounds 2,4-dichlorphenoxyacetic acid (2,4-D), 2,4,5-trichlorophenoxyacetic acid (2,4,5-T), and atrazine (Figure 16.8). Exposed plants exhibit stunted growth and abnormalities including stem twisting and leaf malformations, leaf discoloration (e.g., yellowing, whitening), tissue desiccation, and death. 2,4-D is a synthetic auxin that mimics plant hormones that regulate plant growth, thus stalling plant growth. Atazine is a triazine herbicide that binds to the plastoquinone-binding protein in photosystem II, thus inhibiting photosynthesis and plant growth. Sulfonylurea herbicides (Figure 16.9) include sulfometuron, triasulfuron, tribenuron,

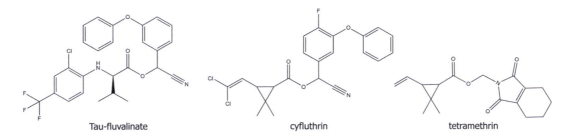

Tau-fluvalinate   cyfluthrin   tetramethrin

Figure 16.6   Structures of pyrethrins.

Figure 16.7 Herbicide products.

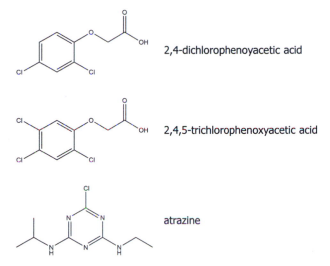

2,4-dichlorophenoyacetic acid

2,4,5-trichlorophenoxyacetic acid

atrazine

Figure 16.8 Structures of chlorinated herbicides.

thifensulfuron, metsulfuron-methyl, and chlorsulfuron; they inhibit acetolactate synthase, a plant enzyme needed for branched chain amino acid synthesis. Other herbicides include glyphosate (N-(phosphonomethyl)glycine) and fluazifop-p-butyl (Figure 16.10). These compounds inhibit amino acid and lipid biosynthesis pathways. In summary, herbicides mimic auxin plant growth hormones to limit plant growth and act as inhibitors in amino acid synthesis, photosynthesis, growth regulation, lipid biosynthesis, and cell division, as shown in Table 16.1. Some species or plant families are not susceptible to some herbicides because they use different biochemical pathways than those targeted or have slightly different enzyme structures, or isozymes. Animals and humans are typically unaffected by treatment as the chemicals target pathways found only in plants.

The United Kingdom was the first to use herbicides for military purposes. It used 2,4,5-T, 2,4-D, and trichloroacetic acid to defoliate and destroy crops in Malaysia in the 1950s. In chemical warfare, "Agent Orange," "Agent White," and "Agent Blue," among others, are mixtures of herbicides that have been employed to destroy the enemy's crops (e.g. crop depletion) and clear or reduce foliage to improve the maneuverability of troops. These were used by the US military in World War II and the Vietnam War and were released by C-123 aircraft. They were so named for the color of the labels (orange, white, blue, etc.) on the barrels that the US military used to store them. "Agent Orange" and "Agent White" were used for deforestation and "Agent Blue" was used to destroy crops.

FUNGICIDES

Agricultural fungicides include propiconazole, maneb, zineb, and benzene hexachloride, to name a few. Propiconazole (1-[[2-(2,4-dichlorophenyl)-4-propyl-1,3-dioxolan-2-yl]methyl]-1,2,4-triazole) is a triazole fungicide

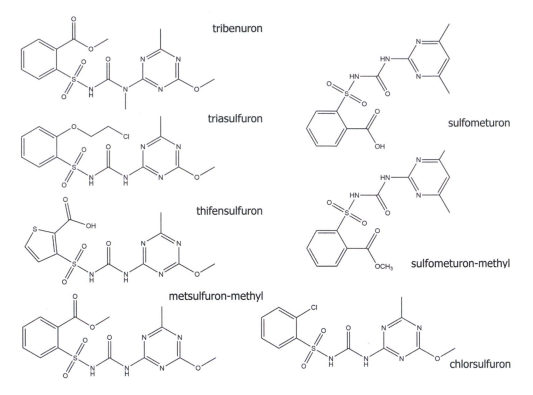

Figure 16.9 Structures of sulfonylurea herbicides.

that inhibits the 14-alpha demethylase enzyme so that an ergosterol precursor cannot be demethylated to form the ergosterol needed for fungal cell membranes. Propiconazole is used to control fungal growth on apricot, lemon, nectarine, peach, plum, and prune fruit plants and staple grains including corn, oats, rice, sorghum, and wheat. It is also used to control fungal growth on nut plants including almond, peanut, and pecan. Maneb ([[2-[(dithiocarboxy) amino]ethyl]carbamodithioato]](2-)-kS,kS']manganese) and zineb (zinc ethane-1,2-diylbis(dithiocarbamate)) are metalloorganic dithiocarbamate fungicides used on many of the same crops listed previously. Benzene hexachloride (hexachlorobenzene) is an organochlorine fungicide. Structures of selected agricultural fungicides are shown in Figure 16.11.

ANTIMICROBIALS

Antimicrobials include both antibiotics that kill bacteria and antifungals that kill fungi. Antifungal antimicrobials kill athlete's foot, candidiasis (thrush), and ringworm infections, among others. Antifungals include polyene and imidazole antifungals. *Polyene* antifungals include amphotericin B, candicidin, filipin, hamycin, natamycin, nystatin, and rimocidin. *Imidazole* antifungals include clotrimazole and related compounds (Figure 16.12).

There are dozens of antibiotics. Antibiotics include sulfa drugs such as sulfanilamide, sulfapyridine, sulfacetamide, and sulfamethizole; penicillins including benzylpenicillin; peptide antibiotics such as gramicidin S; aminoglycosides such as streptomycin; tetracyclines such as chlortetracycline and oxytetracycline; amphenicols such as chloramphenicol; and glycopeptides such as vancomycin, to name a few. Structures of selected antibiotics are shown in

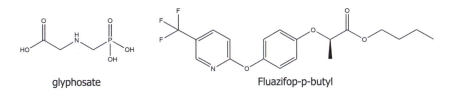

glyphosate   Fluazifop-p-butyl

Figure 16.10 Structures of glyphosate and fluazifop-p-butyl herbicides.

Table 16.1 Classification of herbicides by inhibition effect

Inhibition effect	Description	Chemical families	Examples
Amino acid synthesis	Inhibits amino acid biosynthesis causing growth deformities by targeting acetolactate synthetase, glutamine synthetase, or 5-enoylpyruvyl-shikimate-3-phosphate synthetase	Sulfonylureas, imidazolinones, triazolopyrimidines, glyphosate, glufosinate	Glyphosate, imazethapyr, thifensulfuron, phosphinothricin
Photosynthesis	Inhibits photosynthesis Photosytem II pathway electron transport from photoquinone (Q) Q_A to Q_B and Photosystem I needed for sugar production and growth	Triazines, bromoxynil, bipyridiniums, paraquats, diquats	Atrazine, ametryne, cyanazine, prometryn, simazine
Auxin growth regulation	Causes uncontrolled growth of meristems	Benzoic acids	2,4-D, dicamba
Cell division	Inhibits root cell division and growth	Pyridines, chloroacetanilides	Trifluralin, pendimethalin, metachlor
Lipid biosynthesis	Inhibits synthesis of lipids required for cell membranes by inhibiting acetyl coenzyme A carboxylase	Aryloxyenoxypropionates, cyclohexanediones	Fluzaifop-p-butyl, sethoxydim

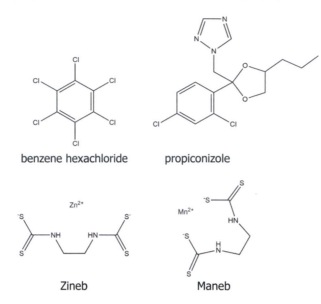

Figure 16.11 Structures of selected agricultural fungicides.

Figure 16.13. These may be introduced to water supplies when they are not disposed of properly including disposal by flushing in toilets. Antibiotic growth promoters used in agriculture include bambermycin, lasalocid, monensin, salinomycin, virginiamycin, bacitracin, carbadox, laidlomycin, lincomycin, neomycin, penicillin, roxarsone, and tylosin. Of these, agricultural antibiotics are most likely to be used in large quantities and enter water supplies and ground soil.

IN THE ENVIRONMENT

Pesticides and herbicides dissipate and exhibit movement by spray drift, volatilization to a gas, movement such as suspension in surface or subsurface runoff into surface or groundwater, leaching into water through the soil, immobilization or adsorption onto soil particles (due to the NOM and metal ion bridges), absorption or uptake by non-susceptible plants and degradation by microorganisms, sunlight (photodegradation), and chemical and biochemical decomposition. The mobility of herbicides and pesticides increases with water solubility. Insoluble compounds adsorb quickly into soils and penetrate tissues more readily. Persistence in the soil depends on the compound's half-life and functionalization, but also soil pH, metal content, temperature and surface area, dissolved organic matter content, weather, and vegetation. Water persistence depends on water pH and dissolved organic matter but also functionalization. Salts and acids remain soluble in water until they are degraded whereas esters precipitate onto surfaces in the water.

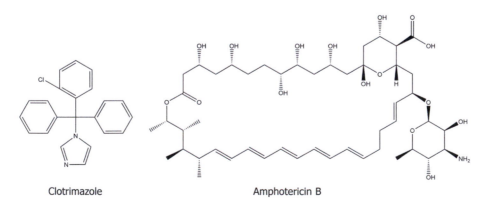

Figure 16.12 Structures of selected antifungals.

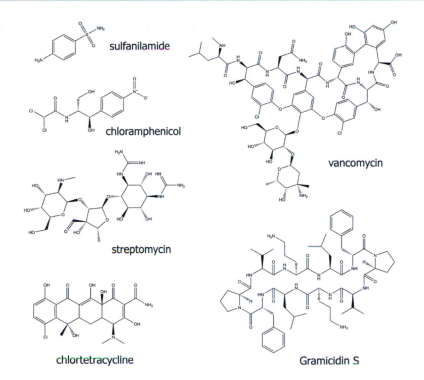

Figure 16.13 Structures of selected antibiotics.

Pesticides are classified by the extent of hazard they present as shown in Table 16.2, their EPA classification. Aldicarb and parathion are two of the most hazardous pesticides and are listed in Class 1A. Carbofuran is a member of Class 1B of highly hazardous poisons. Copper oxychloride (Class II) is a moderately dangerous, inorganic compound, dicopper chloride trihydroxide ($Cu_2(OH)_3Cl$), that is used as a fungicidal spray on important crop plants including orange, coffee, cardamom, cotton, grape, rubber, and tea. Carboxin is also a fungicide and a member of Class III (slightly hazardous). Bromacil (5-bromo-3-(butan-2-yl)-6-methylpyrimidine-2,4(1H,3H)-dione or 5-bromo-3-sec-butyl-6-methyluracil) is a herbicide pesticide used on citrus and pineapple plants that acts by inhibiting photosynthesis. It is listed in Class U, unlikely to cause a hazard.

Table 16.2 Classification of pesticides based on toxicity by the EPA

Class	Label	Examples	Risks	Safety
1A	Extremely hazardous, danger—poison	Aldicarb, parathion	Headache, nausea and vomiting, sweating, diarrhea, loss of coordination, and death	Wear goggles/face shield, rubber gloves, rubber apron
1B	Highly hazardous, danger—poison	Carbofuran	Weakness, nausea and vomiting, sweating, abdominal pain, and blurred vision	Wear goggles/face shield, rubber gloves, rubber apron
II	Moderately hazardous, danger —corrosive	Copper oxychloride	Irreversible eye or skin burns	Wear gloves, googles, apron
III	Slightly hazardous, warning	Carboxin	May be fatal or harmful if swallowed/ inhaled/absorbed dermally, can cause burns or skin/eye irritation	Wear gloves
IV	Very low hazard, caution		Mild skin/eye irritation, mildly toxic when swallowed/inhaled/absorbed dermally	
U	Unlikely to present hazards	Bromacil	Slight hazard near heat or flame	

Source: EPA, https://www.epa.gov/ingredients-used-pesticide-products/ddt-brief-history-and-status.

EXAMPLES OF DETECTION, IDENTIFICATION, AND QUANTIFICATION OF HERBICIDES AND PESTICIDES

Several methods including those discussed previously in this book can be used to detect, identify, and quantitate herbicides and pesticides from natural soil and water samples. A few examples of the application of detection methods follow.

Immunoassays, immunosensors, portable gas chromatography-electron capture detector (GC-ECD) and photoionization detection instruments, and portable Fourier transform–infrared (FT-IR) and Raman spectrometers are available for on-site detection. For example, the SensioScreen TR5000 can be used to detect triazine herbicides in water in only 10 minutes using an enzyme-linked immunosorbent assay method. The triazine herbicides in the evidence sample and a triazine-enzyme conjugate compete for binding sites of membrane-immobilized antibodies. The membrane is rinsed and a blue color signal is produced when the enzyme acts on the substrate. The color intensity is inversely proportional to the concentration of triazines in the sample. Negative controls and negative samples will yield a dark blue color while the positive samples and standards will exhibit a light blue or no color. Portable GC instruments can detect low levels of analyte (ppm) and provide a tentative identification. GC-ECD can be used to detect and identify halogen-containing herbicides and pesticides in a mixture. They can be used with the spectrophotometer instruments equipped with libraries that help identify the questioned substance and the software will provide a list of hits and an associated ranking using an algorithm to aid in identification.

Fluorescence spectrometers can be used to assess herbicide and pesticide binding, sorption, lifetimes, and degradation pathways in the lab. For example, NOM including humic and fulvic acids has been demonstrated to interact with more nonpolar as well as polar herbicides and pesticides. The polar molecules including 2,4-dichlorophenoxyacetic acid and related compounds such as 2,4-dichlorophenoxypropionic acid have been shown to interact with fulvic acids through metal ion bridges with aluminum and other ions soluble in acidic soils and waters (Figure 16.14).

Gas chromatography–mass spectrometry (GC-MS) can be used to detect, identify, and quantitate herbicides and pesticides by comparing the GC retention time of the evidence sample to that of known standards using a method or standard operating procedure and the MS fragmentation pattern and qualifier ions to identify the pesticides. Using optimized methods, 22 organophosphorus pesticides have been shown to be separated using the GC and elute with different retention times. The area under the GC peaks can be used to quantify the substances. GC-MS can also be used to analyze herbicides such as 2,4-D and 2,4,5-T. The mass spectra for 2,4-D and 2,4,5-T are shown in Figure 16.15.

As GC-MS is not a good technique for samples that have a low volatility or that decompose at high temperatures, high-pressure liquid chromatography (HPLC) can be used. For example, in a 1998 case, HPLC was used to detect, identify, and quantitate carbamates. GC is not used for the analysis of carbamates as they decompose at high temperatures. Twenty employees were sickened and hospitalized after a company lunch. One person reported adding black pepper from a can labeled "black pepper" found in the home of a deceased family member to prepare a homemade cabbage salad. HPLC was used to identify and quantitate the contents of the black pepper container. It was determined to be 13.7% aldicarb containing the pesticide TEMIK® and, based on testing, each 6 g portion of salad was estimated to contain 272.6 ppm of aldicarb. The deceased owner of the pepper can had been a crawfish farmer. Aldicarb is a carbamate used on nets for catching crawfish. An investigation by the Louisiana Department of Agriculture and Forestry led to the conclusion that the crawfish farmer had used the pesticide on bait to prevent destruction of his nets by animals. Exposure of carbamates leads to symptoms including gastrointestinal illness, sweating, dizziness, blurred and darkened vision, muscle twitching, and weakness. Other mild symptoms of carbamates can include slowed heart beat and contraction of the pupils. Severe symptoms can include chest tightness, seizures, loss of muscle control, and loss of consciousness. Eight people required and received supportive care and two required and received atropine; all affected persons recovered. In another case, the carbamate insecticide methomyl was analyzed in Soju in the Republic of Korea using stable isotope analysis.

Other methods that can be used to detect and identify herbicides and pesticides include capillary zone electrophoresis and reverse-phase liquid chromatography. Capillary zone electrophoresis and reverse-phase liquid chromatography with UV detection at 254 nm was used to detect sulfonylurea herbicides that cannot be detected using GC due to thermal instability without derivatization. The samples are time consuming to process as they have long retention times

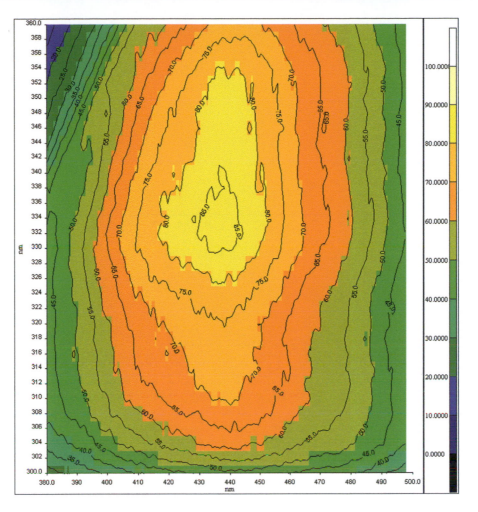

Figure 16.14 Fluorescence EEM plot for the SRFA-Al^{3+}-DCPPA ternary solution prepared at pH 4.0 in 0.1 M NaClO$_4$ buffer containing 15 mg/L SRFA, 0.300 mM Al^{3+}, and 0.715 mM 2,4-dichlorophenoxypropionic acid. The solution was incubated in the dark for 24 hours, then excited from 300 to 360 nm (y-axis), and detected fluorescence emission was plotted from 380 to 500 nm (x-axis). The relative fluorescence units (color) are plotted from 0 to 100 (z-axis).

on the column. In a study, samples were prepared by adding sodium bicarbonate buffer to soil samples and shaking to equilibrate. After separation by centrifugation, the supernatant was recovered and loaded onto a C18 reverse-phase column and eluted with methanol slightly acidified with perchloric acid, HClO$_4$ (to pH 3.5–5.5). Standards need to be run using the same column or capillary, solvent, and protocol for identification using chromatography; benzoic acid was used as the internal standard for quantitation and correlation migration times of the unknowns with elution times of known standards was used for identification. The elution time was inversely correlated with pKa. The sulfonylureas including sulfometuron, triasulfuron, tribenuron, thifensulfuron, metsulfuron-methyl, and chlorsulfuron eluted from 5 to 10 minutes using capillary zone electrophoresis. However, when humic acids found in natural soils and waters were coeluted with the sulfonylureas, a huge and tailed chromatographic peak was observed and made quantification difficult.

Nuclear magnetic resonance spectroscopy (NMR) can provide more specific and *de novo* structure characterization of unknown compounds and detect and identify known compounds. ^{1}H and ^{13}C NMR are both used to detect pesticides and herbicides as most of these compounds contain several of these nuclei. ^{31}P NMR can be used to detect and characterize organophosphates including malathion and parathion, among others. ^{19}F NMR can be used to detect and characterize fluorinated pyrethroids including transfuthrin (C$_{15}$H$_{12}$C$_{12}$F$_4$O$_2$), tefluthrin (C$_{17}$H$_{14}$ClF$_7$O$_2$), silafluofen (C$_{25}$H$_{29}$FO$_2$Si), and flucythrinate (C$_{26}$H$_{23}$F$_2$NO$_4$). NMR can also be used to detect pesticides or herbicides in a mixture or matrix due to the unique peaks of each substance.

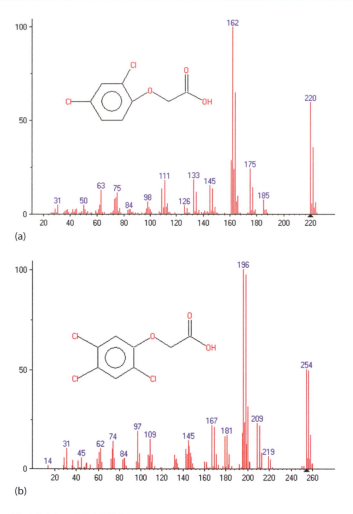

Figure 16.15   Mass spectra of 2,4-D (a) and 2,4,5-T (b).

The identification of a pesticide or herbicide source can be determined based on whether the pesticide or herbicide contains a salt, ester, or adjuvant. Adjuvants are materials that are added to the formulation to increase the efficacy of the herbicide or pesticide. They include surfactants, extenders, activators, compatibility agents, buffers, acidifiers, de-foaming agents, thickeners, dyes, and deposition aids. Salts are water soluble and the salt can dissociate in water. They are less volatile than esters and require use of a surfactant. Esters are not water soluble and require an emulsifying agent. They are more toxic than salts and range in volatility. Volatilization is increased with lower molecular weights, increased temperature, increased soil moisture, and with decreasing content of clay and organic matter.

Chromatographic, spectroscopic, and wet chemistry methods can also be used to detect and quantitate other pollutants and trace materials in the environment including drugs, plastics, toxins, pathogens, and even radioactive material. These materials and their analysis is covered in previous chapters in this book.

QUESTIONS

1. Which method could be used to separate and identify three pesticides?
   a. GC-MS
   b. Infrared spectroscopy
   c. Raman spectroscopy
   d. Scanning electron microscope

2. Which pesticide forms a covalent attachment to acetylcholinesterase?
   a. 2,4,5-T
   b. DDT
   c. Malathion
   d. 2,4-D

3. Name some antidotes to treat pesticide exposure.
    a. Atropine sulfate   b. 2-PAM chloride
    c. Diazepam   d. All of the above

4. The herbicide 2,4-D inhibits:
    a. Amino acid synthesis   b. Auxin growth regulation
    c. Photosynthesis   d. Cell division

5. Which of the following is not a class of pesticides?
    a. Organochlorines   b. Carbamates
    c. Sulfur mustards   d. Pyrethroids

6. What is environmental forensics?

7. How are herbicides and pesticides used in warfare?

8. List some risks of pesticide and herbicide use to humans and the environment.

9. Why was DDT used and why is it banned in the United States?

10. Which pesticides have largely replaced DDT?

Bibliography

Agency for Toxic Substances and Disease Registry (ATSDR). 2002. *Toxicological Profile for DDT, DDE, and DDD*. Atlanta, GA: US Department of Health and Human Services, Public Health Service. www.atsdr.cdc.gov/phs/phs.asp?id=79&tid=20 (accessed January 24, 2018).

Balleteros, B., D. Barceló, A. Dankwardt, P. Schneider, and M.-P. Marcol. Evaluation of a field test kit for triazine herbicides (SensioScreen® TR500) as a fast assay to detect pesticide contamination in water samples. http://digital.csic.es/bitstream/10261/19870/3/Ballesteros_Berta_et_al.pdf (accessed January 24, 2018).

Bradbury, S. 2007. EPA 738-R-07-008: Reregistration eligibility decision for dichlorprop-p (2,4-DP-p). *US EPA*. pp. 1–68.

Carson, R. 1962. *Silent Spring*. Boston, MA: Houghton Mifflin Company.

CDC. 2017. Dichlorodiphenyltrichloroethane (DDT) factsheet. www.cdc.gov/biomonitoring/DDT_FactSheet.html (accessed January 24, 2018).

Chen, Z.L., R.S. Kookanaand, and R. Naidu. 2000. Determination of sulfonylurea herbicides in soil extracts by solid-phase extraction and capillary zone electrophoresis. *Chromatographia* 52(3–4):142–146.

Elkins, K.M., E.M. Traudt, and M.A. Dickerson. 2011. Fluorescence characterization of the interaction suwannee river fulvic acid with the herbicide 2-(2,4-dichlorophenoxy)propionic acid in the absence and presence of aluminum. *J. Inorg. Biochem.* 105:1469–1476.

Elkins, K.M. and D.J. Nelson. 2001. Fluorescence and FT-IR spectroscopic studies of Suwannee River fulvic acid complexation with aluminum, terbium and calcium. *J. Inorg. Biochem.* 87:81–96.

Elkins, K.M. and D.J. Nelson. 2002. Spectroscopic approaches to the study of the interaction of aluminum with humic substances. *Coord. Chem. Rev.* 228:205–225.

Environmental Protection Agency. www.epa.gov/ (accessed January 24, 2018).

EPA. 2017. *DDT—A Brief History and Status*. EPA. www.epa.gov/ingredients-used-pesticide-products/ddt-brief-history-and-status (accessed January 24, 2018).

Farley, T.A. and L. McFarland. 1999. Aldicarb as a cause of food poisoning—Louisiana, 1998. *MMWR* 48(13):269–271. www.cdc.gov/mmwr/preview/mmwrhtml/00056877.htm (accessed January 24, 2018).

Fiedler, H. 2003. *Persistent Organic Pollutants*. Heidelberg: Springer-Verlag.

Hayes, A.W., ed. 2007. *Principles and Methods of Toxicology*, 5th edn. Boca Raton, FL: CRC Press.

Hester, R.E. and R.M. Harrison. 2008. *Environmental Forensics*. London, UK: Royal Society of Chemistry.

Khan, A.I. 2013. Analysis of organophosphorus pesticides by GC. *Thermo Sci*. https://assets.thermofisher.com/TFS-Assets/CMD/Application-Notes/AN-20705-Analysis-Organophosphorus-Pesticides-GC-AN20705.pdf (accessed January 24, 2018).

Larrivee, E.M., K.M. Elkins, S.E. Andrews, and D.J. Nelson. 2003. Fluorescence characterization of the interaction of Al3+ and Pd2+ with Suwannee River fulvic acid in the absence and presence of the herbicide 2,4-dichlorophenoxyacetic acid. *J. Inorg. Biochem*. 97:32–45.

Moreland, D.E. 1967. Mechanisms of action of herbicides. *Annu. Rev. Plant Physiol*. 18(1):365–386.

Morrison, R.D. 1999. *Environmental Forensics*. Boca Raton, FL: CRC Press.

Morrison, R.D. and B.L. Murphy, eds. 2005. *Environmental Forensics: Contaminant Specific Guide*. Cambridge, MA: Academic Press.

Mudge, S.M. 2008. *Methods in Environmental Forensics*. London, UK: Taylor and Francis.

Munnecke, D.M. 1979. Chemical, physical, and biological methods for the disposal and detoxification of pesticides. In *Residue Reviews*, Volume 70, eds. F.A. Gunther and J.D. Gunther. New York, NY: Springer.

Murphy, B.L. and R.D. Morrison, eds. 2014. *Introduction to Environmental Forensics*, 3rd edn. Boca Raton, FL: Elsevier Academic Press.

Nobel Media AB. 2014. The Nobel Prize in Physiology or Medicine 1948, Paul Müller. www.nobelprize.org/nobel_prizes/medicine/laureates/1948/ (accessed January 24, 2018).

NPIC. 2000. DDT technical fact sheet. http://npic.orst.edu/factsheets/archive/ddttech.pdf (accessed January 24, 2018).

O'Sullivan, G. and C. Sandau. 2014. *Environmental Forensics for Persistent Organic Pollutants*. San Diego, CA: Elsevier.

Pesticide Action Network (PAN). The DDT story. www.panna.org/resources/ddt-story (accessed January 24, 2018).

Petrisor, I.G. 2014. *Environmental Forensics Fundamentals: A Practical Guide*. London, UK: Taylor and Francis.

Ritz, K., L. Dawson, and D. Miller. 2008. *Criminal and Environmental Soil Forensics*. New York, NY: Springer.

Rodrigues, M.V.N., F.G.R. Reyes, P.M. Magalhãesa, and S. Rath. 2007. GC-MS determination of organochlorine pesticides in medicinal plants harvested in Brazil. *J. Braz. Chem. Soc*. 18:135–142.

Rogers, P., S. Whitby, and M. Dando. 1999. Biological warfare against crops. *Sci. Am*. 280:70–75.

Song, B. Y., S. Gwak, M. Jung, G. Nam, and N. Y. Kim. Tracing the source of methomyl using stable isotope analysis. Rapid *Commun Mass Spectrom*. 32(3):235–240.

Stellman, J.M., S.D. Stellman, R. Christian, T. Weber, and C. Tomasallo. 2003. The extent and patterns of usage of Agent Orange and other herbicides in Vietnam. *Nature* 422:681–687.

Stenersen, J. 2004. *Chemical Pesticides Mode of Action and Toxicology*. Boca Raton, FL: CRC Press.

Sullivan, P.J., F.J. Agardy, and R.K. Traub. 2001. *Practical Environmental Forensics: Process and Case Histories*. New York, NY: John Wiley & Sons.

Zhao, J. and D.J. Nelson. 2005. Fluorescence study of the interaction of Suwannee River fulvic acid with metal ions and Al3+-metal ion competition. *J. Inorg. Biochem*. 99(2):383–396.

Zhu, Q.-Z., P. Degelmann, R. Niessner, and D. Knopp. 2002. Selective trace analysis of sulfonylurea herbicides in water and soil samples based on solid-phase extraction using a molecularly imprinted polymer. *Environ. Sci. Technol*. 36(24):5411–5420.

Index

AA, *see* Atomic absorption (AA)
Abbe condenser, 35
Abrin, 264, 265, 267, 268, 269
Absorption, 53
Absorption rate, 155
Accelerants, 240–244
Accidents, fire, 239
Accuracy, 7–8
Acetaldehyde dehydrogenases (ALDHs), 155–156
Acetylene, 238
Acrylic, 184
Activation energy, 236
Adamsite, 258
Adhesive, duct tape, 185
Advanced spectroscopy, 78–79; *see also* Spectroscopy
    mass spectrometry, 79–86
    nuclear magnetic resonance spectroscopy, 87–94
AFQAM, *see* Association of Forensic Quality Assurance
        Managers (AFQAM)
Agent Blue, 301
Agent Orange, 301
Agent White, 301
Aiscough, Michael, 252
AK47 weapons, 226, 247
Alcohol, 134, 154–155
    absorption, 155
    accidents and, 157
    in blood, 155
    brain and, 156–157
    detoxification, 156
    ingestion of, 155
    metabolism, 155
Alcohol dehydrogenase (ADH) enzymes, 155
Alcohol deterrent drugs, 156
Alcoholic KOH test, 25–26
Alcohol poisoning, 1–2
ALDHs, *see* Acetaldehyde dehydrogenases (ALDHs)
Alkaloids, 131
Alpha (α) particles, 274
Alterations, handwritten/typewritten documents, 193
Aluminum-27, 276
AMDIS, *see* Automated Mass Spectral Deconvolution and
        Identification System (AMDIS)
American National Accreditation Board (ANAB), 10
American National Standards Institute (ANSI), 10
American Society for Quality (ASQ), 10
American Society of Crime Lab Directors (ASCLD), 8
Ammunition, 224
Amphetamines, 131, 132–133
Amplitude, of wave, 36
ANAB, *see* American National Accreditation
        Board (ANAB)
Anabolic androgenic steroids, 140–141
Anabolic steroids, 140–141

Anagen, growth phase of hair, 177
Analytical scheme, 161
Angel dust, 137
Animal hairs, forensic evidence, 179
Anion-exchange columns, 104
Anisotropy, 37
ANSI, *see* American National Standards Institute (ANSI)
Anthrax, 263, 264, 265
Anthrone spot test, 25
Antianxiety drugs, 134–135
Antibiotics, 305
Antidote, 256
Antimicrobials, 302, 304, 305
Anti-Stokes Raman scattering, 68
Anti-Stokes transitions, 68
Arches, fingerprints pattern, 211, 212
Arm, microscope, 35
Arsine (AsH_3), 254
Arson, 240–244
Asbestos mineral fibers, 178, 179
ASCLD, *see* American Society of Crime Lab
        Directors (ASCLD)
Aspirin, 142
ASQ, *see* American Society for Quality (ASQ)
Assault rifles, 226, 227
Association of Forensic Quality Assurance Managers
        (AFQAM), 8
Aston, Francis, 3, 79
Atomic absorption (AA) spectroscopy, 122–123
Atomic bomb, 278–279
ATR FT-IR, *see* Attenuated total reflectance Fourier
        transform-infrared spectroscopy (ATR FT-IR)
Attenuated total reflectance Fourier transform-infrared
        spectroscopy (ATR FT-IR), 65, 173–175
    of explosives, 246–247
    of hairs, 184
    of natural fibers, 183
    red and green food dyes, spectrum of, 67
    salicylic acid, spectrum of, 68
    of synthetic fibers, 183
Autoignition temperature, 237
Automated Mass Spectral Deconvolution and Identification
        System (AMDIS), 110
Automatic pistols, 225
Automatic rifles, 226

B. anthracis superbug, 286
BAC, *see* Blood alcohol concentration (BAC)
BAC Free app, 158
Backdraft, 240
Background controls, 7
Backing, duct tape, 185
Bacteria, as biological weapons, 265
Ballistics, 227–228

Balmer, J., 117
Banks, Craig, 99
Barbiturates, 135
Barium chloride spot test, 25
Base, microscope, 35
Basecoat, paint, 173
Base peak, 80
Becke line, 169, 171
Becquerel, Henri, 115, 276
Becquerel (Bq), 279
Beer's Law, 56, 57, 119
Belt of stability, 278
Benzodiazepines, 134–135
N-Benzylpiperazine (BZP), 131, 133
Bertillon, Alphonse, 3, 209
Beta (β) particles, 274
Bicomponent fibers, 181, 184
Binary weapons, 286
Biological weapons, 262
    bacteria, 265
    detection and identification, methods of, 272–273
    fungi, 265
    history of use, 262–263
    modern threat classification, 263–264
    protein toxins, 267–271
    recent cases and ongoing threats, 264–265
    small molecule toxins, 271–272
    viruses, 266–267
Bioterrorism, 252
Bioterrorism weapons threats, 286
Birefringence, 37
Bis(dichlorodiphenyl)acetic acid (DDA), 296
Bite mark impressions, 203
Blister agents, 255–256
Bloch, Felix, 3, 87
Blood agents, 253–254
Blood alcohol concentration (BAC), 155
    extrapolation, 153–154
    GC-MS for, 160
    in hours with and without food intake, 155
    risk of accidents and, 157–158
Blowback weapons, 224
Body tube, microscope, 35
Boiling points, 110
Bolt-action rifles, 226
Bombing of Alfred P. Murrah Federal Building in
        Oklahoma city 1995, 235–236
Bombs, radionuclides use in, 278–279
Botulinum toxin, 263, 265, 267, 270, 271
Bragg's Law, 125
Breathalyzer, 159
Breechblock, 224
Brightfield illumination, 38
Broaches, 226
Bromobenzene, 137
Bullet, 224, 225, 229–230
Bullet wipe, 228
Buprenorphine, 139
Butane, heat of combustion, 237, 238
Butler, John, 61
Buttons, 226
BZP, see N-benzylpiperazine (BZP)

Caffeine, 131, 134
Caliber, 227
Calibration standards, 57
Cannabis indica, 136
Cannabis ruderalis, 136
Cannabis sativa plant, see Marijuana
Capillary electrophoresis (CE), 112
Capsicum spray, 258, 259
Carbamates, 297, 298–299
Carbonate sand, 172
Carbon monoxide, 51
    hemoglobin and, 51–52
    poisoning, 51–52, 162
Carboxyhemoglobin (COHb) levels, 52
Carfentanil, 130, 258
    poisoning, 77–78
Carson, Rachel, 3, 295
Cartridge casing, 224, 225, 229
Catagen, growth phase of hair, 177
Catha edulis plant, 141
Cathinones, 131, 133
Cation-exchange column chromatography, 104
Caucasoid hair, 177–178
CBRNE agents, 252
CE, see Capillary electrophoresis (CE)
Cesium-131, 281
Chadwick, James, 276
Chain of custody, 6
Chain reaction, 278
Chelating column chromatography, 104
ChemDraw, 94
Chemical analysis
    of inks and paper, 193–199
    of questioned documents, 192, 193–199
Chemical ionization (CI), 79, 82
Chemical properties, of trace evidence, 169
Chemical shift, 88, 89–93
Chemical tests, 16; see also specific tests
    future of, 27
    for poisons, 24
Chemical warfare agents
    detection and identification methods, 259
    trends in chemical characteristics of, 259
Chemical weapons
    blister agents, 255–256
    blood agents, 253–254
    chemical properties of, 260–261
    incapacitating agents, 257–258
    nerve gases, 256–257
    nettle/urticant agents, 257
    pulmonary/choking agents, 254–255
    riot/tear agents, 258–259
    in terrorism, 253
    vomiting agents, 258, 259
Chen's test, 20
Chlorine, 284
Chlorine gas, 254
Chloroacetophenone, 258
2-Chlorobenzylidene malononitrile, 258, 259
Chloropicrin, 258, 284
Choking agents, 254–255
Chromatic aberrations, lenses, 40

Chromatography, 100
  column chromatography, 103–104
  gas chromatography, 106–111
  paper chromatography, 103
  phases, 100
  thin-layer chromatography, 100, 101–102
  types of, 100
  ultra-performance liquid chromatography, 106
Chromophores, 54, 55
CI, *see* Chemical ionization (CI)
Clandestine labs, 132
Class A fires, 238
Class B fires, 238
Class C fires, 238
Class characteristics, 6
Class D fires, 238
Claviceps purpurea, 137
Clearcoat, paint, 173
Coarse adjustment knob, microscope, 35
Cobalt-60, 275, 277, 281
Cobalt thiocyanate test, 19
Cocaine, 131, 133–134
Codeine, 139
COLA, 10
Colorimetric tests
  for drugs, 16–24
  for explosives, 24–26
Color tests, 6, 15, 16, 17; *see also specific tests*
Column chromatography, 103–104
Combustible materials, 238
Combustion, 237
Comparison microscopes, 43–44, 45
Comparison standards, 7
Compensator, 40
Compound light microscope (CPM), 38–40
Comprehensive Methamphetamine Control Act of 1996, 133
Concentric fractures, 171
Conchoidal lines, 171
Condenser, 35
Conduction, 237
Continental sand, 172
Continuous spectrum, 117
Controlled substances, 130, 131
  abused in sports, 141
  acidic, 145
  anabolic steroids, 140–141
  antianxiety drugs, 134–135
  chemical analysis, 142, 145–147
  chemical and spectroscopic features, 143–144
  chemical structures of, 132
  classification, 130
  control, 130
  depressant drugs, 134–135
  drugs, classes of, 130–131
  hallucinogens, 135–138
  new psychoactive substances, 141–142
  opiates/opioids, 138–140
  scheduling, 130, 131
  stimulants, 131–134
Controlled Substances Act (CSA), 100, 130
Control samples, 7
Convection, 237

Correlation chart, for Raman spectroscopy, 71
Correlation spectroscopy (COSY), 94
Corson, Ben, 258
Cortex, hair, 177
COSY, *see* Correlation spectroscopy (COSY)
Cotton, 181
Coupled, protons, 90
Coupling constant (J), 90
Covert attacks, 252
Crime scene investigation, 4
Crime scene investigators (CSIs), 5
Criminalistics, 4, 5
Critical mass, 278
Cropen test, 26
CSA, *see* Controlled Substances Act (CSA)
CSIs, *see* Crime scene investigators (CSIs)
Cumin, 123
Curie, Frederic Joliot, 276
Curie, Irene Joliot, 276
Curie, Marie Sklodowska, 115, 276
Curie, Pierre, 115, 276
Curie (Ci), 280
Cuticle, hair, 177
Cutters, firearms, 226–227
Cyanogen chloride (NCCl, CK), 253–254
Cyclosarin, 257

Darvon®, *see* Propoxyphene
Data libraries, of mass spectrometry, 85–86
Data precision, 7
Date rape drugs, 135
Datura stramonium, 142
DDA, *see* Bis(dichlorodiphenyl)acetic acid (DDA)
DDD, *see* 1,1-Dichloro-2,2-bis(p-chlorophenyl)ethane (DDD)
DDE, *see* 1,1-Dichloro-2,2-bis(p-dichlorophenyl) ethylene (DDE)
DDT, *see* Dichlorodiphenyltrichloroethane (DDT)
DEA, *see* Drug Enforcement Agency (DEA)
Delta-9-tetrahydrocannabinol (THC), 18, 131, 132, 135
Demerol®, *see* Meperidine
Depleted uranium, 278
Depressant drugs, 134–135
Depth of focus, microscopes, 36
Dermal papillae, skin, 209
Dermis, skin, 209
Designer diseases, 286
Designer superbug weapons, 286
Desorption ionization, 79, 82
Deuterons, 274, 275
Diamagnetic, 87
Dibenzoxazepine, 258, 259
1,1-Dichloro-2,2-bis(p-chlorophenyl)ethane (DDD), 296
1,1-Dichloro-2,2-bis(p-dichlorophenyl) ethylene (DDE), 296
Dichlorodiphenyltrichloroethane (DDT), 295–296, 298–299
Dichroism, 42
Diffraction pattern, 125
Diffuse reflectance infrared Fourier transform spectroscopy (DRIFTS), 64
Dille, James M., 19
Dille–Koppanyi test, 18
Dillinger, John, 212

2-(Dimethylamino)ethyl N,N-dimethylphosphoramidofluoridate, 257
Dimethylheptylpyran, 258
Diphenylamine test, 25
Diphosgene, 254
Diphtheria toxin, 267, 270, 271
Direct analysis in real-time (DART) instrument, 79, 83
Direct flame contact, 237
Dirty bombs, 282
Dispersion, 36
Dispersive magnetic mass analyzers, 83
Disulfur decafluoride (S_2F_{10}), 254
DMT, see N,N-Dimethyltryptamine (DMT)
Document comparison and analysis, 192–193
Dolophine®, see Methadone
Double-action revolvers, 225
Downfield, 89, 90
Doyle, Arthur Conan, 172
DRIFTS, see Diffuse reflectance infrared Fourier transform spectroscopy (DRIFTS)
Dronabinol, 135
The Drug Bible, 160
Drug Enforcement Agency (DEA), 4, 129, 130
Drugfire, 231
Drugs
    abused in sports, 141
    classes of, 130–131
    colorimetric tests for, 16–24
    microcrystalline tests for, 26
    retention time in urine for, 162
    and secondary metabolites, 160
    substances, 16
Drugs abused in sports, 141
Drunkometer, 158
Dual use research, 284–286
Duct tape, 185
Duquenois, Pierre, 3, 17
Duquenois–Levine color test, 9, 17
Dynamite, 245

Ebola, 263, 264, 266, 267
ECD, see Electron capture detectors (ECD)
Ecstasy, see 3,4-methylenedioxymethamphetamine (MDMA)
Edeleano, L., 131
EI, see Electron impact ionization (EI)
Electrocoat primer layer, 173
Electromagnetic spectrum, 53
Electron capture detectors (ECD), 111
Electron impact ionization (EI), 79, 80
Electron multiplier (EM) detectors, 84
Electrons, 116–117
Electrospray ionization (ESI), 79
Emission, 59–60
Emission spectrograph, 117–118
Enamel, 175
Endothermic process, 238
Energy (E), of wave, 36
Environmental forensics, 296
    antimicrobials, 302, 304, 305
    fungicides, 301–302, 304
    herbicides, 297, 300–301
    pesticides, 297–300

Environmental Protection Agency (EPA), 295
EPA, see Environmental Protection Agency (EPA)
Ephedra vulgaris, 131
Ephedrine, 131
Epidermis, skin, 209
Epsilon toxin, 267
Ernst, Richard, 3, 87
Erythroxylum coca plant, 133
Erythroxylum novagranatense plant, 133
ESI, see Electrospray ionization (ESI)
Ethanol, 154–155
    ATR FT-IR of, 159
    to formaldehyde, 156
    IR spectrum of, 159
    structure of, 155
Ethyl alcohol, see Ethanol
Ethyldichloroarsine, 255
Eumelanin, 177
Everspry EverASM Automated Shoe Impression Matcher software, 203
Excitation light sources, 54, 59–60
Excited state, 54, 116–117
Excretion, 156
Expert witnesses, 9
Explosives, 227, 244–245
    chemical structures of, 245
    colorimetric tests for, 24–25
    high, 244–245
    identification of, 246–247
    low, 244
    microcrystalline tests for, 26
    structures of primer, 228
Extinction, 41

Fast Blue B Salt test, 18
Fast-performance liquid chromatography (FPLC), 104, 105
Faulds, Henry, 3, 209
Fax machines, 193
FD&C Yellow 5, 54, 55
Federal Environmental Pesticide Control Act, 1972, 296
Federal Insecticide, Fungicide, and Rodenticide Act, 1947, 296
Federal Insecticide Act, 296
Fenn, John B., 79
Fentanyl, 77, 129, 131, 132, 257, 258
    poisoning, 77–78
Ferric chloride test, 21
Ferric hydroxamate test, 22
Ferric sulfate test, 22
Fibers, as trace evidence, 178–184
    acetate, 182
    acrylic, 182
    ATR FT-IR spectra of, 183
    characteristics, 182
    man-made/synthetic, 178–179, 180
    mineral, 178
    natural, 178–179
    stereomicroscope for evaluation, 179, 181
    types and uses, 180
FID, see Flame ionization detector (FID)
Field desorption ionization, 79
Fine adjustment knob, microscope, 35
Fingerprint impression detection, 205

Fingerprints
    anatomy of, 210
    to certify civil service applications, 209
    classification, 210
    comparison of, 218
    at crime scenes, 213
    evidence, 209–218
    formation, 209
    as identification tool, 209
    latent print development methods, 213–218
    minutiae, 211–212
    on nonporous objects, 216–218
    patterns, 211–212
    physical characteristics, 209
    points for analysis, 213
    position of, 209
    powders, 214–215
    visible prints, 213
Fingerprints, 209
Fingerprints powders, 214–215
Fire, 236–240
    accidents, 239
    classes, 238
    origin of, 241
    stages of, 239–240
    triangle, 237
Firearms, 224–225
    assault rifles, 226, 227
    ballistics, 227–228
    comparisons, 229–231
    evidence handling and labeling, 228–229
    explosives, 227
    handguns, 225
    manufacturing methods, 226–227
    propellants, 227
    rifles, 226
    shotguns, 226, 227
Firing pin, 224
Flame colors, of metals, 118
Flame ionization detector (FID), 110
Flame point, 238
Flame test, 117
Flammable materials, 238, 245
Flashover, 240
Flash point, 237
Flunitrazepam, 135
Fluorescein, 62
Fluorescence, 60
Fluorescence microscopes, 43
Fluorescence spectroscopy, 58–62
    in capillary electrophoresis instruments, 61
    concentration from, 61
    excitation light sources, 59–60
    fluorescence emission and, 60
    in real-time PCR instruments, 61, 63
    schematic of, 61
Fluorimeter, 43
Fluorophores ethidium bromide, 61, 62
FN Five-seveN® semiautomatic pistol, 223–224
Follicles, 177
Follicular tag, 177
Foodborne pathogens, 262

Food Quality Protection Act of 1996, 296
Forensic analysis, 7
Forensic biology, 4
Forensic Biology: A Laboratory Manual, 62
Forensic chemistry, 2–4
    advances, history of, 3
    services providers of, 4
Forensic chemists, 10
Forensic geology, 172
Forensic laboratories, 4
    integrity of, 10
    units of, 4
Forensic odontologist, 203
Forensic science, 2
Forensic scientists, 10
Forensic toxicology, 154
FPLC, *see* Fast-performance liquid chromatography (FPLC)
Fractionation ions, 81
Free-burning, fire stage, 239–240
Frequency
    of light, 36
    of radiation, 87
Froehde test, 23
Fuel cell breath test, 159
Fungi, as biological weapons, 265
Fungicides, 301–302, 304
Furchott, Robert F., 285

GABA, *see* Gamma-aminobutyric acid (GABA)
Gallic acid test, 22
Gallium-67, 281
Galton, Francis, 3, 209
Gamma-aminobutyric acid (GABA), 135
Gamma hydroxybutyrate (GHB), 130, 135
Gamma (γ) radiation, 275
Gas chromatography (GC), 106–111, 147
    of 1-butanol, 242
    detector for forensic applications, 110, 111
    mobile phase, 106
    process, 107–109
    schematic of, 108
    software, 110
    stationary phases, 106, 109
    of unburned gasoline, 243
    of unburned kerosene, 243
    of unburned lamp oil, 243
    of unburned lighter fluid, 242
    of unburned turpentine, 243
    use, 109, 111
Gas chromatography–mass spectrometry (GC-MS), 79, 85, 86, 108
Gasoline, 238
Gas-piston weapons, 225
GC, *see* Gas chromatography (GC)
GC-MS, *see* Gas chromatography–mass spectrometry (GC-MS)
Geiger–Mueller counter, 280–281
Gelb Kreutz, 253
General Rifling Characteristic File, 229
Germ warfare, 252
Gettler, Alexander, 1, 2, 52, 117
GHB, *see* Gamma hydroxybutyrate (GHB)

Glanders, 263, 264, 265
Glass, trace evidence, 169–172
    analysis, 169
    chemical properties of, 169–170
    densities, 170
    fracture examination, 171
    instrumental methods for analysis, 171
    physical features, 169, 170
    refractive index of, 171, 172
    types, 169
Glowing combustion, 239
Glutamate, 135
Gohlke, Roland, 79
GPCR, *see* G protein coupled receptor (GPCR)
G protein coupled receptor (GPCR), 133
Grey (Gy), 280
Griess, Johann Peter, 3
Griess test, 25, 231
GRIM3 software, 171
Grooves, 226, 228
Gross, Hans, 3, 172
Ground state, electrons, 116
Grun Kreutz, 253
G-series agents, 256–257
GSR, *see* Gunshot residue analysis (GSR)
Gun bluing, 217
Gunpowder propellant, 224, 225
Gunshot residue, microscopes to identify, 44
Gunshot residue analysis (GSR), 227, 228, 229, 231

Haber, Fritz, 254
Hahn, Otto, 276
Hairs, forensic evidence, 176–178
    animal, 179
    caucasian *versus* negroid hairs, 179
    characteristics of, 178
    classes, 177–178
    color, 177
    component of, 177
    for DNA analysis, 177
    growth phases of, 177
    layers, 177
    pigments, 177
    root, 177
    schematic and cross section of, 178
    shaft, 177
    shape, 177
    texture, 177
Hair shaft, 177
Half-life, 275–276
Hallucinogens, 135–138
Handbuch für Untersuchungsrichter, 3
Handguns, 225
Handwriting analysis, 192–193
Hanschritt, Konigin, 3
Harrison Narcotic Act of 1914, 133, 139
Hash oil, 136
HCE1, *see* Human carboxylesterase 1 (hCE1)
HCl test, 20
Heat, 237
Heat of combustion, 237, 238
Hemoglobin, 51

Hemp, 136
Henkel, Konrad, 256
Henry, Edward, 209
Henry's Law, 157–158
Herbicides, 297, 300–301
    classification of, 303
    detection, identification, and quantification of, 306–308
    in environment, 304–305
Heroin, 131, 132, 139, 161
Heroin Control Act of 1924, 139
Heteronuclear multiple bond correlation (HMBC), 94
Heteronuclear single quantum coherence (HSQC), 94
Highest occupied molecular orbital (HOMO), 54, 55
High explosives, 244–245
High-performance fibers, 184
High-performance liquid chromatography (HPLC), 58, 104–105, 106, 107
Hillenkamp, Franz, 3, 80
HMBC, *see* Heteronuclear multiple bond correlation (HMBC)
H5N1 flu virus, 286
Hoffman, Felix, 139, 142
Hofmann, Albert, 137
HOMO, *see* Highest occupied molecular orbital (HOMO)
Hondrogiannis, Ellen, 123
Hooke, Robert, 35
Host-swapping diseases, 286
HPLC, *see* High-performance liquid chromatography (HPLC)
HSQC, *see* Heteronuclear single quantum coherence (HSQC)
HU-210, 141
Huffing, 134, 135
Human carboxylesterase 1 (hCE1), 161
Hydrocodone, 139
Hydrogen, 238
Hydrogen cyanide (HCN, AC), 253–254

IAFIS, *see* Integrated Automated Fingerprint Identification System (IAFIS)
IAI, *see* International Association for Identification (IAI)
IBIS, *see* Integrated Ballistic Identification System (IBIS)
Ibogaine, 138
ICP, *see* Inductively coupled plasma emission spectroscopy (ICP)
ICP-MS instrument, 123–124
IEDs, *see* Improvised explosive devices (IEDs)
Ignarro, Louis J., 285
Ignitable liquids, 240–242
Ignition temperature, 236
Illuminators, 35
Imidazole antifungals, 302, 304
Immersion method, for refractive index of substance, 169
Impression evidence, 192, 199–205
Improvised explosive devices (IEDs), 245–246
Incapacitating agents, 257–258
Incendiary weapons, 245–246
Incipient, fire stage, 239
Inductively coupled plasma emission spectroscopy (ICP), 123–124
Influenza, 266, 272
Infrared spectroscopy, 62–68, 197
Infrared vibrations, 64
Ingestion of alcoholic drink, 155
Inidium-192, 281

Inked fingerprints, 213
Inks, chemical analysis of, 193–199
Inorganic chemistry, 116–117
    AA spectroscopy technique in, 122–123
    emission spectrography in, 117–118
    flame test in, 117
    IR spectroscopy in, 120
    Raman spectroscopy in, 120–121
    TLC in, 118
    UV-Vis in, 118–120
    XRF in, 121–122
Inorganic diatomic poisons, 162
Inorganic matter and materials, 116
Instrumental Data for Drug Analysis (IDDA), 160
Integrated Automated Fingerprint Identification System (IAFIS), 7, 209
Integrated Ballistic Identification System (IBIS), 230
Integration, 90
Internal standard, 88
International Association for Identification (IAI), 211
International Organization for Standardization, 10
Intoxilyzer, 159
Iodine-131, 274, 281
Iodine fuming, 217
Ion cyclotron resonance, 83
Ionization, 79, 117
Ionizing radiation, 273–274
Ion trap, 83
Ipomoea purpurea, 142
IR correlation chart, 65
Iris diaphragm, 35
IR spectroscopy, 120
IR spectrum, 63
Isotopes, 125, 278
Isotropic materials, 37

Jablonski diagram of Raman spectroscopy, 68
Janssen, Zacharias, 3, 34
Janssen route, 139
JEOL NMR spectrophotometer, 88
Jeserich, Paul, 3
Johnson, Jean Basset, 138
Johnson, Kirk, 130
JWH-018, synthetic cannabinoid, 99–100, 131, 132

Karas, Michael, 3, 82
Kennedy, Jospeh, 3
Keratin, 177
Ketamine, 130, 138
Kevlar fibers, 184
Klaproth, Martin Heinrich, 276
KN test, *see* Fast Blue B Salt test
Koehler illumination, 40
Koller, Carl, 133
Kolokol-1, 258
Koppanyi, Theodore, 19
Krebs cycle, 155
Kuhn, Richard, 257

Lacassagne, Alexandre, 3
Lacrimator ethyl bromoacetate, 253
La Fon's test, *see* Mecke test

Lands, firearms, 226, 228
Laser ablation-inductively coupled plasma-time of flight-mass spectrometry, 123
Lasers, 53, 60
Latent print development methods, 213–214
Lauterbur, Paul C., 87
LC-MS systems, 85, 86
Lead glass, 170
Leary, Timothy, 137
LEDs, *see* Light-emitting diodes (LEDs)
Lee, Janet, 15
Lenses, 35
    aberrations, 40
    modern, 39, 40
    numerical aperture of, 38–39
Lever-action rifles, 226
Levofentanyl, 257, 258
Lewisite, 255–256
Li, Richard, 62
Liebermann test, 20
Lifting, fingerprint evidence, 218
Light, 36–37
    characteristics, 36
    dispersion, 36
    fluorescence emission of, 60
    polarization of, 36–37
    refraction, 36
    retardation, 37
Light-emitting diodes (LEDs), 53
Light retardation, 37
Line spectra of elements, 119
Liquid chromatography (LC), *see* Column chromatography
Litvinenko, Alexander, 273
Locard, Edmond, 3, 169
Locard's exchange principle, 169
Long-range coupling, 90–91
Loops, fingerprints pattern, 211–212
Lopophora williamsii (Lamaire), 137
Lowest unoccupied molecular orbital (LUMO), 54, 55
Low explosives, 244
LSD, *see* Lysergic acid diethylamide (LSD)
LUMO, *see* Lowest unoccupied molecular orbital (LUMO)
Lysergic acid diethylamide (LSD), 130, 137, 257, 258

Magazine, 224
"Magic" mushrooms, 137
Magnetic field, 87
Magnetic sector mass analyzers, 84
Magnification, 37
Magnifying glass, 34
Ma-huang plant, 131
Malathion, 297
MALDI, 80
Mandelin, K., 3
Mandelin test, 20
Mandrels, firearms, 227
Maneb, 301, 302
Man-made fibers, 178–184
Mansfield, Peter, 87
Marijuana, 135–136
Marijuana Tax Act, 135
Marquis test, 17

Mars Exploration Rover Spirit, 122
Marsh, James, 2, 3
Marsh Test, 2
Martin, John Porter, 3
Mass analyzer, 83
M-16 assault rifle, 226
Mass filter, 83
Mass spectrometers, 110
Mass spectrometry (MS), 79–86
    of acetone, 80
    analyte fragmentation pattern and, 80
    of caffeine, 86
    components, 79
    data libraries of, 85–86
    detectors, 84, 111
    of dodecane found in kerosene, 244
    fractionation ions in, 81
    for methamphetamine, 146
    Nobel Prizes awarded for advances in, 79
    operating modes, 84
    for phentermine, 147
    of phentermine, 85
    of 1-propanol, 82
    of 2-propanol, 82
    of propionaldehyde, 80
    of salicylic acid, 85, 86
Mass spectrum, 80
Match point, 169
Mayer's test, 18
McLafferty, Fred, 79
McMillan, Edwin, 3
McMillan, Edwin Mattison, 276
MDMA, *see* 3,4-Methylenedioxymethamphetamine (MDMA)
Mechanical parts of microscopes, 35, 36
Mechoulam, Raphael, 141
Mecke test, 20–21
Medulla, 177
Meitner, Lise, 276
Meperidine, 139
Mercury (Hg_2), 255
Merremia tuberosa, 142
Mescaline (3,4,5-trimethoxyphenethylamine), 137
Metabolites, 160
Methadone, 139
Methamphetamine, 131, 132
Methanol, 1
Methyldichloroarsine, 255
3,4-Methylenedioxymethamphetamine (MDMA), 130, 138
Methyl fluorophosphoryl homocholine iodide (MFPhCh), 256
Methylphosphonic acid (MPA), 257
Michel-Levy Chart, 37
Michelson, Albert Abraham, 53
Microanalysis, 168
Micro-chemistry of Poisons, 2, 3
Microcrystalline tests, 16
    for drugs, 26
    for explosives, 26
Microfiber, 181
Micrographia, 35
Microscopes, 34
    comparison, 43–44, 45
    CPM, 38–40

depth of focus, 36
    early, 34–35
    fluorescence, 43
    to identify gunshot residue, 44
    magnification, 37
    mechanical parts of, 35, 36
    optical parts of, 35
    parts of, 35–36
    phase contrast, 42
    polarizing light, 40–42
    resolution, 37, 38
    SEM, 44, 45–46
    TEM, 46–47
    working distance of, 36
Microscopy, 34
Microspectrophotometer, 43, 70–71
Mineral fibers, 178
Minerals, 172
Minutiae, 211–212
Mitragyna speciosa, 142
Mobile phases
    column chromatography, 103
    GC, 106
    HPLC, 105
    TLC, 101
Modus operandi, 240
Molecular ion peak, 80
Molly, *see* 3,4-Methylenedioxymethamphetamine (MDMA)
Molotov cocktail, 245–246
Molybdenum disulfide (MoS_2), 217
Mongoloid hair, 177–178
Morphine, 139
MPA, *see* Methylphosphonic acid (MPA)
MS, *see* Mass spectrometry (MS)
Muller, Paul, 295
Multiplicity, 90
Murad, Ferid, 285
Mycotoxins, 265, 271–272

NAA, *see* Neutron activation analysis (NAA)
Naloxone, 139
Naltrexone, 139
NanoDrop UV-Vis spectrophotometer, 57, 58
Nanoparticles, in latent print development, 215–216
Narcan™, *see* Naloxone
Narcotest®, 17
Narcotics, 139
National Fire Protection Association (NFPA), 7, 8
National Institutes of Standards and Technology (NIST), 7
National Integrated Ballistics Information Network, 231
Natural disasters, 252
Natural fibers, 178–184
NCI, *see* Negative chemical ionization (NCI)
Negative chemical ionization (NCI), 82
Negative control, 7
Negroid hair, 177–178
Nelson, Rogers, 129–130
Neptunium, 277
Nerve gases, 256–257
Nessler's test, 24
Nettle agents, 257
Neutron activation analysis (NAA), 125

Neutrons, 274
New psychoactive substances (NPS), 141–142
Newton, Isaac, 53
NFPA, *see* National Fire Protection Association (NFPA)
Nichol, William, 3
Nieman, Albert, 133
NIK® Polytesting System, 17
Ninhydrin, 217
NIST, *see* National Institutes of Standards and
        Technology (NIST)
Nitric acid test, 17
Nitric oxide (NO), 255
Nitrogen mustards, 256
Nitroglycerin, 284–285
Nixon, Richard, 252
NMR spectroscopy, *see* Nuclear magnetic resonance
        spectroscopy (NMR spectroscopy)
N,N-Dimethyltryptamine (DMT), 138
Nobel, Alfred, 3, 245
NOESY, *see* Nuclear Overhauser effect spectroscopy and
        experiments (NOESY)
Nonsteroidal anti-inflammatory drug (NSAID), 142
Norepinephrine, 133
Normal-phase column chromatography, 104
Norris, Charles, 52
NPS, *see* New psychoactive substances (NPS)
NSAID, *see* Nonsteroidal anti-inflammatory drug (NSAID)
Nuclear bomb detonation, 284
Nuclear chemistry, 273
    history of, 276–277
    reactions, 274–275
Nuclear fission reaction, 274, 278
Nuclear fusion reaction, 274, 278
Nuclear magnetic resonance spectroscopy (NMR
        spectroscopy), 87–94
    of carbon atom, 90–91, 92
    ephedrine in D_2O, 91, 92
    of hydrogen atoms, 89–90, 93
    internal standard, 88
    of lidocaine in D_2O, 89
    liquid phase, 93
    modern, 93
    multidimensional, 94
    Nobel Laureates in field of, 87
    radio waves in, 87
    software, 94
    solid-phase, 93
Nuclear Overhauser effect spectroscopy and experiments
        (NOESY), 94
Nuclear power plants
    meltdown, 284
    radioisotopes in, 283–284
Nuclear weapons, 273–284
Numerical aperture (N.A.) of lens, 38–39
Nylon, 184

Obliterations, handwritten/typewritten documents, 193
Ocean floor sand, 172
Ocean Optics spectrophotometers, 57
Ogden Baine, J., 2
Oil Red O, 217
Omelanin, 177

Opiates, 138–140
Opioids, 138–140
Optical interferometers, 53
Optical parts of microscopes, 35
Orbitrap, 83
Organization of Scientific Area Committees (OSACs), 9
Organizations, for forensic profession, 10
Organochlorines, 297–299
Organophosphates, 257, 297, 298, 299
Origin of fire, 241
OSACs, *see* Organization of Scientific Area
        Committees (OSACs)
Overt attacks, 252
Oxidized, 237
Oxycodone, 139

Paint Data Query (PDQ), 173
Paint forensic evidence, 173–175
    ATR FT-IR for, 173–175
    chemistry of, 173
    chips, 173
    coatings, 173
    evaluation, 174
    samples, 173
    smears, 173
Papaver somniferum, 138
Paper chromatography, 103
Papers, chemical analysis of, 193–199
Parathion, 297
Pascal's Triangle, 91
PCC, *see* 1-Piperidinocyclo-hexanecarbonitrile (PCC)
PCI, *see* Positive chemical ionization (PCI)
PCP, *see* 1-1-Phenylcyclohexyl piperidine hydrochloride (PCP)
PD, *see* Plasma desorption (PD)
PDQ, *see* Paint Data Query (PDQ)
Pelargonic acid vanillylamide, 258
Percocet®, *see* Oxycodone
Perfluoroisobutene, 255
Perfluorotributylamine (PFTBA), 85
Performance-enhancing drugs, 140–141
Permanganate test, 22
Personal protective equipment (PPE), 7
Pesticides, 297–300
    classification of, 305
    detection, identification, and quantification of, 306–308
    in environment, 304–305
Peyote, 137
PFTBA, *see* Perfluorotributylamine (PFTBA)
Phase, light, 36
Phase contrast microscopy, 42
Phencyclidine, 137
Phenobarbital, 131, 132
1-1-Phenylcyclohexyl piperidine hydrochloride (PCP), 137
1-(1-Phenylcyclohexyl)pyrrolidine (PHP), 137
Phenyldichloroarsine, 255
Phenylethylamines, 131
Phosgene, 254
Phosgene oxime, 257
Phosphate test, 22–23
Phosphorus-30, 276
Phosphorus-32, 281
Photoelectric effect, 121

Photoionization (PI), 79
Photons, 116–117
PHP, *see* 1-(1-Phenylcyclohexyl)pyrrolidine (PHP)
Physical analysis, of questioned documents, 192–193
Physical developer, 218
Physical evidence, 4–10
  advantages, 6
  class characteristics, 6
  collection devices and containers, 4–5
  comparison standards of, 7
  control samples, 7
  individual characteristics, 6
  packaging, 4
  presumptive test for, 6
  reference samples, 7
Physical properties, of trace evidence, 169
PI, *see* Photoionization (PI)
Pipe bomb, 245
1-Piperidinocyclo-hexanecarbonitrile (PCC), 137
Pistols, 225, 226
Plasma desorption (PD), 79
Plastic polymers
  chemical structures of, 176
  uses and recycling code, 175
Plastic prints, 213
Plutonium-239, 277, 278, 279, 283–284
Poisons
  chemical structures of, 162
  chemical tests for, 24
  heavy metal, 163
Polarization, of light, 36–37
Polarizing light microscopes, 40–42
Polyene antifungals, 302
Polymers forensic evidence, 175–176
  plastic, 175–176
  pyrolysis gas chromatography for, 176
Popp, Georg, 172
Positive chemical ionization (PCI), 82
Positive controls, 7
Positrons, 274, 275
Post-absorption period, 155
PPE, *see* Personal protective equipment (PPE)
Precision, 7–8
Presumptive test, 6
Primer, 224, 225
Primer surfacer layer, 173
Principe de l'echange, 169
Prism, 53
Propane, 238
Propellants, 227
Propiconazole, 301–302
Propoxyphene, 139
Protein toxins, as biological weapons, 267–271
Protons, 274
Psilocybe cubensis, 137
Psilocybe cyanescens, 137
Psilocybin, 137–138
Pulmonary agents, 254–255, 256
Purcell, Edward Mills, 3, 87
Pure Food and Drug Act of 1906, 133
Pyrethroids, 297, 300, 307
Pyrolysis, 237–238

QA, *see* Quality assurance (QA)
QC, *see* Quality control (QC)
QNB, *see* 3-Quinuclidinyl benzilate (QNB)
Quadrupole mass analyzers, 83
Quality assurance (QA), 8
Quality control (QC), 5
Questioned documents
  chemical analysis of, 192, 193–199
  definition, 192
  examples of, 192
  features for testing, 192
  investigation methods, 192
  physical analysis of, 192–193
3-Quinuclidinyl benzilate (QNB), 257, 258

Rabi, Isidor I., 87
Radial fractures, 171
Radiation, 237
  poisoning, treatment of, 282–283
  types of, 274–275
Radioactive decay, 273, 274, 275
  modes of, 277–278
Radioactive material
  detection and identification of, 280–281
  for dirty bombs, 282
Radioactivity
  history of, 276–277
  units, 279–280
Radioisotopes, 276
  in nuclear power plants, 283–284
Radiological material
  cases of accidental poisoning with, 281–282
  uses, 281
Radionuclides, 273, 276
  use in bombs, 278–279
Radium, 115–116
Radium-226, 116
Radium Dial Company, 116
Radon-222 gas, 116
Rad (radiation absorbed dose), 280
Raman, C. V., 3, 68
Raman, Chandrasekhara Venkata, 53
Raman detector, 69
Raman scattering, 68
Raman spectroscopy, 68–70
  advantages of, 70
  correlation chart for, 71
  disadvantages of, 70
  versus infrared spectroscopy, 70
  in inorganic chemistry, 120–121
  Jablonski diagram of, 68
  salicylic acid, spectrum of, 69
Rayleigh scattering, 68
Rayon, 178
Real *versus* virtual image, 35
Recoil weapons, 225
Red Dye 40, 54, 55
Red food coloring, 56, 57
Red phosphorus, 255
Reduced, oxygen, 237
Reference samples, 7
Reflected ultraviolet imaging system (RUVIS), 213

Refraction, 36
Refractive index, 169, 171
    for glass, 172
    for liquids and oils, 171
Reinsch test, 24, 25
Rem, 280
Remifentanil, 258
Replicates, 7
Resolution, 37, 38, 84
Resolving power, 37, 84
Retention factor (Rf) value, 102
Retention time, 161, 162
Reverse-polarity stationary phases, 104
Revia®, *see* Naltrexone
Ricin, 263, 264, 267, 268
Rifles, 226
Riot/tear agents, 258–259
Roentgen, Wilhelm Conrad, 121, 276
ROESY, *see* Rotating frame nuclear Overhauser effect
            spectroscopy (ROESY)
Rohypnol, 130, 135
Rollover, 239–240
Roosevelt, Theodore, 2, 3
Rotating frame nuclear Overhauser effect spectroscopy
            (ROESY), 94
RTX, *see* Ruthenium tetroxide (RTX)
Ruhemann's purple, *see* Ninhydrin
Ruthenium tetroxide (RTX), 217
Rutherford, Ernest, 276
RUVIS, *see* Reflected ultraviolet imaging system (RUVIS)

Safety data sheet (SDS), 7
Salvia divinorum, 138
Samantha Reid Date Rape Prohibition Act of 200, 135
Sarin, 253, 256
Scan mode (SCAN), 84–85
Scanning electron microscopy (SEM), 44, 45–46, 125
Schulenberg, Michael, 130
Scientific Working Group for the Analysis of Seized Drugs
            (SWGDRUG), 9, 79
    analytical techniques, categories of, 9
    Category A methods, 162
    minimum testing criteria, 142
Scientific Working Groups (SWGs), 9
Scintillation counter, 280–281
Scott test, 19
Screening and confirmatory methods, 161
Scrim, duct tape, 185
SDS, *see* Safety data sheet (SDS)
Seaborg, Glenn Theodore, 3, 276, 277
Segre, Emilio, 3
SEM, *see* Scanning electron microscopy (SEM)
Semiautomatic pistols, 225
Semiautomatic rifles, 226
Shielded, 90
Shoeprint Image Capture and Retrieval (SICAR 6), 203
Shoe tread analysis, 204
Shotguns, 226, 227
Siegfried synthesis route, 139
Sievert (Sv), 280
Silent Spring, 3, 295
Silica glass, 170

Silver nitrate, 218
Silver nitrate spot test, 25
Silver nitrate test, 23
Simon test, 23
Simon with acetone test, 23
Single-action revolvers, 225
Single-barrel rifles, 226
Single ion monitoring (SIM) mode, 84–85
Single-shot pistols, 225
Size-exclusion chromatography, 103
Slide/pump-action rifles, 226
Slurry, 103
Small molecule toxins, 271–272
Smallpox, 263, 264, 266, 267
Smoldering, fire stage, 240
Soda-lead glass, 170
Soda-lime glass, 169–170
Sodium nitroprusside test, 23
Soil
    particles by size, 173
    as trace evidence, 172–173
Soldier attacks soldiers readying to deploy at Fort Hood US
            army base, 223–224
Solid phase microextraction fibers, 241
Soman, 256
SOPs, *see* Standard operating procedures (SOPs)
Spark ionization, 79
Spectroscope, 117, 118
Spectroscopy; *see also* Advanced spectroscopy
    fluorescence, 58–62
    infrared, 62–68
    instruments, 53
    methods, 53
    Nobel Prize Laureates in field of, 54
    Raman, 68–70
    UV-Vis, 53–58
Spectrum, 53
Spherical aberrations, lenses, 40
Spin-spin splitting, 90
Spontaneous combustion, 238
SRMs, *see* Standard reference materials (SRMs)
Stage, microscope, 35
Stain, 175
Standard operating procedures (SOPs), 8
Standard reference materials (SRMs), 7
Staphylococcal enterotoxin B, 267, 270
Stationary phases
    column chromatography, 103
    GC, 106, 109
    HPLC, 105
    TLC, 101
Stealth viruses, 286
Stereomicroscopes, 38, 39
Stereomicroscopy, 38
Stern, Otto, 87
Stewart, William, 252
Stibine (SbH$_3$, SA), 254
Stimulants, 131–134
Stipa robusta seeds, 137
Stippling, 227
Stokes radiation, 68
Stokes shift, 60

Stoughton, Roger, 258
Straight-pull bolt-action rifles, 226
Strassmann, Fritz, 276
Strobl, A. Quinn, 129
Sudan Black, 216
Sudden sniffing death syndrome, 134
Sulfate test, 23
Sulfur mustard, 256
Sulfur trioxide, 255
Super Glue® (cyanomethacrylate) fuming, 216
SWGDRUG, *see* Scientific Working Group for the Analysis of Seized Drugs (SWGDRUG)
SWGs, *see* Scientific Working Groups (SWGs)
SYBR Green I, 62, 63
Synge, Richard Laurence Millington, 3
Synthetic Drug Prevention Act of 2012, 100
Synthetic fibers, *see* Man-made fibers

TAAR1, *see* Trace amine-associated receptor 1 (TAAR1)
Tabernaemontana undulata plant, 138
Tabernanthe iboga, 138
Tabernanthe manii, 138
Tabun, 256
Tanaka, Koichi, 79, 83
Tandem mass spectrometers, 83
TCD, *see* Thermal conductivity detectors (TCD)
Technical disasters, 252
Technical Working Groups (TWGs), 9
Technitium-99, 281
Tellurium-201, 281
Telogen, growth phase of hair, 177
TEM, *see* Transmission electron microscopy (TEM)
Temperature, 238
Tempered glass, 170
TEPP, *see* Tetraethyl pyrophosphate (TEPP)
Terrorism, 252
    chemical weapons in, 253
Test fire, 229
Test impressions, 204
Testosterone, 131, 132
Tetraethyl pyrophosphate (TEPP), 297
Tetramethylsilane (TMS), 88–89
THC, *see* Delta-9-tetrahydrocannabinol (THC)
Thermal conductivity detectors (TCD), 110–111
Thermospray ionization (TS), 79
Thin-layer chromatography (TLC), 101–102, 118
    mobile phase, 101
    plate analysis, 102
    process, 101–102
    stationary phase, 101
    use, 102
Thomson, J. J., 3, 79
Time-of-flight secondary ion mass spectrometry (TOF-SIMS), 168
Time-of-flight (TOF) mass detectors, 83
Tip of hair, 177
Tire tread analysis, 204
Titanium tetrachloride, 255
TLC, *see* Thin-layer chromatography (TLC)
TMS, *see* Tetramethylsilane (TMS)
TOCSY, *see* Total correlation spectroscopy (TOCSY)
TOF, *see* Time-of-flight (TOF)

TOF-SIMS, *see* Time-of-flight secondary ion mass spectrometry (TOF-SIMS)
Tool mark impressions, 204
Total correlation spectroscopy (TOCSY), 94
Toxic industrial chemicals, 259
Toxicologists, 154
Trace amine-associated receptor 1 (TAAR1), 133
Trace elements, 116
Trace evidence, 168–169
    analysis tools, 169
    fiber, 178–184
    glass, 169–172
    hairs, 176–178
    materials, 168
    other materials, 184–185
    paint, 173–175
    physical and chemical properties of, 169
    polymers, 175–176
    sample, 168–169
    soil, 172–173
Trace metals, 116
Transmission electron microscopy (TEM), 46–47
Transmitting terminal identifier (TTI), 193
Tryptophan, 61
TS, *see* Thermospray ionization (TS)
Tswett, M. S., 3
TTI, *see* Transmitting terminal identifier (TTI)
Tufa sand, 172
Turn-bolt rifles, 226
Turret, microscope, 35
TWGs, *see* Technical Working Groups (TWGs)
Typewritten documents analysis, 193

Ultra fine metal plus liquid matrix method, 83
Ultra-performance liquid chromatography (UPLC), 106
Ultraviolet-visible spectroscopy (UV-Vis), 53–58
    absorption from HOMO to LUMO energy level, 54, 55
    bovine hemoglobin, absorption spectrum of, 59
    cuvette, 54
    DNA, absorption spectrum of, 54, 56
    in inorganic chemistry, 118–120
    light path, schematic of, 54
    for methamphetamine, 145
    NanoDrop, 57, 58
    Ocean Optics, 57
    for phentermine, 146
    to quantify drugs, 56
    quantitation of iron using, 119
    red food coloring, absorption spectrum of, 56
    salicylic acid, absorption spectrum of, 59
United States Poppy Control Act of 1942, 139
United States Pure Food and Drugs Act (1906), 2
Upfield, 90
UPLC, *see* Ultra-performance liquid chromatography (UPLC)
Uranium, 276
Uranium-235, 278–279
Uranium-238, 283–284
Urticant agents, 257
US Declaration of Independence, 191
US Federal Pesticide Act, 296
US Patriot Act, 132
US Toxic Substances Control Act, 1976, 296

Vacuum metal deposition, 205
Van Leeuwenhoek, Antioni, 34–35
Van Urk's test, 18
Varnish, 175
Vicodin®, *see* Hydrocodone
Vineland Map, 121
Viruses, as biological weapons, 266–267
Visible/patent prints, fingerprints, 213
Visible spectroscopy, 57, 59
Vitali–Morrin test, 21
Voacanga Africana, 138
Volstead Act, 1
Vomiting agents, 258, 259
Von Bayer, Adolph, 135, 297
V-series agents, 256–257
"V"-shaped pattern, fire, 239

WADA, *see* World Anti-Doping Agency (WADA)
Wagner test, 21–22
Wahl, Arthur, 3
Wallner lines, 171
Walsh, Alan, 3
Wavelength, 36
Weapons of mass destruction (WMD)
    attacks, 252
    CBRNE for, 252
    definition, 251–252
    disasters, 252
Weedkillers, *see* Herbicides

Wheatstone, 3
White cross, *see* Xylyl bromide
White phosphorus, 255
Whorls, fingerprints pattern, 211, 212
WMD, *see* Weapons of mass destruction (WMD)
Wood alcohol, 1
Working distance, of microscopes, 36
World Anti-Doping Agency (WADA), 140
Wormley, Theodore, 2, 3
Writing analysis, 192–193
Wuthrich, Kurt, 87

Xenon-131, 274
Xenon arc lamps, 60
X-ray crystallography, 124–125
X-ray diffraction, 124–125
X-ray fluorescence (XRF), 121–122
X-rays, 276, 285
XRF, *see* X-ray fluorescence (XRF)
Xylyl bromide, 253

Yarn, 185

Zeidler, Othmar, 295, 297
Zimmerman test, 23–24
Zinc oxide, 255
Zineb, 301, 302
Zwikker test, 21
Zyklon B, 254